VOLUME SIX HUNDRED AND SEVENTY THREE

METHODS IN ENZYMOLOGY

Helicase Enzymes Part B

METHODS IN ENZYMOLOGY

Editors-in-Chief

ANNA MARIE PYLE
Departments of Molecular, Cellular and Developmental Biology and Department of Chemistry
Investigator, Howard Hughes Medical Institute
Yale University

DAVID W. CHRISTIANSON
Roy and Diana Vagelos Laboratories
Department of Chemistry
University of Pennsylvania
Philadelphia, PA

Founding Editors

SIDNEY P. COLOWICK and NATHAN O. KAPLAN

VOLUME SIX HUNDRED AND SEVENTY THREE

Methods in ENZYMOLOGY

Helicase Enzymes Part B

Edited by

MICHAEL A. TRAKSELIS
Department of Chemistry and Biochemistry,
Baylor University, Waco, TX, United States

ELSEVIER

ACADEMIC PRESS
An imprint of Elsevier

Academic Press is an imprint of Elsevier
50 Hampshire Street, 5th Floor, Cambridge, MA 02139, United States
525 B Street, Suite 1650, San Diego, CA 92101, United States
The Boulevard, Langford Lane, Kidlington, Oxford OX5 1GB, United Kingdom
125 London Wall, London, EC2Y 5AS, United Kingdom

First edition 2022

Notices

Knowledge and best practice in this field are constantly changing. As new research and experience broaden our understanding, changes in research methods, professional practices, or medical treatment may become necessary.

Practitioners and researchers must always rely on their own experience and knowledge in evaluating and using any information, methods, compounds, or experiments described herein. In using such information or methods they should be mindful of their own safety and the safety of others, including parties for whom they have a professional responsibility.

To the fullest extent of the law, neither the Publisher nor the authors, contributors, or editors, assume any liability for any injury and/or damage to persons or property as a matter of products liability, negligence or otherwise, or from any use or operation of any methods, products, instructions, or ideas contained in the material herein.

ISBN: 978-0-323-99773-7
ISSN: 0076-6879

For information on all Academic Press publications
visit our website at https://www.elsevier.com/books-and-journals

Publisher: Zoe Kruze
Developmental Editor: Federico Paulo S. Mendoza
Production Project Manager: James Selvam
Cover Designer: Greg Harris

Typeset by STRAIVE, India

Contents

Contributors

Clara Aicart-Ramos
Department of Macromolecular Structures, Centro Nacional de Biotecnología, Consejo Superior de Investigaciones Científicas, Madrid, Spain

Jean-François Allemand
Institut de Biologie de l'Ecole Normale Supérieure (IBENS), Ecole normale supérieure, CNRS, INSERM, PSL Research University; Laboratoire de Physique de L'Ecole Normale Supérieure de Paris, CNRS, ENS, Université PSL, Sorbonne Université, Université de Paris, Paris, France

Alexander S. Baier
Program in Molecular Medicine, Morningside Graduate School of Biomedical Sciences; Medical Scientist Training Program, T.H. Chan School of Medicine, University of Massachusetts Chan Medical School, Worcester, MA, United States

Darby Ball
Department of Chemistry and Biochemistry, The University of Texas at Dallas, Richardson, TX, United States

Matthew L. Bochman
Molecular & Cellular Biochemistry Department, Indiana University, Bloomington, IN, United States

Timothy A. Bolger
Department of Molecular and Cellular Biology, University of Arizona, Tucson, AZ, United States

Sara B. Carey
Department of Molecular and Cellular Biology, University of Arizona, Tucson, AZ, United States

Carl C. Correll
Center for Proteomics and Molecular Therapeutics, Chicago Medical School, Rosalind Franklin University of Medicine and Science, North Chicago, IL, United States

Vincent Croquette
Institut de Biologie de l'Ecole Normale Supérieure (IBENS), Ecole normale supérieure, CNRS, INSERM, PSL Research University; Laboratoire de Physique de L'Ecole Normale Supérieure de Paris, CNRS, ENS, Université PSL, Sorbonne Université, Université de Paris; ESPCI Paris, Université PSL, Paris, France

Sheena D'Arcy
Department of Chemistry and Biochemistry, The University of Texas at Dallas, Richardson, TX, United States

Subrata Debnath
Department of Biochemistry and Molecular Biology, College of Medicine, Howard University, Washington, DC, United States

Joshua M. Denson
Department of Chemistry & Biochemistry, Utah State University, Logan, Utah, United States

Mark S. Dillingham
DNA:Protein Interactions Unit, School of Biochemistry, University of Bristol, Bristol, United Kingdom

Stephen N. Floor
Department of Cell and Tissue Biology, University of California; Helen Diller Family Comprehensive Cancer Center, University of California, San Francisco, San Francisco, CA, United States

Harshad Ghodke
Molecular Horizons and School of Chemistry and Molecular Bioscience, University of Wollongong; Illawarra Health & Medical Research Institute, Wollongong, NSW, Australia

Mark D. Gold
Department of Chemistry & Biochemistry, Utah State University, Logan, Utah, United States

Spencer J. Gray
Molecular & Cellular Biochemistry Department, Indiana University, Bloomington, IN, United States

Rong Guo
Department of Chemistry, Yale University, New Haven, CT, United States

Anna E. Helmers
Department of Molecular Biosciences, University of Texas at Austin, Austin, TX, United States

Silvia Hormeno
Department of Macromolecular Structures, Centro Nacional de Biotecnología, Consejo Superior de Investigaciones Científicas, Madrid, Spain

Chunyi Hu
Department of Molecular Biology and Genetics, Cornell University, Ithaca, NY, United States

Inga Jarmoskaite
Department of Molecular Biosciences, University of Texas at Austin, Austin, TX, United States

Arlen W. Johnson
Department of Molecular Biosciences and Institute for Cellular and Molecular Biology, The University of Texas at Austin, Austin, TX, United States

Sean J. Johnson
Department of Chemistry & Biochemistry, Utah State University, Logan, Utah, United States

Sreeya Kairamkonda
Department of Biochemistry and Molecular Biology, Colorado State University, Fort Collins, CO, United States

Katrin Karbstein
Department of Integrative Structural and Computational Biology; The Skaggs Graduate School of Chemical and Biological Sciences, Scripps Florida, Jupiter, FL; HHMI Faculty Scholar, Chevy Chase, MD, United States

Ailong Ke
Department of Molecular Biology and Genetics, Cornell University, Ithaca, NY, United States

Dagmar Klostermeier
University of Muenster, Institute for Physical Chemistry, Muenster, Germany

Hervé Le Hir
Institut de Biologie de l'Ecole Normale Supérieure (IBENS), Ecole normale supérieure, CNRS, INSERM, PSL Research University, Paris, France

Christopher D. Lima
Structural Biology Program, Sloan Kettering Institute, Memorial Sloan Kettering Cancer Center; Howard Hughes Medical Institute, New York, NY, United States

Ran Lin
Department of Molecular Biosciences and Institute for Cellular and Molecular Biology, The University of Texas at Austin, Austin, TX, United States

Xing Lu
Department of Biochemistry and Molecular Biology, College of Medicine, Howard University, Washington, DC, United States

Fernando Moreno-Herrero
Department of Macromolecular Structures, Centro Nacional de Biotecnología, Consejo Superior de Investigaciones Científicas, Madrid, Spain

Phoebe Nguyen
Molecular & Cellular Biochemistry Department, Indiana University, Bloomington, IN, United States

Trang Nguyen
Department of Chemistry and Biochemistry, The University of Texas at Dallas, Richardson, TX, United States

David G. Nickens
Molecular & Cellular Biochemistry Department, Indiana University, Bloomington, IN, United States

Jimmy Ouellet
Depixus SAS, 3-5 impasse Reille, Paris, France

Craig L. Peterson
Program in Molecular Medicine, Morningside Graduate School of Biomedical Sciences, University of Massachusetts Chan Medical School, Worcester, MA, United States

M. Rhyan Puno
Structural Biology Program, Sloan Kettering Institute, Memorial Sloan Kettering Cancer Center, New York, NY, United States

Anna Marie Pyle
Department of Chemistry; Department of Molecular, Cellular, and Developmental Biology, Yale University, New Haven, CT; Howard Hughes Medical Institute, Chevy Chase, MD, United States

Prasangi Rajapaksha
Molecular & Cellular Biochemistry Department, Indiana University, Bloomington, IN, United States

Martin Rieu
Institut de Biologie de l'Ecole Normale Supérieure (IBENS), Ecole normale supérieure, CNRS, INSERM, PSL Research University; Laboratoire de Physique de L'Ecole Normale Supérieure de Paris, CNRS, ENS, Université PSL, Sorbonne Université, Université de Paris, Paris, France

Nadia Ruiz-Gutierrez
Institut de Biologie de l'Ecole Normale Supérieure (IBENS), Ecole normale supérieure, CNRS, INSERM, PSL Research University, Paris, France

Rick Russell
Department of Molecular Biosciences, University of Texas at Austin, Austin, TX, United States

Grant D. Schauer
Department of Biochemistry and Molecular Biology, Colorado State University, Fort Collins, CO, United States

Sudha Sharma
Department of Biochemistry and Molecular Biology; National Human Genome Center, College of Medicine, Howard University, Washington, DC, United States

Alisa E. Shaw
Department of Biochemistry and Molecular Biology, Colorado State University, Fort Collins, CO, United States

Robert H. Simmons III
Molecular & Cellular Biochemistry Department, Indiana University, Bloomington, IN, United States

David J. Sun
Molecular & Cellular Biochemistry Department, Indiana University, Bloomington, IN, United States

Srivats Venkataramanan
Department of Cell and Tissue Biology, University of California, San Francisco, CA, United States

Kevin C. Wilkins
Department of Cell and Tissue Biology, University of California; Graduate Division, University of California, San Francisco, San Francisco, CA, United States

Oliver J. Wilkinson
DNA:Protein Interactions Unit, School of Biochemistry, University of Bristol, Bristol, United Kingdom

Matthew K. Yim
Department of Chemistry & Biochemistry, Utah State University, Logan, Utah, United States

Naifu Zhang
Department of Chemistry and Biochemistry, The University of Texas at Dallas, Richardson, TX, United States

Preface

Nucleic acid polymers comprised of DNA or RNA have a spectrum of secondary structure stabilities including extremely dynamic and transient species, helices, loops and stems, and various more stable G-quadruplexes. The range of properties may have been critical to the evolution and selection of nucleic acids as early biological materials, as duplex stability is important for the protection of the genome, while other highly variant structures are utilized for specific purposes within the central dogma. A consequence of these secondary and tertiary nucleic acid structures is that transient separation of the base pairing interactions is required for various informational transactions that occur within the cell. This is accomplished by active and passive helicases, translocases, and ATP-dependent motor proteins and enzymes that act to partially or fully separate or remodel DNA or RNA strands. Nucleic acid remodeling processes are required for DNA replication, DNA repair, DNA recombination, transcription, ribosome biogenesis, translation, RNA splicing, RNA editing, RNA transport, RNA degradation, bacterial conjugation, chromosome segregation, and viral packaging and infection. Therefore, it is not surprising that many inherited human diseases including Bloom syndrome, Werner's syndrome, Rothmund-Thomson syndrome, Warsaw breakage syndrome, Fanconi anemia, Cockayne syndrome, xeroderma pigmentosum, trichothiodystrophy, Crohn's disease, dyskeratosis congenita, and ophthalmoplegia have been linked to defects in specific helicases. Moreover, defects or dysregulation in any 1 of 100 helicases in humans can lead to genomic instability, resulting in increased incidences of cancer, ovarian deficiency, infertility, aging, and neurodegeneration. Furthermore, many viruses encode helicases that play critical roles in viral life cycles, which have become targets for antiviral therapies.

While gel-based unwinding methods have provided for the bulk of early studies on the mechanisms of helicase action, the development and expansion of more recent technologies including single-molecule analyses, chemical biology tools, multiplex fluorescence approaches, advances in structural biology, in vivo DNA labeling and tracking, more advanced genetic methods, various next-generation sequencing isolation methods, and integrated mass spectrometry have greatly contributed to a broader spectrum of interdisciplinary approaches. The outcomes of these approaches have greatly expanded our mechanistic understanding of a variety of helicases, translocases, and remodeling/motor enzymes. Given the diversity of enzymes, their substrates, and the many information processing pathways within a cell, the variety of tools and techniques developed for the examination of helicase/translocase function has become extensive. The breadth of these techniques, as they have been applied to both DNA (Volume 672) and RNA (Volume 673) helicase/translocase substrates, is presented in this two-volume collection of *Methods in Enzymology*.

Volume 672 includes a collection of recent methods and developments that have been primarily utilized to examine DNA helicases and translocases. Chapter 1, by Bryd and colleagues, kicks it off by multiplexing DNA unwinding through multiple fluorophores simultaneously detecting intermediates and products in a single reaction through resolving and imaging of a gel. In Chapter 2, Raney and coworkers utilize methods to investigate the assembly of interacting multiple helicase subunits on a template to enhance unwinding. Chapter 3, by Keck and Duckworth, combines chemical biology tools for incorporating unnatural amino acids with mapping of DNA-binding sites by photoactivated crosslinking. Chapters 4 and 5 examine the assemblies and actions of other replisome proteins in context with hexameric replication helicases by rapid kinetic methods from the Patel laboratory and structural biology methods from the Gao laboratory. Chapter 6, from my own laboratory, takes advantage of TUNEL methodologies to examine DNA gaps and repair processes in bacteria enhanced by targeted genetic mutations in the replication helicase, dnaB. The loading and activation of hexameric helicases is a complex process, and Chapter 7 from the Berger laboratory utilizes biochemical tools to quantify these events. The Brosh laboratory, in Chapter 8, has detailed methods that have become popular recently to quantify DNA fiber lengths in cells to examine helicase-associated DNA replication or repair processes. The design of DNA substrates and utilization of various nucleotide substrates have become more important to trap and solve higher-order helicase structures, and the O'Donnell laboratory has outlined their strategies that have provided for great successes in Chapter 9. Speaking of high-resolution snapshots of hexameric helicase loading, the Costa laboratory has been leading this effort and highlights in silico methods to extract structural intermediates from CryoEM grids in Chapter 10. Chapters 11–13 by Remus, Paeschke, and Myong examine the unique and difficult process of melting very stable G-quadruplex structures, which are only just being recognized for their importance in controlling a variety of genetic information processes.

Chapter 14, by the van Oijen laboratory, details their method of creating extremely long DNA substrates with site-specific lesions for single-molecule analysis of coupled replication and repair events directed by helicases. Replication fork collapse from helicase blocks or translocation of single-stranded nicks can be studied using novel plasmids and by examining the consequences in the robust *Xenopus* egg extract system as outlined by the Dewar laboratory in Chapter 15. Genetic approaches have been useful in teasing out subpathways of DNA recombination, and Ira, Malkova, and colleagues highlight yeast reporter systems to quantify BIR outcomes in Chapter 16. Finally, Volume 672 concludes with Chapter 17 that describes a novel assay developed by the Vindigni laboratory to detect topoisomerase-trapped intermediates within the whole genome, which may be the result of specific inhibitors or defects in repair processing.

Volume 673 focuses primarily on the methods and techniques developed and adapted to examine the mechanisms and actions of RNA helicases, translocases, and remodelers. In Chapter 1, the Peterson laboratory describes quantitative fluorescence approaches to monitor enzymes that can remodel nucleosomes within a chromatin architecture. Katrin Karbstein outlines strategies and approaches for tackling elusive DEAD-box ATPases that act on many different RNA substrates including ribosome assembly in Chapter 2. Chapter 3, from the Pyle laboratory, takes a traditional ATPase assay and adapts it for lower volumes and higher throughput using the Rig-I helicase involved in adaptive immunity as an example. Assays to examine various DEAD-box RNA helicases continue in Chapters 4–7, with Russell, Johnson, Correll, Bolger, and Floor further detailing ATPase, RNA-folding intermediates, protein purifications, assessing cellular phenotypes in yeast, and utilizing human cells and extracts for translational luciferase read outs. RecQ and Pif1 family helicases are utilized in many aspects of DNA replication progression to resolve transcription conflicts or G-quadruplex structures, and the methods showing their activities on properly designed templates in bulk and single-molecule methods are presented in Chapters 8 and 9 by Bochman and Schauer, and genomic binding and transcriptional profiling related to disease in Chapter 10 by Sharma. Single-molecule FRET assays have been a hallmark of helicase studies for several decades now, and Klostermeier details the approaches for proper protein labeling and RNA substrate design in Chapter 11. Longer custom-designed DNA substrates are required for both optical and magnetic tweezer experiments with helicases, and the creation and validation of these are expertly laid out in Chapters 12 and 13 by Dillingham,

Moreno-Herrero, Croquette, and Le Hir. CRISPR-Cas systems also rely on helicase-nuclease enzymes to achieve adaptive immunity, and Ke outlines the methods to reconstitute the biochemistry of a type 1A system and utilize it for gene editing in human cells in Chapter 14. The final three chapters (Chapters 15–17) of Volume 673, by Johnson, Lima, and D'Arcy, detail the purification, reconstitution, substrate design, and HDX-MS structural characterization of RNA helicases and their substrates associated with exosomes.

Let me take this opportunity to extend my sincere thanks to all the authors in these two volumes for providing insights and explicit instructions on how to utilize their methods to examine diverse helicase mechanisms. I commend Paulo Mendoza and the Elsevier staff for their administrative help and guidance. I am grateful to Anna Pyle (editor-in-chief of *Methods in Enzymology*) for the invitation and for being a strong representative within the helicase field. My hope is that other laboratories will find these methods helpful as a starting point to adapt for their own enzymes of interest or extend these methods further for more in-depth understanding of helicase function. As I reflect on these chapters, it is a real honor to be the editor of both volumes, where long-standing leaders in the field, whom I have looked up to, and eager younger investigators bound to take on the mantel someday continue to provide inspiration to me and others working on this broad family of enzymes.

If you enjoyed these two volumes or found them helpful for your own research, I encourage you to consider joining us at the FASEB conference "Helicases and nucleic acid-based machines: Structure, mechanism, regulation, and roles in human diseases" in July 2023, where many of these authors will be attending and presenting and further refining their methods.

Michael A. Trakselis
Editor

CHAPTER ONE

Fluorescence approaches for biochemical analysis of ATP-dependent chromatin remodeling enzymes

Alexander S. Baier[a,b] **and Craig L. Peterson**[a,*]
[a]Program in Molecular Medicine, Morningside Graduate School of Biomedical Sciences, University of Massachusetts Chan Medical School, Worcester, MA, United States
[b]Medical Scientist Training Program, T.H. Chan School of Medicine, University of Massachusetts Chan Medical School, Worcester, MA, United States
*Corresponding author: e-mail address: craig.peterson@umassmed.edu

Contents

Abstract

The dynamic nature of chromatin is an essential mechanism by which gene expression is regulated. Chromatin is comprised of nucleosomes, an octamer of histone proteins wrapped by DNA, and manipulation of these structures is carried out by a family of proteins known as ATP-dependent chromatin remodeling enzymes. These enzymes carry out a diverse range of activities, from appropriately positioning and adjusting the

Methods in Enzymology, Volume 673
ISSN 0076-6879
https://doi.org/10.1016/bs.mie.2022.02.024

density of nucleosomes on genes, to installation and removal of histones for sequence variants, to ejection from DNA. These activities have a critical role in the proper maintenance of chromatin architecture, and dysregulation of chromatin remodeling is directly linked to the pathophysiology of various diseases. Mechanistic understanding of chromatin remodeling enzymes is therefore desirable, both as the drivers of this essential cellular activity and as potentially novel therapeutic targets in disease. In this chapter we cover our current methods for characterization of remodeler substrate binding affinity and catalytic activity, leveraging fluorescence polarization and Förster resonance energy transfer assays.

1. Introduction

The regulation of gene expression is an essential process by which cellular activity is directed and specialized. Regulation takes place from the level of individual gene transcription to genome wide organization and dynamics, and aberrant gene regulation underlies the pathophysiology of many diseases including cancers and developmental disorders (El Hadidy & Uversky, 2019; Lai & Wade, 2011; Mayes, Qiu, Alhazmi, & Landry, 2014; Mirabella, Foster, & Bartke, 2016; Wilson & Roberts, 2011; Wu et al., 2017). One important modality for the regulation of gene expression is chromatin architecture, or the spatial organization of the genome. Each human cell contains approximately two meters of DNA which is constrained within the nucleus, only 5 μm in diameter (Mirny, 2011). This astounding feat of compaction must also ensure the accessibility of various regions of DNA for transcription according to cell type, and this is accomplished through the dynamic nature of chromatin. At its fundamental level, chromatin consists of approximately 147 base pairs of DNA wrapped around an octamer of histone proteins, termed a nucleosome. DNA sequences that promote transcription are typically flanked or occluded by positioned nucleosomes, which can dramatically impact the accessibility of the general transcription machinery. Modulation of nucleosome occupancy, positioning, and composition, is regulated by ATP-dependent chromatin remodeling enzymes which play central roles in transcriptional regulation (Clapier, Iwasa, Cairns, & Peterson, 2017; Swygert & Peterson, 2014).

Chromatin remodeling enzymes can be broadly characterized into four subfamilies, the imitation switch (ISWI), chromodomain helicase DNA-binding (CHD), switching defective/sucrose non-fermentable (SWI/SNF), and inositol-requiring 80 (INO80) subfamilies (Bao & Shen, 2007; Clapier et al., 2017; Corona & Tamkun, 2004; Marfella & Imbalzano, 2007; Mohrmann & Verrijzer, 2005). All chromatin remodelers contain a conserved

ATPase domain within the catalytic subunit, placing these subfamilies in the RNA/DNA helicase superfamily 2, while subfamilies are defined by differing domains flanking the conserved ATPase (Clapier & Cairns, 2009). The critical role of these enzymes is underscored by the sequalae of their mutations, which can be found in many cancers and can cause other heritable diseases (El Hadidy & Uversky, 2019; Greenberg, Long, Swigut, & Wysocka, 2019; Vitobello et al., 2019). Thus understanding their function is desirable, and has historically been achieved through classical techniques such as DNAseI footprinting, electrophoretic mobility shift (EMSA), and restriction enzyme accessibility assays (Bartholomew, 2014; Côté, Quinn, Workman, & Peterson, 1994; Logie & Peterson, 1999; Watanabe et al., 2015).

In this chapter we discuss strategies that use fluorescent probes to investigate the biophysical properties of chromatin remodeling enzymes. The widespread availability of fluorescent dyes has revolutionized our ability to interrogate the activity of chromatin remodeling enzymes, and here we discuss our use of fluorescence polarization (FP) to investigate substrate binding affinities and Förster resonance energy transfer (FRET) to probe catalytic rates of nucleosome re-positioning. The techniques described are broadly applicable to chromatin remodeling enzymes and pose a low technical barrier to adoption.

2. Determining substrate affinity utilizing fluorescence polarization

Fluorescence polarization is an intrinsic property of fluorophores, as when excited by polarized light, a fluorophore will emit light at specific orientations relating to its dipole. As a fluorophore tumbles in space rapidly during the period it is excited, the emission will become less polarized since the emitted light will be detected in all planes. The degree of polarization can be described by the equation:

$$P = \frac{I_{\parallel} - I_{\perp}}{I_{\parallel} + I_{\perp}}$$

where P is an arbitrary and dimensionless unit of polarization, $I_{\parallel}$ is emitted light parallel to the source of illumination, and $I_{\perp}$ is emitted light perpendicular to the source of illumination. The rotational Brownian motion of a spherical molecule in a homogenous solution can be roughly approximated by Perrin's derivation of the Einstein-Stokes relation for a value ρ, defined as the rotational relaxation time, the time for a spherical molecule to rotate through 68.5°:

$$\rho = \frac{3\eta V}{RT}$$

where η is the viscosity of the fluid, V is the volume, R is the universal gas constant, and T is temperature (Perrin, 1926; Watanabe, 2012). Thus, the rotation of a fluorophore is directly proportional to the size of the molecule, and as the rotational relaxation time increases, the value of P will also increase. Armed with this knowledge, the change in P can be used to model binding kinetics as the interaction of a large molecule with a fluorophore-tagged substrate of interest should significantly impact ρ, assuming temperature and solution viscosity remain constant (Fig. 1).

2.1 Before you begin

2.1.1 Timing: 1 week

Consider the type of chromatin remodeling enzyme being studied and reconstitute nucleosomes for use in this assay as appropriate. Our lab studies chromatin remodeling enzymes from yeast and as such we use nucleosomes consisting of a core tetramer of xenopus H3 and H4 with yeast H2A and H2B dimers. In our observations, remodeler substrate affinities range in the tens of nanomolar, thus a starting stock concentration equal to or greater than 2 μM of the remodeler of interest ensures adequate datapoints for a binding curve fit. The following materials should be prepared in advance, using the example of SWR1 complex (SWR1C) from *Saccharomyces cerevisiae*

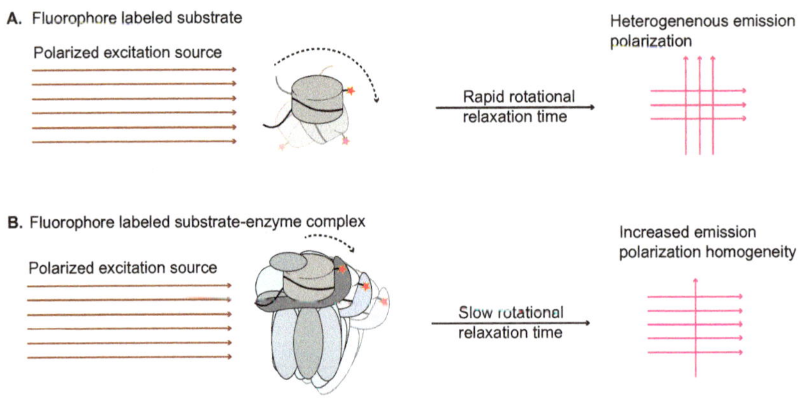

Fig. 1 Schematic of fluorescence polarization experimental setup. (A) shows the polarized excitation of a rapidly tumbling small fluorophore labeled nucleosome substrate with subsequent emission detected in multiple planes. (B) shows the polarized excitation of that same substrate bound to a large chromatin remodeler, resulting in a slower tumbling and an increase in emission polarization homogeneity.

as the chromatin remodeler of interest for the purpose of this protocol. All experiments were conducted at 20 °C.

- Assay buffer
  - o 25 mM HEPES-KOH pH 7.6
  - o 70 mM KCl
  - o 0.2 mM EDTA
  - o 5 mM $MgCl_2$
  - o 1 mM DTT
  - o 0.1 mg/mL Bovine Serum Albumin
- ADP (100 mM stock solution)
- SWR1C (2 μM stock solution)
  - o We purified SWR1C from *S. cerevisiae* engineered with FLAG-tag Swr1 as previously described (Mizuguchi et al., 2004).
- 77N4-ATTO647 labeled nucleosomes
  - o ATTO647 labeled nucleosomes were prepared using a commercial, end-labeled primer to generate a 77N4-ATTO647 601 positioning sequence via Taq PCR amplification, followed by salt dialysis nucleosome reconstitution from recombinant histone octamers as previously described (Luger, Rechsteiner, & Richmond, 1999; Singh et al., 2019). When reconstituted, this end-positioned nucleosome is flanked by 77 bp and 4 bp linker DNAs, with an ATTO647 dye on the 3′ end of the 4 bp linker (Fig. 2A).

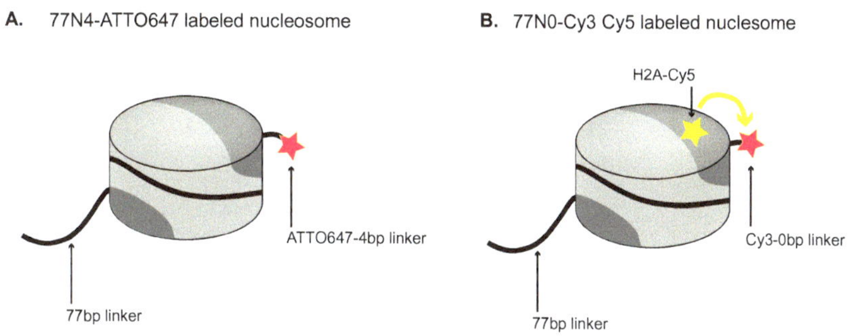

Fig. 2 Design of fluorophore labeled nucleosome substrates. Dark gray regions represent *Saccharomyces cerevisiae* H2A/H2B dimers; light gray regions represent *X. laevis* H3/H4 tetramers. (A) shows the nucleosome labeling scheme used in fluorescence polarization experiments, with ATTO647 label added via labeled primer to distal linker DNA. (B) shows the nucleosome labeling scheme used in FRET experiments, with Cy3 label added via labeled primer to distal linker DNA, and Cy5 added to the octamer via maleimide based labeling at H2A120C (only one label is shown for simplicity, both H2A subunits are labeled with Cy5).

Alternatives: For investigating the binding affinity of a chromatin remodeling enzyme that slides DNA, we recommend using a fluorophore labeled histone for fluorescence polarization experiments to avoid the introduction of noise from a mobile fluorophore. Labels can be added by site directed mutagenesis of a single residue to a cysteine followed by maleimide based labeling as has been previously described (Zhou & Narlikar, 2016). Choice of fluorophore is inconsequential so long as it is relatively photostable and has a convenient excitation and emission wavelength for the configuration of the microplate reader available to the investigator. This strategy has been successfully used for FP experiments with a chromatin remodeler (Markert, Zhou, & Luger, 2021).

2.2 Materials and equipment

- Tecan Spark multimode microplate reader
- 384-well black round bottom microplate
- Adjustable 10 μL 12-tip multichannel (capable of collecting from a 96 well plate or PCR strip format and pipetting into a 384 well plate)

Alternatives: Any microplate reader capable of measuring fluorescence polarization can be used to perform this experiment.

2.3 Step-by-step method details

2.3.1 Setup and instrument calibration

Timing: 20 min

1. Dilute 77N4-ATTO647 nucleosomes to 20 nM in assay buffer on ice, with 120 μL of volume required for each pipetting replicate. For separate conditions, for example apo versus the presence of an ATP analogue such as ADP, prepare separate stock solutions
2. From the apo stock solution prepared in step 1, dilute 12 μL with an additional 12 μL of assay buffer, for a small volume of 10 nM nucleosomes. Add 20 μL to A1 in the microplate and 20 μL of assay buffer alone to A2 and incubate at room temperature for 15 min. These will be used for instrument calibration
3. Using a strip of 12 PCR tubes on ice or a 96-well plate, prepare an 11 step 2-fold serial dilution of the 2 μM stock of SWR1C in assay buffer from right to left such that the rightmost tube contains 2 μM SWR1C, followed by 1 μM SWR1C, ..., to 1.9 nM SWR1C. In the 12th and leftmost tube, have assay buffer alone (Fig. 3A). Each set of conditions will require 10 μL, so adjust the total volume accordingly

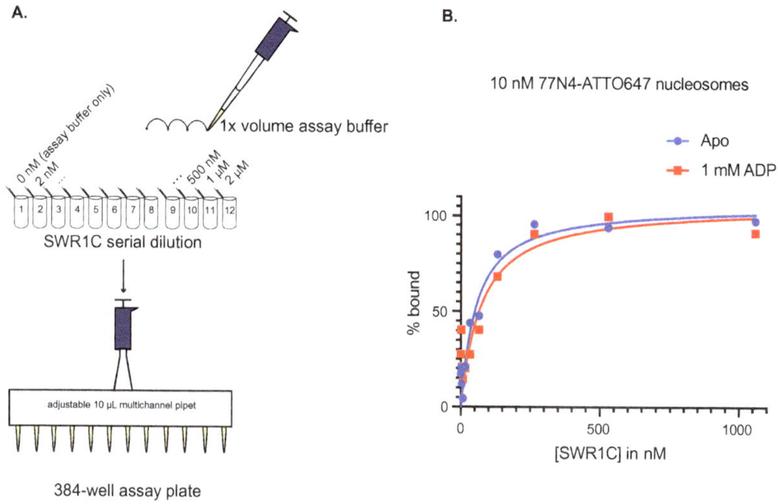

Fig. 3 (A) demonstrates the serial dilution setup for conducting a fluorescence polarization experiment. Briefly, 2 μM SWR1C is placed into tubes 11 and 12. Tube 11 is then diluted with 1 × volume of assay buffer, with half transferred to tube 10. This is then diluted with 1 × volume of assay buffer, continuing serially to tube 2. Tube 1 is filled with assay buffer alone. A multichannel pipettor is then used to transfer the appropriate amount of enzyme to the 384-well microplate. (B) shows fluorescence polarization for 10 nM 77N4-ATTO657 nucleosomes in the presence (red squares) or absence of ADP (blue circles) plotted as a function of SWR1C concentration. Polarization values are normalized to a percentage, and a best fit line derived from the Morrison equation is shown.

4. *Optional:* Prepare an additional serial dilution of the buffer containing the chromatin remodeling enzyme, should that buffer be dramatically different from the assay buffer (i.e. much greater viscosity, etc.).
5. Turn on the Tecan Spark and identify the microplate type in the method editor. Select the fluorescence polarization setting under detection and set the excitation and emission wavelengths to 631 and 686 nm respectively, or to appropriate wavelengths if using a different fluorophore. Set number of flashes to 20
6. After the incubation in Step 2 has finished, load the microplate into the Tecan Spark. Under advanced settings, adjust the *Z*-height of the plate to maximize signal. Calibrate the grating factor (G-Factor) by setting well A1 (containing the 77N4-ATTO647 nucleosomes) as the reference and A2 (containing assay buffer alone) as the blank

2.3.2 Experimental setup

Timing: 45 min

7. Pipet 10 μL of 20 nM 77N4-ATTO647 stock solution from step 1 in rows of 12 starting with B1 on the 384 well microplate
8. *Optional:* Multiple rows can be done per stock solution condition for pipetting replicates
9. Using the multichannel, pipet 10 μL of the serial dilution of SWR1C prepared in step 3 into each row such that lane 1 receives assay buffer alone and lane 12 receives 2 μM SWR1C (Fig. 3A).
10. Tap plate gently to ensure all liquid is at bottom of well. If you prepared a buffer control (optional step 4), include here
11. Incubate plate at room temperature for 15 min
12. In the method editor, highlight all lanes containing sample and press run

2.3.3 Expected outcomes

The Tecan software will generate an excel file containing polarization values for each flash for each well reported in mP. There should be a quantifiable increase in the mP value as the concentration of the chromatin remodeling enzyme increases. The mP values for each well will remain constant over the course of the experiment, indicating that equilibrium has been reached. If this is not the case, incubate for a longer time at Step 10.

2.4 Quantification and statistical analysis

For a single experiment, the mP values for each well should be averaged together prior to analysis. Pipetting replicates (Step 8), if included, are useful for assessing reproducibility between reads but cannot be analyzed to calculate standard errors, as they are not biological replicates. In Prism9, plot the mP value for each condition in the Y column against the concentrations of SWR1C and normalize such that Bmax is 100%. Under the analysis function in Prism9, determine the K_d using the Morrison equation:

$$Y = Bmax\left(\frac{\left([F] + [E] + K_{d,app}\right) - \sqrt{\left([F] + [E] + K_{\mathrm{d,app}}\right)^2 - 4[F][E]}}{2[F]}\right)$$

where Bmax is maximum observed binding, [F] is the concentration of the ATTO647 labeled nucleosomes, [E] is the concentration of SWR1C, and $K_{d,app}$ is the apparent dissociation constant. Example binding curves for SWR1C in the presence of ADP and in the apo state are shown (Fig. 3B).

2.5 Advantages

Determining binding affinities using microplate fluorescence polarization yields many advantages over other traditional techniques. Classical binding assays are performed with radioligands, which are expensive single use reagents, and pose safety and disposal challenges. FP is also a non-destructive assay, meaning conditions can be modified to observe changes in substrate affinity throughout an experiment. This allows for rapid and high throughput investigation of enzyme-substrate interaction under various conditions with relatively low protein requirements, especially when compared to other techniques such as isothermal calorimetry (ITC) or titrations in a cuvette-based fluorimeter. Lastly, the equipment for carrying out FP experiments, namely plate readers, are versatile in their function and broadly available in academic settings. Other techniques such as radioligand investigations and ITC rely on single purpose equipment, which can be a barrier to smaller labs.

2.6 Limitations

The sensitivity of FP assays relies on a directly proportional relationship between the volume of the labeled molecule and the rotational relaxation time. Thus, larger substrates such as nucleosomes may be difficult to interrogate using this technique if the binding partner is comparatively small. In the case of SWR1C, which is >1 MDa in size, this issue doesn't arise. Smaller single subunit remodelers may require optimization of this protocol however, this is not an insurmountable barrier and has been previously demonstrated (Markert et al., 2021). Additionally, the minimum concentration of substrate is limited by the sensitivity of the detector to the fluorophore. In this experiment our ATTO647 labeled substrate was held at 10 nM with the concentration of SWR1C titrated from an order of magnitude lower to two orders of magnitude higher. For substrate-ligand interactions with extremely high affinities, a detectable concentration of substrate may be much greater than the K_d resulting in an artificially inflated $K_{d,app}$. An adequate titration of enzyme concentration above and below the substrate concentration will reveal whether this is the case. For a given fluorophore, titration of labeled substrate compared to a buffer control will allow for the determination of minimum detectable concentration which may vary by reaction volume and detector manufacturer. Substrate-ligand interactions with extremely low affinities may require prohibitive amounts of enzyme to achieve an appropriate titration spectrum (Jarmoskaite, AlSadhan, Vaidyanathan, & Herschlag, 2020).

3. Determining remodeling rates using Förster resonance energy transfer (FRET)

Another useful property of fluorophores can be found in FRET pairs, fluorophores in which the emission of one excited fluorophore excites the other fluorophore. The efficiency of this energy transfer (E) is described the equation:

$$E = \frac{1}{1 + (r/R_o)^6}$$

where r is the distance between the pair and R_o is the empirically determined Förster radius, the distance at which the energy transfer efficiency is 50%. As can be seen from this equation, the energy transfer efficiency is roughly proportional to $1/r^6$, making FRET extremely sensitive to changes on distance. This sensitivity can be leveraged to examine various actions of chromatin remodeling enzymes on nucleosomes, for example sliding, via the careful placement of FRET pairs on a nucleosome (Fig. 4).

To illuminate how the distance between FRET pairs in proportion to the Förster radius effects FRET efficiency, take the example FRET pair of Cyanine3 (Cy3) and Cyanine5 (Cy5) discussed in this protocol. The Förster radius of Cy3/Cy5 is 5.2 nm (Olejko & Bald, 2017), and from this starting radius, an atomic level movement of $\pm$ 1 Å from R_o results in a ΔE of

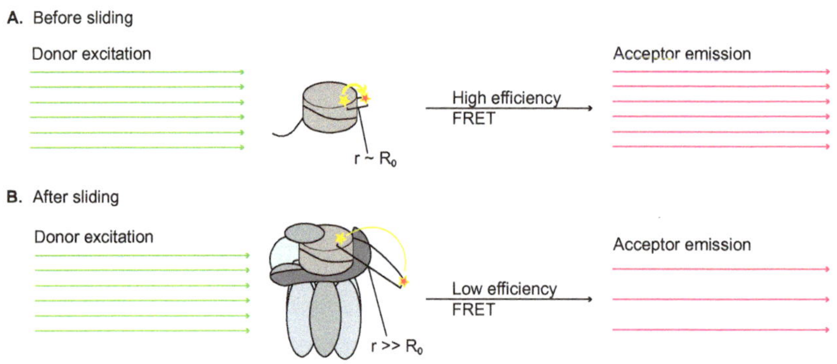

Fig. 4 Schematic of Förster resonance energy transfer based chromatin remodeler sliding assay experimental setup. (A) shows a FRET-pair labeled nucleosome substrate with the distance between labels close to the Förster radius, with excitation of the donor label resulting in high acceptor emission. (B) shows that same substrate after DNA sliding has occurred due to a chromatin remodeler enzyme, resulting in an increased distance between the donor and acceptor fluorophore leading to a decrease in acceptor emission.

5.8%. Thus, careful placement of FRET pair on the nucleosome coupled with informed decision of FRET pair choice based on Förster radius allows for real time observations of exquisite sensitivity.

3.1 Before you begin

3.1.1 Timing: 1 week

Much like with fluorescence polarization, first consider the species of chromatin remodeling enzyme being studied and reconstitute nucleosomes appropriately. The following materials should be prepared in advance, using the example of INO80 complex (INO80C) from *S. cerevisiae* as the chromatin remodeler of interest for the purpose of this protocol. All experiments were conducted at 20 °C. To achieve reaction saturation for any given ATP-dependent chromatin remodeling enzyme, consider adjusting the temperature and concentrations used in this protocol.

- Assay buffer
  - o 25 mM HEPES-KOH pH 7.6
  - o 70 mM KCl
  - o 0.2 mM EDTA
  - o 5 mM $MgCl_2$
  - o 1 mM DTT
  - o 0.1 mg/mL Bovine Serum Albumin
- ATP and AMP-PNP (100 mM stock solutions)
- INO80C (750 nM stock solution)
  - o We purified INO80C from *S. cerevisiae* engineered to have an Ino80-TAP tag as previously described (Sinha, Watanabe, Johnson, Moazed, & Peterson, 2009).
- 77N0-Cy3 Cy5 labeled nucleosomes
  - o Yeast xenopus hybrid octamers were labeled with Cy5 as previously described (Zhou & Narlikar, 2016). In brief, histone subunits were prepared separately via recombinant expression. Yeast H2A position 120 was mutated to a cysteine, followed by maleimide based Cy5 labeling. Free dye was quenched with Tris(2-carboxyethyl)phosphine and then depleted via serial concentration with 3 kDa-cutoff protein concentrators. Labeled 77N0-Cy3 was made using a Cy3 labeled primer coupled with an unlabeled primer to PCR from a 601 positioning sequence plasmid and nucleosomes were then reconstituted by salt dialysis as previously described (Luger et al., 1999). When reconstituted, this end-positioned nucleosome is flanked by 77 bp and 0 bp linker DNAs (Fig. 2B).

3.2 Materials and equipment

- Tecan Spark multimode microplate reader
- 96-well black microplate with clear flat bottom
- 10 μL multichannel pipette

Alternatives: Any microplate reader with appropriate lasers and filters for the given FRET pair can be used to perform this experiment. The 96-well plate format allows for higher volume reactions to maximize signal to noise for exquisite sensitivity while maintaining low sample input requirements and requiring minimal or no optimization, however smaller volumes in a 384-well microplate are tractable for difficult to produce enzymes.

3.3 Step-by-step method details

3.3.1 Setup and instrument calibration

Timing: 20 min

1. Dilute labeled nucleosomes in assay buffer to a total volume of 127 μL such that increasing the volume to 150 μL would result in a final concentration of 10 nM nucleosomes
2. In a black 96 well microplate with a clear flat bottom, pipet 50 μL of assay buffer into well A1 and 42.33 μL of labeled nucleosomes into A2. Add 7.66 μL of assay buffer to the nucleosome wells to bring the volume up to 50 μL
3. Turn on the Tecan Spark and identify the microplate type in the method editor. Select the FRET setting under detection and set the excitation wavelength to 530 nm or to the appropriate wavelength if using a different FRET pair. Manually adjust the gain to 160 and the mode to fluorescence bottom reading
4. Load the microplate into the Tecan Spark. Select the wells with sample in the method editor and perform a fluorescence scan from 560 nm to 700 nm to confirm the absence of signal in the assay buffer alone and good FRET signal from the labeled nucleosomes. The initial fluorescence scan should show a large Cy5 peak and a smaller Cy3 peak (Fig. 5A). *Note:* For novel conditions, single-labeled substrate controls should be investigated to determine if enzyme binding has any quenching effect on fluorescence

3.3.2 Experimental setup

Timing: 20 min

5. In fresh wells, load 42.33 μL of each nucleosome condition, in duplicate. In this experiment we examined INO80C sliding in the presence of ATP or AMP-PNP, and thus loaded four wells

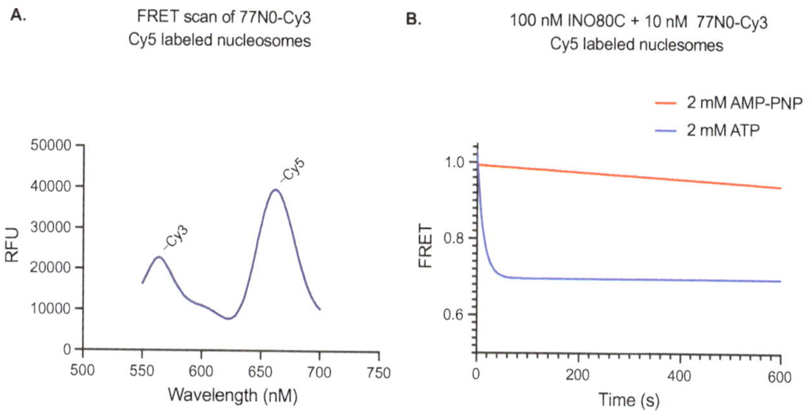

Fig. 5 (A) shows a FRET scan of a 77N0-Cy3 Cy5 labeled nucleosome, plotted as relative fluorescence units (RFUs) over wavelength. The emission peaks of Cy3 and Cy5 are labeled. (B) shows emission at 670 nm of 77N0-Cy3 Cy5 labeled nucleosomes excited with a 530 nm laser in the presence of INO80C and AMP-PNP (red) or ATP (blue). In the presence of ATP, INO80C slides the end positioned nucleosome to a center position, resulting in a loss of FRET. In the presence of AMP-PNP, no sliding occurs, and the emission remains constant. Data is normalized to a fraction of the starting FRET signal.

6. Add 6.66 μL of the INO80C stock such that the final concentration in each well when the volume is brought to 50 μL will be 100 nM
7. Incubate INO80C and nucleosomes for 15 min
8. Prepare a kinetic measurement in the method editor of the Tecan Spark, with 300 cycles at an interval of 2 s, and 35 flashes. Set the excitation wavelength to 530 nm and the emission wavelength to 670 nm. The gain and mode should be the same as listed in Step 3
9. Highlight the wells containing sample in the method editor such that the program is ready to go. Place the microplate on the tray
10. Using the multichannel, initiate the assay by adding 1 μL of 100 mM ATP or AMP-PNP for a final concentration of 2 mM and a total condition volume of 50 μL. Immediately start the program to measure the change in FRET signal over time. With the program queued, time from reaction initiation to measurement is on the order of 1–2 s

3.3.3 Expected outcomes

The Tecan software will generate an excel sheet of relative fluorescence units (RFU) values, an arbitrary and dimensionless value, at the given time interval for the course of the experiment. Values from biological replicates can be averaged and plotted over time. In this experiment, in the presence of ATP the measured emission at 670 nm (Cy5) is expected to decrease after the addition of INO80C, as INO80C slides the end positioned nucleosome to a center position and the Cy3 label on the DNA is slid away from the Cy5 labeled histones. No change in FRET will be observed in the presence of a non-hydrolysable analogue such as AMP-PNP.

3.4 Quantification and statistical analysis

For this experiment, pipetting replicates should be averaged together prior to analysis to generate a list of RFU values over time. In Prism9, enter the RFU values for biological replicates in a single group as a function of time, with separate groups for each condition. In the AMP-PNP condition, no change in FRET is observed and the data shows a linear correlation with minimal slope. The ATP condition can be fit to a single-phase decay using the Prism9 analysis function with the equation:

$$Y = \left(Y_i - Y_p\right)^{-kt} + Y_p$$

where Y is the RFU value, Y_i is the starting RFU, Y_p is the RFU value at the plateau, k is the rate constant, and t is time (Fig. 5B). Other reaction conditions (for example, higher enzyme concentration or temperature) may result in a better fit by two phase decay:

$$Y = \left(P_f\left(Y_i - Y_p\right)\right)^{-k_f t} + \left(P_s\left(Y_i - Y_p\right)\right)^{-k_s t} + Y_p$$

where Y is the RFU value, Y_i is the starting RFU, Y_p is the RFU value at the plateau, P_f is the percent fast phase expressed as a fraction, P_s is the percent slow phase expressed as a fraction, k_f is the rate constant of the fast phase, k_s is the rate constant of the slow phase, and t is time. Other chromatin remodeling enzyme reactions may exhibit behaviors better described by other statistical models.

3.5 Advantages

FRET is an inexpensive and versatile tool for examining chromatin remodeling enzyme kinetics. Through careful placement of FRET pairs, canonical activities such as sliding and dimer exchange can be observed through gain or loss of FRET signal. Older techniques for measuring remodeling activity such as gel shift assays introduce non-native environments and can be difficult to coordinate on an appropriate time scale. FRET additionally allows for rapid and high throughput iteration of experimental conditions with minimal sample requirements. The application of FRET continues to grow, with newer studies leveraging single molecule and overlapping

FRET pair approaches to investigate chromatin remodeling behaviors masked by bulk ensemble assays (Fan, Moreno, Baier, Loparo, & Peterson, 2022; Sabantsev, Levendosky, Zhuang, Bowman, & Deindl, 2019).

3.6 Limitations

The most challenging aspect of FRET assays is generally achieving an acceptable level of signal to noise. This problem can arise due to label concentration and photostability, as well as instrument sensitivity and FRET pair properties. Mitigation strategies include choosing fluorophores with acceptably high photostability and low photoblinking properties, as well as adjusting fluorophore label placement for better initial or final FRET efficiency. FRET pairs with low spectral overlap are also desirable. At Step 4, should a custom assay buffer have high background at the FRET acceptor emission wavelength, the measured emission wavelength can be adjusted to an area of lower noise that still has an acceptable level of FRET acceptor signal as determined by the FRET scan, the FRET pair can be changed, or the buffer reformulated. Additionally, at step 10, there is dead time of 1–2 s between reaction initiation and measurement. For reactions with extremely fast kinetics, this dead time can obscure the initial rate. Strategies to slow the reaction such as lower temperatures, ATP concentrations, and/or enzyme concentration can help mitigate this problem.

4. Summary and conclusions

This chapter demonstrates the use of fluorescent probes to investigate the substrate binding affinity and catalytic activity of selected ATP-dependent chromatin remodeling enzymes from *S. cerevisiae*. We demonstrate that by leveraging the properties of fluorescence polarization and Förster resonance energy transfer, the biophysical properties of ATP-dependent chromatin remodelers can be easily and reproducibly assayed. Furthermore, the protocols outlined here are broadly applicable to both canonical and noncanonical ATP-dependent chromatin remodeling enzymes, of any species, providing a framework for future studies of remodelers that remain poorly characterized.

Acknowledgments

We thank past and present members of the Peterson laboratory for their contributions to standardizing these protocols. This work was supported by NIGMS grant 5R35GM122519-05 awarded to C.L.P. Additional support was provided to A.S.B. from NIGMS grant 5T32GM107000-09.

References

Bao, Y., & Shen, X. (2007). INO80 subfamily of chromatin remodeling complexes. *Mutation Research, Fundamental and Molecular Mechanisms of Mutagenesis, 618*(1–2), 18–29. https://doi.org/10.1016/J.MRFMMM.2006.10.006.

Bartholomew, B. (2014). ISWI chromatin remodeling: One primary actor or a coordinated effort? *Current Opinion in Structural Biology, 24*. https://doi.org/10.1016/j.sbi.2014.01.010.

Clapier, C. R., & Cairns, B. R. (2009). The biology of chromatin remodeling complexes. *Annual Review of Biochemistry, 78*, 273–304. https://doi.org/10.1146/ANNUREV.BIOCHEM.77.062706.153223.

Clapier, C. R., Iwasa, J., Cairns, B. R., & Peterson, C. L. (2017). Mechanisms of action and regulation of ATP-dependent chromatin-remodelling complexes. *Nature Reviews Molecular Cell Biology, 18*(7), 407–422. https://doi.org/10.1038/nrm.2017.26.

Corona, D. F. V., & Tamkun, J. W. (2004). Multiple roles for ISWI in transcription, chromosome organization and DNA replication. *Biochimica et Biophysica Acta (BBA) - Gene Structure and Expression, 1677*(1–3), 113–119. https://doi.org/10.1016/J.BBAEXP.2003.09.018.

Côté, J., Quinn, J., Workman, J. L., & Peterson, C. L. (1994). Stimulation of GAL4 derivative binding to nucleosomal DNA by the yeast SWI/SNF complex. *Science, 265*(5168), 53–60. https://doi.org/10.1126/SCIENCE.8016655.

El Hadidy, N., & Uversky, V. N. (2019). Intrinsic disorder of the BAF complex: Roles in chromatin remodeling and disease development. *International Journal of Molecular Sciences, 20*(21). https://doi.org/10.3390/IJMS20215260.

Fan, J., Moreno, A. T., Baier, A. S., Loparo, J. J., & Peterson, C. L. (2022). H2A.Z deposition by SWR1C involves multiple ATP-dependent steps. *BioRxiv*. https://doi.org/10.1101/2022.01.11.475888. 2022.01.11.475888.

Greenberg, R. S., Long, H. K., Swigut, T., & Wysocka, J. (2019). Single amino acid change underlies distinct roles of H2A.Z subtypes in human syndrome. *Cell, 178*(6), 1421–1436.e24. https://doi.org/10.1016/j.cell.2019.08.002.

Jarmoskaite, I., AlSadhan, I., Vaidyanathan, P. P., & Herschlag, D. (2020). How to measure and evaluate binding affinities. *eLife, 9*. https://doi.org/10.7554/eLife.57264.

Lai, A. Y., & Wade, P. A. (2011). Cancer biology and NuRD: A multifaceted chromatin remodelling complex. *Nature Reviews. Cancer, 11*(8), 588–596. https://doi.org/10.1038/NRC3091.

Logie, C., & Peterson, C. L. (1999). Purification and biochemical properties of yeast SWI/SNF complex. *Methods in Enzymology, 304*, 726–741. https://doi.org/10.1016/S0076-6879(99)04044-6.

Luger, K., Rechsteiner, T. J., & Richmond, T. J. (1999). Preparation of nucleosome core particle from recombinant histones. *Methods in Enzymology, 304*, 3–19. https://doi.org/10.1016/S0076-6879(99)04003-3.

Marfella, C. G. A., & Imbalzano, A. N. (2007). The Chd family of chromatin remodelers. *Mutation Research, Fundamental and Molecular Mechanisms of Mutagenesis, 618*(1–2), 30–40. https://doi.org/10.1016/j.mrfmmm.2006.07.012.

Markert, J., Zhou, K., & Luger, K. (2021). Smarcad1 is an atp-dependent histone octamer exchange factor with de novo nucleosome assembly activity. *Science Advances*, 7(42). https://doi.org/10.1126/SCIADV.ABK2380/SUPPL_FILE/SCIADV.ABK2380_REFERENCE_MATERIAL.ZIP.

Mayes, K., Qiu, Z., Alhazmi, A., & Landry, J. W. (2014). ATP-dependent chromatin remodeling complexes as novel targets for cancer therapy. *Advances in Cancer Research, 121*, 183. https://doi.org/10.1016/B978-0-12-800249-0.00005-6.

Mirabella, A. C., Foster, B. M., & Bartke, T. (2016). Chromatin deregulation in disease. *Chromosoma, 125*(1), 75. https://doi.org/10.1007/S00412-015-0530-0.

Mirny, L. A. (2011). The fractal globule as a model of chromatin architecture in the cell. *Chromosome Research*, *19*(1), 37. https://doi.org/10.1007/S10577-010-9177-0.

Mizuguchi, G., Shen, X., Landry, J., Wu, W. H., Sen, S., & Wu, C. (2004). ATP-driven exchange of histone H2AZ variant catalyzed by SWR1 chromatin remodeling complex. *Science*, *303*(5656), 343–348. https://doi.org/10.1126/SCIENCE.1090701/SUPPL_FILE/MIZUGUCHI.SOM.PDF.

Mohrmann, L., & Verrijzer, C. P. (2005). Composition and functional specificity of SWI2/SNF2 class chromatin remodeling complexes. *Biochimica et Biophysica Acta (BBA) - Gene Structure and Expression*, *1681*(2–3), 59–73. https://doi.org/10.1016/J.BBAEXP.2004.10.005.

Olejko, L., & Bald, I. (2017). FRET efficiency and antenna effect in multi-color DNA origami-based light harvesting systems. *RSC Advances*, 7(39), 23924–23934. https://doi.org/10.1039/C7RA02114C.

Perrin, F. (1926). Polarisation de la lumière de fluorescence. Vie moyenne des molécules dans l'etat excité. *Journal de Physique et le Radium*, 7(12), 390–401. https://doi.org/10.1051/jphysrad:01926007012039000.

Sabantsev, A., Levendosky, R. F., Zhuang, X., Bowman, G. D., & Deindl, S. (2019). Direct observation of coordinated DNA movements on the nucleosome during chromatin remodelling. *Nature Communications*, *10*(1). https://doi.org/10.1038/S41467-019-09657-1.

Singh, R. K., Fan, J., Gioacchini, N., Watanabe, S., Bilsel, O., & Peterson, C. L. (2019). Transient kinetic analysis of SWR1C-catalyzed H2A.Z deposition unravels the impact of nucleosome dynamics and the asymmetry of histone exchange. *Cell Reports*, *27*(2), 374–386.e4. https://doi.org/10.1016/j.celrep.2019.03.035.

Sinha, M., Watanabe, S., Johnson, A., Moazed, D., & Peterson, C. L. (2009). Recombinational repair within heterochromatin requires ATP-dependent chromatin remodeling. *Cell*, *138*(6), 1109–1121. https://doi.org/10.1016/J.CELL.2009.07.013.

Swygert, S. G., & Peterson, C. L. (2014). Chromatin dynamics: Interplay between remodeling enzymes and histone modifications. *Biochimica et Biophysica Acta, Gene Regulatory Mechanisms*, *1839*, 728–736. https://doi.org/10.1016/j.bbagrm.2014.02.013.

Vitobello, A., Perner, J., Beil, J., Zhu, J., Del Río-Espínola, A., Morawiec, L., et al. (2019). Drug-induced chromatin accessibility changes associate with sensitivity to liver tumor promotion. *Life Science Alliance*, *2*(5). https://doi.org/10.26508/lsa.201900461.

Watanabe, F. (2012). In T. T. Ngo (Ed.), *Nonisotopic immunoassay* Plenum Press. https://doi.org/10.1007/978-1-4684-5468-0.

Watanabe, S., Tan, D., Lakshminarasimhan, M., Washburn, M. P., Hong, E. J. E., Walz, T., et al. (2015). Structural analyses of the chromatin remodelling enzymes INO80-C and SWR-C. *Nature Communications*, *6*(1), 1–10. https://doi.org/10.1038/ncomms8108.

Wilson, B. G., & Roberts, C. W. M. (2011). SWI/SNF nucleosome remodellers and cancer. *Nature Reviews. Cancer*, *11*(7), 481–492. https://doi.org/10.1038/NRC3068.

Wu, Q., Lian, J. B., Stein, J. L., Stein, G. S., Nickerson, J. A., & Imbalzano, A. N. (2017, June 1). The BRG1 ATPase of human SWI/SNF chromatin remodeling enzymes as a driver of cancer. *Epigenomics*, *9*, 919–931. https://doi.org/10.2217/epi-2017-0034.

Zhou, C. Y., & Narlikar, G. J. (2016). Analysis of nucleosome sliding by ATP-dependent chromatin remodeling enzymes. *Methods in Enzymology*, *573*, 119–135. https://doi.org/10.1016/BS.MIE.2016.01.015.

CHAPTER TWO

Attacking a DEAD problem: The role of DEAD-box ATPases in ribosome assembly and beyond

Katrin Karbstein[a,b,c,*]

[a]Department of Integrative Structural and Computational Biology, Scripps Florida, Jupiter, FL, United States
[b]HHMI Faculty Scholar, Chevy Chase, MD, United States
[c]The Skaggs Graduate School of Chemical and Biological Sciences, Scripps Florida, Jupiter, FL, United States
*Corresponding author: e-mail address: kkarbst@scripps.edu

Contents

Abstract

DEAD-box proteins are a subfamily of ATPases with similarity to RecA-type helicases that are involved in all aspects of RNA Biology. Despite their potential to regulate these processes *via* their RNA-dependent ATPase activity, their roles remain poorly characterized. Here I describe a roadmap to study these proteins in the context of ribosome assembly, the process that utilizes more than half of all DEAD-box proteins encoded in the yeast genome.

1. Introduction

DEAD-box proteins are a subfamily of ATPases, containing the namesake D-E-A-D amino acid signature in their Walker B motif. In addition to this characteristic eponymous sequence, DEAD-box proteins have 11 additional characterizing sequence features (reviewed in Fairman-Williams, Guenther, & Jankowsky, 2010), some of which are shared with other

Methods in Enzymology, Volume 673
ISSN 0076-6879
https://doi.org/10.1016/bs.mie.2022.03.033

ATPases (such as the Walker A motif). DEAD-box proteins are ubiquitous regulators of RNA biology (Bourgeois, Mortreux, & Auboeuf, 2016; Jarmoskaite & Russell, 2014): the 26 DEAD-box proteins encoded in the yeast genome are involved in transcription (Dbp2), splicing (Prp5, Prp28 and Sub2), RNA export (Sub2, Dbp5), translation (eIF4A (two isoforms), Ded1 (two isoforms), Dhh1). Notably, the largest group is involved in ribosome assembly (14 total: Dbp3, Dbp4, Dbp6, Dbp7, Dbp8, Dbp9, Dbp10, Drs1, Fal1, Has1, Mak5, Rok1, Rrp3, Spb4).

Because of their potential to utilize their ATPase activity to assume key regulatory roles, these factors have garnered a lot of interest. This is especially true in ribosome assembly, not just because of the outsize role this pathway plays in the biology of these proteins, but also because there are surprisingly few energy-utilizing enzymes involved in this highly regulated, hierarchical process (Strunk & Karbstein, 2009), rendering DEAD-box ATPases even more conspicuous as potential regulators of this essential assembly line.

DEAD-box proteins are highly similar, on both a sequence and structural level to the related subclass of DEAH box proteins. As result, in the past, these two classes of proteins have often been referred to together as DEAD/H-box proteins, and it was assumed that they would function similarly. Moreover, a large body of beautiful work from several different labs has shown that DEAH-box proteins can unwind RNA duplexes (Jankowsky, Gross, Shuman, & Pyle, 2000; Jia, Wang, Anderson, & Jankowsky, 2012; Lebaron et al., 2009), similar to DNA (and RNA) helicases, which share structural similarity with both DEAD and DEAH-box proteins. As a result, both protein families were often referred to as RNA helicases. Nonetheless, more recently, mechanistic differences between these two subclasses of RNA-dependent ATPases have become clear (Gilman, Tijerina, & Russell, 2017). Detailed mechanistic analyses have shown that DEAD-box proteins are not processive helicases and can only separate small segments of double stranded RNA (Chen et al., 2008; Liu, Putnam, & Jankowsky, 2008). However, it was shown that DEAD-box proteins can also anneal duplexes, and, critically, dissociate RNA-protein complexes (Jankowsky, 2011; Linder & Jankowsky, 2011; Putnam & Jankowsky, 2013). I suspect that this misnaming of DEAD-box proteins as RNA helicases, instead of more neutrally as RNA-dependent ATPases, or ATP-dependent RNA binding proteins, has contributed to our failure to better understand their functions in biological systems, as essentially, we may have been looking for the wrong thing all along. The goal of this review is to provide a blueprint for (re-) defining the function of DEAD-box proteins in ribosome assembly and beyond.

2. DEAD-box proteins tend to remodel RNA-protein interactions *in vivo*

As discussed above, I believe that part of the reason why the role of DEAD-box proteins in ribosome assembly has been slow to emerge is that we have been looking for roles for RNA helicases, even though DEAD-box proteins are not good unwinding enzymes. While this is a warning against biasing our thinking (too much), it is helpful to consider the known roles of DEAD-box proteins in splicing and ribosome assembly as a basis for further study of the roles of DEAD-box proteins in ribosome assembly and other processes.

Pre-mRNA splicing involves the assembly and remodeling of numerous RNA duplexes and RNA-protein complexes to ensure the high fidelity of the process and the integrity of the mRNA (Wilkinson, Charenton, & Nagai, 2020; Yan, Wan, & Shi, 2019). Not surprisingly, several DEAD-box and DEAH-box proteins are involved in the process, and their roles have been characterized in beautiful molecular detail. In general, it seems that DEAH-box proteins remodel RNA duplexes, *e.g.*, the removal of the lariat by Prp43 (Arenas & Abelson, 1997; Martin, Schneider, & Schwer, 2002; Toroney, Nielsen, & Staley, 2019) or release of the ligated exons by Prp22 (Company, Arenas, & Abelson, 1991; Schwer, 2008). They appear to function "at a distance" by tightly lodging the helicase against the complex and allowing it to then "pull out" the RNA in a winch-type mechanism (Bai et al., 2021; Semlow, Blanco, Walter, & Staley, 2016; Toroney et al., 2019; Wan, Yan, Bai, Lei, & Shi, 2017; Zhang et al., 2019). Notably, a similar role for DEAH-box proteins in ribosome assembly is consistent with the available structures of pre-ribosomes bound to Dhr1 and Mtr4/exosome (Cheng et al., 2020; Du et al., 2020; Lau et al., 2021; Singh, Vanden Broeck, Miller, Chaker-Margot, & Klinge, 2021). In contrast, the DEAD-box proteins (Sub2, Prp5 and Prp28) appear to remove other proteins, typically also leading to larger rearrangements of the intermediates, as described below for the characterized examples (Fig. 1A).

During spliceosome assembly, the splice sites are identified in two-step mechanisms, where the branchpoint is first located by the Mud2/Msl5 complex, which is later replaced with the U2 snoRNA within a larger RNA-protein complex (RNP). Similarly, the 5′-splice-site is first recognized by the U1 snRNP, which is later replaced by U6 snRNP, within the tri-snRNP (U4/U5/U6 snRNPs). DEAD-box proteins play key roles

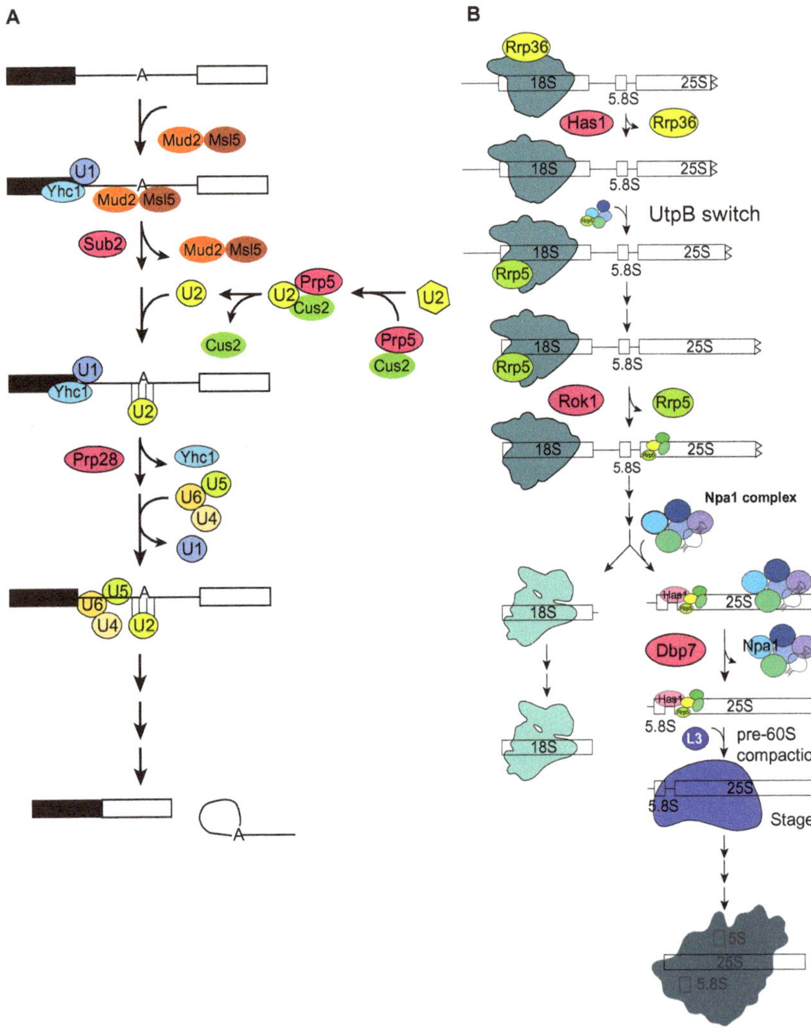

Fig. 1 The characterized roles of DEAD-box proteins in pre-mRNA splicing (A) and ribosome assembly (B). (A) Simplified schematic of pre-mRNA splicing, focusing on steps that involve DEAD-box proteins (shown in magenta). The 5′- and 3′-exons are shown with black and white boxes, respectively, the intron with a line. The branchpoint adenosine is denoted with an A. Early in spliceosome assembly the Mud2/Msl5 heterodimer binds to the branchpoint, while the U1 snRNP, which includes the Yhc1 protein, binds the 5′-splice site. The DEAD-box protein Sub2 released Mud2/Msl5, likely by directly acting on Mud2. Cus2 is required for a conformational rearrangement in U2, required for its binding to the branchpoint. Cus2 release from U2 is necessary to enable the formation of all base pairs between U2 and the branchpoint. To recruit the U4/U5/U6 tri-snRNP, which contains the latent active site, Yhc1 must be released from the 5′-splice-site, a reaction promoted by the DEAD-box protein Prp28. For more details see the text. (B) Simplified scheme of ribosome assembly, focusing on the role of the DEAD-box proteins Has1, Rok1 and Dbp7 (in magenta). Early in 40S assembly Has1 gates the formation

in organizing these RNP-remodeling steps (reviewed in De Bortoli, Espinosa, & Zhao, 2021; Liu & Cheng, 2015).

Genetic data support a role for the DEAD-box protein Sub2 in removal of the Mud2/Msl5 heterodimer from the branchpoint during spliceosome assembly (Fig. 1A). First, deletion of Mud2 rescues Sub2 inactivation or deletion (Kistler & Guthrie, 2001). Similarly, mutations in the RNA binding or Mud2 binding surfaces of Msl5 rescue Sub2 deletion (Jacewicz, Chico, Smith, Schwer, & Shuman, 2015; Wang, Zhang, Lynn, & Rymond, 2008). Intriguingly, Sub2 also has roles in mRNA export (Xie & Ren, 2019).

Cus2 promotes the refolding of an RNA structure within U2 snRNA (Hilliker, Mefford, & Staley, 2007; Perriman & Ares, 2007; Rodgers et al., 2016), thereby preparing U2 for binding to the branchpoint. Nonetheless, Cus2 must be removed to allow for the formation of all base pairs between U2 and the branch point (Talkish et al., 2019). Genetic and biochemical studies suggest strongly that its removal is the role of the DEAD-box protein Prp5 (Fig. 1A). Cus2 deletion, or mutation of the interface with its binding partner Hsh155 abrogate the requirement for ATP hydrolysis by Prp5 in genetic experiments, *in vivo* reporter assays and for spliceosome assembly *in vitro*, as expected if the role for Prp5 was the dissociation of Cus2 (Perriman & Ares, 2000; Perriman, Barta, Voeltz, Abelson, & Ares, 2003; Talkish et al., 2019). Moreover, structural analyses also position Prp5 directly adjacent to Cus2/Hsh155 (Zhang et al., 2021). Interestingly, the activities of Sub2 and Prp5 appear to be interlinked, with human Mud2 binding Hsh155 (Gozani, Potashkin, & Reed, 1998), and formation of the Cus2/Prp5-mediated U2/branchpoint base pairing requiring the release of Mud2/Msl5 (Kao, Cao, Wai, & Cheng, 2021).

Finally, Prp28, which is involved in remodeling of the 5′-splice-site to release the U1 snRNP and incorporate the U4/U5/U6 tri-snRNP, can be inactivated when Yhc1 protein is deleted (Chen et al., 2001). Importantly, Yhc1 binds and stabilizes the U1/5′-splice-site duplex, and moreover

---

of the processosome, by promoting the release of Rrp36, which enables a switch in the UtpB subcomplex, and the recruitment of a set of assembly factors. It also allows for binding of Rrp5 to the platform. After 5′-end processing of the nascent 18S rRNA Rok1 releases Rrp5 from the 40S subunit, allowing for its relocation to the nascent 60S subunit. The nascent 60S subunit is already bound by several assembly factors (in shades of green) but binding of the Npa1 complex (in blue) is required for chaperoning the assembly of domain V. Dbp7 releases this complex, which might lead to the binding of Rpl3, and thus ultimately the compaction of the nascent 60S subunit into the earliest intermediate (called stage A). For more details see the text.

mutation of residues that interact with U1 RNA also allows for bypass of Prp28 (Schwer & Shuman, 2014). Together, these data show that Prp28 is required for the destabilization of the U1–5′-splice-site duplex. Importantly, release of U1 *via* a methylated oligonucleotide that competes with the 5′-splice-site cannot bypass the need for Prp28 *in vitro* (Boesler et al., 2016), and furthermore, mutation of the interface between SmD3 and Yhc1 also allows for bypass of Prp28 (Schwer & Shuman, 2014; Schwer & Shuman, 2015). Together, these observations suggest that Prp28 does not simply disrupt the U1/5′-splice-site duplex but are instead consistent with a model where the DEAD-box protein removes Yhc1, weakening the U1/5′-splice-site duplex, and allowing for accommodation of the U6 within the tri-snRNP in the location that Yhc1 was blocking (Fig. 1A).

Ribosome biogenesis is a dynamic and hierarchical process that allows for the co-transcriptional modification and folding of the rRNA, the latter stabilized by the binding of ribosomal proteins (RPs). Assembly is facilitated and regulated *via* a large machinery, which—among other roles—helps to establish the hierarchy of the process (Huang, Parker, & Karbstein, 2022; Sturm, Cheng, Bassler, Beckmann, & Hurt, 2017), chaperones rRNA folding (Huang & Karbstein, 2021), and enables productive assembly while discouraging dead-end pathways (Liu, Huang, & Karbstein, 2021). This machinery, together with the nascent subunits to which it binds, is dynamically remodeled with factors substituted for other factors or ribosomal proteins. In some cases, these changes are integrated with larger conformational changes in the subunit, or, especially for the later stages, integrated into quality control cycles (Ghalei et al., 2017; Huang, Ghalei, & Karbstein, 2020). How these remodeling events occur remains poorly understood, but because they involve changes in RNA-protein as well as RNA-RNA interactions, and because there are few other ATP or GTP-utilizing enzymes, DEAD-box proteins are prime candidates for regulating and facilitating these events (Strunk & Karbstein, 2009). Indeed, we have previously shown that Rok1 releases Rrp5 from nascent 40S subunits to enable its function in 60S assembly (Fig. 1B, (Khoshnevis et al., 2016)). This conclusion is based on *in vitro* experiments that directly demonstrate this ability for Rok1, and *in vivo* experiments that show that disruption of the Rok1/Rrp5 contact by mutations in Rrp5 phenocopies the effect from Rok1 depletion and leads to the accumulation of Rrp5 on nascent 40S subunits and its depletion from nascent 60S subunits. Recent structural work indicates that Rrp5 release occurs after 5′-end processing of 18S rRNA is complete, but prior to the separation of the RNAs destined for the 40S and 60S subunits (Cheng et al., 2020).

More recently we have shown that even earlier in assembly a different DEAD-box protein, Has1, regulates the repositioning of Rrp5 from the assembling body to the platform (where later it will interact with Rok1) (Liu et al., 2021). Mechanistically, Has1 releases Rrp36, which tethers Rrp5 to the body *via* a direct interaction, from the nascent subunit, thereby enabling its repositioning (Fig. 1B).

Within the 60S assembly pathway, Dbp7 was recently linked to removal of the Npa1 complex, which besides Npa1 includes Npa2, Nop8, Rsa3, Dbp6 and snR190. This complex is required for very early 60S processing steps and appears to chaperone the folding of domain V of 25S rRNA. Mutations within the rRNA and inactivation of snR190 partially rescue the deletion of Dbp7 (Aquino et al., 2021), and snR190, snR34 and the Npa1 complex accumulate in pre-60S intermediates when Dbp7 is deleted (Aquino et al., 2021; Jaafar et al., 2021). However, because snR190 inactivation weakens the association of the Npa1 complex, and because the rescue from rRNA mutation or snR190 inactivation was not complete, and because deletion of the nearby snR34 also partially rescued Dbp7 deletion, the data cannot distinguish between models where Dbp7 removes snR190, or a member of the Npa1 complex, which then weakens the association of the rest of the complex. Nonetheless, these data provide elegant support for a role for Dbp7 in release of a single factor from the Npa1 complex (RNA or protein), which is then linked to larger rearrangements in the nascent 60S subunit (Fig. 1B). Interestingly, the location of the Npa1 complex is likely to overlap the binding site for the large subunit protein Rpl3 (Aquino et al., 2021), which is required for the early compaction of the nascent subunit (Gamalinda et al., 2014). Thus, release of the Npa1 complex may be required for binding of Rpl3 and early compaction of the 5′- and 3′-most domains.

Dbp5 remodels mRNA-protein complexes as they exit the nuclear pore, and also has additional, perhaps related, roles in tRNA export (Lari et al., 2019) and in the nucleus, where it may be involved in the resolution of R-loops (Adams & Wente, 2020). Dbp5 is one of the best-characterized DEAD-box proteins, where biochemical data are closely linked to biology *via* the characterization of point mutations both *in vivo* and *in vitro*. That work has demonstrated that Dbp5 releases the nuclear polyA binding protein Nab2 and the export adaptors Mex67 and Mtr2 from mRNAs as they emerge from the nuclear pore. Biochemical data from *in vitro* reconstituted systems demonstrate that Nab2 release is direct (Tran, Zhou, Corbett, & Wente, 2007). Moreover, genetic data show that a weakly-binding Nab2 mutant can fully rescue a temperature sensitive allele in Dbp5 (Lund & Guthrie, 2005; Tran et al., 2007). Furthermore, a Mex67 mutation that disrupts its interaction with Mtr2 also rescues Dbp5 inactivation. Because Nab2 appears to be an adaptor for Mex67/Mtr2 (Adams & Wente, 2020), it is possible that Dbp5 acts directly on Nab2, with the release of Mex67/Mtr2 an indirect consequence.

Together, these examples show that the characterized roles of DEAD-box proteins in biology involve release of proteins from RNA substrates, a function that DEAD-box proteins can carry out independent of RNA duplex unwinding (Bowers et al., 2006; Fairman et al., 2004; Jankowsky, Gross, Shuman, & Pyle, 2001). Moreover, the data also indicate that these local events are often linked to larger rearrangements. Thus, DEAD-box proteins seem to broadly regulate conformational switches, just not by remodeling RNA duplexes. Finally, the data also demonstrate the importance of combining biochemical and genetic data to fully understand the role of DEAD-box proteins. *E.g.*, for many DEAD-box proteins mutations or deletion of multiple proteins rescue the inactivation of the DEAD-box protein. Additional biochemical and/or genetic data are required to differentiate direct from indirect effects. *E.g.*, the suppression of inactive Dbp5 mutations by Mex67 mutation can be explained by cooperative assembly of the Nab2-Mex67-Mtr2 complex and Dbp5 removing Nab2 as demonstrated. These insights will be used below to suggest a "script" that could be considered when trying to decipher the roles of DEAD-box proteins in specific biological systems. Of course, what is described is a "best-case" scenario, and not all experiments might be possible. For example, if the RNA substrate for a DEAD-box protein is unknown, release assays are very difficult. This roadmap is specifically written for considering roles of DEAD-box proteins in ribosome assembly, however, it can easily be adapted to other biological processes.

3. A roadmap for the discovery of biological roles for DEAD-box proteins

3.1 Characterize the effects on rRNA processing from mutants

As a starting point for thinking about the function of DEAD-box proteins, it is useful to first roughly characterize the effect on the RNA biology under investigation. For this purpose, it is often easiest to investigate RNA processing and this has been done for all the DEAD-box proteins involved in ribosome assembly (Bernstein, Granneman, Lee, Manickam, & Baserga,

2006; Granneman, Bernstein, Bleichert, & Baserga, 2006). It is important though to stress the importance of investigating the effects from inactivating mutations and not the absence of the protein. For many proteins, not just DEAD-box proteins, it has been shown that the proteins often bind before they function, sometimes recruiting other proteins, or merely protecting the RNA from degradation. As a result, deletion (or depletion) affects steps that are beyond the function of the protein in remodeling. A good example is Has1, which has a well-characterized binding site on nascent 60S subunits (Sanghai et al., 2018; Zhou et al., 2019), which may function to stabilize the nascent subunit against degradation. However, its ATPase activity is required only for the assembly of 40S subunits (Dembowski, Kuo, & Woolford Jr., 2013).

3.2 Purify intermediates for mass-spectrometry analyses

A possible next step to undertake could be the isolation of assembly intermediates from cells expressing wild type and inactive versions of the DEAD-box protein of interest, for analysis using mass spectrometry. This standard approach relies on tandem affinity purification (TAP) tags. Thus, one important consideration is that the tagged-protein should be present in the intermediates both before and after the DEAD-box protein dependent step. This sounds trivial, but because DEAD-box protein-mediated remodeling is often linked to larger rearrangements, as outlined above, it is useful to give this point careful consideration, or, even better, ensure that the result does not depend on where one puts the affinity tag. If there is little information on this, one might need to first isolate several different intermediates and determine which of these contain the DEAD-box protein, as in the recent study of Dbp7 (Aquino et al., 2021; Jaafar et al., 2021). Conversely, it might also be useful to pick a protein for tagging that has a somewhat limited temporal association with the nascent ribosomes that is centered around the time of action of the DEAD-box protein. The reason for this is that otherwise the intermediates that accumulate in wild type cells, might be too different from the intermediates that accumulate in mutant cells. Enp1 is a good example for this (Fig. 2). It is recruited after a switch in the UtpB complex (Chaker-Margot, Hunziker, Barandun, Dill, & Klinge, 2015; Hunziker et al., 2019), very early in 40S assembly, and then remains bound until the very latest stages in assembly, dissociating immediately prior to the formation of 80S-like ribosomes, quality control intermediates late in assembly (Fig. 2A, (Huang et al., 2020)). Thus, a number of

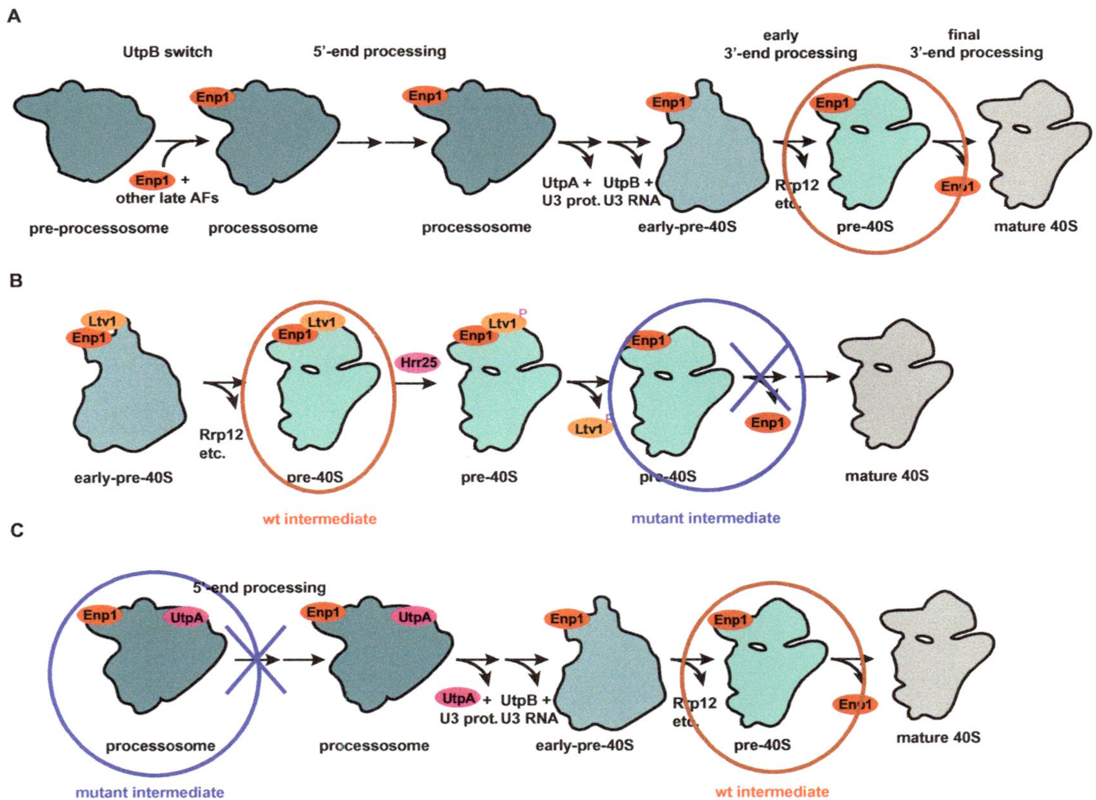

Fig. 2 See figure legend on opposite page.

intermediates contain this protein, but in wild type cells the most abundant one is a largely assembled cytoplasmic intermediate that is the substrate for Hrr25-dependent removal of Enp1's binding partner Ltv1 (Ghalei et al., 2015; Schafer et al., 2006; Schafer, Strauss, Petfalski, Tollervey, & Hurt, 2003). Thus, if one isolates assembly intermediates *via* Enp1-TAP from wild type and mutant cells, one will always compare the mutant intermediates to these late cytoplasmic assembly intermediates. This might be very useful, when the steps immediately prior or after are considered, as the intermediates are very similar and differences limited to the step under investigation (Fig. 2B); however, when one wants to dissect very early steps immediately after Enp1 recruitment, then one compares the early nucleolar intermediate that accumulates in the mutant to a cytoplasmic intermediate (Fig. 2C). Thus, the differences comprise all subsequent assembly steps, making it very difficult to dissect which changes are due to DEAD-box activity, and which are occurring afterwards, and are thus very indirect consequences of DEAD-box activity. In that case, the use of an alternative tag, *e.g.*, on a UtpA component, could refocus the analysis to the change that is under investigation. Note that these two examples also illustrate that a purified intermediate that accumulates with a mutation could be before or after the intermediate purified from wild type cells, something not immediately intuitive.

Finally, the affinity tag should not be differentially obstructed in different intermediates. While genetic complementation is one hint that the tag does not get in the way of additional proteins to grossly perturb conformational

Fig. 2 The importance of finding the correct purification tag. (A) Simplified schematic showing assembly intermediates containing Enp1, a factor present throughout most of 40S ribosome assembly. The most stable intermediate in wild type cells is indicated with the red circle. (B) Focused schematic on late steps in 40S assembly that illustrate how tagged Enp1 can be useful for focusing on intermediates close (and thus similar to) the most stable intermediate. Mutations that block the release of Enp1, like those described in Huang et al. (2020), will lead to the accumulation of the intermediate highlighted with the blue circle. Comparison, by mass-spectrometry or structural tools, of intermediates purified *via* Enp1-TAP from wild type or mutant cells, will compare these two intermediates. (C) Focused schematic on early steps in assembly that illustrate how tagged-Enp1 is not useful to identify changes early in 40S assembly. A mutation early in assembly will lead to the accumulation of the intermediate in the blue circle, which is very different from the intermediate in wild type cells, shown in the red circle. In this case, using a different affinity tag, *e.g.*, one on a UtpA component will help, as UtpA dissociates soon after the change under investigation, refocusing on intermediates around the time of interest.

changes, it merely indicates whether there is enough space, not whether the tag is accessible. *E.g.*, we have observed that Tsr1-TAP only pulls down 80S-like ribosomes, late 40S assembly intermediates, when accessibility of the N-terminally located IgG portion of the tag is required. The calmodulin tag does not appear to be accessible in 80S-like ribosomes (H. Ghalei & K. Karbstein, unpublished observations). This might explain a number of synthetic genetic interactions between Tsr1-TAP and other mutations that impair the formation of 80S-like ribosomes that we have observed (H. Huang & K. Karbstein, unpublished data).

When analyzing such mass-spectrometry data, one might encounter different scenarios. *E.g.*, if the DEAD-box protein releases another protein, the simplest outcome might be that the only protein that is changed is just that protein, which accumulates in intermediates from the inactive mutant. While this scenario would be easy to interpret, it seems unlikely, or at least rare.

The more likely outcome, based on a survey of the literature, is that multiple proteins are changed. Again, one possibility might be that a small number of proteins accumulates in the DEAD-box mutant, which would indicate that a protein complex is released as a function of the DEAD-box activity. A possible example could be Dbp5-mediated remodeling of mRNAs, which should release Nab2/Mex67/Mtr2.

However, as described above, DEAD-box activity is often linked to larger changes in the RNA-protein complex. This will most likely lead to the accumulation of a set of early binding/acting proteins, and depletion of a set of later binding proteins. Because structural and biochemical analyses have now delineated many individual intermediates and their compositional changes, this makes the interpretation of such mass-spectrometry data simpler. *E.g.*, if factors from an earlier stage are accumulated in response to DEAD-box protein inactivation, while later factors are depleted, that suggests that the DEAD-box activity is linked, directly or indirectly, to the transition between these stages. A good example of this finding is the accumulation of very early 60S assembly factors in the absence of Dbp7, while factors for the first structurally characterized intermediate (stage A) and all later-binding factors are depleted ((Aquino et al., 2021), Fig. 1B). Thus, Dbp7 is required for a transition between an early, cotranscriptional assembly intermediate containing few 60S factors that is likely highly mobile and has not yet been isolated, or structurally investigated, and a compacted (stage A) intermediate.

3.3 Confirm candidates in genetic and biochemical assays

Mass spectrometry and analyses of RNA processing should narrow the timing of DEAD-box activity considerably and suggest very specific models for its activity. *E.g.*, if one finds accumulation and depletion of entire groups of intermediates that delineate different already characterized intermediates, one could conclude that the DEAD-box activity is linked to a conformational transition between these intermediates. How, it does so, will then remain to be determined. In an even better example, there will be a small set of proteins that accumulate, which would provide a model for the DEAD-box protein to remove one of these factors. To test these models directly, a combination of genetic and biochemical experiments is recommended.

In the simplest case, the mass-spectrometry data provide a single candidate (or small set of candidates) that might be removed by the DEAD-box activity. In that case, a relatively easy confirmatory experiment would be to test whether deletion of the candidate(s) rescues the growth defect observed from inactivation of the DEAD-box protein, as observed for Cus2 and Prp5. Unfortunately, most ribosome assembly factors are essential, and it is thus more likely than not that this experiment is not possible in its easiest rendition. In that case, one wants to weaken the interactions between the candidate protein and the rest of the ribosome assembly machinery. If there are structures available for the candidate bound to pre-ribosomes, then the simplest thing to do is to disturb its RNA binding interface, or the interface with other assembly factors by site-directed mutagenesis. If such a structure is not available, or the resolution too limited or restricted to a small part of the protein, then one could build a structural model using similarity to other proteins. This is well-implemented on a number of websites, including Jpred and Swiss-model (Drozdetskiy, Cole, Procter, & Barton, 2015; Waterhouse et al., 2018). Often, RNA-binding sites can be identified by their overall positive charge. Conversely, to identify protein-protein interaction sites, we generally look for conserved surface-exposed hydrophobic residues. Such candidates can then be screened in growth assays to identify suppressors of the growth defect arising from the inactive DEAD-box protein. Groups better versed in genetic screens than we are might even carry out an unbiased screen for mutants in the candidate proteins.

If genetic experiments as above identify suppressors, then it should be verified that the suppressor indeed acts as intended. If the mutation is designed to disrupt a protein-protein interface, then yeast-two-hybrid assays or pulldowns with recombinant proteins are useful to demonstrate that indeed the suppressor mutation perturbs the intended protein–protein interaction. If the suppressor mutations are in an RNA binding domain, then either weaker RNA binding or weaker association with pre-ribosomes would be good experimental data to support the suggested mode of suppression. Weaker association with pre-ribosomes is shown most easily in gradient centrifugation experiments (Ghalei et al., 2017; Huang et al., 2020; Rai et al., 2021), or, alternatively, in pulldowns.

As described above, DEAD-box protein activity is often linked to more extensive remodeling events. As also discussed above, evidence for such remodeling events can be garnered from mass-spectrometry experiments, interpreted with the help of structural information. Genetic experiments can also be designed to support a suspected or suggested role for the DEAD-box ATPase in these remodeling events. *E.g.*, mutations in protein-protein or RNA-protein interfaces that are only observed prior to the switch should demonstrate positive genetic interactions with DEAD-box mutations, and might even rescue such mutations, as they destabilize the early structure, rendering DEAD-box activity less necessary. In contrast, mutations in interfaces that are only observed after the switch should demonstrate negative genetic interactions with DEAD-box mutations, as they further disadvantage the switch.

3.4 RNA structure probing experiments

RNA structure probing experiments using small molecules, such as DMS, or SHAPE-reagents, or even hydroxylradicals, can be tools to further dissect the role of DEAD-box proteins, but must be carefully analyzed, keeping in mind the direct changes arising from DEAD-box protein activity, and the indirect changes that can result from the release of a simple ligand as described in the cases above. In that regard, it is especially interesting that formation of the earliest compacted intermediates (the processosome and stage A intermediates for nascent 40S and 60S intermediates, respectively, Fig. 1B) is gated by DEAD-box proteins.

DMS or SHAPE reagents preferentially react with nucleotides that are not engaged in base pairing interactions, and thus can be used to identify secondary structure changes. However, such changes seem to be surprisingly limited, as even the structures of the earliest 40S subunit assembly intermediates show few secondary structure changes relative to the mature subunit, the exception being the U3 binding sites, h28 and h18 (Barandun et al., 2017; Cheng et al., 2020; Du et al., 2020; Kornprobst et al., 2016).

While there are large regions of the RNA that are not resolved in the structures of the earliest assembly intermediates, that does not necessarily mean that they have not yet adopted native secondary structure. In fact, their relative inaccessibility to DMS reagents (Liu et al., 2021) would suggest that they have, as expected because the formation of helices and hairpins is rapid. Instead, it appears most likely that these secondary structure elements are not yet fixed in space *via* tertiary interactions with other RNAs or proteins. However, DMS, SHAPE reagents and hydroxylradicals also require solvent accessibility, and that is much more likely to change as a function of assembly, either because proteins bind or dissociate that might protect an RNA, or because proteins become engaged more stably as has been shown for ribosomal proteins as assembly progresses (Ferreira-Cerca et al., 2007). Furthermore RNA-protein contacts may change if at least one of the partners changes its (tertiary) structure. These changes can be captured in RNA structure probing experiments (Huang & Karbstein, 2021; Liu et al., 2021), and their interpretation is greatly aided by the large amount of structural information already available on ribosome assembly intermediates. Thus, when RNA structure probing data are interpreted with the help of these structures, they can provide further evidence of large-scale changes, as this should affect preferentially the neighborhoods of assembly factors that are recruited or released, or which change. For more details of using structure probing on purified assembly intermediates and a protocol, please consult reference (Liu, Huang, & Karbstein, 2022).

4. Summary and outlook

DEAD-box proteins are key regulators in RNA splicing, mRNA export and ribosome biogenesis, where they induce large changes in the RNA-protein complexes by promoting simple protein dissociation events. While dissecting their function has been slow, we hope that this distillation of the literature and the successful approaches within will help interested researchers to develop a plan that can lead to the successful identification of the roles of more DEAD-box proteins.

Acknowledgments

I would like to thank the people in my lab, especially Xin Liu, for many helpful discussions on ribosome biogenesis in general, and DEAD-box proteins in particular. I would also like to thank Jason Talkish and the members of my lab for comments on the manuscript. Our work is supported by NIH grant R35-GM136323, and HHMI Faculty Scholar grant 55108536.

References

Adams, R. L., & Wente, S. R. (2020). Dbp5 associates with RNA-bound Mex67 and Nab2 and its localization at the nuclear pore complex is sufficient for mRNP export and cell viability. *PLoS Genetics*, *16*(10), e1009033.

Aquino, G. R. R., Hackert, P., Krogh, N., Pan, K. T., Jaafar, M., Henras, A. K., et al. (2021). The RNA helicase Dbp7 promotes domain V/VI compaction and stabilization of inter-domain interactions during early 60S assembly. *Nature Communications*, *12*(1), 6152.

Arenas, J. E., & Abelson, J. N. (1997). Prp43: An RNA helicase-like factor involved in spliceosome disassembly. *Proceedings of the National Academy of Sciences of the United States of America*, *94*(22), 11798–11802.

Bai, R., Wan, R., Yan, C., Jia, Q., Lei, J., & Shi, Y. (2021). Mechanism of spliceosome remodeling by the ATPase/helicase Prp2 and its coactivator Spp2. *Science*, *371*(6525).

Barandun, J., Chaker-Margot, M., Hunziker, M., Molloy, K. R., Chait, B. T., & Klinge, S. (2017). The complete structure of the small-subunit processome. *Nature Structural & Molecular Biology*, *24*(11), 944–953.

Bernstein, K. A., Granneman, S., Lee, A. V., Manickam, S., & Baserga, S. J. (2006). Comprehensive mutational analysis of yeast DEXD/H box RNA helicases involved in large ribosomal subunit biogenesis. *Molecular and Cellular Biology*, *26*(4), 1195–1208.

Boesler, C., Rigo, N., Anokhina, M. M., Tauchert, M. J., Agafonov, D. E., Kastner, B., et al. (2016). A spliceosome intermediate with loosely associated tri-snRNP accumulates in the absence of Prp28 ATPase activity. *Nature Communications*, 7, 11997.

Bourgeois, C. F., Mortreux, F., & Auboeuf, D. (2016). The multiple functions of RNA helicases as drivers and regulators of gene expression. *Nature Reviews. Molecular Cell Biology*, *17*(7), 426–438.

Bowers, H. A., Maroney, P. A., Fairman, M. E., Kastner, B., Luhrmann, R., Nilsen, T. W., et al. (2006). Discriminatory RNP remodeling by the DEAD-box protein DED1. *RNA*, *12*(5), 903–912.

Chaker-Margot, M., Hunziker, M., Barandun, J., Dill, B. D., & Klinge, S. (2015). Stage-specific assembly events of the 6-MDa small-subunit processome initiate eukaryotic ribosome biogenesis. *Nature Structural & Molecular Biology*.

Chen, Y., Potratz, J. P., Tijerina, P., Del Campo, M., Lambowitz, A. M., & Russell, R. (2008). DEAD-box proteins can completely separate an RNA duplex using a single ATP. *Proceedings of the National Academy of Sciences of the United States of America*, *105*(51), 20203–20208.

Chen, J. Y., Stands, L., Staley, J. P., Jackups, R. R., Jr., Latus, L. J., & Chang, T. H. (2001). Specific alterations of U1-C protein or U1 small nuclear RNA can eliminate the requirement of Prp28p, an essential DEAD box splicing factor. *Molecular Cell*, 7(1), 227–232.

Cheng, J., Lau, B., La Venuta, G., Ameismeier, M., Berninghausen, O., Hurt, E., et al. (2020). 90S pre-ribosome transformation into the primordial 40S subunit. *Science*, *369*(6510), 1470–1476.

Company, M., Arenas, J., & Abelson, J. (1991). Requirement of the RNA helicase-like protein PRP22 for release of messenger RNA from spliceosomes. *Nature*, *349*(6309), 487–493.

De Bortoli, F., Espinosa, S., & Zhao, R. (2021). DEAH-box RNA helicases in pre-mRNA splicing. *Trends in Biochemical Sciences*, *46*(3), 225–238.

Dembowski, J. A., Kuo, B., & Woolford, J. L., Jr. (2013). Has1 regulates consecutive maturation and processing steps for assembly of 60S ribosomal subunits. *Nucleic Acids Research*, *41*(16), 7889–7904.

Drozdetskiy, A., Cole, C., Procter, J., & Barton, G. J. (2015). JPred4: A protein secondary structure prediction server. *Nucleic Acids Research*, *43*(W1), W389–W394.

Du, Y., An, W., Zhu, X., Sun, Q., Qi, J., & Ye, K. (2020). Cryo-EM structure of 90S small ribosomal subunit precursors in transition states. *Science*, *369*(6510), 1477–1481.

Fairman, M. E., Maroney, P. A., Wang, W., Bowers, H. A., Gollnick, P., Nilsen, T. W., et al. (2004). Protein displacement by DExH/D "RNA helicases" without duplex unwinding. *Science*, *304*(5671), 730–734.

Fairman-Williams, M. E., Guenther, U. P., & Jankowsky, E. (2010). SF1 and SF2 helicases: Family matters. *Current Opinion in Structural Biology*, *20*(3), 313–324.

Ferreira-Cerca, S., Poll, G., Kuhn, H., Neueder, A., Jakob, S., Tschochner, H., et al. (2007). Analysis of the in vivo assembly pathway of eukaryotic 40S ribosomal proteins. *Molecular Cell*, *28*(3), 446–457.

Gamalinda, M., Ohmayer, U., Jakovljevic, J., Kumcuoglu, B., Woolford, J., Mbom, B., et al. (2014). A hierarchical model for assembly of eukaryotic 60S ribosomal subunit domains. *Genes & Development*, *28*(2), 198–210.

Ghalei, H., Schaub, F. X., Doherty, J. R., Noguchi, Y., Roush, W. R., Cleveland, J. L., et al. (2015). Hrr25/CK1δ-directed release of Ltv1 from pre-40S ribosomes is necessary for ribosome assembly and cell growth. *Journal of Cell Biology*, *208*(6), 745–759.

Ghalei, H., Trepreau, J., Collins, J. C., Bhaskaran, H., Strunk, B. S., & Karbstein, K. (2017). The ATPase Fap7 tests the ability to carry out translocation-like conformational changes and releases Dim1 during 40S ribosome maturation. *Molecular Cell*, *67*(6). 990–1000 e3.

Gilman, B., Tijerina, P., & Russell, R. (2017). Distinct RNA-unwinding mechanisms of DEAD-box and DEAH-box RNA helicase proteins in remodeling structured RNAs and RNPs. *Biochemical Society Transactions*, *45*(6), 1313–1321.

Gozani, O., Potashkin, J., & Reed, R. (1998). A potential role for U2AF-SAP 155 interactions in recruiting U2 snRNP to the branch site. *Molecular and Cellular Biology*, *18*(8), 4752–4760.

Granneman, S., Bernstein, K. A., Bleichert, F., & Baserga, S. J. (2006). Comprehensive mutational analysis of yeast DEXD/H box RNA helicases required for small ribosomal subunit synthesis. *Molecular and Cellular Biology*, *26*(4), 1183–1194.

Hilliker, A. K., Mefford, M. A., & Staley, J. P. (2007). U2 toggles iteratively between the stem IIa and stem IIc conformations to promote pre-mRNA splicing. *Genes & Development*, *21*(7), 821–834.

Huang, H., Ghalei, H., & Karbstein, K. (2020). Quality control of 40S ribosome head assembly ensures scanning competence. *The Journal of Cell Biology*, *219*(11).

Huang, H., & Karbstein, K. (2021). Assembly factors chaperone ribosomal RNA folding by isolating helical junctions that are prone to misfolding. *Proceedings of the National Academy of Sciences of the United States of America*, *118*(25).

Huang, H., Parker, M. D., & Karbstein, K. (2022). The modifying enzyme Tsr3 establishes the hierarchy of Rio kinase activity in 40S ribosome assembly. *RNA*.

Hunziker, M., Barandun, J., Buzovetsky, O., Steckler, C., Molina, H., & Klinge, S. (2019). Conformational switches control early maturation of the eukaryotic small ribosomal subunit. *eLife*, *8*.

Jaafar, M., Contreras, J., Dominique, C., Martin-Villanueva, S., Capeyrou, R., Vitali, P., et al. (2021). Association of snR190 snoRNA chaperone with early pre-60S particles is regulated by the RNA helicase Dbp7 in yeast. *Nature Communications*, *12*(1), 6153.

Jacewicz, A., Chico, L., Smith, P., Schwer, B., & Shuman, S. (2015). Structural basis for recognition of intron branchpoint RNA by yeast Msl5 and selective effects of interfacial mutations on splicing of yeast pre-mRNAs. *RNA*, *21*(3), 401–414.

Jankowsky, E. (2011). RNA helicases at work: Binding and rearranging. *Trends in Biochemical Sciences*, *36*(1), 19–29.

Jankowsky, E., Gross, C. H., Shuman, S., & Pyle, A. M. (2000). The DExH protein NPH-II is a processive and directional motor for unwinding RNA. *Nature*, *403*(6768), 447–451.

Jankowsky, E., Gross, C. H., Shuman, S., & Pyle, A. M. (2001). Active disruption of an RNA-protein interaction by a DExH/D RNA helicase. *Science, 291*(5501), 121–125.

Jarmoskaite, I., & Russell, R. (2014). RNA helicase proteins as chaperones and remodelers. *Annual Review of Biochemistry, 83*, 697–725.

Jia, H., Wang, X., Anderson, J. T., & Jankowsky, E. (2012). RNA unwinding by the Trf4/Air2/Mtr4 polyadenylation (TRAMP) complex. *Proceedings of the National Academy of Sciences of the United States of America, 109*(19), 7292–7297.

Kao, C. Y., Cao, E. C., Wai, H. L., & Cheng, S. C. (2021). Evidence for complex dynamics during U2 snRNP selection of the intron branchpoint. *Nucleic Acids Research, 49*(17), 9965–9977.

Khoshnevis, S., Askenasy, I., Johnson, M. C., Dattolo, M. D., Young-Erdos, C. L., Stroupe, M. E., et al. (2016). The DEAD-box protein Rok1 orchestrates 40S and 60S ribosome assembly by promoting the release of Rrp5 from pre-40S ribosomes to allow for 60S maturation. *PLoS Biology, 14*(6), e1002480.

Kistler, A. L., & Guthrie, C. (2001). Deletion of MUD2, the yeast homolog of U2AF65, can bypass the requirement for sub2, an essential spliceosomal ATPase. *Genes & Development, 15*(1), 42–49.

Kornprobst, M., Turk, M., Kellner, N., Cheng, J., Flemming, D., Kos-Braun, I., et al. (2016). Architecture of the 90S pre-ribosome: A structural view on the birth of the eukaryotic ribosome. *Cell, 166*(2), 380–393.

Lari, A., Rajan, A. A. N., Sandhu, R., Reiter, T., Montpetit, R., Young, B. P., et al. (2019). A nuclear role for the DEAD-box protein Dbp5 in tRNA export. *Elife, 8*.

Lau, B., Cheng, J., Flemming, D., La Venuta, G., Berninghausen, O., Beckmann, R., et al. (2021). Structure of the maturing 90S pre-ribosome in association with the RNA exosome. *Molecular Cell, 81*(2). 293–303 e4.

Lebaron, S., Papin, C., Capeyrou, R., Chen, Y. L., Froment, C., Monsarrat, B., et al. (2009). The ATPase and helicase activities of Prp43p are stimulated by the G-patch protein Pfa1p during yeast ribosome biogenesis. *The EMBO Journal, 28*(24), 3808–3819.

Linder, P., & Jankowsky, E. (2011). From unwinding to clamping—The DEAD box RNA helicase family. *Nature Reviews. Molecular Cell Biology, 12*(8), 505–516.

Liu, Y. C., & Cheng, S. C. (2015). Functional roles of DExD/H-box RNA helicases in Pre-mRNA splicing. *Journal of Biomedical Science, 22*, 54.

Liu, X., Huang, H., & Karbstein, K. (2021). Blocking a dead-end assembly pathway creates a point of regulation for the DEAD-box ATPase Has1 and prevents platform misassembly. *biorxivs*.

Liu, X., Huang, H., & Karbstein, K. (2022). Using DMS-MaPseq to uncover the roles of DEAD-box proteins in ribosome assembly. *Methods* (submitted).

Liu, F., Putnam, A., & Jankowsky, E. (2008). ATP hydrolysis is required for DEAD-box protein recycling but not for duplex unwinding. *Proceedings of the National Academy of Sciences of the United States of America, 105*(51), 20209–20214.

Lund, M. K., & Guthrie, C. (2005). The DEAD-box protein Dbp5p is required to dissociate Mex67p from exported mRNPs at the nuclear rim. *Molecular Cell, 20*(4), 645–651.

Martin, A., Schneider, S., & Schwer, B. (2002). Prp43 is an essential RNA-dependent ATPase required for release of lariat-intron from the spliceosome. *The Journal of Biological Chemistry, 277*(20), 17743–17750.

Perriman, R., & Ares, M., Jr. (2000). ATP can be dispensable for prespliceosome formation in yeast. *Genes & Development, 14*(1), 97–107.

Perriman, R. J., & Ares, M., Jr. (2007). Rearrangement of competing U2 RNA helices within the spliceosome promotes multiple steps in splicing. *Genes & Development, 21*(7), 811–820.

Perriman, R., Barta, I., Voeltz, G. K., Abelson, J., & Ares, M., Jr. (2003). ATP requirement for Prp5p function is determined by Cus2p and the structure of U2 small nuclear RNA.

Proceedings of the National Academy of Sciences of the United States of America, 100(24), 13857–13862.

Putnam, A. A., & Jankowsky, E. (2013). DEAD-box helicases as integrators of RNA, nucleotide and protein binding. *Biochimica et Biophysica Acta, 1829*(8), 884–893.

Rai, J., Parker, M. D., Huang, H., Choy, S., Ghalei, H., Johnson, M. C., et al. (2021). An open interface in the pre-80S ribosome coordinated by ribosome assembly factors Tsr1 and Dim1 enables temporal regulation of Fap7. *RNA, 27*(2), 221–233.

Rodgers, M. L., Tretbar, U. S., Dehaven, A., Alwan, A. A., Luo, G., Mast, H. M., et al. (2016). Conformational dynamics of stem II of the U2 snRNA. *RNA, 22*(2), 225–236.

Sanghai, Z. A., Miller, L., Molloy, K. R., Barandun, J., Hunziker, M., Chaker-Margot, M., et al. (2018). Modular assembly of the nucleolar pre-60S ribosomal subunit. *Nature, 556*(7699), 126–129.

Schafer, T., Maco, B., Petfalski, E., Tollervey, D., Bottcher, B., Aebi, U., et al. (2006). Hrr25-dependent phosphorylation state regulates organization of the pre-40S subunit. *Nature, 441*(7093), 651–655.

Schafer, T., Strauss, D., Petfalski, E., Tollervey, D., & Hurt, E. (2003). The path from nucleolar 90S to cytoplasmic 40S pre-ribosomes. *The EMBO Journal, 22*(6), 1370–1380.

Schwer, B. (2008). A conformational rearrangement in the spliceosome sets the stage for Prp22-dependent mRNA release. *Molecular Cell, 30*(6), 743–754.

Schwer, B., & Shuman, S. (2014). Structure-function analysis of the Yhc1 subunit of yeast U1 snRNP and genetic interactions of Yhc1 with Mud2, Nam8, Mud1, Tgs1, U1 snRNA, SmD3 and Prp28. *Nucleic Acids Research, 42*(7), 4697–4711.

Schwer, B., & Shuman, S. (2015). Structure-function analysis and genetic interactions of the Yhc1, SmD3, SmB, and Snp1 subunits of yeast U1 snRNP and genetic interactions of SmD3 with U2 snRNP subunit Lea1. *RNA, 21*(6), 1173–1186.

Semlow, D. R., Blanco, M. R., Walter, N. G., & Staley, J. P. (2016). Spliceosomal DEAH-box ATPases remodel pre-mRNA to activate alternative splice sites. *Cell, 164*(5), 985–998.

Singh, S., Vanden Broeck, A., Miller, L., Chaker-Margot, M., & Klinge, S. (2021). Nucleolar maturation of the human small subunit processome. *Science, 373*(6560), eabj5338.

Strunk, B. S., & Karbstein, K. (2009). Powering through ribosome assembly. *RNA, 15*(12), 2083–2104.

Sturm, M., Cheng, J., Bassler, J., Beckmann, R., & Hurt, E. (2017). Interdependent action of KH domain proteins Krr1 and Dim2 drive the 40S platform assembly. *Nature Communications, 8*(1), 2213.

Talkish, J., Igel, H., Hunter, O., Horner, S. W., Jeffery, N. N., Leach, J. R., et al. (2019). Cus2 enforces the first ATP-dependent step of splicing by binding to yeast SF3b1 through a UHM-ULM interaction. *RNA, 25*(8), 1020–1037.

Toroney, R., Nielsen, K. H., & Staley, J. P. (2019). Termination of pre-mRNA splicing requires that the ATPase and RNA unwindase Prp43p acts on the catalytic snRNA U6. *Genes & Development, 33*(21 – 22), 1555–1574.

Tran, E. J., Zhou, Y., Corbett, A. H., & Wente, S. R. (2007). The DEAD-box protein Dbp5 controls mRNA export by triggering specific RNA:protein remodeling events. *Molecular Cell, 28*(5), 850–859.

Wan, R., Yan, C., Bai, R., Lei, J., & Shi, Y. (2017). Structure of an intron lariat spliceosome from *Saccharomyces cerevisiae*. *Cell, 171*(1). 120–s132 e12.

Wang, Q., Zhang, L., Lynn, B., & Rymond, B. C. (2008). A BBP-Mud2p heterodimer mediates branchpoint recognition and influences splicing substrate abundance in budding yeast. *Nucleic Acids Research, 36*(8), 2787–2798.

Waterhouse, A., Bertoni, M., Bienert, S., Studer, G., Tauriello, G., Gumienny, R., et al. (2018). SWISS-MODEL: Homology modelling of protein structures and complexes. *Nucleic Acids Research, 46*(W1), W296–W303.

Wilkinson, M. E., Charenton, C., & Nagai, K. (2020). RNA splicing by the spliceosome. *Annual Review of Biochemistry*, *89*, 359–388.

Xie, Y., & Ren, Y. (2019). Mechanisms of nuclear mRNA export: A structural perspective. *Traffic*, *20*(11), 829–840.

Yan, C., Wan, R., & Shi, Y. (2019). Molecular mechanisms of pre-mRNA splicing through structural biology of the spliceosome. *Cold Spring Harbor Perspectives in Biology*, *11*(1).

Zhang, X., Zhan, X., Yan, C., Zhang, W., Liu, D., Lei, J., et al. (2019). Structures of the human spliceosomes before and after release of the ligated exon. *Cell Research*, *29*(4), 274–285.

Zhang, Z., Rigo, N., Dybkov, O., Fourmann, J. B., Will, C. L., Kumar, V., et al. (2021). Structural insights into how Prp5 proofreads the pre-mRNA branch site. *Nature*, *596*(7871), 296–300.

Zhou, D., Zhu, X., Zheng, S., Tan, D., Dong, M. Q., & Ye, K. (2019). Cryo-EM structure of an early precursor of large ribosomal subunit reveals a half-assembled intermediate. *Protein & Cell*, *10*(2), 120–130.

CHAPTER THREE

Monitoring functional RNA binding of RNA-dependent ATPase enzymes such as SF2 helicases using RNA dependent ATPase assays: A RIG-I case study

Rong Guo[a] and Anna Marie Pyle[a,b,c,*]

[a]Department of Chemistry, Yale University, New Haven, CT, United States
[b]Department of Molecular, Cellular, and Developmental Biology, Yale University, New Haven, CT, United States
[c]Howard Hughes Medical Institute, Chevy Chase, MD, United States
*Corresponding author: e-mail address: anna.pyle@yale.edu

Contents

Abstract

The highly conserved Superfamily 1 (SF1) and Superfamily 2 (SF2) nucleic acid-dependent ATPases, are ubiquitous motor proteins with central roles in DNA and RNA metabolism (Jankowsky & Fairman, 2007). These enzymes require RNA or DNA binding to stimulate ATPase activity, and the conformational changes that result from this coupled behavior are linked to a multitude of processes that range from nucleic acid unwinding to the flipping of macromolecular switches (Pyle, 2008, 2011). Knowledge about the relative affinity of nucleic acid ligands is crucial for deducing mechanism and understanding biological function of these enzymes. Because enzymatic ATPase activity is directly coupled to RNA binding in these proteins, one can utilize their ATPase activity as a simple reporter system for monitoring functional binding of RNA or DNA to an SF1 or SF2 enzyme. In this way, one can rapidly assess the relative impact of mutations in the protein or the nucleic acid and obtain parameters that are useful for setting up more quantitative direct binding assays. Here, we describe a routine method for employing NADH-coupled enzymatic

Methods in Enzymology, Volume 673
ISSN 0076-6879
https://doi.org/10.1016/bs.mie.2022.03.064

ATPase activity to obtain kinetic parameters reflecting apparent ATP and RNA binding to an SF2 helicase. First, we provide a protocol for calibrating an NADH-couple ATPase assay using the well-characterized ATPase enzyme hexokinase, which a simple ATPase enzyme that is not coupled with nucleic acid binding. We then provide a protocol for obtaining kinetic parameters (K_mATP, V_{max} and K_mRNA) for an RNA-coupled ATPase enzyme, using the double-stranded RNA binding protein RIG-I as a case-study. These approaches are designed to provide investigators with a simple, rapid method for monitoring apparent RNA association with SF2 or SF1 helicases.

1. Introduction

DNA and RNA metabolism involve the activity of numerous ATPase enzymes that require DNA or RNA binding to become activated and to undergo ATP hydrolysis (Bleichert & Baserga, 2007; Jankowsky & Fairman, 2007; Pyle, 2008; Sloan & Bohnsack, 2018). In many such cases, the ATPase active-site is not even structurally assembled or correctly aligned in the absence of RNA or DNA (Pyle, 2008, 2011). These proteins, which fall into two phylogenetically-determined superfamilies (SF1 and SF2) (Gorbalenya & Koonin, 1993), have many functions, including duplex unwinding and translocation, while some members catalyze more local changes in macromolecular structure (Jankowsky, Gross, Shuman, & Pyle, 2001; Pyle, 2008, 2011). For example, many SF2 proteins function as "smart" RNA binding proteins that use ATP binding and hydrolysis to trigger single conformational changes that lead to rearrangement of large ribonucleoprotein complexes, while others use ATPase activity to facilitate dissociation from a nucleic acid ligand structure (Dumont et al., 2006; Jankowsky et al., 2001; Pyle, 2008, 2011).

Because ATPase activity is selectively triggered by nucleic acid (NA) binding in these enzymes, one can utilize ATPase enzymatic activity as a "handle," or proxy, for monitoring NA association with the enzyme (Beran & Pyle, 2008; Ding, Kohlway, & Pyle, 2011). There are several advantages to this: (a). DNA or RNA-dependent ATPase activity reflects *functional binding* of NA to the protein, which can therefore be distinguished from nonspecific binding that frequently occurs between NA and highly basic members of the SF1 and SF2 families. Direct binding assays cannot distinguish functional and nonspecific NA binding. (b). Because NA-dependent ATPase activity reflects functional binding, it enables the investigator to conduct structure/function studies to elucidate the ideal shape or sequence of the stimulatory NA ligand, and thereby learn about the biologically relevant NA target molecule. Alternatively, it can be used to obtain a quick readout on the effects of protein mutations. (c). NA-dependent ATPase assays are quick and simple. They do not require specialized equipment or the optimization of a direct binding assay in order to get an informative initial understanding of protein-NA affinity.

While NA-dependent ATPase assays are simple and highly informative, it is ideal to conduct them together with properly-designed direct assays of NA binding to the protein of interest. This is because NA-dependent ATPase assays do not provide a true K_d value. Rather, they provide a K_mRNA value, which should only be considered an upper limit on the estimated value of the dissociation constant, K_d. In other words, RNA or DNA binding may be much stronger (and K_d lower) than it appears from the apparent K_mRNA values obtained using this approach. This is because the sensitivity of ATPase assays depends on the amount of ATP hydrolysis that can be measured, and at very low enzyme concentrations, the total amount of ATP hydrolysis may be too small to detect using the coupled assay we describe below. Other assays, such as those using isotopically-labeled ATP, can be more sensitive and provide a K_mRNA value that is closer to the true K_d, and the reader is directed to other publications where this method is described (Beran, Lindenbach, & Pyle, 2009; Beran, Serebrov, & Pyle, 2007; Ding et al., 2011). In any of these assays, if the actual K_d is less than (tighter than) the concentration of enzyme needed to monitor ATPase activity, then the NA-dependent activity of the enzyme will be in the stoichiometric binding regime and cannot be considered a proxy for the actual dissociation constant, K_d. It can, however, provide useful information on relative NA binding or stoichiometry under the conditions of the assay, and it can be considered an upper bound on the relative affinity between WT enzyme and NA ligand. For example, in the RIG-I case study provided below, RNA-dependent ATPase assays can accurately be used to monitor dsRNA binding to the enzyme if the RNA is a weak ligand (in this case, a blunt duplex with a 5'-OH terminus, Kd $\sim$ 20 nM) (Ren, Linehan, Iwasaki, & Pyle, 2019; Vela, Fedorova, Ding, & Pyle, 2012). However, this assay cannot be used to accurately measure binding of the highest-affinity RIG-I targets (blunt RNA duplexes terminated with a 5′-triphosphate, $K_d \sim 10$ pM) (Ren et al., 2019; Vela et al., 2012), which have K_d values that are several logs lower than non-phosphorylated dsRNAs.

Given its simplicity and ease of implementation, we present the following protocol for employing an enzymatically coupled ATPase assay established (De La Cruz, Sweeney, & Ostap, 2000; Imamura, Tada, & Tonomura, 1966; Lindsley, 2001; Luo et al., 2011) to monitor NA binding to a NA-dependent ATPase enzyme. To facilitate the use of this assay, we also describe the assay calibration protocol using a relatively cheap, commercially available non-nucleic acid dependent ATPase. Although we use this approach to monitor dsRNA binding by a SF2 enzyme below, one can easily adapt it to monitor NA interactions with SF1 enzymes. While the example enzyme described below (RIG-I) does not display ATPase activity in the absence of dsRNA, it is still possible to use this assay for monitoring NA binding by an enzyme with background levels of unstimulated ATPase activity (such as the NS3 enzyme from Hepatitis C Virus). In those cases, the results must simply be normalized for background hydrolysis by the enzyme (Beran et al., 2007, 2009; Ding et al., 2011).

2. NADH-coupled ATPase assay to measure RNA binding affinity

The NADH-coupled enzyme assay has been widely applied to characterize steady-state kinetics for ATPases (Beisenherz, Bucher, & Garbade, 1955; Bucher & Pfleiderer, 1955; De La Cruz et al., 2000; Imamura et al., 1966; Lindsley, 2001; Luo et al., 2011; Montpetit, Seeliger, & Weis, 2012). Instead of measuring inorganic phosphate-release, the NADH-coupled assay monitors ADP production via a multi-enzyme ATP regeneration system (Scheme 1). In this system, the ATPase of interest converts ATP into inorganic phosphate and ADP. Pyruvate kinase (PK) then converts ADP back to ATP while phosphoenolpyruvate (PEP) is transformed to pyruvate. The pyruvate is reduced to lactate by lactate dehydrogenase (LDH), which couples to the NADH oxidation reaction where NADH (which absorbs at 340 nm) is oxidized to NAD^+ (which absorbs at 260 nm). The loss in absorbance at 340 nm is proportional to the quantity of ATPase hydrolysis and can be monitored via spectrometric methods. In our protocol, we adapted the NADH-coupled assay to a facile, low-volume

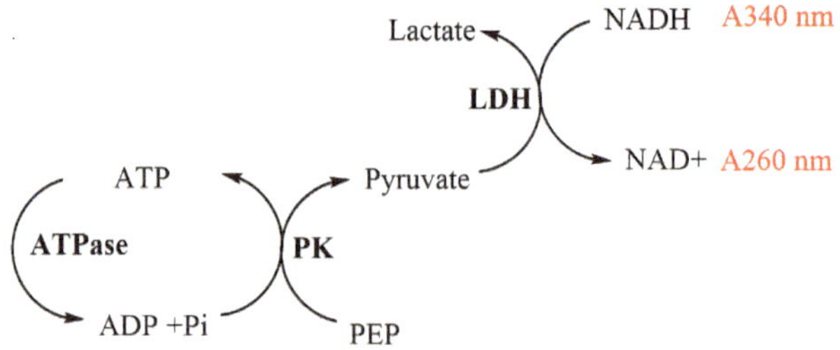

Scheme 1 The ATPase hydrolysis reaction is coupled to the oxidation of NADH via a series of enzymatic reactions.

system (50 μL total volume) enabling measurement of ATPase hydrolysis rate in a conventional plate reader. Time-dependence measurements of ATP hydrolysis determines reaction velocities. In turn, ATP and RNA-concentration dependence measurements of velocity provide kinetic parameters, including the apparent binding constants or K_M values, and k_{cat} or the turnover number for the enzyme.

2.1 NADH-coupled reaction preparation

To set up a 50 μL NADH-coupled reaction in a 96-well plate format, it is important to make the 5 × enzyme mix containing PEP, PK, LDH and NADH (see Table 1). Then the 10 μL 5 × enzyme mix is incubated with the SF2 helicase of interest, together with increasing concentrations of substrates in each well at room temperature for over 1 h before the addition of ATP to initiate the reaction. The reaction details are shown below:

2.1.1 Equipment

- Synergy Neo2 Multi-Mode Plate Reader
- 96-well plates

2.1.2 Reagents and buffers

- ATPase Assay Buffer: 25 mM HEPES pH 7.4, 150 mM NaCl and 1.5 mM $MgCl_2$, 5 mM DTT
- NADH (Sigma N8129), dissolved in water
- Lactic dehydrogenase (LDH, Sigma L-1254), stored in 50% glycerol
- Pyruvate kinase (PK, Sigma 1012816300), stored in 50% glycerol
- Phosphoenolpyruvic acid (PEP, Sigma, P-7127), stored in water, pH is adjusted to 7.0

  Note: All enzymes should be saved in −20 °C to prevent degradation, assay buffer should be stored in 4 degree.

Table 1 The 5 × NADH cocktail component.

	Stock	Final concentration
LDH	4000 U/mL	100 U/mL
PK	2000 U/mL	500 U/mL
PEP	100 mM	2.5 mM
NADH	10 mM	1 mM
ATPase buffer	1 ×	

2.1.3 ATPase reaction setup and procedure

The coupled enzymes are mixed into the 5 × cocktail solution showed below (Table 1). Note that the cocktail mix should be made fresh before each use. SF2 helicase and RNA will be diluted to the concentrations as needed and incubated with the 5 × mix (Table 2) for at least 1 h. ATP is then added to initiate the ATPase reaction. Centrifuge the plate for 3 min at 3000 rpm and start to monitor the absorbance at 340 nm over at least 15 min at 20 s intervals.

2.1.4 Data analysis

- Obtain the slope of NADH absorbance vs time
- Convert the units from absorbance to concentration of NADH (Imamura et al., 1966) using the extinction coefficient for NADH ($\varepsilon_{340} = 6220\,\text{M}^{-1}\,\text{cm}^{-1}$) via the Beer- Lambert Law (Eq. 1).

$$[NADH] = \frac{Abs340}{\varepsilon NADH * pathlength} \tag{1}$$

where ε is the extinction coefficient of NADH; pathlength of a plate is given or calculated from vendor's handbook

- The linear change in [NADH] vs time is equivalent to the rate of ATP hydrolysis and is in units of molarity per second.
- The ATPase hydrolysis velocity is plotted as a function of the concentration of the ligand of interest (RNA or ATP in the case of SF2 helicase). The data can be analyzed using various graphing software, such as GraphPad Prism.

Table 2 Reaction setup (50 μL final volume).

		Stock	Desired concentration	Volume
Incubate for at least 1 h	5 × cocktail mix	5 ×	1 ×	10 μL
	ATPase(SF2 helicase)	Vary	2–50 nM[a]	Vary
	Substrate (RNA)	Vary	vary	Vary
	ATPase Buffer	1 ×	1 ×	Add up to 45 μL
	ATP	Vary	Vary	5 μL

[a]The lowest concentration of SF2 helicase needs to be determined experimentally.

- The data will be fitted to the quadratic Briggs-Haldane equation (Eq. 2) and Michaelis Menten curve can be plotted and K_m and V_{max} can be extrapolated from the curve.

$$y = yo + (amp) \times \frac{x + p + Km - \sqrt{(x + p + Km)^2 - 4xp}}{2p} \tag{2}$$

where y = catalytic velocity; y_o = background catalytic velocity without ATPase; amp = Vmax-y_o = k_{cat}; x = total substrate concentration, p = total ATPase concentration.

2.1.5 Notes

- For each ATPase reaction, a control reaction under the same conditions without ATPase and nucleic acid addition is needed to ensure the proper function of the assay. Low to no ATPase activity is expected for the controls
- Plots of NADH versus time can be nonlinear, with a slope that can change at the beginning of the reaction due to the turbulence of adding ATP (red portion in Fig. 1) and it should become linear after a short period of time. Only use the linear region for determining velocity

When the NADH is depleted and detection limit of the plate reader has been reached, the Abs340 nm vs time will stop changing and slope becomes flat

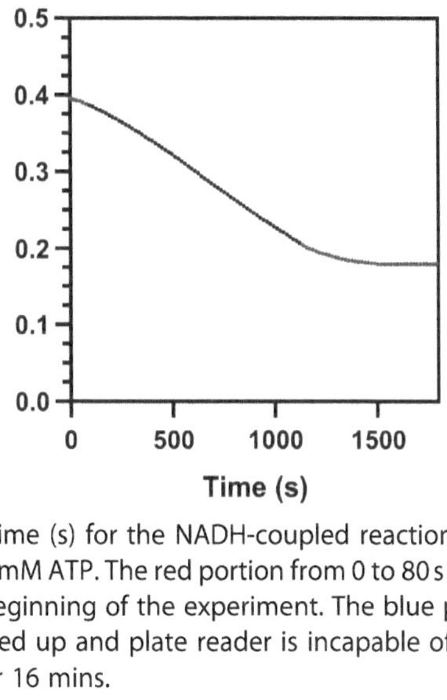

Fig. 1 Abs340 nm vs Time (s) for the NADH-coupled reaction containing RIG-I and 5′ OHSLR14, initiated by 5 mM ATP. The red portion from 0 to 80 s represents the instability of absorbance at the beginning of the experiment. The blue portion of the curve represents the NADH is used up and plate reader is incapable of reading the changes in NADH absorbance after 16 mins.

(see blue portion of the curve, Fig. 1). Only the linear range of the Abs340 nm vs Time(s) should be used for calculation of ATPase rate.

2.2 Calibrating the ATPase assay using hexokinase

Due to the complexity of this multienzyme assay, it is important to validate it under local laboratory conditions using a well-established ATPase enzyme before applying an RNA-coupled ATPase of interest. Here, we use hexokinase (HK), which is a model enzyme that converts glucose to glucose-6-phosphate with ATP hydrolysis, to calibrate the ATPase assay, and we compare our kinetic parameters with values reported in the literature. It is important to note that hexokinase exists in three isozymes and the kinetic parameters of HK might vary from different sources (recombinant vs. commercially available brands). To standardize the protocol, we used the yeast hexokinase isozyme suspension from Sigma Aldrich (SKU: 11426362001). The reference values for $K_{m,\ ATP}$ and $K_{m,\ glucose}$ are 0.27 ± 0.02 mM and 0.14 ± 0.002 mM (Hillard & Stewart, 1998), which are in agreement with the past literature values (Fromm & Zewe, 1962; Gao & Leary, 2003).

2.2.1 Reaction setup

The reaction setup is based on that described in Section 2.1, with the following modification: ATPase buffer is changed to 50 mM MOPS pH7.4, 1 mM $MgCl_2$ to keep consistent with the past HK kinetic studies (Hillard & Stewart, 1998).

1. 5× NADH enzyme cocktail is constituted as Table 1.

To obtain Km for ATP

- Mix 0.25 U/mL (9.7 nM) HK (sigma, SKU: 11426362001) with 5 mM Glucose and 10 μL 5× NADH enzyme mix. Combine the reaction components to a total volume to 45 μL per well and centrifuge at 3000 rpm for 10 min to ensure proper mixing. Incubate at room temperature for 1 h.
- Initiate the reaction with 5 μL 10× ATP with 11 different concentrations (listed as final concentrations in reaction, μM): 0, 4.8, 9.8, 19.6, 39, 78.1, 156.3, 312.5, 625, 1250 and 2500.

To obtain Km for HK substrate: Glucose

- Mix 0.25 U/mL (9.7 nM) HK (sigma, SKU: 11426362001) with 5 μL 10× 11 different concentrations of glucose (listed as final concentrations, μM): 0, 4.8, 9.8, 19.6, 39, 78.1, 156.3, 312.5, 625, 1250 and 2500.

- Add the assay buffer and 10 μL 5 × NADH enzyme mix to the total volume of 45 μL. Mix the reaction by centrifugation and incubate at room temperature for 1 h.
- Initiate the reaction with 5 μL 10 × ATP with final concentration of 5 mM ATP.

2. Centrifuge the plate for 3 min at 3000 rpm and start monitor the plates at Abs340 nm for over 15 min at 20s interval.
3. The data obtained were fitted to the quadratic Briggs-Haldane Eq. 2 and analyzed as described in Section 2.1.4.

2.2.2 Sample data

Here, we show the Michaelis–Menten curve of ATPase rate versus ATP and Glucose (Fig. 2) obtained using the described protocol in Section 2.2.1 and the analysis described in Section 2.1.4. The ATPase reactions without HK show no ATPase activity (Fig. 2A). The $K_{m,ATP}$ fitted from the curve is 0.266 ± 0.03 mM which matches well with the literature value of 0.269 ± 0.02 mM (Hillard & Stewart, 1998). Similarly, the $K_{m,\ Glucose}$ is fitted to be 0.151 ± 0.02 mM which is also consistent with the reference data (Hillard & Stewart, 1998).

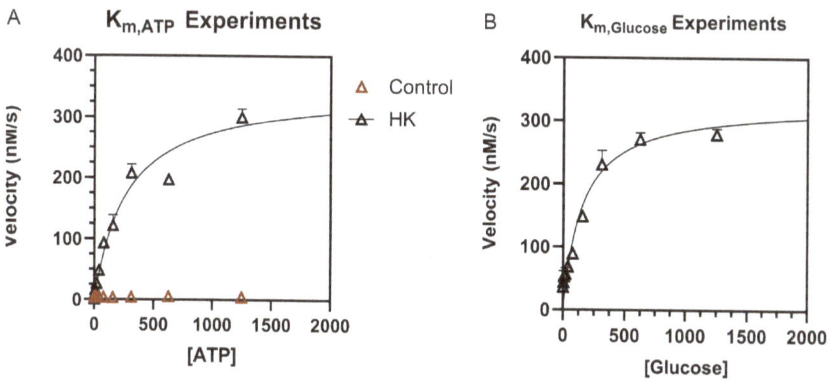

Fig. 2 Example Michaelis–Menten curve for HK. Steady state ATP hydrolysis by HK stimulated with varying concentrations of (A) ATP; (B) Glucose. In panel (A), the control reaction without HK stimulated with different concentrations of ATP shows no catalytic activity. The extrapolated binding constant for ATP ($K_{m,\ ATP}$) is 0.266 ± 0.03 mM and V_{max} are 346 ± 12 nM/s. $K_{m,\ Glucose}$ is fitted to be 0.151 ± 0.02 mM and V_{max} is 326 ± 12 nM/s. Plotted values are mean ± SD (n = 3).

2.2.3 *Troubleshooting tips*

The calibration of the NADH-coupled ATPase assay should provide K_m values comparable with those cited in the literature. Here are some common tips if the assay is not behaving well:

- If the NADH versus time is not linear and slope is changing, it is likely due to improper mixing or no mixing after ATP addition. Centrifuge or mixing is needed before reading the plate
- If the slope of NADH versus time is shallow, it could be caused by (1) low concentration or loss of catalytic activity of the ATPase; (2) The pathlength of the plate is small. An Increase ATPase concentration and/or change the plate with larger pathlength to decrease signal to noise ratio.
- If the slope of NADH versus time is still shallow after increasing ATPase concentration, consider use new enzymes for constituting 5 × enzyme mix.
- If the replicability of the data is not consistent, consider adding low amount of detergents such as 0.01% to 0.1% triton to stabilize the reaction system.

2.3 Determine RNA binding affinities for SF2 protein: A RIG-I case study

Despite the diversity of SF2 enzyme biological mechanisms, all SF2 enzymes have nucleic acid dependent ATPase activity (Jankowsky & Fairman, 2007). An important topic in the field is to find the natural or optimal ligands for a given SF2 enzyme, from which the ligand recognition and activation mechanism can be further explored. To this end, a facile method that reports on the binding affinity of different nucleic acids for SF2 proteins is useful for mechanistic studies. As a case study, we use this method to determine the apparent K_m value of short, double-stranded RNA molecule binding to RIG-I, which is an important member of the SF2 family (Vela et al., 2012). Here, we measure the binding affinity between 5' OH stem-loop RNA containing 14 base pairs (5' OH SLR 14), which has been studied extensively in the past as a weak immunomodulatory ligand (Linehan et al., 2018; Ren et al., 2019).

2.3.1 *Reaction setup*

1. The reaction setup is described in Section 2.1. In general: the 5 × NADH enzyme cocktail is constituted as shown in Table 1. We will first determine the ATP saturation concentration.

To determine the K_m for ATP

- Mix 50 nM WT RIG-I with 500 μM 5'OH SLR14 and 10 μL 5 × NADH enzyme mix. The total mixture volume should be 45 μL per well. Centrifuge the plate at 3000 rpm for 10 min to ensure proper mixing. Incubate at room temperature for 1 h.
- Initiate the reaction with 5 μL 10 × ATP with 11 different concentrations serial diluted from 5000 μM across the plate (listed as final concentrations, μM): 0, 19.6, 39, 78.1, 156.3, 312.5, 625, 1250, 2500 and 5000.

To obtain Km for 5' OH SLR14

- Mix 50 nM WT RIG-I, 10 μL 5 × NADH enzyme mix and 5 μL 10 × 9 different concentrations of 5'OH SLR14 across the plate (listed as final concentrations, nM): 0, 7.81, 15.63, 31.25, 62.5, 125, 250, 500 and 100.
- Add ATPase buffer to a total volume of 45 μL per well and centrifuge at 3000 rpm for 10 mins to ensure proper mixing. Incubate at room temperature for 1 h.
- Initiate the reaction with 5 μL 10 × ATP with final concentration of 5 mM ATP, where V max is reached informed by Michaelis–Menten curve of Km,ATP (Fig. 3A).

2. Centrifuge the plate for 3 min at 3000 rpm and start monitor the plates at Abs340 nm for over 15 min at 20s interval.

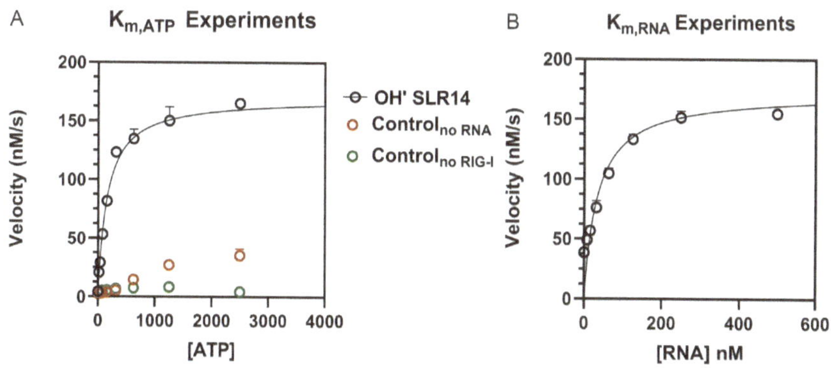

Fig. 3 Michaelis–Menten curve for WT RIG-I versus ATP and RNA using 5' OH SLR14. Steady state ATP hydrolysis by wild type RIG-I proteins stimulated with varying concentrations of (A) ATP; (B) RNA. This experiment shows that without RIG-I or RNA, there is little detectable catalytic activity. The $K_{m,\ ATP}$ is 156 ± 11 μM and $K_{m,\ 5'\ OH\ SLR14}$ is 34.1 ± 5.7 nM. V_{max} determined from ATP and RNA experiments is 169 ± 3.0 nM/s and is 172 ± 7.2 nM/s, respectively. Plotted values are mean ± SD (n = 3).

2.3.2 Sample data

The ATPase data were analyzed as described in Section 2.1.4 and were fitted to a Michaelis–Menten curve across ATP and RNA with increasing concentration (Fig. 3). From the ATPase rate versus ATP curve, we obtained the K_m for ATP and determined the saturation ATP concentration at 5 mM, where the velocities plateau. This high concentration of ATP is then used for stimulating ATPase activity in determining K_m for RNA experiments. The K_m for ATP is determined to be 156 ± 11 μM (Fig. 3A) and K_m for 5' OH SLR14 is 34.1 ± 5.7 nM (Fig. 3B). The obtained $K_{m,5'\ OH\ SLR14}$ agrees with the previous directing binding data on the 14 bp blunt duplet with a 5' OH terminus where K_d is around 20 nM (Ren et al., 2019; Vela et al., 2012).

3. Summary

Here, we provide a stepwise NADH-coupled assay where RNA binding affinity for RIG-I can be readily evaluated. The continuous measurement of absorbances using plate reader allows us to monitor ATP hydrolysis under various conditions simultaneously. The assay we present can serve as a model for accessing the affinity of different nucleic acid substrates for SF1 and SF2 proteins and may provide insights into their activation mechanisms.

Acknowledgments

A.M.P. is an investigator of the Howard Hughes Medical Institute. R.G. is supported by Colton Center for Autoimmunity at Yale. We thank Dr. Dave Rawling for consultation of NADH-coupled ATPase assay and Wenshuai Wang for SF2 helicase discussion.

References

Beisenherz, G., Bucher, T., & Garbade, K. H. (1955). Alpha-glycerophosphate dehydrogenase from rabbit muscle. *Methods in Enzymology*, *1*, 391–397. https://doi.org/10.1016/0076-6879(55)01063-X.

Beran, R. K., Lindenbach, B. D., & Pyle, A. M. (2009). The NS4A protein of hepatitis C virus promotes RNA-coupled ATP hydrolysis by the NS3 helicase. *Journal of Virology*, *83*(7), 3268–3275. https://doi.org/10.1128/JVI.01849-08.

Beran, R. K. F., & Pyle, A. M. (2008). Hepatitis C viral NS3-4A protease activity is enhanced by the NS3 helicase. *Journal of Biological Chemistry*, *283*(44), 29929–29937. https://doi.org/10.1074/jbc.M804065200.

Beran, R. K., Serebrov, V., & Pyle, A. M. (2007). The serine protease domain of hepatitis C viral NS3 activates RNA helicase activity by promoting the binding of RNA substrate. *The Journal of Biological Chemistry*, *282*(48), 34913–34920. https://doi.org/10.1074/jbc.M707165200.

Bleichert, F., & Baserga, S. J. (2007). The long unwinding road of RNA helicases. *Molecular Cell*, *27*(3), 339–352. https://doi.org/10.1016/j.molcel.2007.07.014.

Bucher, T., & Pfleiderer, G. (1955). Pyruvate kinase from muscle. *Methods in Enzymology*, *1*, 435–440. https://doi.org/10.1016/0076-6879(55)01071-9.

De La Cruz, E. M., Sweeney, H. L., & Ostap, E. M. (2000). ADP inhibition of myosin V ATPase activity. *Biophysical Journal*, *79*(3), 1524–1529. https://doi.org/10.1016/S0006-3495(00)76403-4.

Ding, S. C., Kohlway, A. S., & Pyle, A. M. (2011). Unmasking the active helicase conformation of nonstructural protein 3 from hepatitis C virus. *Journal of Virology*, *85*(9), 4343–4353. https://doi.org/10.1128/Jvi.02130-10.

Dumont, S., Cheng, W., Serebrov, V., Beran, R. K., Tinoco, I., Jr., Pyle, A. M., et al. (2006). RNA translocation and unwinding mechanism of HCV NS3 helicase and its coordination by ATP. *Nature*, *439*(7072), 105–108. https://doi.org/10.1038/nature04331.

Fromm, H. J., & Zewe, V. (1962). Kinetic studies of yeast hexokinase. *The Journal of Biological Chemistry*, *237*, 3027–3032.

Gao, H., & Leary, J. A. (2003). Multiplex inhibitor screening and kinetic constant determinations for yeast hexokinase using mass spectrometry based assays. *Journal of the American Society for Mass Spectrometry*, *14*(3), 173–181. https://doi.org/10.1016/S1044-0305(02)00867-X.

Gorbalenya, A. E., & Koonin, E. V. (1993). Helicases–amino-acid-sequence comparisons and structure-function-relationships. *Current Opinion in Structural Biology*, *3*(3), 419–429. https://doi.org/10.1016/S0959-440x(05)80116-2.

Hillard, S. W., & Stewart, K. K. (1998). A bypass trapped-flow analysis system evaluation of enzyme kinetic parameters with a coupled enzyme assay and fluorescence detection. *Talanta*, *45*(3), 507–512. https://doi.org/10.1016/s0039-9140(97)00180-x.

Imamura, K., Tada, M., & Tonomura, Y. (1966). The pre-steady state of the myosin–adenosine triphosphate system. IV. Liberation of ADP from the myosin–ATP system and effects of modifiers on the phosphorylation of myosin. *Journal of Biochemistry*, *59*(3), 280–289.

Jankowsky, E., & Fairman, M. E. (2007). RNA helicases—one fold for many functions. *Current Opinion in Structural Biology*, *17*(3), 316–324. https://doi.org/10.1016/j.sbi.2007.05.007.

Jankowsky, E., Gross, C. H., Shuman, S., & Pyle, A. M. (2001). Active disruption of an RNA-protein interaction by a DExH/D RNA helicase. *Science*, *291*(5501), 121–125. https://doi.org/10.1126/science.291.5501.121.

Lindsley, J. E. (2001). Use of a real-time, coupled assay to measure the ATPase activity of DNA topoisomerase II. *Methods in Molecular Biology*, *95*, 57–64. https://doi.org/10.1385/1-59259-057-8:57.

Linehan, M. M., Dickey, T. H., Molinari, E. S., Fitzgerard, M. E., Potapova, O., Iwasaki, A., et al. (2018). A minimal RNA ligand for potent RIG-I activation in living mice. *Science Advances*, *4*(2), e1701854. https://doi.org/10.1126/sciadv.1701854.

Luo, D., Ding, S. C., Vela, A., Kohlway, A., Lindenbach, B. D., & Pyle, A. M. (2011). Structural insights into RNA recognition by RIG-I. *Cell*, *147*(2), 409–422. https://doi.org/10.1016/j.cell.2011.09.023.

Montpetit, B., Seeliger, M. A., & Weis, K. (2012). Analysis of dead-box proteins in mRNA export. *RNA Helicases*, *511*, 239–254. https://doi.org/10.1016/B978-0-12-396546-2.00011-5.

Pyle, A. M. (2008). Translocation and unwinding mechanisms of RNA and DNA helicases. *Annual Review of Biophysics*, *37*, 317–336. https://doi.org/10.1146/annurev.biophys.37.032807.125908.

Pyle, A. M. (2011). RNA helicases and remodeling proteins. *Current Opinion in Chemical Biology*, *15*(5), 636–642. https://doi.org/10.1016/j.cbpa.2011.07.019.

Ren, X., Linehan, M. M., Iwasaki, A., & Pyle, A. M. (2019). RIG-I selectively discriminates against 5'-monophosphate RNA. *Cell Reports*, *26*(8), 2019–2027 e2014. https://doi.org/10.1016/j.celrep.2019.01.107.

Sloan, K. E., & Bohnsack, M. T. (2018). Unravelling the mechanisms of RNA helicase regulation. *Trends in Biochemical Sciences*, *43*(4), 237–250. https://doi.org/10.1016/j.tibs.2018.02.001.

Vela, A., Fedorova, O., Ding, S. C., & Pyle, A. M. (2012). The thermodynamic basis for viral RNA detection by the RIG-I innate immune sensor. *The Journal of Biological Chemistry*, *287*(51), 42564–42573. https://doi.org/10.1074/jbc.M112.385146.

CHAPTER FOUR

Measurement of ATP utilization in RNA unwinding and RNA chaperone activities by DEAD-box helicase proteins

Inga Jarmoskaite[†], Anna E. Helmers, and Rick Russell*
Department of Molecular Biosciences, University of Texas at Austin, Austin, TX, United States
*Corresponding author: e-mail address: rick_russell@cm.utexas.edu

Contents

Abstract

RNA helicase proteins perform coupled reactions in which cycles of ATP binding and hydrolysis are used to drive local unwinding of double-stranded RNA (dsRNA). For some helicases in the ubiquitous DEAD-box family, these local unwinding events are integral to folding transitions in structured RNAs, and thus these helicases function as RNA chaperones. An important measure of the efficiency of the helicase-catalyzed reaction is the ATP utilization value, which represents the average number of ATP molecules hydrolyzed during RNA unwinding or a chaperone-assisted RNA structural rearrangement. Here we outline procedures that can be used to measure the ATP utilization value in

[†] Present address: Department of Genetics, Stanford University, Stanford, CA, United States.

Methods in Enzymology, Volume 673
ISSN 0076-6879
https://doi.org/10.1016/bs.mie.2022.04.004

RNA unwinding or folding transitions. As an example of an RNA folding transition, we focus on the refolding of the *Tetrahymena thermophila* group I intron ribozyme from a long-lived misfolded structure to its native structure, and we outline strategies for adapting this assay to other RNA folding transitions. For a simple dsRNA unwinding event, the ATP utilization value provides a measure of the coupling between the ATPase and RNA unwinding activities, and for a complex RNA structural transition it can give insight into the scope of the rearrangement and the efficiency with which the helicase uses the energy from ATPase cycles to promote the rearrangement.

1. Introduction

Helicase proteins function throughout RNA biology in diverse roles. A common role for RNA helicases is to use ATP to promote RNA structural transitions—folding steps and conformational changes—by accelerating unwinding of local secondary structure (*i.e.*, short RNA helices). This ATP-dependent RNA chaperone activity is required for complex RNA folding processes, including those of the ribosomal subunits and the spliceosome (Jarmoskaite & Russell, 2014; Staley & Guthrie, 1998; Xing, Ma, & Tran, 2019), and is thought to underlie the ability of cellular mRNAs to exist at a steady state in structural ensembles that are less structured than at equilibrium under similar solution conditions *in vitro* (Ding et al., 2014; Herschlag, 1995; Rouskin, Zubradt, Washietl, Kellis, & Weissman, 2014). Similarly, ATP-dependent disruption of intermolecular contacts may underlie the ability of RNA helicases to mediate RNA transport into and out of intracellular condensates (Van Treeck & Parker, 2018; Weis & Hondele, 2022).

DEAD-box proteins comprise the largest family of RNA helicases, with dozens of members present in eukaryotes, including humans (Fairman-Williams, Guenther, & Jankowsky, 2010; Linder & Jankowsky, 2011). DEAD-box helicases rely on an ATP-dependent mechanism of dsRNA unwinding activity that differs substantially from conventional helicase activity but is exceptionally well-suited for efficient disruption of short RNA helices like those found commonly in structured RNAs (Hilbert, Karow, & Klostermeier, 2009). In the ATP-bound form, DEAD-box proteins can bind directly to segments of dsRNA, which produces local unwinding of the two strands and tight binding of the helicase core to one of the strands in a conformation that is not compatible with helix formation (Del Campo & Lambowitz, 2009b; Mallam, Del Campo, Gilman, Sidote, & Lambowitz, 2012; Sengoku, Nureki, Nakamura, Kobayashi, & Yokoyama, 2006) (Fig. 1A). Thus, this ATP-dependent binding produces

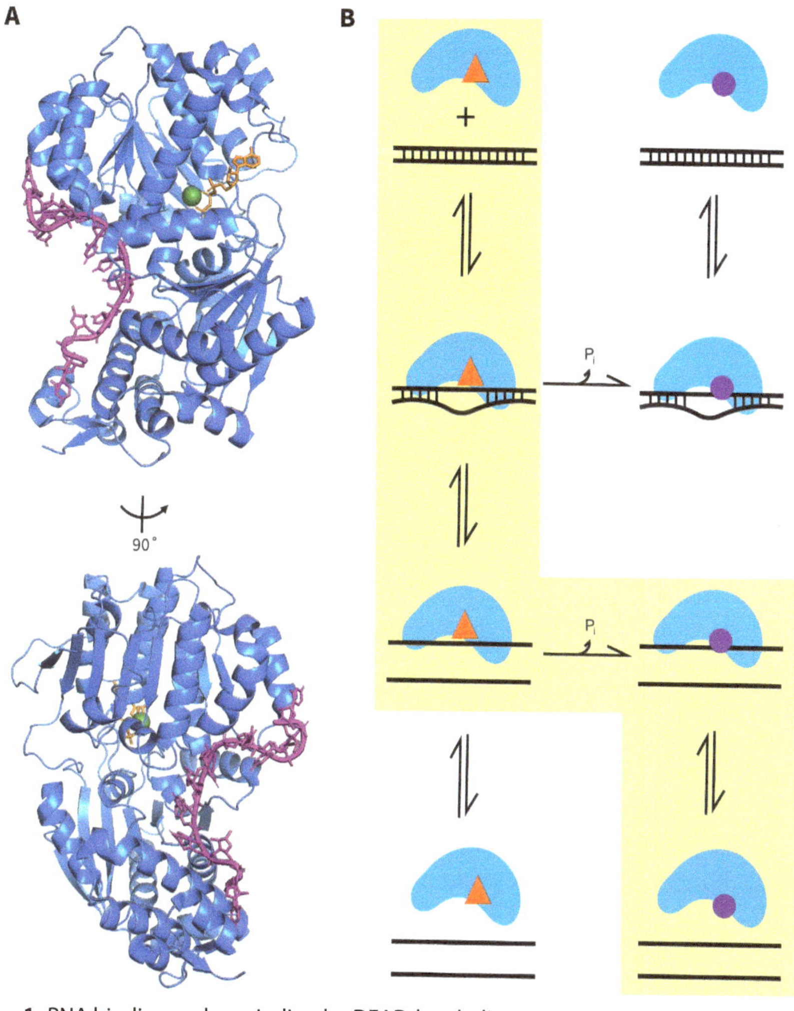

Fig. 1 RNA binding and unwinding by DEAD-box helicase proteins. (A), Structural views of ssRNA binding by the helicase core, as shown for Mss116 (Del Campo & Lambowitz, 2009b). The ssRNA (purple) binds to Mss116 (blue) in a crimped conformation, incompatible with a standard duplex geometry. Also shown are a bound ATP analog (orange) and a coordinated Mg^{2+} ion (green). (B) ATP-dependent RNA unwinding by DEAD-box helicases. In the dominant pathway, highlighted in yellow, the ATP-bound form of the helicase binds dsRNA and produces local unwinding (starting from top left; ATP is indicated by a red triangle). This unwinding reaches completion, resulting in the dissociation of one of the strands, producing the ssRNA-bound form shown in panel A. ATP hydrolysis and release of P_i weaken binding of ssRNA (bound ADP is indicated by a purple circle), resulting in ssRNA dissociation (bottom right). Thus, this pathway results in complete helix unwinding with hydrolysis of one ATP. Less populated pathways produce ATPase activity without complete duplex unwinding (top two rows, counterclockwise) and complete duplex unwinding without ATP hydrolysis (left column, top to bottom).

local RNA unwinding (Chen et al., 2008; Liu, Putnam, & Jankowsky, 2008). Subsequently, ATP hydrolysis and the release of the inorganic phosphate (P_i) product favor a helicase core conformation with reduced affinity for single-stranded (ss) RNA, permitting release of the helicase from the unwound RNA strand and completing the cycle of linked ATPase and RNA unwinding activities (Fig. 1B) (Cao et al., 2011). Because this RNA unwinding activity stems from a single cycle of RNA binding and release and does not involve translocation, unwinding of an entire short helix (6–11 base pairs) typically involves hydrolysis of just a single ATP (Chen et al., 2008; Henn, Cao, Hackney, & De La Cruz, 2008; Liu et al., 2008). This stoichiometric ATP consumption is much lower than that for conventional, directional helicase activity, which generally requires hydrolysis of one ATP per base pair unwound, with limited variability noted for individual helicases (Lohman, Tomko, & Wu, 2008; Martinez-Senac & Webb, 2005; Pyle, 2008; Rajagopal, Gurjar, Levin, & Patel, 2010).

The seemingly simple mechanism of ATP-dependent RNA unwinding activity by DEAD-box helicases belies the complexity of the RNA structural transitions that rely on this activity. For complex RNA folding steps or conformational rearrangements, one might imagine that many local unwinding events and ATPs are likely to be required, both because of a potential requirement for the unwinding of multiple RNA helices and because the same RNA helix may need to be unwound multiple times before the transition progresses fully and successfully.

In this chapter, we outline methods to measure the amount of ATP consumed during a DEAD-box protein-mediated RNA unwinding event or an RNA chaperone-mediated structural transition. The basic strategy is to measure the rate of ATP hydrolysis and the RNA structural transition under identical conditions and to compare the amounts of the products from each reaction. Such measurements have revealed an ATP utilization value of approximately one for unwinding of short RNA helices (6–11 base pairs); *i.e.*, one ATP per completely unwound helix, for several DEAD-box proteins (Chen et al., 2008), as shown in the major pathway for helix unwinding in Fig. 1B. Although we expect this result to be general for DEAD-box helicases, provided that the duplex being measured is short enough to fit inside the binding site of the protein or with little enough extruded that the rest of the duplex unwinds spontaneously, much remains unknown. Alternative pathways exist, which give ATPase activity without unwinding or RNA unwinding without ATPase activity (Fig. 1B), and these pathways may be engaged to different degrees with changes in solution conditions, protein concentrations, and duplex length. Most generally, it is unclear whether the stoichiometric consumption of ATP is a universal property of DEAD-box helicases or whether there is additional variability in the strictness of the coupling between the ATPase and RNA unwinding cycles. Even less is known about the ATP utilization by DEAD-box helicases during complex RNA structural transitions (Jarmoskaite, Tijerina, & Russell, 2021).

The chapter comprises two main experimental sections. In Section 2, we focus on ATP utilization during unwinding of short RNA helices, and in Section 3, we outline a largely analogous strategy to determine ATP utilization during a large-scale RNA structural rearrangement. We expect Section 2 to be broadly applicable across DEAD-box proteins and different dsRNAs with minimal changes. Section 3 is geared toward a narrower set of helicases with known chaperone functions and with RNA targets whose folding is reasonably well understood. Here, we specifically describe methods used to measure ATP utilization by the DEAD-box helicase CYT-19 during native folding of a group I self-splicing intron from a discrete misfolded conformation. We also outline general strategies to adapt the method to new RNA folding transitions.

2. ATP utilization in RNA unwinding reactions

The central idea of the method is to measure the rates of ATP hydrolysis and RNA unwinding under identical conditions (Fig. 2). Although this is a simple idea, a bit of thought (and possibly some trial and error) is required to choose concentrations of the RNA and protein that will maximize the experimental signal-to-noise ratio and allow a successful measurement.

To make the RNA unwinding detectable, a pulse-chase protocol is used, in which dsRNA is first formed by mixing the labeled strand with an excess of the complementary strand (the pulse; Fig. 2A). After the dsRNA is formed, an excess amount of an unlabeled version of the labeled strand is added (the chase). A large excess of the chase strand (5–10-fold), *relative to the complementary strand*, is necessary for a large signal for RNA unwinding because the signal arises from the labeled strand being outcompeted by the chase strand for binding to the relatively small, limiting amount of the complementary strand (Fig. 2B). However, a large excess of the chase strand will increase the background in the ATPase measurement, and therefore the noise, because ssRNA binding by DEAD-box proteins stimulates ATP hydrolysis—although to a lesser degree than dsRNA (Fig. 2C). This background ATPase activity should be measured in a control reaction with

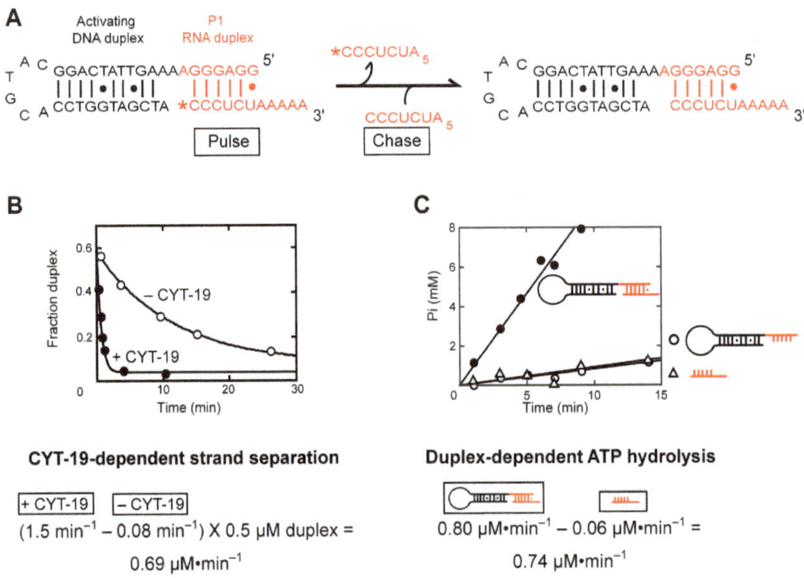

Fig. 2 Results of ATP utilization measurement in RNA unwinding by the DEAD-box helicase CYT-19 (Chen et al., 2008). (A) RNA unwinding substrate and labeling strategy. The RNA helix (red) was appended to a flanking DNA helix, which was shown to increase CYT-19 activity (Tijerina, Bhaskaran, & Russell, 2006). In the pulse phase, the RNA helix was formed by adding labeled $CCCUCUA_5$ (labeling indicated by asterisk), and unwinding of this helix was trapped in the chase phase by excess unlabeled $CCCUCUA_5$. Results are shown for RNA unwinding measurements (B) and ATPase measurements (C). For both sets of experiments, CYT-19 was 2 μM and the duplex was 0.5 μM. For ATPase measurements, excess of the substrate strand was present (1 μM). For RNA unwinding measurements, this strand was increased to 5 μM in panel (B), which was shown to increase the signal for RNA unwinding without changing the observed rate constant. Conditions for all measurements were 25 °C, 10 mM Mg^{2+}, and 50 μM ATP-Mg^{2+}. These measurements gave an ATP utilization value of 1.1 ± 0.1.

the ssRNA only, and subtracted, but the larger the value of the background is relative to the signal, the greater will be the noise after the subtraction.

The concentrations of the helicase and the dsRNA should both be moderately high (hundreds of nanomolar to low micromolar), unlike conventional single-turnover or steady-state experiments where very low concentrations are used of the dsRNA substrate or the helicase, respectively. The reason for choosing a relatively high concentration of the dsRNA substrate is to increase the signal in the ATPase measurement relative to the background rate in the absence of any RNA or with the excess ssRNA only. The reason for choosing a relatively high protein concentration is to increase the signal in the RNA unwinding experiment, keeping in mind that the short dsRNA helices that are efficiently unwound by DEAD-box helicases will also have a detectable spontaneous rate of unwinding, and this rate needs to be subtracted from the observed rate in the presence of the helicase to determine the rate of helicase-dependent unwinding.

Within these constraints, the concentrations of the protein, substrate, and ATP should be varied to establish the robustness of the measurement. In addition, the solution conditions should be varied to determine how such changes impact the coupling between ATPase activity and RNA unwinding, as read out in the ATP utilization value. The length and sequence of the RNA helix should also be varied to examine how changes in the substrate impact the ATP utilization.

2.1 Equipment

- Gel electrophoresis system for non-denaturing polyacrylamide gels (used in a cold room or with a connected cooling system)
- Chamber for thin-layer chromatography
- Spectrophotometer for protein and RNA concentration determinations
- Typhoon™ gel imager system (phosphor or fluorescence) (Cytiva)
- Beta radiation shield and Geiger counter (for radioactivity-based measurements).
- Fluorometer (for fluorescence-based ATPase assays)

2.2 Reagents

- Purified DEAD-box helicase protein; for purification protocols, see, *e.g.*, (Del Campo & Lambowitz, 2009a; Pan et al., 2014)
- RNA oligonucleotides to form the dsRNA substrate (purchased from Dharmacon or Integrated DNA Technologies (IDT))
- RNase-free water
- RNase AWAY (Invitrogen) or equivalent surface decontaminant
- ATP (premixed with equimolar $MgCl_2$). Pre-mixing ATP with Mg^{2+} ions to form a stoichiometric complex is essential to prevent Mg^{2+} depletion from the reaction solution, which could destabilize dsRNA and influence the affinity of the helicase protein for RNA.
- For radioactive labeling of RNA:
  - T4 polynucleotide kinase (New England Biolabs (NEB))
  - [γ-^{32}P]ATP, 150 μCi/μL (6000 Ci/mmol; Perkin-Elmer, NEG035C)

- For radioactivity-based measurements of ATPase activity:
  - [γ-^{32}P]ATP, 10 μCi/μL (3000 Ci/mmol; Perkin-Elmer, BLU003H250UC)

    Note: This less-concentrated ATP is better suited for ATPase measurements than the product used for oligonucleotide labeling, because of the much lower background contamination level of Pi (the product of ATP hydrolysis).
  - 150 mM EDTA (pH 8.0) for quenching ATPase reactions
- For fluorescence-based unwinding assays:
  - one of the RNA oligonucleotides with a fluorescent label (purchased from, *e.g.*, IDT or labeled in-house (Greenfeld & Herschlag, 2013))
- For fluorescence measurements of ATPase activity
  - Coumarin-modified phosphate binding protein (Thermo Fisher Scientific, PV4406)
- Buffer, monovalent, and divalent salts for reactions; prepared with RNase free water and filter-sterilized (0.22 μm) or autoclaved
- Reagents for 20% non-denaturing polyacrylamide gel electrophoresis
- PEI-cellulose thin-layer chromatography (TLC) sheets (J.T.Baker)
- TLC mobile phase, containing 1 M formic acid and 0.5 M LiCl

2.3 Procedure

2.3.1 ATPase measurements

The experiment is performed with a moderate concentration of the dsRNA substrate (hundreds of nanomolar) and a moderate excess of one of the strands (2–10-fold). Under these conditions, unwinding of the dsRNA is rapidly reversed by binding of the strand in excess to its complement. Thus, the concentration of the dsRNA substrate remains essentially constant during the measurement, and the steady-state rate of ATP hydrolysis is dominated by interactions of the helicase with the dsRNA substrate. In the description below, we use radioactively labeled ATP and RNA as the default labeling strategy, and we outline adjustments for fluorescence-based assays in Section 2.3.3. An experimental workflow outlining the experimental steps is shown in Fig. 3A.

1. Form the dsRNA by incubating the two strands briefly in reaction buffer (with a modest excess of the strand that will not be labeled in the RNA unwinding measurements). For oligonucleotides with the potential to form competing self-structure, it may be necessary to perform a cycle of brief heating and cooling.

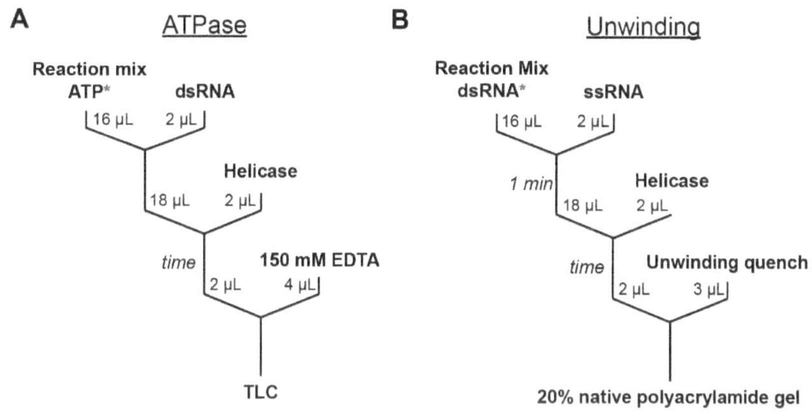

Fig. 3 Experimental workflows for measurement of ATP utilization during helicase-promoted RNA unwinding. Workflows are shown for ATPase (A) and RNA unwinding (B) rate measurements. Asterisks indicate radioactively labeled components (ATP or RNA).

2. Prepare reaction pre-incubation solution (16 μL) consisting of: Na-MOPS, pH 7.0, $MgCl_2$, KCl, ATP-Mg^{2+}, and a trace amount of [γ-^{32}P]ATP. Use concentrations that will result in desired concentrations after the volume is adjusted to 20 μL with addition of dsRNA and protein.
3. Add 2 μL of the dsRNA from step 1.
4. Initiate reaction with protein addition (2 μL). Use a final protein concentration of 1–5 μM.
5. At various times, quench aliquots (2 μL) by adding them to 4 μL of 150 mM EDTA solution.
6. Separate ATP and P_i using thin-layer chromatography. Cut each TLC sheet (20 x 20 cm) into three equally sized strips and, using a pencil and a ruler, gently draw a line 1 cm from the long edge of each strip. Pipet a small volume from each quenched time point (1–2 μL) along this line, leaving a 1-cm space between the centers of adjacent aliquots. Separate ATP and P_i by placing the TLC sheet upright into a glass chamber, with the mobile phase covering the bottom of the chamber, until the mobile phase migrates up the sheet and reaches the top.
7. Dry the TLC sheet (we use a hair dryer), cover it with plastic wrap, and expose a phosphor screen to the TLC sheet for 1–24 h (depending on the level of [γ-^{32}P]ATP used).
8. Scan the phosphor screen with a phosphorimager and quantify the band intensities using ImageQuant or ImageJ to determine the fraction of

ATP hydrolyzed at each time point ($P_i/(P_i + ATP)$). Plot this fraction against reaction time and perform a linear fit. Use the baseline P_i level in the radioactive ATP stock as the y-intercept (determined by analyzing an aliquot of the reaction mix without added helicase protein).

9. Determine the ATPase rate by multiplying the slope by the total unlabeled ATP concentration (see Fig. 2C). The units of the ATPase rate are concentration/time.
10. Perform an experiment that is exactly the same except that it lacks the dsRNA (*i.e.*, the reaction includes only the ssRNA at a concentration equivalent to the excess ssRNA). For example, if the standard measurement includes 0.5 μM of one strand and 1.2 μM of the other strand, the control reaction should have 0.7 μM of the strand in excess.

2.3.2 RNA unwinding measurements

The RNA unwinding measurement is performed using a pulse-chase procedure under conditions that are identical to those of the ATPase measurement, but the RNA unwinding measurement also includes a trace amount of a labeled oligonucleotide of the same sequence as the strand present in excess. A reaction workflow showing the experimental steps is shown in Fig. 3B.

1. Form the dsRNA in the "pulse" phase by incubating the labeled strand with a small excess of the complementary strand in reaction buffer. It may be helpful to mix the labeled strand with a low concentration of the same oligonucleotide unlabeled, and then add that mixture to a small excess of the complementary strand. Typical concentrations of the strands are in the low μM range. For oligonucleotides with the potential to form competing self-structure, it may be necessary to perform a cycle of brief heating and cooling.
2. Prepare the reaction pre-incubation solution (16 μL) of: Na-MOPS, pH 7.0, $MgCl_2$, KCl, labeled dsRNA from step 1 above, and ATP-Mg^{2+}. Use concentrations that will result in desired concentrations after volume adjustment to 20 μL with addition of unlabeled chase ssRNA and protein.
3. Add 2 μL of unlabeled ssRNA chase. The final concentration of this ssRNA should be the same as above in the ATPase measurement.
4. After a brief incubation of a defined time (typically 1 min), initiate the reaction by adding 2 μL of helicase protein. Mix well by flicking the tube.

5. At various times, quench aliquots (2 μL) by adding them to 3 μL of solution containing 50 mM $MgCl_2$, 0.1 mg/mL proteinase K, 30% glycerol, 0.4 mg/mL xylene cyanol, and 0.4 mg/mL bromophenol blue.
6. To determine the spontaneous unwinding rate, perform the same experiment without added helicase. For this reaction, add an equivalent volume (2 μL) of the helicase storage or dilution buffer to ensure that the solution conditions are identical.
7. Separate labeled ssRNA and dsRNA using 20% non-denaturing polyacrylamide gel electrophoresis at 4–6 °C.
8. Dry the gel and expose a phosphor screen (1–24 h, depending on the level of radiolabeled oligonucleotide used).
9. Scan the phosphor screen by using a phosphorimager and quantify the band intensities using ImageQuant or ImageJ to determine the fraction of the RNA duplex that remains intact at each time point (Duplex/(Duplex + ssRNA)).
10. Fit the time course of RNA unwinding to a single exponential equation to determine the rate constant (see Fig. 2B). For comparison to the ATPase rate, determine the initial steady-state rate of RNA unwinding by multiplying the rate constant by the concentration of dsRNA, which is equivalent to the concentration of the strand complementary to the labeled strand. This steady-state unwinding rate can also be estimated from the initial slope of the curve multiplied by the dsRNA concentration.

2.3.3 ATPase and RNA unwinding measurements with fluorescently labeled molecules

Methods for fluorescence-based measurements of ATP hydrolysis by helicase proteins have been described previously (Özeş, Feoktistova, Avanzino, Baldwin, & Fraser, 2014; Toseland & Webb, 2010). Briefly, coumarin-modified phosphate binding protein (MDCC-PBP) is a commercially available phosphate biosensor that enables real-time observation of ATP hydrolysis with minimal perturbation of the reaction conditions (Toseland & Webb, 2010). MDCC-PBP binds the ATP hydrolysis product P_i, resulting in a protein conformational change and a fluorescence increase. By calibrating the fluorescence signal with a series of known concentrations of P_i, the ATPase rate can be quantitatively determined (Özeş et al., 2014). While avoiding radioactive labeling, these measurements require a fluorometer or a fluorescence plate reader that is capable of sensitive, real-time kinetic measurements. Potential effects of PBP on the observed helicase activity should be assessed by performing RNA unwinding reactions with and without PBP.

Unwinding measurements with a fluorescently labeled dsRNA can be carried out with minimal modification from the protocol above, provided that a fluorescence scanner is available. High-quantum yield fluorescent dyes such as Cy3b are recommended to improve the signal-to-noise ratio. A simple fluorescence labeling strategy has been previously described in case pre-labeled oligonucleotides are unavailable or cost-prohibitive (Greenfeld & Herschlag, 2013).

As an alternative to gel-based unwinding measurements, real-time RNA unwinding measurements can also be performed using a fluorometer (Özeş et al., 2014).

2.4 Notes

1. To calculate the ATP utilization value per RNA unwinding event, the ATPase rate is divided by the unwinding rate. To account for the low level of ATPase stimulation by the excess ssRNA, the ATPase rate measured with ssRNA only ($V_{\mathrm{ATPase,ss}}$) should be subtracted from that measured during the unwinding reaction (V_{ATPase}) (Fig. 2C). The unwinding rate should likewise be adjusted by subtracting the spontaneous unwinding rate ($V_{\mathrm{unwind-}}$) from the unwinding rate measured with the helicase and ATP present ($V_{\mathrm{unwind+}}$) (Fig. 2B):

$$\#\mathrm{ATPs} = \frac{V_{ATPase} - V_{ATPase,ss}}{V_{unwind+} - V_{unwind-}} \tag{1}$$

2. Note that the protein concentration is not involved in the calculation of the ATPase rate or the RNA unwinding rate. Thus, the possibility of a fraction of inactive protein does not affect these rates or the ATP utilization value.

3. ATP utilization in RNA chaperone/refolding reactions

Regardless of the scale of the RNA structural rearrangement, the strategy for measuring ATP utilization remains essentially the same. The basic approach is to measure the kinetics of ATP hydrolysis and helicase-assisted RNA refolding under identical conditions. The ratio of the ATPase and refolding rates yields the total number of ATP molecules consumed per refolding event. Additional steps can be taken to further parse the ATP utilization into productive vs. non-productive ATP hydrolysis events and to identify factors that determine ATP consumption.

A requirement for measuring ATP utilization is that conditions exist under which the rate of the RNA folding transition is substantially increased by the ATP-dependent activity of the helicase. A range of reaction conditions should be surveyed to identify those that give at least a three-to-five-fold increase in the RNA refolding rate upon addition of the helicase and ATP. To this end, one should systematically vary the concentrations of ATP and the helicase and measure the rates of the RNA refolding event using a quantitative folding assay, which we have done by using the catalytic activity of the *Tetrahymena* ribozyme to measure the fraction of the ribozyme present in the native state (Potratz & Russell, 2012; Wan, Mitchell, & Russell, 2009). The helicase concentration should be varied in the presence of one or more constant ATP concentrations. Analogously, the ATP concentration should be varied in the presence of a constant helicase concentration that gives a substantial refolding rate enhancement and falls within a range in which the refolding rate increases with helicase concentration. The RNA concentration should be kept constant and, as in duplex unwinding measurements, should typically be in the hundreds-of-nanomolar range, comparable to the helicase concentration. Ultimately, the ideal starting conditions for ATP utilization are ones where both the helicase and ATP are sub-saturating, such that there is robust but slightly below-maximal enhancement of refolding relative to the rate without helicase and ATP.

Note: In our experience with CYT-19, high helicase concentrations can inhibit refolding, resulting in a plateau or even a decrease in the observed refolding rates with further increases in protein concentration (Jarmoskaite et al., 2021). High CYT-19 concentrations also show substantial ATP-independent activity, probably by binding and stabilizing unfolded RNA conformations (Bhaskaran & Russell, 2007). These observations emphasize the importance of systematically pre-screening the reaction conditions to find a regime in which the signal, *i.e.,* the ATP-dependent, helicase-dependent acceleration of the folding transition, is large enough to provide for a robust measurement of the ATP utilization.

As in duplex unwinding measurements, it is critical that the ATPase activity of the helicase be RNA-dependent. Ideally, the RNA-independent ATPase activity, measured under identical conditions to the refolding reaction but without added RNA, should be negligible relative to the ATPase rates in the presence of the RNA of interest. If significant

RNA-independent ATPase activity is observed, it is important to rule out co-purified cellular RNA in the helicase protein preparation. Contaminating RNA can be detected in a UV absorption spectrum by the presence of a peak or a prominent "shoulder" at 260 nm. Poly(ethyleneimine) (PEI) precipitation during protein purification is an effective way to remove cellular RNA (Atkinson & Jack, 1973). If RNA contamination is ruled out and the helicase still displays significant RNA-independent ATPase, the RNA-independent ATPase rate should be subtracted before calculating ATP utilization.

3.1 Equipment

- Gel electrophoresis system for denaturing polyacrylamide gels
- TLC chamber
- Typhoon™ phosphor imager (Cytiva)
- Beta radiation shield and Geiger counter

3.2 Reagents

- Purified DEAD-box helicase protein (for purification of the DEAD-box RNA chaperone CYT-19, see (Pan et al., 2014)).
- In-vitro transcribed, purified ribozyme. (We prepare the *Tetrahymena* L-21 ribozyme by transcription with T7 RNA polymerase (Zaug, Grosshans, & Cech, 1988) and purify it with Qiagen RNeasy kits.)
- For radioactive labeling of ribozyme substrate oligonucleotide:
  - T4 polynucleotide kinase (New England Biolabs (NEB)).
  - [γ-^{32}P]ATP, 150 μCi/μL (6000 Ci/mmol; Perkin-Elmer, NEG035C).
- For ATPase assays:
  - [γ-32P]ATP, 10 μCi/μL (3000 Ci/mmol; Perkin-Elmer, BLU003H250UC).
  - 150 mM EDTA (pH 8.0).
  - PEI-cellulose TLC sheets (J.T.Baker).
  - TLC mobile phase: 1 M formic acid and 0.5 M LiCl.
- RNase-free water, buffer, monovalent, and divalent salts for reactions.
- ATP (premixed with equimolar $MgCl_2$). Pre-mixing ATP with Mg^{2+} ions to form a stoichiometric complex is essential to prevent Mg^{2+} depletion from the reaction solution, which could destabilize the RNA structure and influence the affinity of the helicase protein for RNA.

- Reagents for 20% denaturing polyacrylamide gel electrophoresis.
- Reagents for a folding assay to monitor the folding transition of interest (*e.g.*, as described in (Potratz & Russell, 2012; Wan et al., 2009)).

3.3 Procedure

We describe the experiments below based on the specific example of the well-characterized group I intron ribozyme from *Tetrahymena thermophila* and the refolding of its misfolded conformation to the native conformation promoted by the DEAD-box protein CYT-19 (Jarmoskaite et al., 2021). After addressing the general considerations above, we describe some steps and conditions to consider for adapting the method to other RNA and helicase systems.

The workflow is schematically depicted in Fig. 4. Refolding measurements should be performed first, with the objective of establishing conditions at which refolding is greatly accelerated by the helicase. When a range of suitable conditions has been identified, the refolding and ATPase measurements can be performed in parallel, potentially even using the same reaction (Section 3.3.3).

3.3.1 ATPase measurements

1. The workflow is shown in Fig. 4A. Prepare misfolded *Tetrahymena* ribozyme by adding 10 mM $MgCl_2$ to unfolded RNA and incubating at 25 °C for 5 min. We use 2 μM ribozyme, 10 mM $MgCl_2$, and 50 mM Na-MOPS (pH 7.0) at this step (final concentrations), in a 20 μL volume or enough for the desired number of ATPase and refolding reactions, each using 2 μL. Store the misfolded RNA on ice.
2. Prepare the reaction mix of: Na-MOPS, pH 7.0, $MgCl_2$, KCl, ATP-Mg^{2+}, and a trace amount of [γ-^{32}P]ATP. The reaction mix should be prepared at 1.25-fold the desired final concentration to account for the dilution after the addition of 2 μL of protein and 2 μL of ribozyme to 16 μL of reaction mix. For reaction components that are already present in the prefolded ribozyme (such as $MgCl_2$ and buffer), concentrations in the reaction mix should be lowered accordingly; *e.g.*, adding the prefolded ribozyme will contribute 1 mM Mg^{2+} to the final Mg^{2+} concentration, so for a total Mg^{2+} concentration of 2 mM, the reaction mix should only contain 1.25 mM Mg^{2+}.

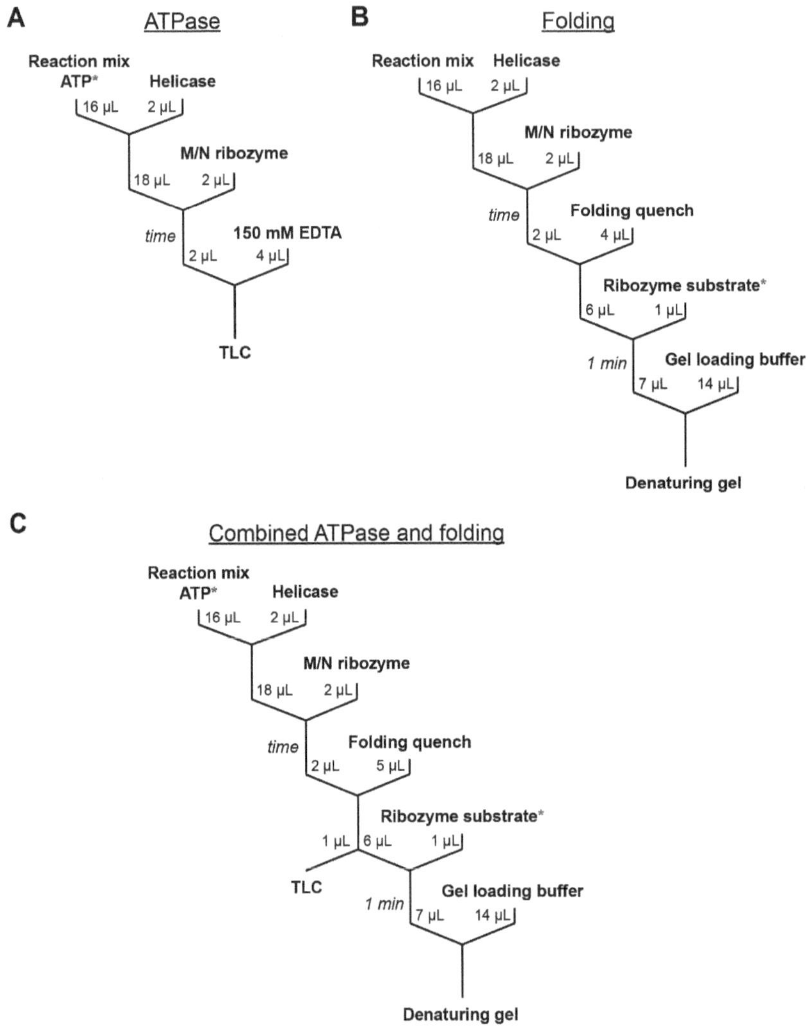

Fig. 4 Experimental workflows for measurement of ATP utilization during a helicase-promoted RNA folding transition. Workflows are shown for ATPase (A), RNA refolding (B), and combined ATPase and refolding (C) rate measurements. Asterisks indicate radioactively labeled components (ATP and/or ribozyme substrate oligonucleotide). In (C), the concentrations of folding quench components should be adjusted to be 80% lower than those listed in Section 3.3.2 (step 3) to account for the slightly larger volume required to provide sufficient sample for both TLC and gel analysis.

3. Aliquot 16 μL of reaction mix to a microcentrifuge tube for each reaction. A few minutes before starting the reaction, equilibrate the solution at the desired temperature.
4. Prepare one or more helicase dilutions at 10-fold above the final concentrations. Keep these protein solutions on ice.

5. Aliquot 150 mM EDTA (ATPase quench solution) into a number of tubes corresponding to the number of time points planned, 4 μL per tube. Keep at room temperature.
6. To start the reaction, pipette 2 μL of the helicase dilution into the reaction mix. Immediately add 2 μL of the ribozyme and note the time. The time of adding the ribozyme serves as reference (zero time point) for calculating the time elapsed at each subsequent time point. Mix well by flicking the tube or by pipetting up and down (with the volume set to at least half the reaction volume).
   - The helicase is added, followed by the ribozyme, to prevent the possibility of any rapid folding transitions of the pre-folded ribozyme under the reaction conditions before the helicase is added. By initiating the reaction with the addition of pre-folded ribozyme, we ensure that all of the reactions start with the same ensemble of ribozyme conformations.
7. Immediately begin taking time points. To determine the initial ATPase rate, collect a series of time points at 15–30 s intervals (or as quickly as possible) by transferring 2 μL aliquots to the tubes containing 4 μL of ATPase quench solution. To measure ATPase activity throughout the full time course of refolding, collect additional time points (at longer intervals) up to the time corresponding to 3–5 half-lives of the refolding reaction (measured in Section 3.3.2).
8. Comparing the ATPase activity during the folding transition to the ATPase stimulation by the native (correctly folded) ribozyme can provide useful insights into the extent of productive vs. non-productive ATP hydrolysis (see Section 3.4). Therefore, for every set of conditions (including each helicase, ATP-Mg^{2+}, and Mg^{2+} concentration), we also perform an identical experiment with the native ribozyme instead of misfolded ribozyme. To prepare the native *Tetrahymena* ribozyme, add 10 mM $MgCl_2$ to unfolded RNA and incubate at 50 °C for 45 min. Store the native ribozyme on ice.
9. For a subset of conditions (or, in case of substantial RNA-independent ATPase activity, all conditions), also measure the ATPase activity without added RNA. Carry out the reaction identically but use 2 μL of a ribozyme-free solution of 10 mM $MgCl_2$ and 50 mM Na-MOPS (pH 7.0).
10. Determine the baseline level of P_i in the [γ-^{32}P]ATP stock by quenching an aliquot of the reaction mix (without helicase and ribozyme). This value will represent the amount of P_i hydrolyzed at the zero time point (*i.e.*, the y-intercept value of the ATPase time course plot).

This baseline value should be determined in every experiment because it can vary considerably between [γ-^{32}P]ATP stocks and can increase with prolonged storage.

11. Process the quenched aliquots using TLC and determine the initial rates, as described in Section 2.3.1.

3.3.2 Refolding measurements

The refolding reaction is performed analogously to the ATPase reaction until the quenching step, upon which an additional experimental step is required to determine the fraction of native ribozyme. The two-stage (discontinuous) folding assay using ribozyme catalytic activity has been described in detail previously (Potratz & Russell, 2012; Wan et al., 2009). These references should be consulted for adapting the assay to other ribozymes.

1. The workflow is shown in Fig. 4B. Prepare misfolded ribozyme and helicase dilutions as in Section 3.3.1.
2. Prepare reaction mix as in Section 3.3.1 but omit [γ-^{32}P]ATP.
3. Prepare a folding quench solution containing a high Mg^{2+} concentration (to block further refolding and promote catalysis), guanosine (required for catalysis), and proteinase K (to degrade the helicase). The quench should be prepared at 1.5-fold the final concentration and, for the *Tetrahymena* ribozyme, contains the following: 75 mM $MgCl_2$ (or lower if substantial Mg^{2+} is contributed by the refolding reaction), 750 μM guanosine, 1.5 mg/mL proteinase K, and 50 mM Na-MOPS (pH 7.0). Aliquot 4 μL of quench solution per tube, with one tube for each planned time point.
4. Prepare a dilution of ^{32}P-labeled oligonucleotide substrate in TE buffer, calculated to provide sufficient signal at the gel electrophoresis step ($\geq$1000 cpm per lane). The volume should be sufficient to provide 1 μL per time point. The substrate concentration should be much lower than the ribozyme concentration during the substrate cleavage reaction. An upper limit on the concentration of the radiolabeled substrate stock can be conveniently calculated from the amount used in the labeling reaction and the final volume of the storage solution.
5. Start the reaction and collect time points as in Section 3.3.1. During wait times between longer time points, or between reactions, process the quenched time points by adding 1 μL of labeled substrate and quenching the substrate cleavage reaction exactly 1 min later with 14 μL of denaturing gel loading buffer (72% formamide (vol/vol), 100 mM EDTA,

0.4 mg/mL xylene cyanol, and 0.4 mg/mL bromophenol blue). The substrate binds with comparable rate constants to the misfolded and native ribozyme, but after 1 minute, only the substrate that initially bound to the native ribozyme is cleaved. The fraction of cleaved substrate therefore indicates the fraction of native ribozyme in the sample (Wan et al., 2009). The appropriate cleavage time should be determined empirically for other systems.

6. For every set of conditions, measure the helicase-independent refolding rate by replacing the helicase with 2 μL of buffer that was used to prepare helicase dilutions. Also measure the ATP-independent effects of the helicase on refolding by omitting ATP from the reaction mix.
7. Separate the cleaved and uncleaved substrate on a 20% denaturing polyacrylamide gel. After drying, expose a phosphor screen to the dried gel overnight, scan the screen using a phosphorimager, and quantify the fraction of cleaved substrate with ImageQuant or ImageJ.
8. Plot the exponential refolding curve. Determine the initial rate as described for the duplex unwinding measurements, with one of the following options: (1) fit the refolding curve to a single exponential equation and multiply the rate constant by the total ribozyme concentration; (2) fit the earliest time points (with the extent of refolding not exceeding ~20% of the amplitude of the refolding curve) to a line, with the slope, multiplied by the ribozyme concentration, indicating the initial refolding rate. Proceed to the calculations in Section 3.4.

3.3.3 *Combined ATPase and refolding measurements*

If desired, the ATP utilization measurements can be simplified by using a single reaction that includes trace [γ-^{32}P]ATP and is quenched in the usual way with the folding quench solution (Section 3.3.2; Fig. 4C). Because the folding quench contains proteinase K, it also stops ATP hydrolysis by degrading the helicase. Aliquots of the quenched time point can then be independently analyzed by TLC and by the oligonucleotide cleavage assay.

3.4 Notes

1. The total number of ATPs hydrolyzed per refolding event is most simply calculated as the ratio of the initial ATPase rate (measured in the presence of the misfolded ribozyme) and initial refolding rate:

$$\#\text{ATPs} = \frac{V_{ATPase}}{V_{refolding}} \tag{2}$$

2. If the spontaneous refolding rate is non-negligible relative to helicase-assisted refolding (*i.e.*, not all of the refolding activity can be assumed to occur due to the helicase activity), the helicase-dependent refolding rate should be estimated by subtracting the helicase-independent refolding rate ($V_{\text{refolding}-}$) from the refolding rate in the presence of the helicase ($V_{\text{refolding}+}$):

$$\#\text{ATPs} = \frac{V_{ATPase}}{V_{refolding+} - V_{refolding-}} \tag{3}$$

3. In the case of the *Tetrahymena* ribozyme, we know that the misfolded ribozyme has a very similar overall structure to the native ribozyme (Mitchell, Jarmoskaite, Seval, Seifert, & Russell, 2013; Russell et al., 2006). As a result, robust ATPase activity is observed both with the misfolded ribozyme (refolding of which is accelerated by CYT-19 and ATP) and with the native ribozyme (which CYT-19 does not detectably unfold under the conditions of refolding measurements), although the ATPase rate is higher in the presence of misfolded RNA (Fig. 5A) (Jarmoskaite et al., 2021). To narrow down the number of productive ATP hydrolysis events, we assume that (1) all ATP hydrolysis with native ribozyme is non-productive for unfolding of the ribozyme structure; (2) all interactions with native ribozyme are also formed with misfolded ribozyme. Under these assumptions, the ATPase rate measured with native ribozyme can be subtracted as non-productive background:

$$\#\text{ATPs} = \frac{V_{ATPase,M} - V_{ATPase,N}}{V_{refolding+} - V_{refolding-}} \tag{4}$$

4. A method for determining the ATP utilization from the full ATPase time course instead of initial rates is depicted in Fig. 5B. This approach yields consistent results to the initial-rate approach (Jarmoskaite et al., 2021). Nevertheless, note that because of the long time course, this approach is more prone than initial rates to potential errors from inhibition of the helicase by accumulated ADP and general loss of protein activity.
5. To further dissect what determines the ATP utilization value, portions of the ribozyme can be mutated, truncated, or extended. The stability of the RNA can be systematically varied through mutations or by varying the Mg^{2+} concentration.

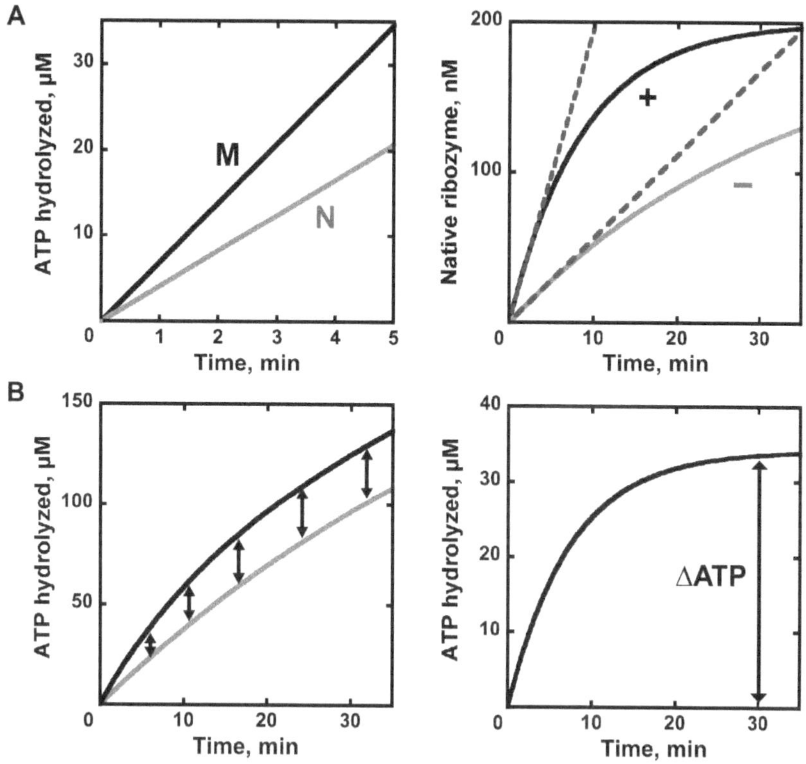

$$\#ATPs = \frac{\Delta ATP}{[Ribozyme]}$$

Fig. 5 Determination of the ATP utilization value for ribozyme refolding. (A) The initial-rate method. Left: To obtain the ATP hydrolysis rate associated with refolding, the initial ATPase rates (slopes of linear fits) are determined during refolding (black; M) and in the presence of the pre-folded native ribozyme (grey; N). All panels depict simulated data based on previous results for CYT-19 and the *Tetrahymena* ribozyme (Jarmoskaite et al., 2021). To narrow down the number of productive ATP hydrolysis events, for the *Tetrahymena* ribozyme, we subtract the background ATPase activity seen with N ribozyme from the ATPase rate measured during the refolding of M ribozyme (see notes in Section 3.4 for the underlying assumptions). Right: The initial refolding rate is determined both in the presence (+) and absence (–) of helicase and ATP, and the spontaneous rate is subtracted. The ATP utilization value is calculated using Eq. (4). (B) The full time-course approach. Left: The ATPase activity is recorded until the completion of the refolding reaction, *i.e.*, until at least ~90% of the ribozyme has reached the native state (based on the refolding data). The same time points as those taken in the refolding reaction (black) are collected in the presence of pre-folded native ribozyme (gray). At each time point, the difference in the amount of hydrolyzed ATP is calculated between the two time courses (arrows) and the resulting values are plotted (right). The amplitude of the resulting curve (arrow) corresponds to the total amount of productive ATP hydrolysis. Division by the ribozyme concentration yields the number of ATPs hydrolyzed per ribozyme.

4. Conclusion

In this chapter, we have described methods for quantifying the coupling efficiency between ATP hydrolysis and RNA structural rearrangements by DEAD-box RNA helicases. Applications of these or analogous methods to several DEAD-box proteins have advanced our understanding of the RNA unwinding mechanism by these ubiquitous helicase proteins and have provided mechanistic insights into the RNA chaperone function of the DEAD-box helicase CYT-19. We hope that this chapter will spur further investigation of ATP utilization values across functionally diverse DEAD-box proteins, their RNA targets, and factors that regulate DEAD-box protein activity.

Acknowledgments

Research in the Russell lab is supported by NIH grant R35 GM131777.

References

Atkinson, A., & Jack, G. W. (1973). Precipitation of nucleic acids with polyethyleneimine and the chromatography of nucleic acids and proteins on immobilised polyethyleneimine. *Biochimica et Biophysica Acta (BBA) - Nucleic Acids and Protein Synthesis, 308*(1), 41–52.

Bhaskaran, H., & Russell, R. (2007). Kinetic redistribution of native and misfolded RNAs by a DEAD-box chaperone. *Nature, 449*(7165), 1014–1018.

Cao, W., Coman, M. M., Ding, S., Henn, A., Middleton, E. R., Bradley, M. J., et al. (2011). Mechanism of Mss116 ATPase reveals functional diversity of DEAD-box proteins. *Journal of Molecular Biology, 409*(3), 399–414.

Chen, Y., Potratz, J., Tijerina, P., Del Campo, M., Lambowitz, A., & Russell, R. (2008). DEAD-box proteins can completely separate an RNA duplex using a single ATP. *Proceedings of the National Academy of Sciences of the United States of America, 105*(51), 20203–20208.

Del Campo, M., & Lambowitz, A. M. (2009a). Crystallization and preliminary X-ray diffraction of the DEAD-box protein Mss116p complexed with an RNA oligonucleotide and AMP-PNP. *Acta Crystallographica Section F, 65*(8), 832–835.

Del Campo, M., & Lambowitz, A. M. (2009b). Structure of the yeast DEAD box protein Mss116p reveals two wedges that crimp RNA. *Molecular Cell, 35*(5), 598–609.

Ding, Y., Tang, Y., Kwok, C. K., Zhang, Y., Bevilacqua, P. C., & Assmann, S. M. (2014). In vivo genome-wide profiling of RNA secondary structure reveals novel regulatory features. *Nature, 505*(7485), 696–700.

Fairman-Williams, M. E., Guenther, U.-P., & Jankowsky, E. (2010). SF1 and SF2 helicases: family matters. *Current Opinion in Structural Biology, 20*(3), 313–324.

Greenfeld, M., & Herschlag, D. (2013). Chapter fifteen—Fluorescently labeling synthetic RNAs. In J. Lorsch (Ed.), *Vol. 530. Methods in enzymology* (pp. 281–297). Academic Press.

Henn, A., Cao, W., Hackney, D., & De La Cruz, E. (2008). The ATPase cycle mechanism of the DEAD-box rRNA helicase, DbpA. *Journal of Molecular Biology, 377*(1), 193–205.

Herschlag, D. (1995). RNA chaperones and the RNA folding problem. *Journal of Biological Chemistry*, *270*(36), 20871–20874.

Hilbert, M., Karow, A. R., & Klostermeier, D. (2009). The mechanism of ATP-dependent RNA unwinding by DEAD box proteins. *Biological Chemistry*, *390*(12), 1237–1250.

Jarmoskaite, I., & Russell, R. (2014). RNA helicase proteins as chaperones and remodelers. *Annual Review of Biochemistry*, *83*, 697–725.

Jarmoskaite, I., Tijerina, P., & Russell, R. (2021). ATP utilization by a DEAD-box protein during refolding of a misfolded group I intron ribozyme. *Journal of Biological Chemistry*, *296*.

Linder, P., & Jankowsky, E. (2011). From unwinding to clamping—The DEAD box RNA helicase family. *Nature Reviews Molecular Cell Biology*, *12*(8), 505–516.

Liu, F., Putnam, A., & Jankowsky, E. (2008). ATP hydrolysis is required for DEAD-box protein recycling but not for duplex unwinding. *Proceedings of the National Academy of Sciences of the United States of America*, *105*(51), 20209–20214.

Lohman, T. M., Tomko, E. J., & Wu, C. G. (2008). Non-hexameric DNA helicases and translocases: mechanisms and regulation. *Nature Reviews Molecular Cell Biology*, *9*(5), 391–401.

Mallam, A. L., Del Campo, M., Gilman, B., Sidote, D. J., & Lambowitz, A. M. (2012). Structural basis for RNA-duplex recognition and unwinding by the DEAD-box helicase Mss116p. *Nature*, *490*(7418), 121–125.

Martinez-Senac, M. M., & Webb, M. R. (2005). Mechanism of translocation and kinetics of DNA unwinding by the helicase RecG. *Biochemistry*, *44*(51), 16967–16976.

Mitchell, D., Jarmoskaite, I., Seval, N., Seifert, S., & Russell, R. (2013). The long-range P3 helix of the *Tetrahymena* ribozyme is disrupted during folding between the native and misfolded conformations. *Journal of Molecular Biology*, *425*(15), 2670–2686.

Özeş, A. R., Feoktistova, K., Avanzino, B. C., Baldwin, E. P., & Fraser, C. S. (2014). Real-time fluorescence assays to monitor duplex unwinding and ATPase activities of helicases. *Nature Protocols*, *9*(7), 1645–1661.

Pan, C., Potratz, J. P., Cannon, B., Simpson, Z. B., Ziehr, J. L., Tijerina, P., et al. (2014). DEAD-box helicase proteins disrupt RNA tertiary structure through helix capture. *PLoS Biology*, *12*(10), e1001981.

Potratz, J. P., & Russell, R. (2012). RNA catalysis as a probe for chaperone activity of DEAD-box helicases. *Methods Enzymology*, *511*, 111–130.

Pyle, A. M. (2008). Translocation and unwinding mechanisms of RNA and DNA helicases. *Annual Review of Biophysics*, *37*(1), 317–336.

Rajagopal, V., Gurjar, M., Levin, M. K., & Patel, S. S. (2010). The protease domain increases the translocation stepping efficiency of the hepatitis C virus NS3-4A helicase. *Journal of Biological Chemistry*, *285*(23), 17821–17832.

Rouskin, S., Zubradt, M., Washietl, S., Kellis, M., & Weissman, J. S. (2014). Genome-wide probing of RNA structure reveals active unfolding of mRNA structures in vivo. *Nature*, *505*(7485), 701–705.

Russell, R., Das, R., Suh, H., Travers, K., Laederach, A., Engelhardt, M., et al. (2006). The paradoxical behavior of a highly structured misfolded intermediate in RNA folding. *Journal of Molecular Biology*, *363*(2), 531–544.

Sengoku, T., Nureki, O., Nakamura, A., Kobayashi, S., & Yokoyama, S. (2006). Structural basis for RNA unwinding by the DEAD-box protein *Drosophila* Vasa. *Cell*, *125*(2), 287–300.

Staley, J. P., & Guthrie, C. (1998). Mechanical devices of the spliceosome: Motors, clocks, springs, and things. *Cell*, *92*(3), 315–326.

Tijerina, P., Bhaskaran, H., & Russell, R. (2006). Nonspecific binding to structured RNA and preferential unwinding of an exposed helix by the CYT-19 protein, a DEAD-box RNA chaperone. *Proceedings of the National Academy of Sciences of the United States of America*, *103*(45), 16698–16703.

Toseland, C. P., & Webb, M. R. (2010). Fluorescence tools to measure helicase activity in real time. *Methods*, *51*(3), 259–268.

Van Treeck, B., & Parker, R. (2018). Emerging roles for intermolecular RNA-RNA interactions in RNP assemblies. *Cell*, *174*(4), 791–802.

Wan, Y., Mitchell, D., & Russell, R. (2009). Catalytic activity as a probe of native RNA folding. *Methods Enzymology*, *468*, 195–218.

Weis, K., & Hondele, M. (2022). The role of DEAD-Box ATPases in gene expression and the regulation of RNA–protein condensates. *Annual Review of Biochemistry*, *91*.

Xing, Z., Ma, W. K., & Tran, E. J. (2019). The DDX5/Dbp2 subfamily of DEAD-box RNA helicases. *Wiley Interdisciplinary Reviews: RNA*, *10*(2), e1519.

Zaug, A. J., Grosshans, C. A., & Cech, T. R. (1988). Sequence-specific endoribonuclease activity of the *Tetrahymena* ribozyme: enhanced cleavage of certain oligonucleotide substrates that form mismatched ribozyme-substrate complexes. *Biochemistry*, *27*(25), 8924–8931.

CHAPTER FIVE

In vitro characterization of Dhr1 from *Saccharomyces cerevisiae*

Ran Lin[a], Carl C. Correll[b], and Arlen W. Johnson[a,*]

[a]Department of Molecular Biosciences and Institute for Cellular and Molecular Biology, The University of Texas at Austin, Austin, TX, United States
[b]Center for Proteomics and Molecular Therapeutics, Chicago Medical School, Rosalind Franklin University of Medicine and Science, North Chicago, IL, United States
*Corresponding author: e-mail address: arlen@austin.utexas

Contents

Methods in Enzymology, Volume 673
ISSN 0076-6879
https://doi.org/10.1016/bs.mie.2022.03.058

Abstract

The RNA helicase Dhr1 from *S. cerevisiae* is an essential enzyme required for the assembly of the cytosolic small ribosomal subunit (SSU). A critical feature of the SSU is the central pseudoknot, an RNA fold that organizes the overall architecture of the subunit and connects all four domains of the 18S ribosomal RNA (rRNA). The initial folding of rRNA is guided, in part, by the U3 small nucleolar RNA, which base-pairs with the pre-rRNA in such a way as to preclude premature formation of the central pseudoknot. Thus, the essential role of Dhr1 is the unwinding of U3 from the pre-rRNA to allow folding of the central pseudoknot. Enzymes of the DEAH/RNA helicase A-like (RHA) family, to which Dhr1 belongs, are involved in splicing and ribosome biogenesis. They typically unwind RNA duplexes by translocation along a single strand of RNA in a 3′ to 5′ direction, driven by ATP hydrolysis. The substrate specificity of these enzymes requires tight regulation of their activity, by restricting access to their substrates, requiring adaptors to recruit them to their substrates and mechanisms of inhibiting and activating their activity. Purified Dhr1 is an active RNA-dependent ATPase with specific unwinding activity. Here, we provide detailed protocols for its purification and assays for its ATPase and unwinding activities.

1. Introduction

The DEAH helicase Dhr1 from *S. cerevisiae*, also identified as Ecm16, is an essential 145 kDa protein that is required for assembly of the small ribosomal subunit (SSU). In eukaryotic cells, the bulk of ribosome synthesis is carried out in the nucleolus, a membraneless subcompartment of the nucleus. Ribosome assembly is initiated by the transcription of the ribosomal DNA (rDNA) repeats by RNA polymerase I. In yeast, RNA polymerase I synthesizes a 35S primary transcript that contains three rRNAs, 18S of the small subunit, and 5.8S and 25S of the large subunit, flanked by external transcribed spacers (ETSs) and interrupted by two internal transcribed spacers (ITSs). The spacer sequences are removed by endo and exonucleases during ribosome assembly. Within a single rDNA repeat, the 5′-ETS is transcribed first, followed by 18S and ITS1 sequences with subsequent sequences devoted to assembly of the large subunit synthesized later. The initial assembly of ribosomes is co-transcriptional with the small subunit assembling before the large subunit due to the order of sequences within the 35S transcript. Early rRNA processing events include extensive rRNA modification guided by small nucleolar RNAs (snoRNAs) with box C/D snoRNAs directing ribose 2′-OH methylation and box H/ACA snoRNA directing pseudouridylation. Concomitant with RNA modification, the transcript begins folding into domains that will eventually constitute the RNA domains of the ribosomal subunits. Folding of the rRNA and loading of ribosomal proteins are coordinated by a large host of assembly factors. For general reviews about ribosome biogenesis in yeast see (Baßler & Hurt, 2019; Klinge & Woolford, 2019; Woolford & Baserga, 2013).

A critical and conserved step in small subunit biogenesis is the assembly of the SSU processome, a large complex of pre-rRNA, assembly factors and a subset of ribosomal proteins. Indeed, early electron micrographs of rDNA transcription revealed the presence of terminal knobs (Miller & Beatty, 1969), which were subsequently recognized to be the assembling SSU processome (Dragon et al., 2002; Grandi et al., 2002). Assembly of the SSU processome is scaffolded on the 5′-ETS of the pre-rRNA as well as the 333 nucleotide (nt) U3 box CD snoRNA (Fig. 1A) (Barandun et al., 2017; Chaker-Margot, Hunziker, Barandun, Dill, & Klinge, 2015; Cheng et al., 2019; Du et al., 2020; Sun et al., 2017; Zhang, Wu, Cai, Chen, & Ye, 2016). Unlike other box CD snoRNAs, U3 does not direct methylation but rather acts as an RNA chaperone, helping to spatially organize the pre-rRNA and prevent premature formation of a central tertiary contact. Early genetic experiments identified three U3-pre-rRNA interactions: U3 box A/A' with 6–22 of 18S (designated U3-18S duplex), U3 5′ hinge with 470–479 of ETS (designated U3-ETS1) and U3 3′ hinge with 281–291 of ETS (designated U3-ETS2) (Hughes, 1996; Kornprobst et al., 2016; Marmier-Gourrier, Cléry, Schlotter, Senty-Ségault, & Branlant, 2011; Sharma & Tollervey, 1999). A fourth interaction, between U3 and nts 1139–1143, was also proposed (Hartl et al., 2013). Recent cryo-EM structures of the SSU processome confirm the presence of the first three duplexes with some subtle differences (Sun et al., 2017). However, the fourth proposed duplex was not observed and instead, a duplex between U3 and nts 1111–1123 was observed.

The transition from the SSU Processome to the mature 40S proceeds through multiple intermediates (Cheng et al., 2020) and involves rRNA and protein rearrangements that result in compaction of the overall assembly (Fig. 1). A critical element of the mature 40S subunit is the central pseudoknot, comprising nts 4–20 and 1138–1144 of 18S rRNA. The central pseudoknot is located at the nexus of, and organizes, the four rRNA domains of the small subunit thereby establishing the decoding center (Fig. 1C) (Ben-Shem et al., 2011). Notably, U3-hybridization with nts 9–16 of 18S in the SSU processome precludes premature formation of the central pseudoknot of 18S rRNA. Consequently, U3 must be unwound from 18S rRNA to allow folding of the central pseudoknot. The transition

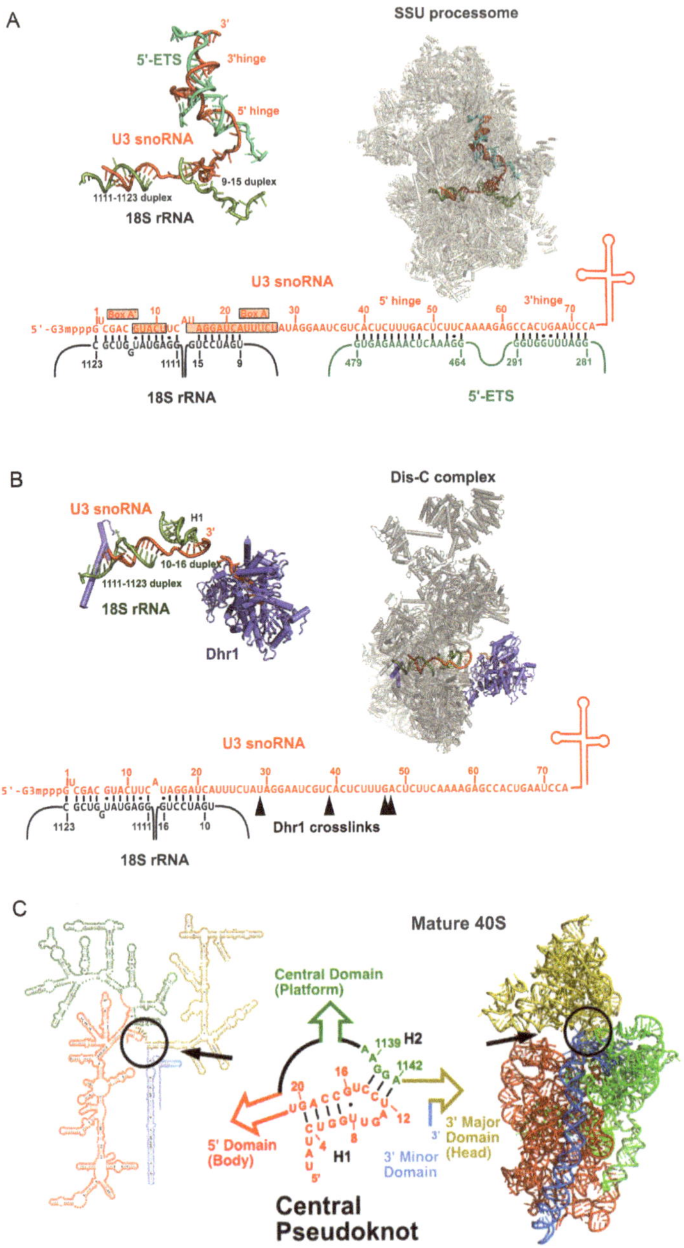

Fig. 1 See figure legend on opposite page.

from SSU processome to pre-40S also requires pre-rRNA cleavages to liberate the nascent pre-40S from the assembling pre-60S downstream of the SSU processome on the same primary transcript. Endonuclease cleavages at sites A0 and A1 release the 5′-ETS, which is degraded by the nuclear exosome (Allmang et al., 1999; Thoms et al., 2015). A third cleavage at site A2, within the internal transcribed spacer separating the 18S and 5.8S/25S rRNAs, releases the small subunit precursor from the large subunit precursor. In addition to its role in SSU processome assembly, U3 snoRNA is also required for cleavage at sites A0, A1 and A2 (Dragon et al., 2002; Hughes & Ares, 1991), presumably by establishing the correct RNA conformation to promote recognition by the endonucleases that cleave at these sites.

Genetic and biochemical studies identified Dhr1 as the RNA helicase that unwinds U3 from 18S rRNA. Dhr1 is a member of the SF2 superfamily and classified as a DEAH/RNA helicase A-like (RHA) enzyme based on sequence similarity to other known RNA helicases. These enzymes typically display 3′ to 5′ unwinding activity (Hamann, Enders, & Ficner, 2019; Ozgur et al., 2015). Initial characterization of Dhr1 function *in vivo* revealed that it was required for cleavage at sites A1 and A2 and that it stably associated with U3 sno-RNA suggesting a role for Dhr1 in rRNA rearrangements necessary for forming the central pseudoknot (Colley, Beggs, Tollervey, & Lafontaine, 2000). Further work showed that overexpression of mutants with substitutions in motif I (nucleotide binding), II or III (nucleotide hydrolysis) were dominant lethal and resulted in inhibition of cleavage at sites A0, A1 and A2 (Granneman, Bernstein, Bleichert, & Baserga, 2006). Subsequently, we used *in vivo* protein-RNA crosslinking to show that Dhr1 binds to U3 snoRNA, primarily nucleotides 22–63 of the BoxA and the 5′ hinge. Considering that Dhr1 likely translocates in a 3′ to 5′ direction on its RNA target while unwinding RNAs, this crosslinking analysis places Dhr1 immediately downstream of the U3:18S heteroduplexes

Fig. 1 Overview of U3 snoRNA and 18S transactions during SSU assembly. (A) U3 snoRNA (red) scaffolds assembly of the SSU processome by base pairing with 18S rRNA (olive) and the 5′ ETS (cyan). U3 (nts 1–76) based paired with 18S and 5′-ETS (upper left) extracted from the intact SSU processome (PDB: 5wlc) (upper right). BoxA′ and BoxA regions are highlight (lower). (B) U3 snoRNA (red) base paired with 18S rRNA (olive) and the 5′ ETS (cyan) with Dhr1 (blue) extracted from the Dis-C transition intermediate (PDB: 6zqg) (upper right). In the Dis-C complex, 5′ ETS sequences have been enzymatically removed (lower). The positions of Dhr1 crosslinks to U3 are shown (black arrows, lower). (C). The central pseudoknot (middle) links all four rRNA domains of the SSU. Left, secondary structure of 18S from *S. cerevisiae*. Right, tertiary structure (PDB: 3u5b).

(Fig. 1B). We also showed that a catalytic mutant of Dhr1, Dhr1-K420A, fails to release U3 RNA *in vivo* (Sardana et al., 2015) and prevents the folding of the central pseudoknot, as assayed by chemical probing with DMS. Furthermore, mutations in residues of U3 that reduce base-pairing with 18S rRNA suppress a cold-sensitive mutant of Dhr1 (Sardana et al., 2015). Lastly, recent structural studies of transition intermediates of the SSU processome reveal Dhr1 positioned adjacent to the U3:18S duplexes prior to their unwinding in the DisC intermediate (Cheng et al., 2020). These results identify the native substrate of Dhr1 to be the U3:18S duplexes. A similar conclusion has been reached for DHX37, the mammalian homolog of Dhr1 (Choudhury, Hackert, Memet, Sloan, & Bohnsack, 2019).

Cryo-EM studies of 40S assembly have revealed several intermediate steps in the transition from the SSU Processome to the 40S subunit (Cheng et al., 2020). In particular, the Dis-C complex represents an intermediate in which many of the SSU Processome factors have been released but Dhr1 has not yet unwound U3 from 18S, indicating that the unwinding of U3 is a relatively late step in the transition from the SSU Processome to 40S. Dhr1 is first resolved in cryo-EM structures in the post-A1 intermediate, a state immediately after cleavage at A1. However, Dhr1 is required for A1 cleavage and catalytic mutants block cleavage at A1, suggesting that Dhr1 first assembles into the SSU earlier than the post-A1 state, despite it not being resolved by cryo-EM; presumably it is bound but not ordered sufficiently to determine their structure. This is consistent with studies on the order of assembly of factors into the SSU processome. Dhr1 remains associated with the maturing subunit during the transition from the SSU processome to the Dis-C complex in which Dhr1 is bound to U3 downstream of the U3:18S heteroduplexes (Fig. 1B). Further rRNA rearrangements, triggered by the recruitment of the methyltransferase Bud23, are thought to be necessary to reposition nts 1138–1144 of 18S rRNA proximal to the U3:18S heteroduplexes to promote productive formation of the central pseudoknot upon unwinding of U3 from the rRNA by Dhr1 (Black, Sardana, Elmir, & Johnson, 2020).

The presence of Dhr1 in the SSU processome prior to unwinding U3 suggests that Dhr1 activity is tightly controlled to prevent premature unwinding and ensure productive folding of the central pseudoknot. Control of Dhr1 activity is likely regulated, in part, through access to its substrate. Notably, cleavage at A1 by Utp24 followed by exonucleolytic removal of the 5′-ETS by the nuclear exosome may be necessary for

Dhr1 to engage with U3 snoRNA. Dhr1 is also regulated by additional assembly factors. However, unlike other DEAH RNA helicases, which utilize G-Patch proteins for their activation, no G-Patch protein has been identified as an activator of Dhr1. Instead, we identified the assembly factor Utp14 as an interactor of Dhr1 that activates its unwinding activity *in vivo* and *in vitro* (Zhu, Liu, Anjos, Correll, & Johnson, 2016). Biochemical analysis of mouse DHX37 and Utp14 revealed that Utp14 enhances the RNA binding of DHX37, and may act as a processivity factor for DHX37 unwinding (Boneberg et al., 2019). Lastly, a recent crystal structure of N-terminally truncated Dhr1 revealed the presence of a loop of Dhr1 (aa644 to 728), which sits in the expected RNA channel of the enzyme thereby precluding RNA binding and suggesting a mechanism of autoinhibition (Singh, Vanden Broeck, Miller, Chaker-Margot, & Klinge, 2021) (Fig. 2).

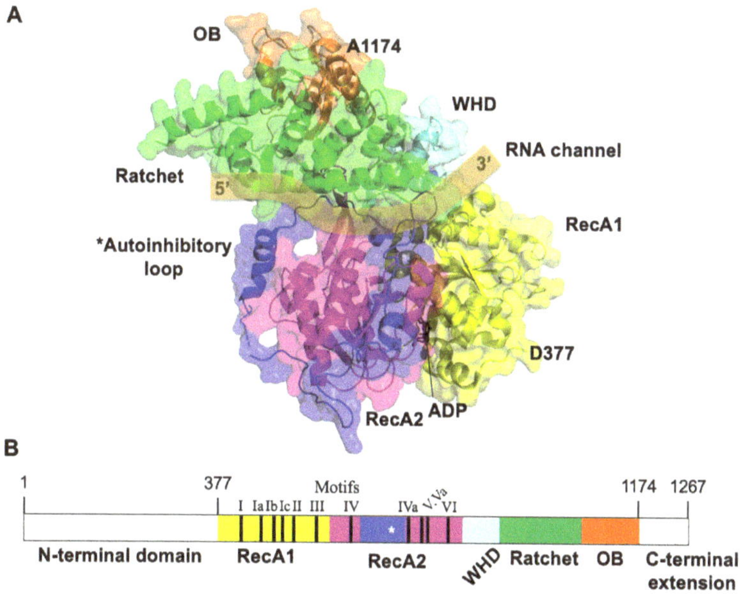

Fig. 2 Dhr1. (A) The crystal structure of Drh1 (amino acids 377–1174, PDB: 7MQJ) colored by functional domains: RecA1 (yellow), RecA2 (magenta with blue autoinhibitory loop), winged helix domain (WHD, cyan), ratchet domain (green), oligonucleotide/oligosaccharide binding domain (OB, orange). Based on superposition (not shown) with co-crystal structure a related helicase Prp43 bound to single stranded RNA (PDB: 5LTA) and putative RNA channel is identified (tan). (B) Schematic of the sequence of Dhr1 with catalytic motifs I through VI marked and color coded as in panel (A).

2. Biochemical characterization of Dhr1

Like other DEAH RNA helicases, Dhr1 displays RNA-dependent ATPase activity. The yeast enzyme is activated by polyA and by U3 snoRNA but, unlike mouse DHX37 (Boneberg et al., 2019), it is not activated by short (10–20 nt) homopolymers (unpublished). This lack of activation by oligonucleotides may be due to the presence of the putative autoinhibitory loop of yeast Dhr1 (described above and (Singh et al., 2021) (Fig. 2)). Substitution of the expected catalytic metal binding carboxylates D516 and E517 of the DEAH motif with alanines is lethal *in vivo* and largely abrogates its ATPase activity.

Whereas most DEAH helicases can unwind generic substrates with an extended 3′-homopolymeric strand, Dhr1 is not active on such substrates. This is likely related to its inability to bind short homopolymers. Consequently, we developed an unwinding substrate for Dhr1 based on native U3 and pre-rRNA sequences. In the SSU Processome, U3 forms two short heteroduplexes with 18S rRNA: the U3 BoxA' duplex with 18S nts 1111–1123 and the U3 BoxA duplex with 18S nts 9–15 are stabilized by interacting proteins. Formation of these U3-18S duplexes requires unfolding of box A/A' stem loop of U3. Without an annealing factor entropy favors reformation of the unfolded U3 stem loop over association and duplex formation with a second RNA. U3 also forms 2 duplexes with 5′-ETS RNA. Formation of the duplex between the 5′ hinge and nts 464–479 of ETS1 occurs *in vitro* but it is short and unstable in the absence of an annealing factor. Thus, while the U3-18S duplexes and the 5′ hinge duplex can be formed *in vitro* they require additional protein factors, and the added equilibria complicate unwinding assays. Only the duplex formed between the 3′ hinge and nts 281–291 of 5′-ETS (denoted here as U3:ETS2) forms spontaneously upon mixing the two RNA molecules and thus we have routinely used this duplex to measure Dhr1 unwinding activity (Sardana et al., 2015; Zhu et al., 2016) (and see Section 6). Although this substrate may not reflect the *in vivo* substrate of Dhr1, the enzyme recognizes and unwinds this substrate and thus provides a useful model for studying Dhr1 unwinding activity *in vitro*. Interestingly, although the D516A E517A mutant of Dhr1 displays reduced ATPase activity and is inactive in a multiple turnover unwinding assay, the mutant does support unwinding in pre-steady state conditions (Sardana et al., 2015). Although this could suggest that Dhr1 employs a mechanism akin to that of DEAD box proteins which do not unwind in a processive manner, we believe a more likely explanation is that the U3:ETS2 substrate is a special case in which a single binding event is sufficient to displace ETS2 from U3.

3. Expression and purification of full-length Dhr1 from yeast

We have previously reported the purification and characterization of full length Dhr1 from bacterial cells (Sardana et al., 2015). Here, we present an alternative purification of the full length protein from yeast as a cleavable fusion to maltose binding protein (MBP).

3.1 Equipment

- Cabinet incubator for 30 °C
- Shaking incubator (Innova 44, New Brunswick Scientific)
- FPLC (AKTA)
- Cryo-mill (Retsch MM400) fitted with 50 mL stainless steel chambers
- Centrifuge (Beckman J25)
- Centrifuge rotor JA-10.5 with bottles (Beckman)
- Centrifuge rotor JA-25.50 with tubes (Beckman)
- Refrigerated microfuge 5427R (Beckman)
- 18 × 150 mL culture tubes and 250 and 4000 mL flasks
- Dialysis tubing (12–14 kDa MWCO)
- Vivaspin 2.0100 kDa filter

3.2 Reagents

- Synthetic media master mix: Combine 2 g each: adenine, histidine, lysine, methionine, tryptophan and uracil; 8 g each: alanine, arginine, asparagine, aspartic acid, cysteine, glutamic acid, glutamine, glycine, inositol, isoleucine, phenylalanine, proline, serine, threonine, tyrosine, valine; and 0.8 g para-amino benzoic acid
- Synthetic media lacking leucine (SD-Leu): for 1 L: combine 6.7 g yeast nitrogen base without amino acids, 2 g synthetic media master mix, and 950 mL H_2O. pH to ~5.8 with 1 N NaOH. Sterilize by autoclaving. Add 50 mL of sterile 40% (*w*/*v*) dextrose. For solid media, add 20 g of agar before autoclaving
- Sterile 20% galactose

- pAJ4911: high-copy yeast shuttle vector with the prototrophic LEU2 selectable marker. The vector expresses MBP-10xhistidine-TEV-Dhr1 under control of the galactose-inducible *GAL1* promoter
- Yeast strain 334 (MATalpha pep4–3 prb1–1122 ura3–52 leu2–3, 112 reg1–501 gal1) (Hovland, Flick, Johnston, & Sclafani, 1989). This strain is deficient for glucose repression of the galactose regulon and allows induction of genes under control of the galactose regulon in the presence of glucose, avoiding the need for cumbersome shifts in carbon source. This strain is also deficient for the vacuolar proteases Pep4 and Prb1, facilitating purification of protease-sensitive proteins
- Buffer-X (50 mM Tris•HCl, pH 7.5, 500 mM NaCl, 10% glycerol, 1 mM each PMSF and benzamidine and 1 μM each leupeptin and pepstatin).
- Wash buffer (50 mM Tris-HCl, pH 7.5, 500 mM NaCl, 10% glycerol, 5 mM BME, 1 mM each PMSF and 1 μM each leupeptin and pepstatin)
- 5 mL amylose resin (New England Biolabs)
- 1 mL HisTrap FF (Cytiva)
- Tobacco etch virus (TEV) protease (New England Biolabs)

3.3 Procedure

Cell Growth:

1. Streak out yeast strain 334 transformed with pAJ4911 from −80 °C glycerol stock onto a SD-Leu plate and incubate at 30 °C for 2–3 days
2. Late afternoon, use a sterile pipet tip to inoculate a single colony into 2 mL of liquid SD-Leu media in a 18 × 150 mm culture tube and incubated overnight in a shaker at 30 °C
3. Late afternoon, subculture the 2 mL overnight culture into 50 mL of the same media in a 250 mL flask and incubated at 30 °C with shaking overnight
4. Set two 4 L flasks, each with 1 L of SD-Leu medium, at 30 °C
5. ~9 AM. Subculture the overnight culture into the two prewarmed flasks of SD-Leu medium to give an OD of 0.1. This should require ~20 mL of overnight culture per liter. Incubate with shaking at 30 °C until OD ~0.4–0.5 which should take approximately 8 h
6. Add 50 mL of sterile 20% (*w*/*v*) galactose to induce expression and continue growth overnight (approx. 16 h) at 30 °C with shaking
7. Prior to harvesting cells, prechill 20 mL of Buffer-X and ensure that you have liquid nitrogen available

8. Harvest cells by centrifugation 5000 × *g* for 5 min. Decant supernatant and place cells on ice
9. Wash cells by resuspending in 20 mL of ice-cold Buffer-X and centrifuge at 5000 × *g* for 5 min. Yield should be ~10 g for 2 L of culture
10. Resuspend cells in 0.3 mL Buffer-X per gram of cell pellet. Drip the cell suspension from a serological pipette into liquid nitrogen in a 50 mL conical tube to make "popcorn."
11. Loosely cap to allow liquid nitrogen to vent
12. After all liquid nitrogen has evaporated, tighten the cap and store the popcorn at −80 °C until further use

Preparation of "Grindate":

13. Prechill the grinding chambers and balls of the ball mill in liquid nitrogen
14. Load popcorn into the chambers, not exceeding 5 g per 50 mL chamber
15. Pulverize with 5 cycles of shaking for 3 min at 15 Hz. Between cycles, remove chambers and submerg in liquid nitrogen to maintain the sample in a frozen state
16. Transfer the grindate to prechilled 50 mL conical tubes without thawing and stored at −80 °C until further use

Amylose affinity chromatography:

(All subsequent steps are performed at 0–4 °C.)

17. Thaw the grindate and add an equal volume of cold Buffer-X. Mix well by inversion
18. Clarify the suspension by centrifugation for 10 min at 5000 × *g*
19. Transfer the supernatant to a clean round-bottom centrifuge tube compatible with high g force and clarify again by centrifugation for 20 min at 50,000 × *g*
20. Remove the clarified extract with a serological pipette, avoiding the lipid layer on the surface and the loose top layer of the pellet
21. Save 40 μL of sample for analysis by SDS-PAGE
22. Apply the clarified extract by gravity flow to a 5 mL column of amylose resin pre-equilibrated in Buffer-X
23. Wash the column with 25 mL of Buffer-X
24. Wash the column with a 25 mL of wash-buffer
25. Elute protein with wash buffer containing 10 mM maltose, collecting 1 mL fractions
26. Analyze fractions by SDS-PAGE
27. Pools fractions containing the highest concentration and purity of the fusion protein

    (See Fig. 3A for an example purification over amylose.)

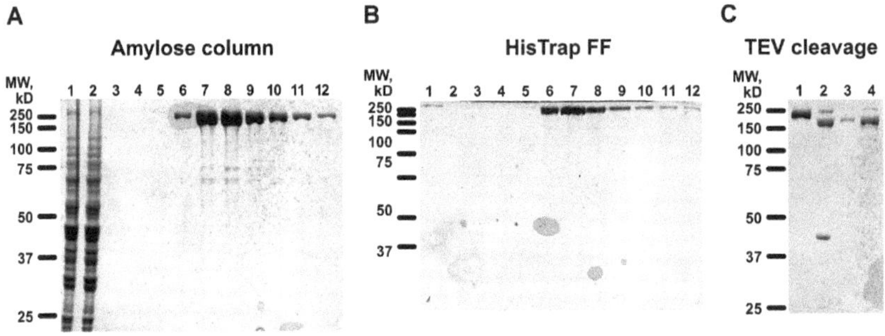

Fig. 3 Example purification of full-length Dhr1 from yeast. (A) Amylose column step. Lane 1, clarified extract; 2, flow through; 3, wash 1; 4 wash 2; lanes 5–12, elution fractions 3–10. Drh1-h. (B) HisTrap FF step. Lane 1, load; 2, flow through; 3, wash 1; 4 wash 2; lanes 5–12, elution fractions 3–10. (C) TEV cleavage and clean up. Lane 1, HisTrap pool pre cleavage; 2, cleaved pool; 4, post cleavage flowthrough and combined washed; 5 concentrated post cleavage. Dhr1 is 146 kDa and His-MBP is ~45 kDa.

Metal affinity chromatography (HisTrap FF)

28. Supplement the pooled fractions with imidazole to 10 mM final concentration
29. Load one half of pooled protein at 1 mL/min onto a prepacked 1 mL HisTrap FF column attached to an FPLC
30. Wash the column with 10 column volumes of wash buffer supplemented with 10 mM imidazole
31. Wash the column with 10 column volumes of wash buffer supplemented with 20 mM imidazole
32. Elute protein in a single step with wash buffer supplemented with 250 mM imidazole, collecting 0.5 mL fractions
33. Repeat steps 28 through 31
34. Analyze fractions by SDS-PAGE
35. Pools fractions containing the highest concentration and purity of the fusion protein

    (See Fig. 3B for an example purification over HisTrap FF.)

Removal of MBP-10xhistidine tag:

36. To cleave the MBP-10xhistidine tag, add fresh BME to the Ni-NTA pool for a final concentration of 5 mM
37. Add 10 units of histidine-tagged TEV protease per mL of sample
38. Load into dialysis tubing (12–14 kDa) and dialyze overnight at 4 °C against Wash Buffer
39. Check the extent of TEV cleavage by SDS-PAGE

40. Pass the cleavage reaction over a 0.5 mL Ni-NTA column by gravity to remove the cleaved MBP-6xhistidine tag, residual uncleaved fusion protein, as well as the histidine-tagged TEV protease. Reload the flow through three times and save the final flow through
41. Wash with four 0.5 mL aliquots of Wash buffer
42. Analyze the flow through and fractions by SDS-PAGE and pool fractions containing the highest concentration of protein
43. Concentrate the sample on a Vivaspin 2.0100 kDa filter
44. Aliquot the concentrated sample, flash freeze in liquid nitrogen, and store at −80 °C

    (See Fig. 3C for an example TEV cleavage.)

3.4 Notes

1. Although this procedure describes the preparation of yeast extract using a ball mill under cryogenic conditions which is optimal for avoiding proteolysis during extract preparation, MBP-Dhr1 appears quite stable and it is likely that the extract can also be made by more conventional disruption with glass beads.
2. Frozen cell "popcorn" can be stored at −80 °C for several months.
3. Cell grindate can be stored at −80 °C in sealed tubes for at least several weeks making it convenient to prepare grindate and store it at −80 °C for protein purification on a subsequent day.

4. Expression and purification of N-terminally truncated Dhr1

Full-length Dhr1 expresses poorly in bacterial cells due to its extended unstructured N-terminus. Consequently, it is convenient to express the protein lacking the N-terminal 375 amino acids which shows improved expression and stability in bacterial cells. Although this protein does not display unwinding activity on substrates that we have tested, including the U3-ETS2 duplex that full length Dhr1 unwinds (Sardana et al., 2015), Dhr1Δ1-375 retains the conserved RecA1 and RecA2 domains and is an active RNA-dependent ATPase making this protein a useful for limited biochemical analysis. This N-terminally truncated protein corresponds to constructs that have been used for crystallization of Dhr1 from yeast (Roychowdhury et al., 2019; Singh et al., 2021) and mouse (Boneberg et al., 2019).

4.1 Equipment

- Cabinet incubator for 30 °C
- Shaking incubator (Innova 44, New Brunswick Scientific)
- Centrifuge (Beckman J25)
- Centrifuge rotor JA-10.5 with bottles (Beckman)
- Centrifuge rotor JA-25.50 with tubes (Beckman)
- Refrigerated microfuge 5427R (Beckman)
- Sonicator fitted with a 0.5 in. disruptor horn (Branson)
- FPLC (AKTA)

4.2 Reagents

- Plasmid pAJ3487
- *E. coli* BL21 CodonPlus (DE3)-RIL cells (Agilent)
- Sterile LB medium (for 1 L, 10 g tryptone, 10 g NaCl and 5 g yeast
- extract). For solid LB medium, add agar to 1.5% (*w/v*) final concentration
- Lysis Buffer (40 mM Tris-HCl, 500 mM NaCl, 10% glycerol, 5 mM BME, 1 mM each PMSF and benzamidine and 1 μM each leupeptin and pepstatin).
- SP buffer (30 mM Tris-HCl pH 8.0, 5 mM NaAcetate, 5% glycerol and 1 mM DTT).
- Dialysis tubing (12–14 kDa MWCO)

4.3 Procedure

Cell growth:

1. Transform *E. coli* BL21 CodonPlus (DE3)-RIL cells with pAJ3487 and selected on LB plates containing ampicillin (50 $\mu g\,mL^{-1}$) and chloramphenicol (25 $\mu g\,mL^{-1}$).
2. Use a single colony to inoculate 10 mL of LB medium containing ampicillin and chloramphenicol in a 50 mL flask and culture at 37 °C overnight with shaking
3. Subculture the entire 10 mL overnight culture into 1 L of the same medium, prewarmed, in a 4 L flask, and grow at 37 °C with vigorous shaking
4. When cells reached an $OD_{600} \sim 0.5$, shift to 16 °C, and add IPTG to a final concentration of 0.5 mM. Continue growth overnight at 16 °C with vigorous shaking
5. Harvest cells by centrifugation at 5000 × *g* for 5 min and place on ice

6. Resuspend the cells in 20 mL of 40 mM Tris-HCl, 500 mM NaCl and 10% glycerol and pelleted again at 5000 × *g* for 5 min
7. Store the cell pellet at −80 °C until further use

Cell lysis:

(All steps were done at 0–4 °C.)

8. Thaw the cell pellet on ice and resuspended in 50 mL of Lysis Buffer
9. Add lysozyme to 0.1 mg/mL and hold cells on ice for 30 min
10. Transfer the cell suspension to a 100 mL beaker set in an ice water bath and disrupt cells by sonication with a 0.5 in. disruptor horn (Branson) at a power of 9, 50% duty in three cycles of 20 pulses followed by 3 min in an ice water bath
11. Add Dnase I to 0.1 mg/mL, RNaseA to 7 u/mL and RNase I to 10 u/mL and incubate on ice for 30 min
12. Clarify by centrifugation for 20 min at 20,000 × *g*

Metal affinity chromatography (Ni-NTA):

13. Add imidazole to the clarified extract to 10 mM final concentration and load onto pre-equilibrated 1.5 mL Ni-NTA column by gravity
14. Wash with 30 CV of lysis buffer supplemented with 10 mM imidazole
15. Wash with 20 CV of lysis buffer supplemented with 20 mM imidazole
16. Elute with lysis buffer supplemented with 250 mM imidazole, collecting 0.5 mL fractions
17. Analyze fractions by SDS-PAGE and pool fractions containing the most protein and free of contaminants
18. Dialyze sample overnight against SP buffer containing 50 mM NaCl

Ion-exchange chromatography (SP HiTrap)

19. Centrifuge the sample at 18,000 × *g* for 10 min to remove any particulate material
20. Pre-equilibrate a 1 mL HiTrap SP HP column attached to an FPLC with SP buffer supplemented with 50 mM NaCl
21. Load onto 1 mL HiTrap SP HP column attached to an FPLC at 1 mL/min
22. Wash with 10 column volumes of SP buffer containing 50 mM NaCl
23. Elute with a 10 column-volume linear gradient from 50 mM to 1 M NaCl in SP buffer
24. Check fractions by SDS-PAGE and pool fractions containing the most protein and free of contaminants
25. Dialyze the sample against SP buffer containing 50 mM NaCl
26. Aliquot and flash freeze aliquots in liquid nitrogen and store at −80 °C

4.4 Notes

1. The metal affinity chromatography can be carried out on an FPLC if preferred

5. Microtiter plate assay for RNA-dependent ATPase activity

The following protocol is based on the method first described by Webb (Webb, 1992). In this assay, free phosphate is utilized by purine nucleoside phosphorylase (PNP) to convert the chromogenic substrate 2-amino-6-mercapto-7-methylpurine ribonucleoside (MESG) to ribose 1-phosphate and the chromophore 2-amino-6-mercapto-7-methylpurine. The reaction is monitored by absorbance at 360 nm. We have adapted this assay to 384 well plates to minimize the consumption of protein.

5.1 Equipment

- Synergy HTX multimode reader (BioTek Instruments, Inc.)
- Microplate, 384 well, flat bottom, clear, non-binding (Greiner Bio-one #E190239B)
- Microplate centrifuge (Fisher Scientific)

5.2 Reagents

- Deionized water
- Reaction buffer (60 mM Tris-HCl, pH 8.0, 40 mM KCl, 15 mM NaCl, 4% [*v*/v] glycerol, 10 mM $MgCl_2$, 2 mM NaOAc and 2.4 mM DTT)
- 20 mM ATP
- 20 mM $MgCl_2$
- 2-amino-6-mercapto-7-methylpurine ribonucleoside (MESG) (1 mM) (ThermoFisher Scientific)
- Purine nucleoside phosphorylase (PNP) (100 U/mL) (ThermoFisher Scientific)
- Pyrophosphate standard solution: 50 mM $Na_4P_2O_7$
- Polyadenylic acid potassium salt [poly(A)] (Sigma, #P9403-25 mg)

5.3 Procedure

Prepare the reaction mixtures

The assay is performed in 384-well plates in a final volume of 20 μL. At a minimum, reactions are performed in triplicate

1. Prepare master mix which contains: Reaction buffer, MESG and PNP. An example master mix setup is shown below:

	Stock	Final Concentration	Volume/reaction(μL)
Reaction buffer	10 ×	1 ×	2
MESG	1 mM	0.2 mM	4
PNP	100 U/mL	1 U/mL	0.2
H_2O	–	–	1.8
Total	–	–	8

2. Prepare stocks for each RNA concentration being tested (with or without RNA or at a varying RNA concentrations). For poly(A) titration of final concentrations of 0, 5, 10, 20, 40 and 100 μM, 5 × stocks of poly (A) concentrations of 0, 25, 50, 100, 200 and 500 μM will be prepared
3. Prepare a 5 × stock for protein condition being tested (with or without protein or at a varying protein concentrations). For an example, a final concentration of 0.5 μM Dhr1ΔN is being tested, a 5 × stock of 2.5 μM Dhr1ΔN will be prepared
4. Prepare ATP mix, containing an equal molar amount of 10 mM ATP and 10 mM $MgCl_2$

The following table is an example to show the RNA titrating of Dhr1ΔN setup, with three replicates.

		Blank x3			With enzyme x3		
PolyA concentration (μM)		**1**	**2**	**3**	**4**	**5**	**6**
0	A						
5	B						
10	C						
20	D						
40	E						
100	F						

Perform the experiment

5. Turn on the Synergy HTX multimode BioTek microplate reader
6. Open the open the Gen5™ program which controls BioTek microplate reader
7. Set the reading absorbance of the samples to 360 nm by clicking: New > Protocol > Procedure > Read > Absorbance > Wavelength: 360 nm
8. Set the program reading frequency every 30 s by clicking: Start Kinetics > Run time: 40 min (A 40-min time window is routinely used for Dhr1) > Interval: 30 s
9. Pipette 8 μL master mix into each well of 384-well microplate
10. Pipette 4 μL different Poly(A) solution into each well of 384-well microplate
11. Pipette 4 μL enzyme solution (2.5 μM) into each well of 384-well microplate
12. Pipette 4 μL ATP solution into each well and pipette up and down to mix the reaction well
13. Centrifuge the microplates in the microplate centrifuge for 20 s to remove all bubbles and make sure the solution in the wells is distributed evenly
14. Place the microplate in the BioTek microplate reader and begin reading

Data analysis

15. Create a standard curve by plotting the absorbance at 360 nm vs phosphate concentration (μM). Based on the standard curve, determine the relationship between Absorbance at 360 nm and free Pi. $x=(y-b)/a$ (x = the concentration of released phosphate, y = absorbance at 360 nm). An example standard curve is shown in Fig. 4A
16. Convert values for absorbance to moles of Pi
17. An example of data generated from Gen5 is shown in Fig. 4B. The data was plotted with absorbance at 360 nm vs time. To calculate the ATPase activity, determine the initial reaction rates. In this example, rates are linear within the first 300 s
18. Using Graphpad Prism 8, plot the released phosphate (mols) vs time (s)
19. In Graphpad Prism 8 use "Simple linear regression" to determine slope to generate initial linear rates
20. Enzyme blank reactions are subtracted from corresponding enzyme-containing reactions
21. Plot the initial rate from step 4 against poly(A) concentration based on "Michaelis-Menten" model using non-linear fit. Fig. 4C is an example plot of the ATPase activity *versus* poly(A) concentration

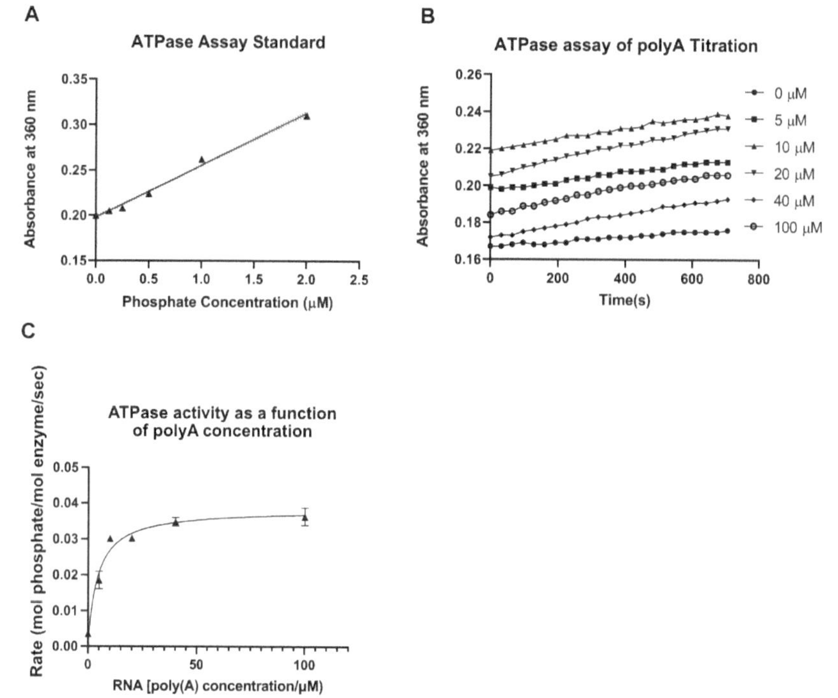

Fig. 4 Example data for polyA-dependent ATPase activity. (A) Example standard curve for free Phosphate. (B) Example data for titration of poly(A) activation of Dhr1ΔN ATPase activity. (C) Example plot of ATPase vs poly(A) concentration based on data from panel (B).

5.4 Notes

1. Because reactions begin upon addition of ATP:$MgCl_2$, one must work relatively quickly to ensure reactions are read during their linear phase. Setting up a manageable number of reactions, determined empirically, in a given cycle is advised. Multiple cycles of reactions can be processed in consecutive wells of a single 384-well plate

6. U3-ETS2 duplex unwinding assays under pre-steady state conditions

In this section, we describe an unwinding assay using the U3:ETS2 substrate described in Section 2 (and see Fig. 5).

6.1 Equipment

- Heating water bath

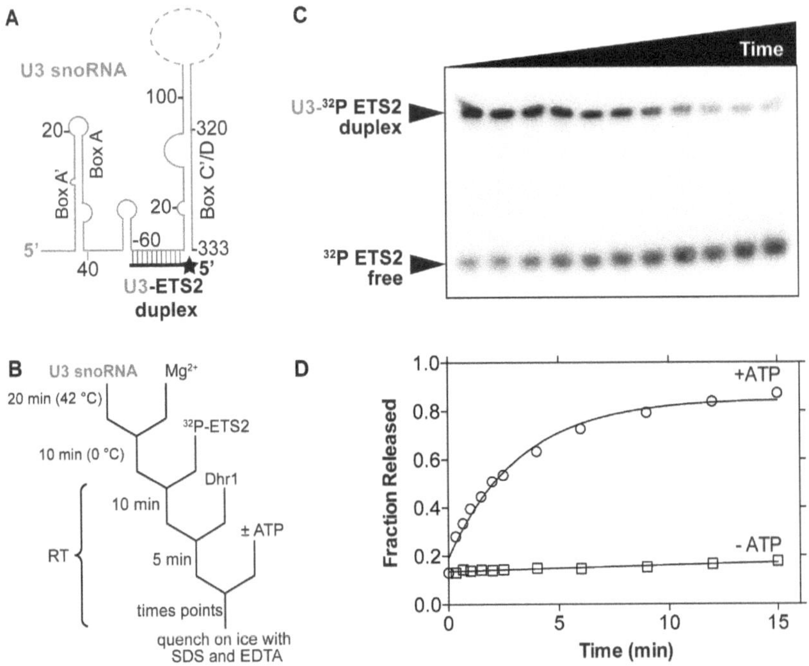

Fig. 5 Assaying *in vitro* unwinding activity of Dhr1. (A) Schematic of the U3-ETS2 substrate. (B) Outline of reaction scheme with results shown in panels (C and D). (C) Representative unwinding data from native gel analysis showing ^{32}P-labeled ETS2 free (unwound) or in duplex with U3 snoRNA. (D) Fraction unwound is plotted as a function of time after addition of either ATP (circles) or mock (squares).

- Native gel apparatus
- Cold room or cold box to run the electrophoretic mobility shift assays (EMSAs)
- Power supply (at least 200 V)
- Gel Drier (BioRad)
- Fuji imaging plate (Bas 2024) and Typhoon 7000FLA (Amersham Biosciences, GE)
- Image Quant 5.0 software quantification of bands on gel

6.2 Reagents

- U3 snoRNA (300 nM stock in water, *in vitro* transcribed from a linearized plasmid (Gerczei & Correll, 2004))
- ETS2 RNA (125 nM stock in water. After shrimp alkaline phosphatase treatment, T4 kinase (Fermentas) was used in the presence of [γ-^{32}P]

ATP (Perkin–Elmer, 150 μCi/μL) to label at the 5′-end of each RNA according to the manufacturer's instructions.)
- Dhr1 (3–5 μM stock in 30 mM Tris [pH 8.0], 150 mM NaCl, 5% (*v*/v) glycerol, 5 mM sodium acetate and 1 mM DTT)
- RNasin (Promega)
- Bovine serum albumin (BSA)

6.3 Procedure

See Fig. 5B for a schematic of the reaction scheme.

1) Refold RNA by incubating U3 snoRNA for 20 min at 42 °C, then add $MgCl_2$ and incubate 10 min on ice and then add ^{32}P ETS2 (final concentration 1 mM $MgCl_2$, 150 nM U3, 62.5 nM ETS2) and incubate 10 min RT
2) Dilute 10-fold and incubate 10 min at RT (final 2.5 mM Mg, 10 mM DTT, 1 mg/mL BSA, 250 mM NaCl)
3) Dilute fivefold, add Dhr1 and mock buffer (20 mM Tris (8.0), 40 mM KCl) and incubate in reaction buffer for 5 min RT
4) Initiate reaction by adding ATP•$MgCl_2$ (final concentrations were 500 nM Dhr1, 1 mM ATP, 1.5 mM $MgCl_2$, 1 nM U3 snoRNA, ≤0.3 nM ^{32}P ETS2, 15 mM Tris (pH 8.0), 23 mM KCl, 70 mM NaCl, 0.7% (*v*/v) glycerol, 2.4 mM DTT, 0.8 units/μL RNasin, and 0.2 mg/mL BSA)
5) Remove aliquots at different times and quench the reaction by addition of one-half volume of stop buffer (150 mM Tris [pH 8.0], 0.3% [*w*/*v*] SDS, and 150 mM EDTA).
6) Prepare a 6% nondenaturing polyacrylamide gel (50:1 cross-linking ratio) with 40 mM Tris-acetate, pH 8.0, 1 mM ethylenediaminetetraacetic acid, 50 mM KCl, and 2.5% (v/v) glycerol
7) Immediately before loading the gel, add an equal volume of 40% (*w*/*v*) sucrose, 50 mM KCl, and 80 mM Tris-acetate, pH 8.0 to samples
8) Load samples and run the gel for 45 min at 125 V at 4 °C using a gel running buffer that is the same as the buffer used to prepare the gel.
9) Image the bound and free ^{32}P-labeled ETS species by autoradiography using a Fuji imaging plate and Typhoon 7000FLA
10) Quantify data using Image Quant 5.0. The fraction of free ETS2 RNA relative to total with background corrected as a function of time was then plotted (Fig. 5C and D).

6.4 Notes

1) Whereas all three preparations of Dhr1 (bacterially expressed C-terminally His-tagged protein, yeast expressed C-terminally His-tagged protein and yeast expressed N-terminally MBP and His tagged protein with a TEV cleavage site, described in Section 3) showed unwinding activity. The published results (Sardana et al., 2015; Zhu et al., 2016) and results shown in Fig. 5used the bacterially expressed His-tagged protein
2) To prevent melting of short thermolabile RNA duplexes during electrophoresis, native gels must be run with active cooling or in a cold room
3) For steady-state U3-ETS2 unwinding assays the reaction was essentially the same as for pre-steady state reactions described above, with the following several differences. Instead of using U3 snoRNA, 5 μM U3 (nucleotides 1–76) was incubated with 0.5 μM ETS2 with trace ^{32}P-ETS2 to form the U3-ETS2 duplex. Non-specific binding to Dhr1 prevented the use of full length U3 snoRNA under these conditions. After pre-incubating the duplex with 0.1 μM Dhr1, reactions were initiated by rapid addition of 5 μM ETS2 and 6 mM mixture of ATP and MgCl2. The remaining buffer conditions are the same as those in the pre-steady reactions

7. Summary and conclusions

Here, we have provided methods for the purification of full-length and truncated Dhr1 proteins from *E. coli* or yeast, as well as established ATPase and RNA unwinding assays. There remain a number of outstanding questions about Dhr1 function. For example, we previously identified Utp14 as an accessory factor of Dhr1 and showed that Utp14 stimulates the unwinding activity of Dhr1 *in vitro* (Zhu et al., 2016). However, we did not observe activation of Dhr1 ATPase activity by Utp14. In contrast, the RNA-dependent ATPase activity of murine DHX37 is stimulated by murine Utp14 (Boneberg et al., 2019). This apparent discrepancy may reflect differences between full-length and N-terminally truncated Dhr1 or differences between the yeast and mammalian proteins. We hope that the methods provided in this chapter will serve as a starting point for understanding the mechanism of regulation of Dhr1 activity and its role in progression of the SSU Processome to a pre-40S particle.

Acknowledgment

This work was funded by NIH grant GM108823 to A.W.J.

Declarations of interest

None.

References

Allmang, C., Kufel, J., Chanfreau, G., Mitchell, P., Petfalski, E., & Tollervey, D. (1999). Functions of the exosome in rRNA, snoRNA and snRNA synthesis. *EMBO Journal*, *18*(19), 5399–5410. https://doi.org/10.1093/emboj/18.19.5399.

Barandun, J., Chaker-Margot, M., Hunziker, M., Molloy, K. R., Chait, B. T., & Klinge, S. (2017). The complete structure of the small-subunit processome. *Nature Structural & Molecular Biology*, *24*(11), 944–953. https://doi.org/10.1038/nsmb.3472.

Baßler, J., & Hurt, E. (2019). Eukaryotic ribosome assembly. *Annual Review of Biochemistry*, 1–26.

Ben-Shem, A., De Loubresse, N. G., Melnikov, S., Jenner, L., Yusupova, G., & Yusupov, M. (2011). The structure of the eukaryotic ribosome at 3.0 Å resolution. *Science*, *334*(6062), 1524–1529. https://doi.org/10.1126/science.1212642.

Black, J. J., Sardana, R., Elmir, E. W., & Johnson, A. W. (2020). Bud23 promotes the final disassembly of the small subunit Processome in Saccharomyces cerevisiae. *PLoS Genetics*, *16*. https://doi.org/10.1371/journal.pgen.1009215.

Boneberg, F. M., Brandmann, T., Kobel, L., van Den Heuvel, J., Bargsten, K., Bammert, L., et al. (2019). Molecular mechanism of the RNA helicase DHX37 and its activation by UTP14A in ribosome biogenesis. *RNA*, *25*(6), 685–701. https://doi.org/10.1261/rna.069609.118.

Chaker-Margot, M., Hunziker, M., Barandun, J., Dill, B. D., & Klinge, S. (2015). Stage-specific assembly events of the 6-MDa small-subunit processome initiate eukaryotic ribosome biogenesis. *Nature Structural & Molecular Biology*, *22*(11), 920–923. https://doi.org/10.1038/nsmb.3111.

Cheng, J., Baßler, J., Fischer, P., Strauss, D., Beckmann, R., Hurt, E., et al. (2019). Thermophile 90S pre-ribosome structures reveal the reverse order of co-transcriptional 18S rRNA subdomain integration article thermophile 90S pre-ribosome structures reveal the reverse order of co-transcriptional 18S rRNA subdomain integration. *Molecular Cell*, *75*, 1–14.

Cheng, J., Lau, B., La Venuta, G., Ameismeier, M., Berninghausen, O., Hurt, E., et al. (2020). 90S pre-ribosome transformation into the primordial 40S subunit. *Science*, *369*(6509), 1470–1476. https://doi.org/10.1126/SCIENCE.ABB4119.

Choudhury, P., Hackert, P., Memet, I., Sloan, K. E., & Bohnsack, M. T. (2019). The human RNA helicase DHX37 is required for release of the U3 snoRNP from pre-ribosomal particles. *RNA Biology*, *16*(1), 54–68. https://doi.org/10.1080/15476286.2018.1556149.

Colley, A., Beggs, J. D., Tollervey, D., & Lafontaine, D. L. (2000). Dhr1p, a putative DEAH-box RNA helicase, is associated with the box C+D snoRNP U3. *Molecular and Cellular Biology*, *20*(19), 7238–7246.

Dragon, F., Gallagher, J. E., Compagnone-Post, P. A., Mitchell, B. M., Porwancher, K. A., Wehner, K. A., et al. (2002). A large nucleolar U3 ribonucleoprotein required for 18S ribosomal RNA biogenesis. *Nature*, *417*(6892), 967–970. https://doi.org/10.1038/nature00769.

Du, Y., An, W., Zhu, X., Sun, Q., Qi, J., & Ye, K. (2020). Cryo-EM structure of 90S small ribosomal subunit precursors in transition states. *Science, 369*(6509), 1477–1481. https://doi.org/10.1126/SCIENCE.ABA9690.

Gerczei, T., & Correll, C. C. (2004). Imp3p and Imp4p mediate formation of essential U3-precursor rRNA (pre-rRNA) duplexes, possibly to recruit the small subunit processome to the pre-rRNA. *Proceedings of the National Academy of Sciences of the United States of America, 101*(43), 15301–15306. https://doi.org/10.1073/pnas.0406819101.

Grandi, P., Rybin, V., Bassler, J., Petfalski, E., Strauss, D., Marzioch, M., et al. (2002). 90S pre-ribosomes include the 35S pre-rRNA, the U3 snoRNP, and 40S subunit processing factors but predominantly lack 60S synthesis factors. *Molecular Cell, 10*(1), 105–115.

Granneman, S., Bernstein, K. A., Bleichert, F., & Baserga, S. J. (2006). Comprehensive mutational analysis of yeast DEXD/H box RNA helicases required for small ribosomal subunit synthesis. *Molecular and Cellular Biology, 26*(4), 1183–1194. https://doi.org/10.1128/mcb.26.4.1183-1194.2006.

Hamann, F., Enders, M., & Ficner, R. (2019). Structural basis for RNA translocation by DEAH-box ATPases. *Nucleic Acids Research, 47*(8), 4349–4362. https://doi.org/10.1093/nar/gkz150.

Hartl, T. A., Ni, J., Cao, J., Suyama, K. L., Patchett, S., Bussiere, C., et al. (2013). Regulation of ribosome biogenesis by nucleostemin 3 promotes local and systemic growth in drosophila. *Genetics, 194*(1), 101–115. https://doi.org/10.1534/genetics.112.149104.

Hovland, P., Flick, J., Johnston, M., & Sclafani, R. A. (1989). Galactose as a gratuitous inducer of GAL gene expression in yeasts growing on glucose. *Gene, 83*(1), 57–64. https://doi.org/10.1016/0378-1119(89)90403-4.

Hughes, J. M. X. (1996). Functional base-pairing interaction between highly conserved elements of U3 small nucleolar RNA and the small ribosomal subunit RNA. *Journal of Molecular Biology, 259*(4), 645–654. https://doi.org/10.1006/jmbi.1996.0346.

Hughes, J. M. X., & Ares, M. (1991). Depletion of U3 small nucleolar RNA inhibits cleavage in the 5′ external transcribed spacer of yeast pre-ribosomal RNA and impairs formation of 18S ribosomal RNA. *EMBO Journal, 10*(13), 4231–4239. https://doi.org/10.1002/j.1460-2075.1991.tb05001.x.

Klinge, S., & Woolford, J. L. (2019). Ribosome assembly coming into focus. *Nature Reviews Molecular Cell Biology, 20*(2), 116–131. https://doi.org/10.1038/s41580-018-0078-y.

Kornprobst, M., Turk, M., Kellner, N., Cheng, J., Flemming, D., Kos-Braun, I., et al. (2016). Architecture of the 90S pre-ribosome: A structural view on the birth of the eukaryotic ribosome. *Cell, 166*(2), 380–393. https://doi.org/10.1016/j.cell.2016.06.014.

Marmier-Gourrier, N., Cléry, A., Schlotter, F., Senty-Ségault, V., & Branlant, C. (2011). A second base pair interaction between U3 small nucleolar RNA and the 5′-ETS region is required for early cleavage of the yeast pre-ribosomal RNA. *Nucleic Acids Research, 39*(22), 9731–9745. https://doi.org/10.1093/nar/gkr675.

Miller, O. L., & Beatty, B. R. (1969). Visualization of nucleolar genes. *Science, 164*(3882), 955–957. https://doi.org/10.1126/science.164.3882.955.

Ozgur, S., Buchwald, G., Falk, S., Chakrabarti, S., Prabu, J. R., & Conti, E. (2015). The conformational plasticity of eukaryotic RNA-dependent ATPases. *FEBS Journal, 282*(5), 850–863. https://doi.org/10.1111/febs.13198.

Roychowdhury, A., Joret, C., Bourgeois, G., Heurgué-Hamard, V., Lafontaine, D. L. J., & Graille, M. (2019). The DEAH-box RNA helicase Dhr1 contains a remarkable carboxyl terminal domain essential for small ribosomal subunit biogenesis. *Nucleic Acids Research, 1–16*. https://doi.org/10.1093/nar/gkz529.

Sardana, R., Liu, X., Granneman, S., Zhu, J., Gill, M., Papoulas, O., et al. (2015). The DEAH-box helicase Dhr1 dissociates U3 from the pre-rRNA to promote formation of the central pseudoknot. *PLoS Biology*, *13*(2), e1002083. https://doi.org/10.1371/journal.pbio.1002083.

Sharma, K., & Tollervey, D. (1999). Base pairing between U3 small nucleolar RNA and the 5′ end of 18S rRNA is required for pre-rRNA processing. *Molecular and Cellular Biology*, *19*(9), 6012–6019. https://doi.org/10.1128/mcb.19.9.6012.

Singh, S., Vanden Broeck, A., Miller, L., Chaker-Margot, M., & Klinge, S. (2021). Nucleolar maturation of the human small subunit processome. *Science*, *373*(6560). https://doi.org/10.1126/science.abj5338.

Sun, Q., Zhu, X., Qi, J., An, W., Lan, P., Tan, D., et al. (2017). Molecular architecture of the 90S small subunit pre-ribosome. *eLife*, *6*. https://doi.org/10.7554/eLife.22086.

Thoms, M., Thomson, E., Bassler, J., Gnadig, M., Griesel, S., & Hurt, E. (2015). The exosome is recruited to RNA substrates through specific adaptor proteins. *Cell*, *162*(5), 1029–1038. https://doi.org/10.1016/j.cell.2015.07.060.

Webb, M. R. (1992). A continuous spectrophotometric assay for inorganic phosphate and for measuring phosphate release kinetics in biological systems. *Proceedings of the National Academy of Sciences of the United States of America*, *89*(11), 4884–4887. https://doi.org/10.1073/pnas.89.11.4884.

Woolford, J. L., Jr., & Baserga, S. J. (2013). Ribosome biogenesis in the yeast Saccharomyces cerevisiae. *Genetics*, *195*(3), 643–681. https://doi.org/10.1534/genetics.113.153197.

Zhang, L., Wu, C., Cai, G., Chen, S., & Ye, K. (2016). Stepwise and dynamic assembly of the earliest precursors of small ribosomal subunits in yeast. *Genes & Development*, *30*(6), 718–732. https://doi.org/10.1101/gad.274688.115.

Zhu, J., Liu, X., Anjos, M., Correll, C. C., & Johnson, A. W. (2016). Utp14 recruits and activates the RNA helicase Dhr1 to undock U3 snoRNA from the Preribosome. *Molecular and Cellular Biology*, *36*(6), 965–978. https://doi.org/10.1128/MCB.00773-15.

CHAPTER SIX

Translational control by helicases during cellular stress

Sara B. Carey and Timothy A. Bolger*
Department of Molecular and Cellular Biology, University of Arizona, Tucson, AZ, United States
*Corresponding author: e-mail address: tbolger@email.arizona.edu

Contents

Abstract

Stress is inevitable, so all organisms have developed response mechanisms to allow for their survival during times of stress. Regulation of gene expression is a critical part of these responses, which allows for the appropriate cohort of proteins to be produced to counter the stress while downregulating others in order to conserve resources. Translation is both highly energy intensive and able to rapidly shift the proteome, thus making it a key target for regulation during stress. Numerous stress pathways converge on translation, and examining the regulatory mechanisms that underlie these pathways is essential for understanding the initial and long-term effects of stress on cells. A number of RNA helicases, including eIF4A, Ded1/DDX3X, and Dhh1/DDX6, have been previously linked to translation, and given their ability to dramatically alter RNA-protein interactions, they are well-positioned to play critical roles in translation regulation during stress. Therefore, assessing the role of helicases in these conditions is vital to the overall understanding of stress. Outlined below are key assays focusing on two areas: assessing cellular phenotypes in growth and survival during stress conditions, and

Methods in Enzymology, Volume 673
ISSN 0076-6879
https://doi.org/10.1016/bs.mie.2022.03.056

analyzing cellular translation in the presence and absence of stress. The combination of these two approaches will begin to establish the function(s) of a given helicase in the overall stress response.

1. Introduction

Whether single-celled microbes or higher eukaryotes, cells frequently face a variety of environmental stresses, and they mount a variety of responses to these stresses in order to continue to proliferate and survive. In particular, adverse conditions, such as nutrient deprivation, osmotic shifts, or temperature shock, induce highly coordinated changes in gene expression at all levels so that the proper cohort of proteins is produced to allow for maintenance of critical cell functions and viability (Albig & Decker, 2001; Liu & Qian, 2014; Saavedra, Tung, Amberg, Hopper, & Cole, 1996). The extensive regulatory mechanisms governing gene expression in the cellular stress response ensure that these changes are implemented with high fidelity (Liu, Beyer, & Aebersold, 2016; Mitchell & Parker, 2014). Although vital for cellular response and organismal survival, many of these mechanisms and their connections to signaling hubs, such as the Target-of-Rapamycin (TOR) pathway, are poorly understood.

Changes in translation play a particularly large role in stress responses. Translation is one of the most energy intensive cellular processes; therefore, in energy limiting conditions, general downregulation of translation is required in order to conserve cellular resources for responding to the stress (Liu & Qian, 2014). Supporting this idea, ribosome footprinting (Ribo-Seq) studies have indicated that massive translational reprogramming occurs under stress with changes in translational efficiency of a third or more of all mRNAs (Gerashchenko, Lobanov, & Gladyshev, 2012; Ingolia, Ghaemmaghami, Newman, & Weissman, 2009). However, this regulation is specific and complex, wherein bulk translation is down-regulated, but translation of some mRNAs is up-regulated (e.g., those that encode proteins needed to respond to the stress) (Crawford & Pavitt, 2019; Liu & Qian, 2014). While some of the regulatory mechanisms controlling translation in stress are relatively well-defined, such as the canonical Integrated Stress Response (ISR) that results in phosphorylation and inhibition of the translation factor eIF2α (Costa-Mattioli & Walter, 2020), others remain much more poorly defined. Underscoring the importance of this research, stress response pathways are often misregulated in disease, including cancer and aging (Gonskikh & Polacek, 2017; Robichaud, Sonenberg, Ruggero, & Schneider, 2019).

In addition to the ISR, another major signaling hub for controlling stress responses is the Target-Of-Rapamycin (TOR) pathway. When active the TOR pathway is responsible for cell proliferation, protein synthesis, and ribosome synthesis, while stresses such as nutrient starvation cause inhibition of TOR and repression of these processes, including translation (Gonzalez & Rallis, 2017; Saxton & Sabatini, 2017). Inhibition of TOR leads to inactivation of a number of translation factors through a variety of mechanisms, including dephosphorylation of ribosomal protein S6, inhibition of eIF4E through 4E-BPs, and degradation of eIF4G and other factors (Berset, Trachsel, & Altmann, 1998; Kelly & Bedwell, 2015; Spriggs, Bushell, & Willis, 2010). We and others have also found that at least two members of the DEAD-box RNA helicase family, eIF4A and Ded1, are part of the translational response to TOR inhibition. The DEAD-box proteins are the largest family of RNA helicases, and they play multiple, critical roles in gene expression, acting to remodel both RNA secondary structures and RNA-protein binding in a largely non-processive manner (Valentini & Linder, 2021). PDCD4, which is downregulated by the TOR-dependent S6 Kinase, is a well-established inhibitor of eIF4A, and eIF4B, a binding partner of eIF4A, is also regulated by TOR (Roux & Topisirovic, 2018; Yang et al., 2003). More recently, we showed that Ded1 appears to switch from promoting translation initiation in pro-growth conditions to down-regulating translation following TOR inactivation, at least in part by promoting dissociation and degradation of eIF4G (Aryanpur et al., 2019). In addition to these direct roles in translation, eIF4A, Ded1 and its human ortholog DDX3X, and Dhh1/DDX6 all play roles in the assembly and disassembly of stress granules and processing bodies, cytoplasmic accumulations of RNA and protein that form in response to stresses (Beckham et al., 2008; Carroll, Munchel, & Weis, 2011; Hilliker, Gao, Jankowsky, & Parker, 2011; Tauber et al., 2020).

Because these regulatory mechanisms are vital for adequate translation of proteins and for coordinated and effective response to stress, understanding how to evaluate RNA helicases for translation abnormalities is crucial for understanding the total cellular response to stress. Outlined in this chapter are specific assays for understanding differences in translation between wild-type cells and genetically altered cells both in normal, pro-growth conditions and during stress. Two sets of assays are described. The first set (Section 3) allow for characterization of growth phenotypes during stress. These growth phenotypes indirectly assess translation but can provide reliable insight into helicase function. For instance, mutating a relevant helicase, such as Ded1, could affect cells' ability to continue to grow and/or survive during stress. Such a result alone does not provide definitive evidence that the helicase is promoting this resistance via a translational response. Assays in the second set (Section 4) can then address this hypothesis more directly by examining cellular translation status under the same conditions as the first set of assays. Combining the results from both sets of assays would thus build confidence that the helicase (or other factor) has an important function in the translational stress response. In essence, the protocols described here provide a roadmap for the initial characterization of a specific factor hypothesized to play a role in these responses.

2. General reagents and conditions

Throughout this chapter are general protocols for assessing the function of a helicase regarding translational control during cellular stress. These protocols are meant to serve as guidelines to assess any helicase in any type of stress but were optimized using the specific conditions and yeast strains that are detailed below.

2.1 Media

2.1.1 YEPD

1. Dextrose stock: 40% dextrose (w/v) in dH_2O. Autoclave.
2. Liquid media (1 L): 10 g yeast extract, 20 g peptone, 200 μL 10 N NaOH in 950 mL dH_2O.
   a. Aliquot media (237.5 mL is a convenient stock volume) as desired and autoclave.
   b. Add dextrose to media to final concentration of 2% (12.5 mL in 237.5 mL media).
3. Plate media (1 L): 10 g yeast extract, 20 g peptone, 20 g agar, 200 μL 10 N NaOH in 950 mL dH_2O in a suitable Erlenmeyer flask.
   a. Add a stir bar to media and autoclave.
   b. Place media on stir plate and stir until mixture has cooled enough to comfortably touch the glass.
   c. Add dextrose to media to final concentration of 2% (50 mL for 1 L media). Add to side of flask to reduce bubbles.
   d. Pour plates: add 20 to 25 mL of media to each 10 cm petri plate. We recommend wrapping the neck of the flask in paper towels to catch drips. A Bunsen burner can be used to pop bubbles.

2.1.2 Minimal/selective media

1. Amino acid drop-out mix: 100 g yeast nitrogen base (YNB) w/o amino acids and w/o ammonium sulfate, 293.11 g ammonium sulfate, 1.2 g adenine sulfate, 1.2 g L-arginine sulfate, 1.2 g L-histidine-HCl, 1.8 g L-isoleucine, 3.6 g L-leucine, 1.8 g L-lysine-HCl, 1.2 g L-methionine, 3.0 g L-phenylalanine, 1.2 g L-tryptophan, 1.8 g L-tyrosine, 1.2 g L-uracil, 9.0 g L-valine, 1.5 g L-aspartic acid, 1.5 g L-glutamic acid, 1.5 g L-threonine, 1.5 g L-serine, 1.5 g L-proline.
   a. The above list contains all necessary amino acids. If selective media is required, leave that amino acid out of the mix.
   b. Thoroughly mix amino acids, then gradually mix in YNB and ammonium sulfate. We recommend a blender if possible.
2. Dextrose stock: 40% dextrose (w/v) in dH_2O. Autoclave.
3. Liquid media (1 L): 7.1 g amino acid drop-out mix, 200 μL 10 N NaOH in 950 mL dH_2O.
   a. If adjustments to the drop-out mix are needed, the desired amino acid can be added in the following amounts (per L): 20 mg histidine, 60 mg leucine, 20 mg uracil, 20 mg tryptophan, 30 mg lysine, 20 mg methionine, 20 mg adenine.
   b. Aliquot media (237.5 mL is a convenient stock volume) as desired and autoclave.
   c. Add dextrose to media to final concentration of 2% (12.5 mL in 237.5 mL media).
4. Plate media (1 L): 7.1 g amino acid drop-out mix, 20 g agar, 200 μL 10 N NaOH in 950 mL dH_2O in a suitable Erlenmeyer flask.
   a. If adjustments to the drop-out mix are needed, add the desired amino acid as above.
   b. Add a stir bar to media and autoclave.
   c. Place media on stir plate and stir until mixture has cooled enough to comfortably touch the glass.
   d. Add dextrose to media to final concentration of 2% (50 mL for 1 L media). Add to side of flask to reduce bubbles.
   e. Pour plates: add 20 to 25 mL of media to each 10 cm petri plate. We recommend wrapping the neck of the flask in paper towels to catch drips. A Bunsen burner can be used to pop bubbles.

2.2 Stress conditions

Table 1 lists several potential stressors along with the treatment conditions that we have tested previously. The user may find that they require further

Table 1 Stressors and recommended conditions.

Stressor	Details of treatment
Rapamycin (TOR inhibition)	200 ng/mL added to (plates or liquid) media
Hydrogen peroxide (oxidative stress)	0.75 mM in (liquid) media. Usage in plates appears to have inconsistent effects. Note that H_2O_2 degrades fairly rapidly, even at 4 °C. Consider replacing stock every 2 to 3 months.
Sodium azide (oxidative stress)	0.5% in (liquid) media. Best used for stress granule induction. Usage in plates appears to have inconsistent effects.
Heat shock	Incubate at 42 °C if performing an assay such as polysome profiles. Cells will not grow appreciably at that temperature, however. If a survival assay is desired, incubate at 50 °C in liquid culture for 1 h and then spread on plates at 30 °C.
Nitrogen withdrawal	Pre-grow cells in minimal media (as in Section 2.1.2) until mid-log phase, then switch into a minimal media without ammonium sulfate or any amino acids added.
Glucose withdrawal	Pre-grow cells in minimal media with 2% glucose until mid-log phase, then switch to a minimal media lacking glucose. Low concentrations of glucose (e.g., 0.2%, 0.05%) can also be used.
Stationary phase	Grow cells to mid-log phase. Spin down and resuspend in minimal media to an OD_{600} of 0.2, and grow at 30 °C (3 to 14 days).
High salt (osmotic shock)	Add 1 M NaCl to (plates or liquid) media

optimization in their hands, however. Of course, other stressors (e.g., DNA damage agents, heavy metal toxicity, etc.) can also be used as the user finds appropriate.

2.3 Strains previously used

Table 2 contains a list of several yeast strains that we have previously tested in many of the above stress conditions (Aryanpur et al., 2019, 2022; Bolger & Wente, 2011; Takahara & Maeda, 2012; Thomas & Rothstein, 1989). We have generally used the W303 yeast background in our studies. Many stress responses (particularly TOR dependent ones) seem to be more pronounced in this background compared to, for example, the S288C/Research Genetics strains used to generate many of the available yeast libraries.

Table 2 A selection of previously used strains.

Strain Name	Genotype	Source
W303	*MATα ade2-1 ura3-1 his3-11,15 leu2-3,112 trp1-1 can1-100*	Thomas and Rothstein, (1989)
TS161	*MATα ura3-52*	Takahara and Maeda (2012)
TS184	*MATα ura3-52* $TOR1^{L2134M}$	Takahara and Maeda (2012)
TBY52	*MATα ded1::KAN ade2-1 ura3-1 his3-11,15 leu2-3,112 trp1-1 can1-100 +pCEN/LEU2/DED1*	Aryanpur et al. (2019)
TBY121	*MATα ded1::KAN ade2-1 ura3-1 his3-11,15 leu2-3,112 trp1-1 can1-100 +pCEN/LEU2/ded1-ΔCT*	Aryanpur et al. (2019)
TBY108	*MATα ded1::KAN ade2-1 ura3-1 his3-11,15 leu2-3,112 trp1-1 can1-100 +pCEN/LEU2/ded1-Δ591-604*	Aryanpur et al. (2019)
SWY4277	*MATα ded1::KAN ade2-1 ura3-1 his3-11,15 leu2-3,112 trp1-1 can1-100 +pCEN/LEU2/ded1-120*	Bolger and Wente (2011)
TBY159	*MATα tif4631::HYG ded1::KAN ade2-1 ura3-1 his3-11,15 leu2-3,112 trp1-1 can1-100 +pCEN/LEU2/DED1*	Aryanpur, Mittelmeier, and Bolger (2022)
TBY165	*MATα tif4631::HYG ded1::KAN ade2-1 ura3-1 his3-11,15 leu2-3,112 trp1-1 can1-100 +pCEN/LEU2/ded1-Δ536-604*	Aryanpur et al. (2022)
TBY118	*MATα ded1::KAN NIP1-PROTA::spHIS5 ade2-1, ura3-1, his3-11,15 leu2-3,112 trp1-1 can1-100 +pCEN/LEU2/DED1*	Aryanpur et al. (2019)
TBY130	*MATα ded1::KAN NIP1-PROTA::spHIS5 ade2-1 ura3-1 his3-11,15 leu2-3,112 trp1-1 can1-100 +pCEN/LEU2/ded1-ΔCT*	Aryanpur et al. (2019)

3. Growth assays to assess stress phenotypes

Yeast growth assays are routinely used to assess growth phenotypes of mutant strains. Perhaps the most well-known qualitative method to assess yeast growth phenotypes is the spot titer method (yeast serial dilution growth assay). Advantages of this method include ease of setup and the ability to qualitatively assess differences in growth between several different strains. However, this method is not well suited for capturing smaller but still significant differences in yeast growth rate and cannot accurately differentiate between different phases of yeast growth (lag, exponential, diauxic, stationary). Furthermore, it is not a good method for screening a larger number of strains in different growth conditions, as making serial dilutions is time-consuming and becomes limiting. Liquid culture methods that measure optical density of cultures (at 600 nm) over a time-course are a more quantitative way to assess yeast growth and can capture more subtle differences in growth rate of yeast strains. Ultimately, this can lead to a more nuanced understanding of how different gene mutations affect yeast growth rate and determine whether specific growth phases are affected. Additionally, liquid growth methods can easily be adapted to screen many yeast strains under different growth conditions and/or drug treatments using 96-well plate micro-cultures. Both protocols have their advantages and disadvantages depending on the nature of the question being addressed.

A central consideration in assessing stress phenotypes is the nature of the stress, as different stressors induce multiple different pathways as part of the stress response. Examples of these include sugar or amino acid deprivation, temperature shock, oxidative stress, and osmotic shock, all of which have been shown to be mediated by helicases (Aryanpur et al., 2019, 2022; Beck et al., 2014; Carroll et al., 2011; Hilliker et al., 2011; Iserman et al., 2020; Seok, Nguyen, Van Nguyen, Lee, & Moon, 2020). Growth assays may be designed for chronic stress, where cells are grown at constant semi-permissive conditions (e.g., <1 mM hydrogen peroxide), for survival/recovery, where cells are subjected to intense stress where no growth occurs followed by incubation in pro-growth conditions (e.g., 50 °C for 1 h followed by 30 °C recovery), or for pharmacological inhibition of specific stress pathways (e.g., 200 ng/mL rapamycin). If a general analysis of phenotypes during cellular stress is desired, we recommend assaying a panel of these different conditions.

3.1 Spot titer growth analysis

The spot titer method is one of the simplest methods for growth analysis of yeast strains. Cells at a pre-determined range of concentrations are "spotted" in a small droplet onto the desired agar media, then incubated until grown sufficiently to discern differences amongst strains. This protocol is for 5 spots of 1:5 dilutions with 100,000, 20,000, 4000, 800, 160 cells per spot. Other dilutions, numbers of cells, and number of spots can also be used; the math and preparation will need to be adjusted to reflect these changes. As mentioned above, the media used will depend on the stress condition tested. General controls will include media lacking the stress (or cells that have not been exposed to it) and ideally, positive and negative control yeast strains.

3.1.1 Equipment and reagents

1. Spectrophotometer or hemacytometer (depending how cell concentration will be calculated)
2. Microcentrifuge
3. Benchtop vortexer
4. Incubator(s)
5. Flatbed scanner with associated software (EPSON Perfection V500)
6. Sterile 1.5 mL tubes
7. Plates with desired media (see Section 2.1 and Table 1)
8. Sterile dH_2O or liquid media
9. Desired yeast strains (see Table 2)

3.1.2 Procedure

1. Grow yeast cultures overnight in 3 mL of appropriate media.
2. Back-dilute cultures and grow to log phase growth (OD600 = 0.2 to 0.8).
3. Set up tubes for serial dilutions:
   a. For each sample to be analyzed collect 5 × 1.5 mL Eppendorf tubes.
   b. Add 400 μL sterile water (or liquid media, see notes) to 4 of the 5 sets of tubes, leaving the fifth tube empty. The fifth tube will be used for step 5.
   c. Warm up plates for 30 to 60 min prior to plating (see notes for plate suggestion).
4. Determine the number of cells/mL for each strain by:
   a. Counting with hemacytometer: follow standard directions, but generally, the average number of cells in the large squares (4 × 4 or 5 × 5) is multiplied by 105 cells/mL.
   b. Measuring OD600 (Note: this is obviously faster but may be inaccurate if one or more of the strains are irregular in shape or size). Multiply OD600 values by 50 × 106 cells/mL.

5. Calculate volume of culture needed for the initial spot (see below). Pipet appropriate number of cells into the empty tube from tube setup (#3).
   a. If plating on YEPD plates, use 6.25×10^6 cells.
   b. For selective plates, use 5×10^6 cells.
6. Equalize sample volumes and cell concentration:
   a. If volumes of all samples are less than 250 μL, add sterile water (or liquid media) up to 250 μL.
   b. If not, centrifuge samples at full speed for 1 min and aspirate supernatant. (See notes about selective media.) Add 250 μL sterile water or liquid media and vortex briefly.
7. Add 100 μL of cell suspensions from this set of tubes to the next set of tubes (with 400 μL water or media already in them).
8. Vortex briefly.
9. Repeat by adding 100 μL of the diluted cells to the next set of tubes to make successive dilutions.
10. Tape down plates on top of pattern template (Fig. 1A).
11. Spot cell suspensions: Take 4 μL (5 μL if selective media) from first dilution, pipet onto plate(s) as a circular spot. Repeat for rest of

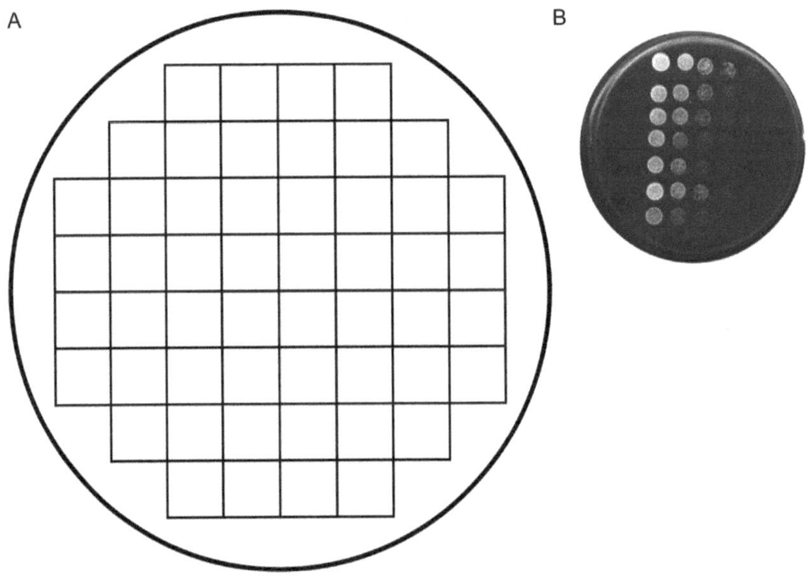

Fig. 1 Spot titer growth analysis. (A) Template for spot titer assay for a standard 10 cm plate (to scale). (B) Example of a spot titer assay result.

dilutions, arranging the serially diluted spots in a line for each strain. Be sure to briefly vortex each dilution before spotting.

12. Wait for spots to dry/absorb into plate, then incubate at desired temperature.
13. Take out plates, allow to equilibrate to room temperature (to avoid condensation during imaging), and use a flatbed scanner to image as desired (Fig. 1B; also see notes for information on growth period).

3.1.3 Notes

1. Unlike liquid phase growth assays, spot titer assays are less sensitive to whether the cells are in early-, mid-, or late-log phase when plated, although the cell concentrations of the strains tested should be relatively close to each other.
2. Diluting in sterile water versus media: while yeast cells can tolerate a temporary suspension in water, it is both nutrient deprivation and an osmotic shock. The user may therefore wish to consider diluting the cells in liquid media similar to that used to induce the stress.
3. Warming up plates: 6 to 8 strains can fit on one plate, depending on how tightly packed the spots are. Use this as a basic metric to determine the number of plates needed.
4. Centrifuging with selective media: cells will spin down more easily in selective media if water or YEPD is added to the tubes prior to centrifugation.
5. Pipeting: if dilutions are made in 96-well plates instead of 1.5 mL tubes, a multichannel pipet can be used for making the dilutions and for spotting cells. Extra care must be taken that equal volumes are pipetted for each channel, so normal micropipets are preferred unless a large number of strains are to be assayed.
6. Growth period: Wild-type cells on dextrose at 30 °C or 37 °C usually take ~1 day, at 25 °C takes ~2 days. Other sugars, drug treatments, or growth at 16 °C can require 5 to 10 days. Once the colonies have grown sufficiently, the apparent colony size will increase more slowly. Strains that grow more slowly can then "catch up," making differences in growth less apparent. Thus, the ideal imaging time must be determined empirically and is before growth has begun to slow.

3.2 96-Well plate growth analysis

This protocol describes the use of 96-well plate micro-cultures and a plate reader to measure OD_{600} over a time course, followed by curve fitting to generate growth curves that can be used for comparison of growth rates of different strains in specific growth phases. This assay has the advantages of allowing comparison between a large number of different combinations of strains and conditions, which is highly useful for stress studies, and providing quantitative measurements of both growth rate and the extent of lag phase. A disadvantage is that the range of concentrations that can be accurately measured (and therefore the potential length of incubations) is fairly small compared to the spot titer assay described above.

3.2.1 Equipment and reagents

1. Visible light plate reader (We use the Molecular Devices VersaMax)
2. Plate reader software (SoftMax Pro)
3. Data analysis software (Excel and Graphpad Prism)
4. Shaking incubator
5. Multi-channel pipet (recommended)
6. Toothpicks (sterile)
7. 96-well plates (flat-bottom)
8. Plates with desired media (see Section 2.1)
9. Liquid media and any desired treatments (see Table 1)
10. Desired yeast strains (see Table 2)

3.2.2 Procedure

1. Preparing strains for inoculation:
   a. Streak out strains of interest for single colonies on either YEPD or selective plates.
   b. Pick individual colonies with toothpick and streak a "strain-track" for each strain onto a fresh YEPD or selective plate using the serial dilution plate template as a guide for spacing (see Fig. 2A for reference). A small amount of yeast on the toothpick is sufficient. Larger amounts may cause uneven growth for later steps.
   c. Incubate plate overnight at 30 °C.
   d. The next morning, check the plates to see if there is a relatively even array of strain-tracks corresponding to the different strains.
2. Inoculating the 96-well plate micro-cultures:
   a. Using proper sterile technique, use a multichannel pipet to dispense 250 μL of liquid media into the wells of a flat-bottom 96-well plate.
   b. Inoculate wells of 96-well plate by picking a small amount of yeast from strain tracks (see notes for suggestions). Inoculate a minimum of 5 replicates per strain track. See Fig. 2B for reference.

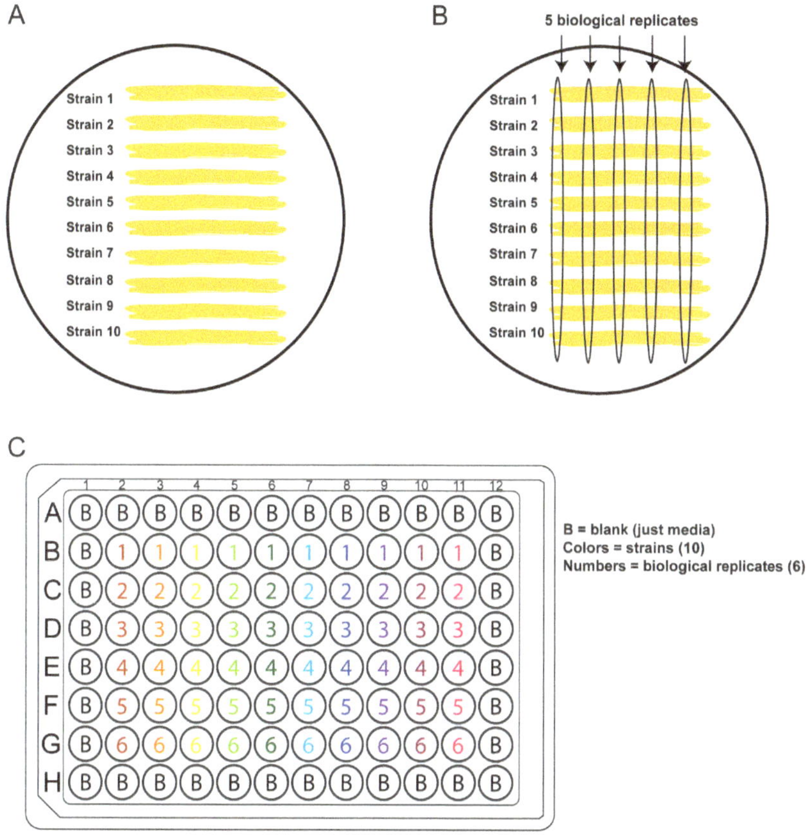

Fig. 2 96-Well plate growth analysis. (A) Layout for streaking individual strains ("strain-tracks") onto a standard 10cm plate for generating growth curves. (B) Diagram for selecting five biological replicates from a "strain-tracks" stock plate. (C) Sample layout for 96-well plate for six biological replicates of 10 strains/conditions. Note that the perimeter is reserved for buffer-only controls to reduce edge effects.

   c. Set shaking incubator to 30 °C and 350 rpm. Put plate with lid on shaker and tape it securely to the shaker. Let the cultures grow overnight.

3. Back diluting the cultures and allowing them to reach exponential growth:

   a. The next morning check the plate to see that each well has grown. The overnight micro-cultures should all be overgrown (in stationary phase) at this point with visible cells settled at the bottom of the plate.

   b. Increase the shaker speed to 450 rpm, re-tape the plate and let it shake for ~5 min to resuspend the cells.

c. Obtain a new 96-well flat-bottom plate and use multichannel pipet to dispense 200 μL of fresh media per well.
d. Inoculate wells of new plate with 25 μL of the overnight micro-culture. Pipet up-and down ~10 × to mix.
e. Turn on the plate reader and ready it to take readings. If using a VersaMax, turn on and open SoftMax Pro software. Set wavelength to 600 nm.
f. Measure OD_{600} using plate reader.
g. Check OD_{600} readings. Each back-diluted culture should have an OD_{600} between 0.2 and 0.3. If cultures are not within this range, adjust accordingly with more cells or more media as needed.
h. Incubate at 30 °C while shaking at 400 rpm.
i. Check OD_{600} after 4 to 6 h using plate reader. ODs should be between 0.4 and 0.7. Adjust volume with appropriate media so all micro-cultures are at a similar density.
j. Dispense 200 μL fresh media into wells of new 96-well flatbottom plate. See notes and Fig. 2C for reference on the well arrangement. Add drug treatment/stress at this stage if doing so.
k. Calculate the volume of cultures to add to each well for and OD_{600} of 0.07 to 0.1. This should be approximately 20 μL of a mid-log culture.
l. Inoculate calculated amount to each well in the new plate. Mix well by pipetting up and down 10 × with multichannel pipet. Make sure to mix well without generating bubbles.
m. Check that OD_{600} values are 0.07–0.1 using plate reader. At this stage the volume of each micro-culture should be the same and OD_{600} measurements should be within 0.02 OD_{600} units from one another.
n. Set shaking incubator to desired temperature and 350 rpm. Put plate with lid on shaker and tape it to the shaker.

4. Time-course OD_{600} measurements:
   a. Measure OD_{600} of wells using the plate reader at pre-determined time-points. At a minimum it is recommended to take one OD_{600} measurement per cell doubling for exponentially growing cultures.
   b. Before taking an OD_{600} measurement, set shaker to 450 rpm for 5 min to make sure cells are resuspended.
   c. Take care to get rid of bubbles with pipet or by tapping on counter before measurement.
   d. Continue taking OD_{600} measurements until all strains reach stationary phase (OD_{600} of ~1.4).

5. Data analysis and curve fitting:

a. Compile all data into a spreadsheet program (e.g., Excel).

b. Calculate average of all background (media only) OD_{600} readings. Subtract background average from values for experimental wells for each time point.

c. Use a graphing/curve fitting program to generate an XY graph using background subtracted measurements. We recommend Graphpad Prism as it has many options but is optimized for biological data sets.

d. It may be helpful to perform a Q test to identify true outliers at this stage. The small scale coupled with exponential growth can result in substantial amplification of small errors. We recommend that exclusion of any outliers be applied to the entire replicate rather than individual time-points.

e. Fit a curve to each strain/condition with a re-parameterized version of the Gompertz growth equation, which has the following form (Zwietering, Jongenburger, Rombouts, & van't Riet, 1990):

$$OD_{600} = Ae^{-e^{\frac{\mu_{max} \cdot e}{A}(\lambda - t) + 1}}$$

Here, t is the incubation time, A represents the asymptotic maximum value (the maximum OD_{600} in stationary phase), μ_{max} represents the maximum growth rate (the tangent of the inflection point), and λ is a measure of the lag time (technically, the x-axis intercept of μ_{max}). See notes for further discussion.

f. If the curve fitting does not seem to describe the data, consider:

i. Judiciously adding constraints, such as setting all A parameters to the same value. Try to avoid overfitting.

ii. Repeating experiments with more time points and/or replicates.

g. Compare curves between different strains and conditions. Following stress treatments, cells will often display a long lag time (λ) before beginning significant growth. The growth rate itself (μ_{max}) may also differ between strains or before/after treatment. Use appropriate statistical tests, such the extra sum-of-squares F test, to determine significance between parameters.

3.2.3 Notes

1. Before starting experiment:

a. Plan the timing of experiment beforehand to minimize the need to perform measurements at odd hours. For long drug treatments or stress conditions the timing of the experiment becomes more important. It may take some trial and error.

b. Pre-plan the layout of the 96-well plate with strains, replicates and experimental conditions. It is helpful to make a template for each experiment (see Fig. 2C). Leave blank wells on the outer edges of the plate; these wells tend to have higher evaporation than the inner wells, throwing off the growth measurements.

c. Always make sure 96-well plate is taped securely onto the shaking incubator.

d. Minimize/eliminate bubbles in micro-cultures prior to OD_{600} measurement, as they can interfere with accurate measurements.

2. For initial inoculation of colonies, single colonies can be hand-picked using a toothpick or sterile pipet tip, or multiple colonies can be selected at once using the multi-channel pipet. The latter is not easy and is dependent on accurate spacing of strain tracks.

3. The Gompertz growth equation is one of several described in microbiology literature (Zwietering et al., 1990). For stress treatments, it has the advantage of including a variable-length lag phase as well as a maximal growth rate, both useful measurements. The usual form of the Gompertz equation is:

$$f(t) = ae^{-e^{b-ct}}$$

However, in this form, the parameters are purely mathematical and do not correspond to biologically-relevant values. It is unlikely that any curve-fitting software on hand will have the re-parameterized form as a default (Prism does not); therefore, it will likely need to be input as a custom equation. A linear version of the re-parameterized form that may be helpful for software entry is:

$$y = A \bullet \exp(-\exp(((u \bullet \exp(1))/A) \bullet (l - x) + 1))$$

4. In general, this protocol is described using software and equipment common to our lab, but equipment and software of your preference can be substituted.

4. Translation assays to assess stress phenotypes

As discussed above, translation regulation is a key part of stress responses, resulting in massive reprogramming of translation and the redistribution of cellular resources (Crawford & Pavitt, 2019; Liu & Qian, 2014). In general, protein synthesis is repressed during stress with translation of most mRNAs, especially those encoding proteins with pro-growth or "housekeeping" functions, being down-regulated. Despite this bulk inhibition, select mRNAs, including those encoding "stress response genes" such as protein chaperones, are translationally upregulated, while others are particularly sensitive to down-regulation, indicating that the translational response is specific and complex (Gerashchenko et al., 2012; Ingolia et al., 2009). Furthermore, after cells adapt to the stress or it is removed, the renewed growth is accompanied by a resumption of bulk translation, which requires an additional layer of regulation (Advani & Ivanov, 2019; Aryanpur et al., 2022). Several mechanisms have been identified that regulate various aspects of this specificity, including the action of RNA helicases such as eIF4A and Ded1/DDX3X, but much remains poorly understood (Aryanpur et al., 2019; Guenther et al., 2018; Tauber et al., 2020; Tsokanos et al., 2016).

In this section, several assays are described that indicate the status of cellular translation: ^{35}S-methionine labeling, luciferase reporters, and polysome profiles/fractionation. One consideration is that ^{35}S-methionine labeling and polysome profiles themselves are read-outs for general/bulk translation, especially for the most abundant and highly translated mRNAs, so regulation of specific mRNAs, such as stress response genes, will likely not be represented (in some cases such as heat shock, expression of individual heat shock proteins can sometimes be observed using ^{35}S-methionine labeling). In contrast, luciferase reporter translation is specific for the particular construct design; thus, different reporters can be used as "generic" indicators for translation or for translation of mRNAs with very specific regulatory mechanisms. Additionally, as described below, fractionation of polysome profiles can be performed to examine the association of proteins and RNAs with ribosomal complexes, which can also be used as a proxy for the translation status of specific mRNAs. The most comprehensive way to examine translation of individual mRNAs is using ribosome footprinting (Ribo-seq), which is discussed elsewhere (e.g., McGlincy & Ingolia, 2017; Powers & Brar, 2021). However, it is worth noting that the normalization procedures for Ribo-seq are often optimized for examining relative differences between mRNAs rather than absolute levels of translation, so this method may not be ideal when bulk translational changes are expected.

A general caveat with these assays is that it can be difficult to distinguish between changes in translation and effects on upstream gene expression processes such as transcription. Especially if characterizing a novel factor or condition, it may be helpful to include additional controls to show that the observed translational effect is not due to indirect effects, for example in the form of mutants in the transcription and/or mRNA export machinery. Additionally, for luciferase assays, RT-PCR can be performed to ensure that reporter mRNA levels are similar, and in polysome profiles, analysis of the profile pattern can sometimes suggest if indirect effects are present.

4.1 Translation status via ^{35}S-methionine labeling

This protocol is used to observe in vivo biosynthesis of proteins. The addition of radioactively labeled methionine allows newly formed proteins to be observed via polyacrylamide gel. It also allows the visualization of translation during stress conditions by displaying all new proteins synthesized after stress conditions have been added. A limitation of this method is that it does not innately allow for characterization of the individual proteins being actively translated other than their apparent molecular weight, and therefore cannot be used to examine specific proteins. It is also important to note that before attempting this protocol, the proper certification must be acquired to handle radioactive materials, although alternative formulations of methionine labeled with fluorescent tags or biotin are also available.

4.1.1 Equipment and reagents

1. Shaking incubator
2. Spectrophotometer
3. Benchtop centrifuge
4. Microcentrifuge
5. Power supply and gel apparatus for SDS-PAGE
6. Gel dryer
7. Film and developer/dark room
8. Desired yeast strains (see Table 2)
9. Liquid media and any desired treatments (see Table 1)
10. Minimal media lacking methionine (see Section 2.1)
11. ^{35}S-methionine, 1000 Ci/mmol (recommend using within one half-life)
12. Lysis buffer: 1.85 M NaOH, 7.4% β-mercaptoethanol
13. 50% Trichloroacetic acid (TCA)
14. 1 M Tris Base (do not adjust pH)
15. SDS sample buffer: 100 mM Tris pH 6.8, 5% SDS, 0.1 M DTT, 0.02% bromophenol blue, 15% glycerol
16. Fixing solution: 40% methanol, 10% acetic acid, 3% glycerol

17. Fluoro-Hance autoradiography enhancer
18. Reagents for SDS-PAGE gels
19. Safe work space, storage, and waste disposal for radioactive materials

4.1.2 Procedure

1. Inoculate 20 mL yeast cultures in a 125 mL flask and grow to mid-log phase. If stresses, drug treatments, etc., are to be used, perform these at this stage and have the cultures ready for step 2.
2. Measure and record OD_{600} of cultures with spectrophotometer.
3. Transfer cells to 50-mL conical tubes and spin down cells in centrifuge for 5 min at 4000 rpm.
4. Discard supernatants.
5. Methionine starvation (see also notes for alternative method):
   a. Re-suspend cells in 10 mL of minimal media lacking methionine.
   b. Pellet cells again by centrifugation and discard supernatant.
   c. Resuspend cells in 3 mL of minimal medial lacking methionine.
6. Add 25 μCi of ^{35}S-methionine per OD_{600} unit (the OD_{600} measurement times the amount of culture (~20 mL) spun down) to each culture. Make sure to follow proper radioactivity safety protocols from this point forward.
7. Incubate cells for 30 to 60 min in 30 °C while shaking. Stress treatments should also be repeated for this incubation if possible.
8. Record OD_{600} of samples.
9. Pellet the cells via centrifugation for 5 min at 4000 rpm.
10. Discard supernatant.
11. Lyse cells (if needed cells can be frozen and stored at this step). Multiple methods can be used, and here we briefly describe an adapted version of the crude cell lysate protocol of Yaffe and Schatz (1984) (also see notes):
    a. Resuspend cells in 1 mL H_2O and transfer to 1.5 mL tubes. Spin 30 s in microcentrifuge and discard supernatant.
    b. Resuspend in 160 μL of freshly prepared lysis buffer. Mix and incubate on ice 10 min.
    c. Add 160 μL 50% TCA. Mix and incubate on ice 10 min.
    d. Spin 2 min in microcentrifuge. Wash pellet (don't resuspend) with 0.5 mL 1 M Tris base.
    e. Resuspend in 150 μL of SDS sample buffer per OD_{600} unit (OD_{600} × culture volume). Alternate heating at 95 °C and vortexing to resuspend the pellet.

   f. Briefly centrifuge to bring all liquid to the bottom of the tube before loading.

12. Load as much sample as possible on a polyacrylamide gel (10% or 4–15% gradient). Run samples as standard SDS-PAGE (see notes).
13. Fix and dry gel:
    a. Remove gel from electrophoresis apparatus and add to fixing solution. Gel should be completely submerged in solution for a minimum of 1 h.
    b. Decant fixing solution and add sufficient Fluoro-Hance to submerge gel. Incubate 20 min.
    c. Dry gel using a gel dryer.
14. Visualize ^{35}S-labeled bands by exposing dried gel to autoradiography film. Exposure times can vary from 1 day to a full week, and the proper exposure time should be determined empirically. Develop film (see Fig. 3 for example).
15. For quantitation, scan film and perform band densitometry using ImageJ (or a similar application) to measure total ^{35}S signal in each lane.

4.1.3 Notes

1. Alternative to methionine starvation: skip this step and just re-suspend cells in 3 mL of YEPD or minimal complete media. Since unlabeled

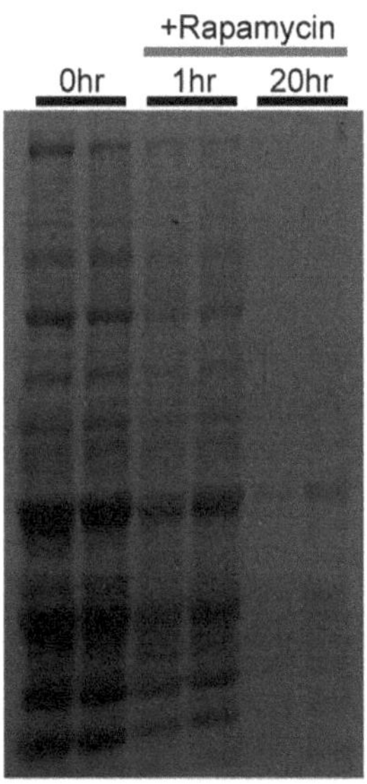

Fig. 3 ^{35}S-methionine labeling. Representative ^{35}S-methionine autoradiogram displaying newly translated proteins 0, 1 h, and 20 h after treatment with rapamycin. Film exposed for 4.5 days and loaded with 10 μL sample.

methionine will be present, this will reduce the labeling of proteins and produce a lower signal, but in many cases the signal will still be sufficient to visualize.
2. Once the ^{35}S-methionine is added in step 6, all samples, reagents, and equipment used are considered radioactively contaminated. Please plan accordingly and follow all appropriate safety protocols.
3. Other extract preparation protocols can be used if preferred.
4. Standard SDS-PAGE conditions are recommended. Details are not described here but can be found in many laboratory protocols.

4.2 Translation status via luciferase reporter

This protocol describes the use of a luciferase assay as a reporter for intracellular translation status. In vitro translation systems with luciferase mRNA are also available, although they are not described here (see Wu, Amrani, Jacobson, & Sachs, 2007). Such systems can be of great utility in studying translation (e.g., Balagopal & Parker, 2011); however, it is not straightforward to study stress-specific functions in vitro, so here we focus on the reporter assays in cells. In these assays, the amount of luciferase signal is dependent on the design of the reporter construct. In theory, any highly expressed promoter and 5' UTR fused to the luciferase coding region could serve as a baseline reporter, and reporters can then be altered to add secondary structure, mutate the start codon, or other changes to test more specific aspects of translation. We recommend pFJZ342 (*CEN/URA3*), which contains the promoter and 5' UTR of *RPL41A* with an insertion of 23 CAA repeats fused to the firefly luciferase coding sequence (Sen, Zhou, Ingolia, & Hinnebusch, 2015). This is part of a set of reporters with 5' UTRs of different lengths and secondary structures, which have been used to characterize the RNA helicases Ded1 and eIF4A (Brown, Vergara, Whelan, Guerra, & Bolger, 2021; Sen et al., 2015).

4.2.1 Equipment and reagents

1. Spectrophotometer
2. Shaking incubator
3. Mini bead-beater (Biospec)
4. Luminometer (Promega Glomax)
5. 1.5 mL screw top tubes
6. 1.5 mL microfuge tubes
7. Acid-washed glass beads (425–600 μm)
8. 1 mL syringes with small gauge needles (recommend 23G or higher)

9. Luciferase lysis buffer (25 mM Tris phosphate (pH 7.8), 0.5% Triton, 2 mM EGTA, 2 mM DTT, 10% Glycerol)
10. Luciferase assay reagent (luciferin). If using Promega reagents, add suggested volume of buffer to lyophilized reagent. If custom-made, resuspend luciferin at 150 μg/mL in buffer containing 15 mM $MgSO_4$, 15 mM K_2HPO_4, 4 mM EGTA, 1 mM ATP, 1 mM DTT). Protect from light. Can aliquot and freeze at −80 °C.
11. Yeast strains transformed with luciferase reporters (see Table 2)

4.2.2 Procedure

1. Sample Preparation:
   a. Inoculate yeast cultures and grow overnight.
   b. Dilute cultures and split into triplicate (3–5 mL each). Grow cultures to mid-log phase ($OD_{600} = \sim 0.5$).
   c. If stresses and/or drug treatments are to be performed, dilute cultures back to OD_{600} of 0.05 to 0.2 depending on length of treatment. Add drug/stress and incubate.
   d. Recover samples and record OD_{600} of each sample.
   e. Transfer up to 1.5 mL of culture to microfuge tubes and spin down in microcentrifuge for 1 min.
   f. Aspirate media and resuspend in 200 μL luciferase lysis buffer.
   g. Add ~100 μL glass beads to screw top tubes, then transfer samples into them.
   h. Shake in mini-bead beater at 4 °C twice for 2.5 min each, with 1 min on ice in between runs (or 4 × 1.25 min with 30 s on ice in between).
   i. Spin 6 min in microcentrifuge at 4 °C.
   j. Using a 1 mL syringe with small-gauge needle or a thin gel-loading tip, transfer supernatants from screw cap tubes to 1.5 mL tubes.
   k. Keep on ice or at −20 °C.
2. Luciferase Assay
   a. Turn on luminometer to warm up at least 15 min before use and thaw luciferase assay reagent. Protect the reagent from light.
   b. Set luminometer to a single injection of 100 μL, minimal delay, and 10 s (or more) integration time.
   c. Prime luciferase assay reagent in injector.
   d. If desired, insert an empty tube in the luminometer and run as a "blank" to check that the injector and detector are working properly (see notes).
   e. Add 5 to 50 μL of sample to a 1.5 mL microfuge tube (or appropriate container for luminometer).

f. Run each sample in the luminometer and record result.
g. Flush the luminometer with water after use.
h. Normalize each sample by dividing the luciferase value by the OD_{600}, and compare values.

4.2.3 Notes

1. This assay is a measure of the amount of luciferase protein present in the cells, so when performing a time-course, residual luciferase may still be present even if translation is repressed. Therefore, we recommend choosing time points on the order of hours rather than minutes.
2. The luciferase value for the blank may be subtracted from the sample values as background, but in our experience, the blanks give a negligible value compared to samples.
3. Luciferase assays have a large dynamic range, eight orders of magnitude or more. The amount of sample added to the assay can be adjusted if needed, however.

4.3 Polysome profiles and fractionation

Polysome profiling utilizes centrifugation of cellular extracts over a sucrose gradient to separate the different types of ribosomal complex (40S, 60S, 80S, and polysomes) by relative density. These profiles can potentially give several different kinds of outputs that provide information on cellular translational status. First, a trace of the absorbance (at 255 nm, a measurement of RNA concentration) along the gradient is generated, marking each of the ribosomal complexes and their relative abundances. By measuring the area under the curve of each peak, the relative amount of translation that is occurring globally can be quantified. With this information, mutations, drug treatments, or stress conditions can be evaluated as to what extent they are able to alter global, or bulk, translation (e.g., Chuang, Weaver, Liu, & Chang, 1997). Generally, profiles would be expected to show translation repression during stress responses, which is then restored as the cells adapt and recover (Aryanpur et al., 2019, 2022; Shenton et al., 2006). Second, fractions can be collected from the gradients and immunoblotted for specific translation factors or other proteins. These results can help reveal the effects of changes in ribosomal complexes as a result of stress or mutations. An additional crosslinking step (see notes) is recommended when performing in-depth protein analysis. A method to use fluorescently-tagged proteins to generate continuous traces of protein signal, similar to the RNA absorbance trace, has also recently been described (Shaffer & Rollins, 2020). Third, RNA can be isolated from the gradient fractions, and quantitative RT-PCR can be performed in order to examine the association of individual mRNAs with more or less active translation complexes (see notes). However, this method is technically challenging and requires additional equipment; thus, other methods may be more accessible for examining translation of individual mRNAs.

4.3.1 Equipment and reagents

1. Small gradient mixer (Hoefer SG30)
2. Stir plate
3. Small peristaltic pump (Fisher mini-pump is sufficient)
4. Small-diameter (~2 mm) tubing, connectors, ~12 cm stainless steel tube (interior diameter ~ 2 mm) and rubber stopper
5. Shaking incubator
6. Spectrophotometer
7. Benchtop centrifuge
8. Vortexer with lid (Turbo-Mix)
9. Microcentrifuge
10. Standard balance
11. Ultracentrifuge (Beckman)
12. SW-41TI ultracentrifuge rotor
13. Density gradient fractionation system (Brandel). See Fig. 4 and notes
    a. Syringe pump with 50 mL glass syringe (Brandel)
    b. Gradient separator (Brandel)

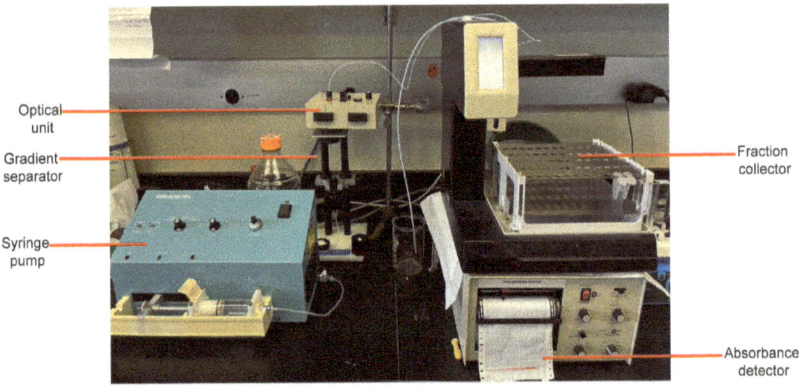

Fig. 4 Density gradient fractionation system (Brandel). Major components are labeled. Flow moves from left to right: Fluorinert is pushed from syringe pump through tubing to gradient separator, pushing sample in gradient tube (not shown) up into optical unit of absorbance detector, through tubing to fraction collector, and into 1.5 mL tubes or waste beaker.

   c. Absorbance detector (Teledyne-ISCO UA-6 with Type 11 Optical Unit)
   d. Fraction collector (Teledyne-ISCO Foxy R1)
   e. Data capture software (Brandel PeakChart)
14. Thinwall polypropylene tubes (recommend Beckman #331372, 14 × 89 mm)
15. 50 mL conical tubes
16. 1.5 mL tubes
17. Acid-washed glass beads (425–600 μm)
18. Cycloheximide (10 mg/mL)
19. Low salt (LS) buffer (20 mM HEPES-KOH, 100 mM KOAc, 5 mM MgOAc, 1 mM EDTA) in RNase-free water (see notes). Add fresh day of: 1 mM DTT, 100 μM PMSF, 1 × protease inhibitor cocktail, 100 μg/mL cycloheximide
20. 7% sucrose (w/v) solution in LS buffer (add DTT and cycloheximide but other additives not needed). See notes.
21. 47% sucrose (w/v) solution in LS buffer (add DTT and cycloheximide but other additives not needed). See notes.
22. Fluorinert FC-40 chase solution
23. RNase-free water
24. Liquid media and any desired treatments (see Section 2.1 and Table 1)
25. Desired yeast strains (see Table 2)
26. Ice-cold ethanol (optional)
27. SDS sample buffer (optional): 100 mM Tris pH 6.8, 5% SDS, 0.1 M DTT, 0.02% bromophenol blue, 15% glycerol

4.3.2 Procedure

1. Pour gradients (can be done 1 day prior and stored at 4 °C, also see notes):
   a. Set up gradient preparation apparatus (see Fig. 5A). Place gradient mixer on stir plate and add a micro stir bar to the front well. Attach tubing from gradient mixer to pump and from pump to stainless tube. Insert tube through rubber stopper and place stopper on top of a polypropylene ultracentrifuge tube so that stainless tube extends inside. Make sure the ultracentrifuge tube is securely held in a vertical position, and position the stainless tube so it does not quite touch the bottom of the ultracentrifuge tube.
   b. Move the stainless tube/stopper to empty into a beaker. Add RNase-free water to gradient mixer wells and turn on cartridge

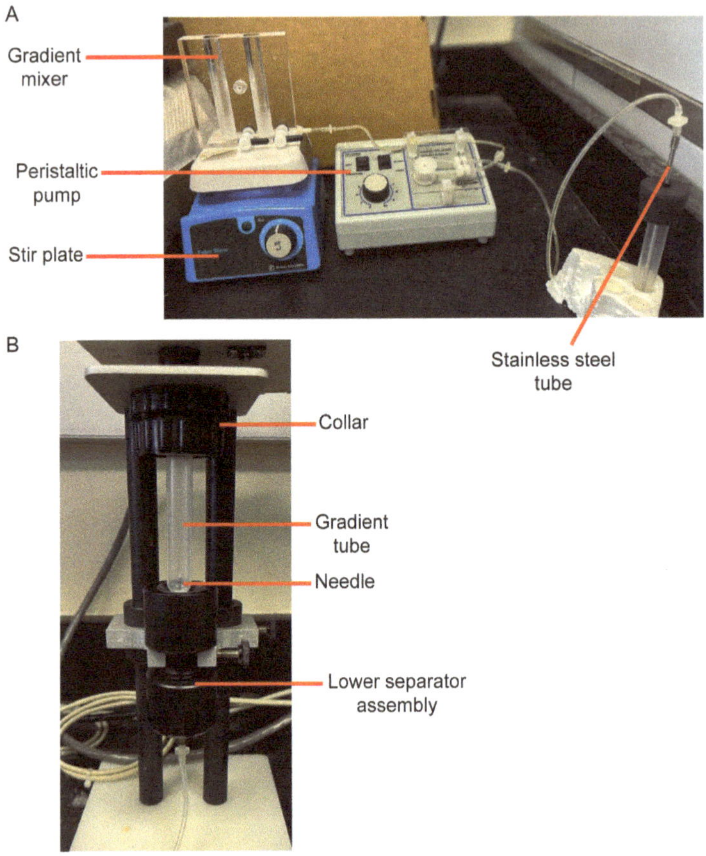

Fig. 5 Gradient pouring setup and gradient separator. (A) Setup for pouring gradients with components labeled. Sucrose solutions are added to the mixer; then are pulled into the tubing, through the pump, and out of the stainless steel tube into the gradient tube. (B) Close-up of the gradient separator with gradient tube and other relevant parts labeled.

pump and stir plate (stir bar should be spinning cleanly). Wash system with water, ethanol, and water again. Run until all liquid has drained.

c. Close valves on gradient mixer.

d. Add 5.7 mL of 47% sucrose solution to rear well of mixer, and 5.7 mL of 7% to front well.

e. Open valve between wells of mixer, then front valve, then turn on pump (use a low setting; it should take 5 to 10 min to pour each gradient). The solution should flow smoothly between the wells and should start moving into the tubing.

f. The progress of the sucrose solution can be tracked through the tubing. When it approaches the stainless tube/stopper, move it so it empties into the pre-positioned polypropylene ultracentrifuge tube.
g. When the solution in the mixer is nearly empty, air bubbles will form in the line. Continue running the pump until bubbles are near/within the stainless tube, then stop it. The gradient solution should be near (but not at) the top of the polypropylene tube (if it is not, discard the gradient and pour again). In one smooth, continuous motion, remove the stainless tube from the ultracentrifuge tube (this minimizes mixing of the gradient in the tube) and place in waste beaker.
h. Restart the pump to run out any last bits. Reset for the next gradient. Washing between each gradient is not usually necessary.
i. When finished, be very gentle moving gradients around to minimize mixing. Wash system extensively (as in step b); the sucrose solutions can leave significant residue if not cleaned.

2. Harvest cells:
   a. Culture 30 to 50 mL of desired yeast strains to mid-log phase in desired conditions. Add drugs and/or stress conditions and treat as desired (see notes). Measure OD_{600} and multiply by culture volume to get "OD units". Protocol works best with 8 to 15 OD units. Generally, this protocol is limited to 6 samples at a time due to the ultracentrifuge rotor.
   b. Add cycloheximide at 100 μg/mL directly to cultures and incubate on ice 5 min. Transfer to 50 mL conical tube.
   c. Spin down cells for 5 min at 3000 × g
   d. Decant media and add 5 mL of LS buffer (+cycloheximide).
   e. Spin down again and remove all supernatant. Can freeze cell pellets overnight at this step if needed.
   f. Resuspend pellets in LS buffer (+additives) at 13.3 μL per OD unit and transfer to 1.5 mL tubes.
   g. Add an approximately equal volume of glass beads to each tube.
   h. Vortex at 4 °C for 5 min. Spin in microcentrifuge at 4 °C for 5 min.
   i. Transfer extracts (supernatants) to fresh tubes.
3. Lyse cells, load gradients, and spin:
   a. Carefully load extracts on the top of each gradient by layering along the side of the tube. We recommend 100 μL (7.5 OD units) per sample.

b. Gently place gradients into the buckets for SW-41 rotor. Balance them as closely as possible: weigh each pair of buckets, and add extra LS Buffer to the top of the lighter sample. Ideally each pair is within 0.02 g.

c. Carefully transport buckets to ultracentrifuge, hang buckets on rotor, and set up ultracentrifuge. Make sure rotor is mounted properly in ultracentrifuge and that vacuum is working properly. Spin at 38,000 rpm for 2.5 h at 4 °C.

4. Polysome fractionator operation protocol (This protocol is specific to using the Brandel density gradient fractionation system, but equivalent steps will need to be performed in other systems):

   a. Turn on the absorbance detector and let warm up for at least 15 min. Open the data collection software application (PeakChart).

   b. Turn on the fraction collector. If desired, set up to collect fractions, and place the appropriate number of open 1.5 mL tubes in the collector (see notes).

   c. Assemble the gradient separator apparatus (see Fig. 5B).

   d. As needed, rinse lines with RNAse-free water, ethanol, and water again, followed by clearing the lines out with air. We recommend adding a three-way valve between the syringe pump and the bottom of the gradient separator and inserting a spare polypropylene tube into the separator. The third outlet of the three-way valve can then be used for pushing water, ethanol, or air through the system via a syringe. Please note that the fraction collector must be started in order to push anything through.

   e. If needed, add Fluorinert to the glass syringe in the syringe pump: Disconnect the tubing to the gradient separator and place in Fluorinert stock. Filling the syringe to 40–50 mL is preferable. Avoid bubbles since they will disturb the gradient if they are pumped through it. The pump can be turned on its side to remove excess air from the system. Reconnect tubing to separator.

   f. Prime the system with Fluorinert: if present, remove the spare tube from the separator. Extend tube piercing needle until the outlet hole is visible, then turn on syringe pump (forward, 1.5 mL per min). Watch Fluorinert travel through tubing until it starts coming out of the needle, then stop pump. Wipe up excess Fluorinert. Retract needle below gasket.

   g. When ultracentrifuge spin is complete, open each bucket and remove the gradients with forceps. Keep at 4 °C until analyzed.

h. Insert a gradient into the gradient separator: Loosen the upper collar in the separator and carefully insert the gradient tube into the collar. Push firmly upwards and tighten the collar until the tube is securely held. Raise the lower separator assembly so that the gradient tube is centered and tight between the top and bottom of the separator, and tighten in place. Extend the needle until it pierces the tube and both black lines are visible in the tube. See Fig. 5B.

i. Set syringe pump to remote start/stop, 1.5 mL per min, forward.

j. Start the pump via the software. In PeakChart, click the green trace button—looks like a marker tip. The fraction collector should start, and the absorbance trace should start in the software (if it does not, stop and troubleshoot, otherwise sample may be lost). Immediately switch fraction collector to waste (trash can icon).

k. Fluorinert should start coming out of the needle hole and filling the bottom of the gradient tube. The gradient should be slowly pushed up into the top of the separator and the UV detector in the optical unit.

l. Adjust baseline of the absorbance trace ("Baseline Adjust" knob on the optical unit) as needed. We find that the baseline reading at the beginning of a trace (in air) should be 0.2 to 0.25 V to avoid bottoming out later in the run.

m. If collecting fractions, when the detector begins detecting sample (trace jumps upward), switch collector back to run (arrow icon) if delay is set. If delay is not used, wait to switch collector until sample is about to come out of the fractionator.

n. Run entire profile (~10 min). At the end, the trace will often spike, then begin acting erratically. Run sample all the way through fractionator and wait until Fluorinert begins dripping out (Fluorinert has less cohesion than an aqueous sample, so the drops are smaller and faster).

o. Stop trace in application (red "x" button in PeakChart). Manually switch the syringe pump to reverse at 6 mL per min, and pump Flourinert back into glass syringe. There may still be some sample at the Fluorinert/air interface, so stop the pump before the Fluorinert reaches the needle hole.

p. Remove the gradient tube from the separator and discard. Close used fraction tubes, remove from fractionator, and store at 4 °C until ready to precipitate protein. Reset the system for the next sample. It is usually not necessary to clean the lines between each sample, but this can be done if lines appear dirty/blocked.

q. When finished, rinse lines as in step (d) with RNAse-free water, ethanol, and water again, followed by clearing the lines out with air. We recommend also disassembling the separator and wiping down its components to minimize sucrose residue.

5. Analysis of profiles (see Fig. 6):

a. To assess the general translation status of samples, first calculate the area under the curve for the peaks from the UV absorbance trace. PeakChart has a tool for calculating this, but instead, we recommend opening the image file in ImageJ/Fiji. Draw a baseline underneath the trace; a common practice is to draw a horizontal line at the lowest point on the trace, usually between the 80S and the disome peak. Draw vertical lines to separate the peaks as needed, and use the measurement tool to calculate the area.

b. The most important comparison is the relative area of the 80S monosome peak versus the polysome peaks. A larger polysome to monosome ratio indicates more active translation, and a lower ratio indicates lower bulk translation. Note that translation repression generally leads to an increased 80S peak and not substantial increases

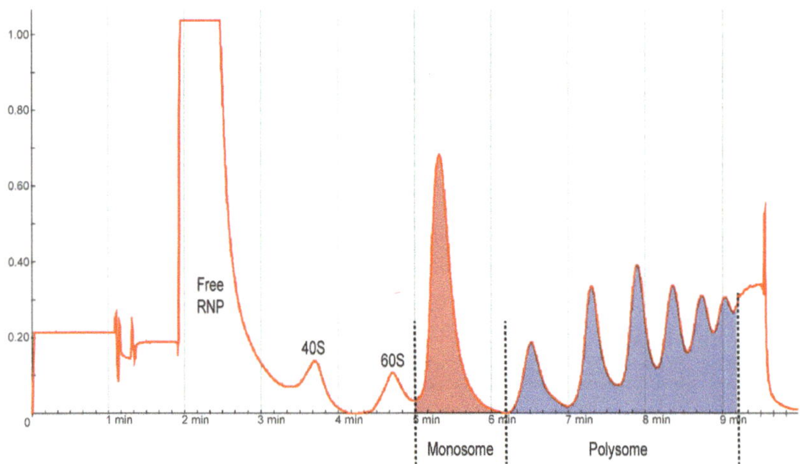

Fig. 6 Sample polysome profile. Output absorbance trace from PeakChart showing a sample profile (of unstressed wild-type yeast cells) with peaks for free RNPs, 40S, 60S, monosome (80S), and polysomes labeled. Note that the sample reaches the detector about 2 min after starting the trace, and the bottom of the gradient is reached after slightly more than 9 min. Shaded regions display the area under the curve that would be calculated for the polysome (blue) to monosome (red) ratio. The baseline (see Section 4.3.2, step 5, note a) is not specifically labeled since it would overlay the x-axis for this profile.

in the "free" 40S and 60S. However, for long time points in stress conditions, ribophagy, degradation of ribosomes, can also occur, resulting in a general loss of signal from the trace. Other types of effects have also been observed on the absorbance traces, such as defects in ribosome biogenesis, which can manifest as changes in the relative amounts of 40S and 60S complexes and/or the appearance of "halfmers," shoulders that appear on each monosome and polysome peak. However, we caution users that these differences are not always reliable and can vary from sample to sample.

c. If fractions have been collected, then the association of proteins with the various complexes can be examined. To do so, first precipitate proteins from the fractions by addition of 0.8 mL ice-cold ethanol to each 0.5 mL fraction, followed by vortexing and incubation overnight at −20 °C. Spin samples in microcentrifuge at 4 °C for 10 min, and remove supernatant. Wash the pellets once with 75% ethanol and then resuspend in SDS sample buffer. Proteins can then be separated by SDS-PAGE and blotted for the desired factors. Association of the target protein with translation complexes can be observed as well as shifts to more or less actively translated mRNAs in different conditions.

d. Fractions can also be examined for individual mRNAs. RNA extraction can be performed via phenol or with a commercial kit as desired. Depending on the technique, the fractions may need to be diluted with additional LS buffer before extraction to decrease the sucrose concentration. Fractions may also be pooled (for example, into "non-translated," "low translated," and "highly translated" pools) for simplicity. Relative RNA levels can then be determined via qPCR or Northern blot and then compared to the absorbance trace, where association with heavier polysomes is evidence of higher translation levels.

4.3.3 Notes

1. This protocol is designed around the Brandel Density Gradient Fractionation system, the only complete system available of which we are aware. However, systems can also be constructed out of the individual components listed. A syringe pump is highly recommended, as other types, such as peristaltic pumps, do not provide the same levels of smooth, continuous action. The gradient separator is fairly specialized but is available from Brandel and others. If accessible to the user,

HPLC or FPLC systems may be able to provide the necessary flow cell, UV absorbance readings, and data capture, while fraction collectors are available from multiple suppliers. If using a custom-assembled system, the operation section of the protocol will need to be modified to fit the system, but generally the same steps will need to be included.

2. Given that this protocol relies on intact RNA-protein complexes, degradation due to RNases is a concern, although ribosomal RNA itself is somewhat RNase-resistant. Therefore, all buffers, tubes, and other reagents that come into contact with the samples following lysis should be RNAse-free. This includes the interior spaces of the fractionation system, so RNase-free water should be used for rinsing the lines. Water can be either DEPC-treated or purchased as RNAse-free.
3. When preparing sucrose solutions, please note that the percentages listed are (solute) weight by (solution) volume. Given the high sucrose percentages, the volume of the sucrose will need to be accounted for, i.e., much less than 100 mL of LS buffer will be needed for 100 mL of 47% sucrose solution.
4. Other methods of generating sucrose gradients exist and can be used if desired, such as layering a step gradient and allowing it to equilibrate overnight. We have found that using a gradient mixer is a more reliable and user-friendly method.
5. When examining translation during stress conditions (or indeed in mutant cells), the user should carefully consider the appropriate time course. Translation is repressed fairly quickly in stress (in minutes rather than hours), and this is reflected in the polysome profiles in 15 min or less of rapamycin treatment, for example (Aryanpur et al., 2019). However, our studies and others have shown that the extent of repression continues to increase for a number of hours, depending on the severity of the stress, while eventually slower processes such as ribophagy can affect the profiles (Aryanpur et al., 2019; Kelly & Bedwell, 2015).
6. If fractions are to be collected, the fraction collector unit must be set up correctly. With the Brandel system collector, the user can choose between collecting each fraction through an equal number of drops (droplet icon) or through an equal time of collection (clock icon). We generally find the timer method to be more consistent. The number of drops or amount of time for each fraction will need to be determined empirically, depending on the number of desired fractions. With either method, a delay (hourglass icon) should be set before the counter/timer begins. The delay will synchronize the fraction collecting with the

absorbance measurement, reflecting the time it takes for the sample to travel from the detector to the fraction collector outlet. The length of the delay will need to be empirically determined but is ~30 s for a typical Brandel setup. Programming in the delay will also locate the fraction markers in the PeakChart software at the correct points on the trace. When using this method, remember to switch the collector to waste immediately after starting the trace and to switch it back to the run once the sample reaches the detector.

7. If in-depth analysis of protein association with translation complexes, particularly pre-initiation complexes, is desired, consider adding a cross-linking step. Formaldehyde crosslinking has been shown to aid in retention of associated proteins during profiling (see Valasek, Szamecz, Hinnebusch, & Nielsen, 2007). We modified this protocol slightly to treat cells with 0.25% formaldehyde for 15 min on ice, followed by quenching with 0.5 mM glycine (Aryanpur et al., 2019). Note that this is a significantly lower formaldehyde concentration, as we found that higher levels introduced artifacts into the profiles.

5. Other general protocols

Above are specific instructions for several protocols associated with analyzing helicases for translation functions in stress. These protocols are essential for understanding the complex phenotypes that are associated with stress in cells, but there are, of course, other assays that may be useful in examining translation during stress that complement the above approaches. Two of these approaches are described in general form below: stress granule analysis and protein half-life evaluation.

The formation of stress granules is another common response to cellular stress. Stress granules are membrane-less cytoplasmic foci of RNA and proteins that form under conditions of translation repression (Buchan & Parker, 2009; Protter & Parker, 2016). They are thought to be sites for the storage of repressed mRNAs that can later be degraded or shifted back into active translation, although their precise function has remained somewhat unclear. Stress granule analysis is immensely useful when analyzing a stress condition that is known to trigger stress granule formation. Given the association of translation repression and stress granules, effects on stress granule dynamics can be evidence of translational changes, changes in stress granules themselves, or both. It should be noted that rapamycin treatment is not known to induce stress granule formation, although many other stresses, such as glucose deprivation and oxidative stress, are great candidates for this type of analysis. In brief, stress granules can be assessed by growing *S. cerevisiae* cells (or others) containing fluorescent stress granule markers to mid-log phase. The most common of these markers is *PAB1* fused to *GFP* or *mCherry* (Buchan, Kolaitis, Taylor, & Parker, 2013). At this point, depending on the type of stress being analyzed, either an addition of drug or a change of media needs to be completed. For most assays there is a time requirement to induce stress granules and this time can vary between types of stress from a few minutes to several hours. Cells can then be collected and concentrated for evaluation using a fluorescence microscope over the appropriate time course. Added attention should be given to the concentration portion of the sample preparation since centrifugation has been shown to induce stress phenotypes independent of initial cell conditions. For more detailed protocols, there are several available (e.g., Aulas et al., 2017; Buchan, Nissan, & Parker, 2010), but in essence the user should follow basic fluorescence microscopy protocols.

Another generalized approach to analyzing phenotypes of stress is to evaluate changes in protein levels. As cells are subjected to stress conditions, particularly for prolonged periods, subsets of proteins, including ribosomes and certain translation factors, begin to be degraded, possibly as a scavenging response in scarce resources (Berset et al., 1998; Kelly & Bedwell, 2015; Kraft, Deplazes, Sohrmann, & Peter, 2008). These degradation mechanisms can be studied through examining protein levels during stress, including in cells with mutations in the factors that promote degradation. Using basic methods of cell collection, cell lysis, and western blotting for specific factors an apparent protein "half-life" in stress can be determined. For example, the apparent half-life of eIF4G1 in rapamycin-treated yeast cells was about 1 h in our hands, but it increased over twofold in cells with mutations in *DED1*, which helps control its degradation (Aryanpur et al., 2019). Understanding how the half-life of proteins change over the course of stress can elucidate basic functions of those proteins and even help to clarify whether those proteins have an active, passive, or non-existent role in the stress response.

These are just two of the many assays that can be used, alongside the more detailed protocols provided, to evaluate stress in cells. It is highly recommended that when evaluating a DEAD-box helicase or other protein for a role in stress responses, effects on basic functions are considered before more complex methods are undertaken. This will help ensure that adequate information is known about the protein before committing to time- and labor-intensive assays.

6. Summary and conclusion

As discussed above, gene expression processes are not static, but rather they flexibly respond to changes in cellular conditions, particularly during stresses like a lack of nutrients or extreme environmental conditions. Translational control is critical during stress both because of the immense energy requirements of translation and because it allows relatively quick alterations to the proteome. Furthermore, given their ability to modulate RNA-RNA and RNA-protein interactions, RNA helicases are a natural regulatory node in these conditions. Described above are two different complementary approaches to addressing helicase (or any protein) function in translation during stress. First, growth assays (Section 3) are a straightforward way to associate a given gene with a stress function. Once growth phenotypes have been identified under stress, more specific assays can be designed. These could include examining not only different stress conditions, but also address whether the phenotypes are due to effects on cell growth, survival, and/or recovery during stress. However, it is difficult to conclude that a factor is specifically affecting translation from growth assays alone. In contrast, translation assays (Section 4) carried out in stress conditions can address whether the translational stress response is affected. It should be noted, however, that it is possible that effects in other processes (e.g., transcription) can have indirect effects on translation. More specific translation assays (e.g., reporters that test a particular aspect of translation) and/or analysis of association with translation complexes can provide additional confidence that the observed translation phenotypes are, in fact, direct. Furthermore, assays that examine a detailed mechanism can also add to our understanding of the complex regulation of the translational response to cellular stress.

Acknowledgments

The authors would like to thank Michael Trakselis, Frederico Paulo Mendoza, and other editors at *Methods in Enzymology* for this opportunity. We would also like to thank current and former members of the Bolger lab for helpful advice and discussion. This work was partly supported by a grant from the National Institute of General Medical Sciences (R01-GM136827).

References

Advani, V. M., & Ivanov, P. (2019). Translational control under stress: Reshaping the Translatome. *BioEssays*, *41*(5), e1900009.

Albig, A. R., & Decker, C. J. (2001). The target of rapamycin signaling pathway regulates mRNA turnover in the yeast Saccharomyces cerevisiae. *Molecular Biology of the Cell*, *12*(11), 3428–3438.

Aryanpur, P. P., Mittelmeier, T. M., & Bolger, T. A. (2022). The RNA helicase Ded1 regulates translation and granule formation during multiple phases of cellular stress responses. *Molecular and Cellular Biology, 42*(1), e0024421.

Aryanpur, P. P., Renner, D. M., Rodela, E., Mittelmeier, T. M., Byrd, A., & Bolger, T. A. (2019). The DEAD-box RNA helicase Ded1 has a role in the translational response to TORC1 inhibition. *Molecular Biology of the Cell, 30*(17), 2171–2184.

Aulas, A., Fay, M. M., Szaflarski, W., Kedersha, N., Anderson, P., & Ivanov, P. (2017). Methods to classify cytoplasmic foci as mammalian stress granules. *Journal of Visualized Experiments, 123.*

Balagopal, V., & Parker, R. (2011). Stm1 modulates translation after 80S formation in Saccharomyces cerevisiae. *RNA, 17*(5), 835–842.

Beck, Z. T., Cloutier, S. C., Schipma, M. J., Petell, C. J., Ma, W. K., & Tran, E. J. (2014). Regulation of glucose-dependent gene expression by the RNA helicase Dbp2 in Saccharomyces cerevisiae. *Genetics, 198*(3), 1001–1014.

Beckham, C., Hilliker, A., Cziko, A. M., Noueiry, A., Ramaswami, M., & Parker, R. (2008). The DEAD-box RNA helicase Ded1p affects and accumulates in Saccharomyces cerevisiae P-bodies. *Molecular Biology of the Cell, 19*(3), 984–993.

Berset, C., Trachsel, H., & Altmann, M. (1998). The TOR (target of rapamycin) signal transduction pathway regulates the stability of translation initiation factor eIF4G in the yeast Saccharomyces cerevisiae. *Proceedings of the National Academy of Sciences of the United States of America, 95*(8), 4264–4269.

Bolger, T. A., & Wente, S. R. (2011). Gle1 is a multifunctional DEAD-box protein regulator that modulates Ded1 in translation initiation. *The Journal of Biological Chemistry, 286*(46), 39750–39759.

Brown, N. P., Vergara, A. M., Whelan, A. B., Guerra, P., & Bolger, T. A. (2021). Medulloblastoma-associated mutations in the DEAD-box RNA helicase DDX3X/DED1 cause specific defects in translation. *The Journal of Biological Chemistry, 100296.*

Buchan, J. R., Kolaitis, R. M., Taylor, J. P., & Parker, R. (2013). Eukaryotic stress granules are cleared by autophagy and Cdc48/VCP function. *Cell, 153*(7), 1461–1474.

Buchan, J. R., Nissan, T., & Parker, R. (2010). Analyzing P-bodies and stress granules in *Saccharomyces cerevisiae. Methods in Enzymology, 470,* 619–640.

Buchan, J. R., & Parker, R. (2009). Eukaryotic stress granules: The ins and outs of translation. *Molecular Cell, 36*(6), 932–941.

Carroll, J. S., Munchel, S. E., & Weis, K. (2011). The DExD/H box ATPase Dhh1 functions in translational repression, mRNA decay, and processing body dynamics. *The Journal of Cell Biology, 194*(4), 527–537.

Chuang, R. Y., Weaver, P. L., Liu, Z., & Chang, T. H. (1997). Requirement of the DEAD-box protein Ded1p for messenger RNA translation. *Science, 275*(5305), 1468–1471.

Costa-Mattioli, M., & Walter, P. (2020). The integrated stress response: From mechanism to disease. *Science, 368*(6489).

Crawford, R. A., & Pavitt, G. D. (2019). Translational regulation in response to stress in Saccharomyces cerevisiae. *Yeast, 36*(1), 5–21.

Gerashchenko, M. V., Lobanov, A. V., & Gladyshev, V. N. (2012). Genome-wide ribosome profiling reveals complex translational regulation in response to oxidative stress. *Proceedings of the National Academy of Sciences of the United States of America, 109*(43), 17394–17399.

Gonskikh, Y., & Polacek, N. (2017). Alterations of the translation apparatus during aging and stress response. *Mechanisms of Ageing and Development, 168,* 30–36.

Gonzalez, S., & Rallis, C. (2017). The TOR signaling pathway in spatial and temporal control of cell size and growth. *Frontiers in Cell and Development Biology, 5,* 61.

Guenther, U. P., Weinberg, D. E., Zubradt, M. M., Tedeschi, F. A., Stawicki, B. N., Zagore, L. L., et al. (2018). The helicase Ded1p controls use of near-cognate translation initiation codons in 5' UTRs. *Nature*, *559*(7712), 130–134.

Hilliker, A., Gao, Z., Jankowsky, E., & Parker, R. (2011). The DEAD-box protein Ded1 modulates translation by the formation and resolution of an eIF4F-mRNA complex. *Molecular Cell*, *43*(6), 962–972.

Ingolia, N. T., Ghaemmaghami, S., Newman, J. R., & Weissman, J. S. (2009). Genome-wide analysis in vivo of translation with nucleotide resolution using ribosome profiling. *Science*, *324*(5924), 218–223.

Iserman, C., Desroches Altamirano, C., Jegers, C., Friedrich, U., Zarin, T., Fritsch, A. W., et al. (2020). Condensation of Ded1p promotes a translational switch from housekeeping to stress protein production. *Cell*, *181*(4), 818–831 e819.

Kelly, S. P., & Bedwell, D. M. (2015). Both the autophagy and proteasomal pathways facilitate the Ubp3p-dependent depletion of a subset of translation and RNA turnover factors during nitrogen starvation in Saccharomyces cerevisiae. *RNA*, *21*(5), 898–910.

Kraft, C., Deplazes, A., Sohrmann, M., & Peter, M. (2008). Mature ribosomes are selectively degraded upon starvation by an autophagy pathway requiring the Ubp3p/Bre5p ubiquitin protease. *Nature Cell Biology*, *10*(5), 602–610.

Liu, Y., Beyer, A., & Aebersold, R. (2016). On the dependency of cellular protein levels on mRNA abundance. *Cell*, *165*(3), 535–550.

Liu, B., & Qian, S. B. (2014). Translational reprogramming in cellular stress response. *Wiley Interdisciplinary Reviews: RNA*, *5*(3), 301–315.

McGlincy, N. J., & Ingolia, N. T. (2017). Transcriptome-wide measurement of translation by ribosome profiling. *Methods*, *126*, 112–129.

Mitchell, S. F., & Parker, R. (2014). Principles and properties of eukaryotic mRNPs. *Molecular Cell*, *54*(4), 547–558.

Powers, E. N., & Brar, G. A. (2021). Performing ribosome profiling to assess translation in vegetative and meiotic yeast cells. *Methods in Molecular Biology*, *2252*, 89–125.

Protter, D. S. W., & Parker, R. (2016). Principles and properties of stress granules. *Trends in Cell Biology*, *26*(9), 668–679.

Robichaud, N., Sonenberg, N., Ruggero, D., & Schneider, R. J. (2019). Translational control in Cancer. *Cold Spring Harbor Perspectives in Biology*, *11*(7).

Roux, P. P., & Topisirovic, I. (2018). Signaling pathways involved in the regulation of mRNA translation. *Molecular and Cellular Biology*, *38*(12).

Saavedra, C., Tung, K. S., Amberg, D. C., Hopper, A. K., & Cole, C. N. (1996). Regulation of mRNA export in response to stress in Saccharomyces cerevisiae. *Genes & Development*, *10*(13), 1608–1620.

Saxton, R. A., & Sabatini, D. M. (2017). mTOR signaling in growth, metabolism, and disease. *Cell*, *168*(6), 960–976.

Sen, N. D., Zhou, F., Ingolia, N. T., & Hinnebusch, A. G. (2015). Genome-wide analysis of translational efficiency reveals distinct but overlapping functions of yeast DEAD-box RNA helicases Ded1 and eIF4A. *Genome Research*, *25*(8), 1196–1205.

Seok, H. Y., Nguyen, L. V., Van Nguyen, D., Lee, S. Y., & Moon, Y. H. (2020). Investigation of a novel salt stress-responsive pathway mediated by Arabidopsis DEAD-box RNA helicase gene AtRH17 using RNA-Seq analysis. *International Journal of Molecular Sciences*, *21*(5).

Shaffer, D., & Rollins, J. A. (2020). Fluorescent Polysome profiling in Caenorhabditis elegans. *Bio-Protocol*, *10*(17), e3742.

Shenton, D., Smirnova, J. B., Selley, J. N., Carroll, K., Hubbard, S. J., Pavitt, G. D., et al. (2006). Global translational responses to oxidative stress impact upon multiple levels of protein synthesis. *The Journal of Biological Chemistry*, *281*(39), 29011–29021.

Spriggs, K. A., Bushell, M., & Willis, A. E. (2010). Translational regulation of gene expression during conditions of cell stress. *Molecular Cell*, *40*(2), 228–237.

Takahara, T., & Maeda, T. (2012). Transient sequestration of TORC1 into stress granules during heat stress. *Molecular Cell*, *47*(2), 242–252.

Tauber, D., Tauber, G., Khong, A., Van Treeck, B., Pelletier, J., & Parker, R. (2020). Modulation of RNA condensation by the DEAD-box protein eIF4A. *Cell*, *180*(3), 411–426 e416.

Thomas, B. J., & Rothstein, R. (1989). The genetic control of direct-repeat recombination in Saccharomyces: The effect of rad52 and rad1 on mitotic recombination at GAL10, a transcriptionally regulated gene. *Genetics*, *123*(4), 725–738.

Tsokanos, F. F., Albert, M. A., Demetriades, C., Spirohn, K., Boutros, M., & Teleman, A. A. (2016). eIF4A inactivates TORC1 in response to amino acid starvation. *The EMBO Journal*, *35*(10), 1058–1076.

Valasek, L., Szamecz, B., Hinnebusch, A. G., & Nielsen, K. H. (2007). In vivo stabilization of preinitiation complexes by formaldehyde cross-linking. *Methods in Enzymology*, *429*, 163–183.

Valentini, M., & Linder, P. (2021). Happy birthday: 30 years of RNA helicases. *Methods in Molecular Biology*, *2209*, 17–34.

Wu, C., Amrani, N., Jacobson, A., & Sachs, M. S. (2007). The use of fungal in vitro systems for studying translational regulation. *Methods in Enzymology*, *429*, 203–225.

Yaffe, M. P., & Schatz, G. (1984). Two nuclear mutations that block mitochondrial protein import in yeast. *Proceedings of the National Academy of Sciences of the United States of America*, *81*(15), 4819–4823.

Yang, H. S., Jansen, A. P., Komar, A. A., Zheng, X., Merrick, W. C., Costes, S., et al. (2003). The transformation suppressor Pdcd4 is a novel eukaryotic translation initiation factor 4A binding protein that inhibits translation. *Molecular and Cellular Biology*, *23*(1), 26–37.

Zwietering, M. H., Jongenburger, I., Rombouts, F. M., & van't Riet, K. (1990). Modeling of the bacterial growth curve. *Applied and Environmental Microbiology*, *56*(6), 1875–1881.

CHAPTER SEVEN

Lysate and cell-based assays to probe the translational role of RNA helicases

Kevin C. Wilkins[a,b], Srivats Venkataramanan[a], and Stephen N. Floor[a,c,*]

[a]Department of Cell and Tissue Biology, University of California, San Francisco, CA, United States
[b]Graduate Division, University of California, San Francisco, San Francisco, CA, United States
[c]Helen Diller Family Comprehensive Cancer Center, University of California, San Francisco, San Francisco, CA, United States
*Corresponding author: e-mail address: stephen@floorlab.org

Contents

Methods in Enzymology, Volume 673
ISSN 0076-6879
https://doi.org/10.1016/bs.mie.2022.03.032

Abstract

Translation initiation is the first step in protein synthesis, during which the small subunit of the ribosome scans the 5′ untranslated region (5′UTR) of an mRNA to identify a start codon and commence translation elongation. By unwinding and modulating secondary structures and other RNA features present in the 5′UTR, RNA helicases can regulate ribosome scanning and start codon selection. This chapter presents an approach to measure the effect of RNA helicases on mRNA translation initiation. 5′UTR luciferase reporters are transcribed *in vitro* and employed in either of two assays. The *in vitro* assay translates the reporters in a cell-free whole-cell lysate system, which allows for greater biochemical manipulation and tighter control over confounding effects. In the alternative cell-based approach, the reporters are transfected and translated in living cells, which provides a more physiological setup. Either method can be used to investigate how the perturbation of a helicase, such as changes in protein levels or mutations, affects translation initiation at the 5′UTR level. The chapter also discusses alternative approaches, troubleshooting, and further applications of these methods. These assays will provide insights on the role of helicases and other translational factors as regulators of the proteome both in physiological and diseased settings.

1. Introduction

1.1 RNA helicases and translation

The translation of messenger RNA (mRNA) into protein is a fundamental and tightly regulated process. Canonical cap-dependent translation starts with the recruitment of the 40S ribosome subunit and associated factors, collectively known as the 43S pre-initiation complex (PIC), to the 5′-cap of an mRNA. Recruitment is mediated by cap-binding protein eIF4E that, together with scaffolding protein eIF4G and RNA helicases, comprises the eIF4F complex. Once bound near the 5′-cap, the 43S PIC travels along the 5′ untranslated region (5′UTR) in the 5′–3′ direction until it reaches a start codon where it is joined by the 60S large ribosomal subunit and translation elongation commences. Known as ribosome scanning, the process of the ribosome moving along a 5′UTRis intensely regulated.

The 5′UTR contains features that regulate protein production from the open reading frame (ORF) by altering the rate of ribosome scanning and start codon selection (Hinnebusch, Ivanov, & Sonenberg, 2016; Leppek, Das, & Barna, 2018). Hairpins, RNA secondary structures, G-quadruplexes, pseudoknots, and other features can alter translation by slowing down ribosomes, preventing them from reaching the main start codon altogether, or promoting use of an alternative start codon (Hinnebusch et al., 2016; Leppek et al., 2018). Upstream open reading frames (uORFs) result from initiation at start codons before the main ORF start codon. Upstream initiation generally decreases but, more infrequently, can also increase protein production at the main ORF depending on the arrangement of the start codons and reading frames (Calvo, Pagliarini, & Mootha, 2009; Hinnebusch et al., 2016). The strength of the Kozak sequence environment around the start codon, both of the main ORF and of the uORFs, can drastically alter their usage (Kozak, 2002). Secondary structures can stimulate usage of a uORF through proximity, which causes ribosomes to dwell longer on a start codon and initiate elongation (Kozak, 1990). Other regulatory features in 5′UTRs include RNA modifications, RNA binding protein (RBPs) binding sites, and long non-coding RNAs (lncRNAs), which can form ribonucleoprotein complexes or interact with other RNA molecules (Leppek et al., 2018). Understanding the complex 5′UTR regulatory network is essential for a complete picture of how protein production is regulated.

One class of proteins that have an important role in modulating and interacting with 5′UTR regulatory features are RNA helicases. The family of RNA helicases contains multiple subfamilies, some of which function as processive helicases and others function as distributive RNA chaperones (Fairman-Williams, Guenther, & Jankowsky, 2010); we refer to this entire group as RNA helicases in this method. eIF4A, a member of the eIF4F complex and one of the canonical initiation factors, is a crucial DEAD-box RNA helicase whose role is to unwind secondary structures in 5′UTR to allow for ribosome progression (Iwasaki, Floor, & Ingolia, 2016; Leppek et al., 2018; Pestova & Kolupaeva, 2002). The related DEAD-box RNA helicase DDX3X is similarly implicated in resolving secondary structures ahead of ribosome scanning (Calviello et al., 2021; Nashchekin, Zhao, Visa, & Daneholt, 2006; Sharma & Jankowsky, 2014; Soto-Rifo et al., 2012). The DExH-box RNA helicase DHX29 has also been shown to be required for proper translation of highly structured 5′UTRs (Parsyan et al., 2009; Pisareva, Pisarev, Komar, Hellen, & Pestova, 2008). Whether DHX29 acts by unwinding mRNA or by modulating the ribosomal RNA or a combination of both is under active investigation (Sweeney et al., 2021). Another example is provided by DHX36, which has been shown to resolve G-quadruplexes within the 5′UTR in order to promote proper translation (Murat et al., 2018; Sauer et al., 2019). Given the importance of tightly regulating protein synthesis, it is unsurprising that many helicases involved in mRNA translation are also implicated in several diseases, particularly in cancer (Cai et al., 2017; Fuller-Pace, 2013; Hao et al., 2020; van Voss, van Diest, & Raman, 2017). Developing methods to functionally interrogate RNA helicases during protein production is an essential step to advancing the understanding of their role in health and disease.

Characterizing the direct role of individual RNA helicases on mRNA translation in mammalian cells presents challenges. Perturbing a given helicase to observe its role on translation in cells can be difficult, since any secondary effects on translation, transcription, or other RNA processing steps can act as confounding factors. Additionally, targeting a helicase that affects translation can alter the proteome which can indirectly affect protein production. Small molecules can be effective ways to perturb activity (Iwasaki et al., 2016; Naineni et al., 2020), but achieving specificity can be hard since there are many similar RNA helicases (Arrowsmith et al., 2015; Calviello et al., 2021; Floor, Barkovich, Condon, Shokat, & Doudna, 2016). These challenges can complicate studies observing the direct effects of the helicase on the mechanism of translation. Reconstituted systems, particularly in yeast, have been fundamental to avoid these issues and have been an excellent source of mechanistic insight into translation initiation (Acker, Kolitz, Mitchell, Nanda, & Lorsch, 2007; Aitken et al., 2016). An alternative option are mRNA reporter-based *in vitro* translation assays which can measure the effects of perturbing a given helicase on translation. Such assays have also been used for other RNA binding proteins and translation factors, such as eIF3 (Lee, Kranzusch, & Cate, 2015), as well as investigating translation by transcript isoforms (Floor & Doudna, 2016).

In this chapter, we present a protocol to measure the effect of helicases on mRNA translation. We describe how to transcribe mRNA luciferase reporters *in vitro* and how to employ them in both an *in vitro* and an *in cell* \translation assay. These assays can be used to observe the effects of the perturbation of a given helicase on protein production. Observing how mutations of a given helicase, as well as alterations of its protein levels, impact translation initiation on 5′UTRs can provide insight on its function as a regulator or protein levels as well as its malfunction in disease systems. We also outline the advantages and the limitations of the *in vitro* and *in cell* systems compared to *in vivo* investigations. Finally, we discuss troubleshooting, adaptations of protocol for further experiments and alternative strategies.

2. Before you begin

2.1 List of reagents

Reagent	Source	Identifier
2′-*O*-methyltransferase (enzyme, buffer, SAM)	NEB	M0366S
Adenosine 5′-triphosphate disodium salt hydrate	Sigma-Aldrich	A7699
Cytidine 5′-triphosphate disodium salt	Sigma-Aldrich	C1506
Guanosine 5′-triphosphate sodium salt hydrate	Sigma-Aldrich	G8877
Uridine-5′-triphosphate trisodium salt	Sigma-Aldrich	U6750
1 mM amino acids	Promega	L4461
ATP	Sigma-Aldrich	A26209
cOmplete™, Mini, EDTA-free Protease-Inhibitor-Cocktail	Merck	4693159001
Creatine Kinase (CK) from rabbit muscle	Merck	10127566,001
Creatine Phosphate	Merck	10621714001
DTT	Gold Bio	DTT500
EDTA 500 mM, pH 8.0	Teknova	E0308
EtOH	VWR	V1101
GlycoBlue™ Coprecipitant	Ambion/ ThermoFisher	AM9516
Hepes pH 7.6	Gold Bio	H-400
KOAc	Sigma-Aldrich	P5708
$MgCl_2$	Sigma-Aldrich	M2670
$MgOAc_2$	Sigma-Aldrich	M5661
Murine RNAse inhibitor	NEB	M0314S
NaOAc pH 5.2	Fisher Scientific	P171–500
NorthernMax™-Gly Sample Loading Dye	Ambion/ ThermoFisher	AM8551
RNase-free DNase	Promega	M6101

Continued

—cont'd

Reagent	Source	Identifier
RNasin Ribonuclease Inhibitor	Promega	N2611
Spermidine	Sigma-Aldrich	S2626
Superase:IN	Ambion/ ThermoFisher	AM2696
TBE 10 × Concentrate	VWR	E442
T7 polymerase (1 mg/mL)		
Tris-Cl, pH 8.1	Gold Bio	T-095-100
Triton X-100	Sigma-Aldrich	T8787
Vaccinia capping enzyme system (enzyme, buffer, SAM, GTP)	NEB	M2080S
Zymo Clean & Concentrator 25 columns	Zymo Research	D4033
Promega Dual Glo	Promega	E2920
Renilla Luciferase Assay System	Promega	E2820

2.2 Equipment

1. Cell culture equipment and reagents
2. GloMax Explorer plate reader (Promega) or equivalent
3. Thermocycler
4. Water bath (can also use thermocycler or thermomixer)
5. Opaque 96-well plates, sterile (PerkinElmer – 6,005,688 or equivalent)

2.3 RNA reporters

The RNA reporters employed in this assay contain a 5′UTR followed by a Renilla luciferase ORF (Calviello et al., 2021; Floor & Doudna, 2016; Iwasaki et al., 2016). Plasmids and full sequences are available on Addgene (Addgene ID 182638 for the minimal 5′UTR reporter and ID 182639 for an example of an experimental 5′UTR reporter). Other luciferases could also be used in place of Renilla. Transcription is driven by a bacteriophage T7 promoter and terminated *via* run-off transcription following 60 adenosine nucleotides encoded in the template downstream of the stop codon to mimic polyadenylation (Fig. 1A). The 5′UTR of each reporter is chosen based on the biological question of the experiment, for example

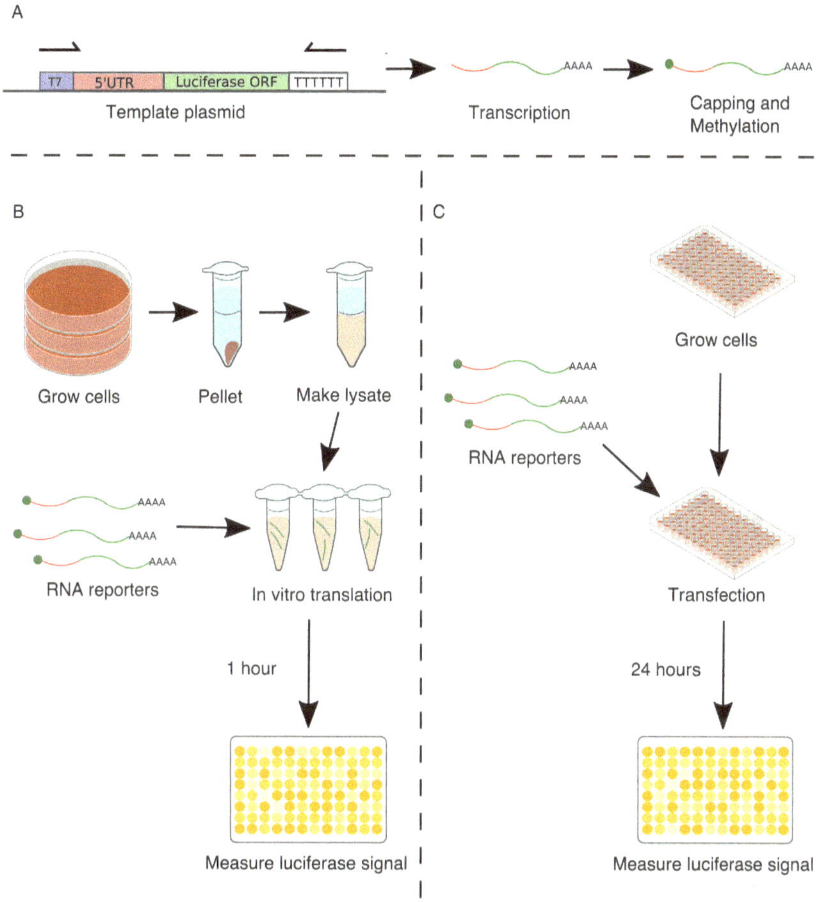

Fig. 1 Schematic of the transcription of the reporters and the *in vitro* and *in vivo* translation assays. (A) *In vitro* transcription of RNA reporters. The template plasmid contains a T7 promoter followed by a 5′UTR (either the control or a 5′UTR of interest), followed by the luciferase open reading frame (ORF) and a string of Adenosines. PCR on this plasmid is used to obtain the template DNA containing the reporter. The DNA is then used to transcribe the RNA reporters. The reporters are then capped and methylated. (B) *In vitro* assay. Cells are grown, pelleted, and lysed to obtain the whole-cell lysate. The lysate is enriched and combined with the mRNA reporters in a 1 h translation reaction. The reactions are moved to an opaque plate and the luciferase signal is measured. (C) *In cell* assay. Cells are grown in an opaque plate. The mRNA reporters are transfected into the cells. After 24 h, the media is removed and the cells are lysed, and the luciferase signal is read on an imager.

5′UTRs that are sensitive to the perturbation of a given helicase or translational factor. In Calviello et al., the 5′UTRs employed were selected among those known to be translationally controlled by DDX3X (Calviello et al., 2021). Additionally, this assay also employs a control reporter, which we refer to here as the minimal 5′UTR reporter. As the name suggests, it possesses a minimal 8 nucleotide 5′UTR. During data analysis, the signal of all other reporters is normalized to the signal of this control reporter to account for variations in the translational activity of the different lysates. This reporter can be modified or adapted for future experiments to obtain a control that is non-sensitive or the least sensitive as possible to the biological perturbation that is investigated.

2.4 Cells and lysates

The *in vitro* assay (step 3.2) measures the luciferase signal from RNA reporters translated for 1 h in an enriched whole-cell lysate reaction (see Fig. 1B). The lysate is obtained by growing and lysing cells and then supplementing it with additional reagents (Lee et al., 2015). This protocol and the papers referenced have been tested and optimized for HEK 293T, HCT-116, and HeLa cells (Calviello et al., 2021; Lee et al., 2015; Venkataramanan, Gadek, Calviello, Wilkins, & Floor, 2021). Other cells may also be employed but may require optimization and may not achieve the desired signal.

The *in cell* assay (step 3.3) measures the luciferase signal from RNA reporters transfected into cells (see Fig. 1C). This assay has been optimized for the same cell lines as step 3.2; using other cell types may require optimization. Because the luciferase measurement protocol requires lysis in a small volume (see step 3.3.1), adherent cells are recommended. Suspension cells may need to be removed from their media for lysis and this may prove unfeasible or compromise the strength of the signal.

Note: The relative advantages and disadvantages of the two assays are discussed in Section 6.

2.5 Perturbation of the helicase

Each experiment should include a control condition and one or more experimental conditions in which the helicase (or the translational factor) of interest is perturbed. For example, the assay can be run with cells expressing wild-type *versus* a mutant helicase, or cells expressing normal protein levels *versus* depletion. Chemical inhibition or inducible degron tags may prove advantageous, when possible, due to the rapid impact on translation before compensatory effects can occur. When using gene knockout or depletion with CRISPRi or siRNA, it is possible that some effects may be secondary in nature.

3. Step-by-step method details

This protocol is composed of three sections. Step 3.1 describes the *in vitro* transcription of mRNA reporters (Rio, 2011). These can be used in either the *in vitro* translation assay (step 3.2) or the *in cell* translation assay (3.3) or other assays (Bai, Zhou, & Doudna, 2013; Calviello et al., 2021). Quantification and statistical analysis for both assays are described in step 4. The advantages and limitations of each assay are described in Section 5.

3.1 *In vitro* transcription, capping, and 2′-O methylation of long RNAs

This protocol is for *in vitro* transcription of long RNAs off plasmid or PCR product templates (see Fig. 1A) (Rio, 2011). The template is purified and the size verified by running an agarose gel, and transcription occurs *via* a T7 promoter with a fixed 3′ polyA tail encoded in the template (typically A60). Following this protocol, the RNA will be ready for transfection into mammalian cells (step 3.3) or *in vitro* translation (step 3.2) (Bai et al., 2013; Calviello et al., 2021). Substantial amounts of template are necessary: at least 1 μg of template per 100 μL reaction; ideally closer to 5 μg of template, especially for very long templates. Use RNase sensitive protocols and reagents for all steps of this procedure.

3.1.1 PCR amplification of template

1. The DNA template is amplified using the following primers:

Primer name	Primer sequence
Reverse template primer	**5′-TTT TTT TTT TTT TTT TTT TTT TTT TTT TTT TTT TTT TTT TTT TTT TTT TTT TTT TTT TTT CTG CAG**
Forward template primer	5′-CGG CCA GTG AAT TCG AGC TCT AAT ACG ACT CAC TAT AGG

Note: The reverse primer ensures 60 thymidine residues are present on the template. This primer is used to amplify any 3′ UTR that has a constant 3′ hexanucleotide region but can be modified for higher efficiency PCR by extending the complementarity region if only one 3′ UTR is being used.

2. Assemble the following PCR reaction. Prepare one reaction (total volume of 100 μL) for each of the RNAs you are transcribing

Reagent	Amount per 100 μL rxn
HF buffer	20 μL
dNTP	2 μL
Forward template primer (100 μM)	0.5 μL
Reverse template primer (100 μM)	0.5 μL
Phusion polymerase	1 μL
Template (50 ng total)	10 μL
Water	66 μL

3. Run PCR with a two-step protocol:
   (1) 98 °C melting for 2 min
   (2) 98 °C melting for 30 s
   (3) 67 °C annealing for 30 s
   (4) 72 °C annealing/extension for 3 min
   (5) Repeat steps 2, 3 and 4 for 30 cycles
   (6) 10 min final extension time

Note: Steps 2 and 3 can be also done with different PCR kits and polymerases. Regardless of which PCR kit is used, step 4 is used to check if you have the right product.

4. Run the PCRs on a native agarose gel, verify the size of the template band(s), and gel extract with a Qiagen DNA gel extraction kit (alternatively, other gel purification kits can be used).

Note: The size of the template PCR product will be the length of the RNA reporter construct including primers. It should be 1098 bp + length of the 5′UTR for the plasmid cited in Section 2.3 but will differ if the luciferase or other transcribed regions change.

5. You have now obtained the DNA template that will be used for RNA transcription

3.1.2 In vitro *transcription of RNA*

Note: This protocol employs purified T7 polymerase at 1 mg/mL (Rio, 2013). It is possible to transcribe the RNA by using alternative kits, such as the MEGAscript™ T7 Transcription Kit (AMB13345) and following manufacturer instructions. If using an alternative kit, assemble transcription and skip to step 5.

1. Assemble a 10 × transcription buffer:

300 mM Tris-Cl, pH 8.1
250 mM MgCl2
0.1% Triton X-100
20 mM spermidine
DTT

2. Assemble ACGU solution by mixing adenosine, cytidine, guanosine, uridine 5′-triphosphate disodium salt hydrates at 25 mM each. **Critical**: Ensure solution of nucleoside triphosphates is adjusted to pH 7

Note: The solution can be stored long term at −20 °C.

3. Warm all reagents for step 4 except T7 polymerase and Superase:IN to room temperature

Note: Adding cold spermidine to template can cause precipitation of the template.

4. Assemble the transcription reaction by mixing the following in a tube at room temperature. If you are transcribing more than one RNA, assemble a master mix, split it into different tubes, and add the templates separately

Note: This reaction can be run in either a 1.5 mL tube or a PCR tube.

Reagent	100 μL reaction (μL each)
Template (from step 3.1.1)	1 to 5 g
ACGU (25 mM)	30 μL
10 × txn buffer (RT)	10 μL
DTT (1 M)	2 μL
MgCl2 (1 M)	1 μL
T7 polymerase (1 mg/mL)	10 μL
Superase:IN	1 μL
RNase-free H2O	Up to 100

5. Incubate at 37 °C in a stable incubator—a water bath, stable dry incubator, or PCR thermocycler. Warm rooms and heat blocks can have higher variability in temperature that may decrease reproducibility
6. Wait 1 hour and check the reactions—they should be somewhat cloudy from pyrophosphate precipitation. If they are, add 1 µL of 1 mg/mL pyrophosphatase to each

   Note: Reactions that are not cloudy might also have worked, although it is less likely. You can proceed with the protocol and check if non-cloudy reactions were successful with a gel in step 3.1.4.
7. Allow the reactions to proceed for 3–4 hours
8. Add 1 µL of RQ1 RNase-free DNase to each reaction and incubate at 37 °C for 30 min to digest the template
9. Precipitate transcription products by adding 200 µL 0.3 M NaOAc pH 5.2, 1 µL GlycoBlue and 750 µL 100% EtOH to each reaction

   Note: If the reactions were run in a 1.5 mL tube, add these directly to the tube. If the reactions were run in a smaller tube, add these to a new 1.5 mL tube and then transfer the reactions into the tube.
10. Incubate at −80 °C for >1 hour to overnight
11. **Pause point**: The samples can be kept at stored at −80 °C for months

3.1.3 Resuspending EtOH precipitated RNA

1. Pellet precipitations by centrifugation at 16,000 × *g* at 4 °C for 25 min
2. Remove supernatant with a pipette and discard
3. Wash with 500 µL −20 °C 70% EtOH
4. Centrifuge at 16,000 × *g* at 4 °C for 5 min
5. Remove supernatant with a pipette and discard
6. Quick-spin and remove residual ethanol with a P20 pipet
7. Allow pellet to air dry for 5 min

   Note: Do not overdry as this can impact recovery.
8. Resuspend pellet in 100 µL 50 mM EDTA
9. Purify RNA with Zymo RNA Clean & Concentrator 25 column following the small RNA removal protocol and elute in 30 µL RNase-free H_2O
10. Measure concentration of RNA on a nanodrop or Qubit or other assay

3.1.4 Quality control: Glyoxal gel

1. RNA molecules will fold intramolecularly and may not run according to their sequence length in a native gel. Chemical modification of adenine, cytidine, and guanosine bases by glyoxal reduces intramolecular

secondary structure, promoting size-dependent migration of RNA molecules without using formamide or other denaturing gels

2. Add 2 μL of the NorthernMax™-Gly Sample Loading Dye to 2 μL of each RNA. Add 5 μL of the dye to 5 μL of the RNA ladder. Incubate reactions for 30 min at 50 °C (Rio, 2015).
3. Run samples on an 1% agarose gel. Make the gel and run it in 0.5 × TBE buffer. **Critical**: Ensure the gel and the running buffer are made from RNase free water to prevent RNase contamination
4. Check size of the products
5. If degradation products are apparent on the agarose gel, use image quantification software to quantify the band corresponding to full-length product and use this to determine concentration in combination with nanodrop readings
6. Alternatively, if significant truncated products are present purify the full-length product by agarose gel purification using the Zymo Agarose Gel Zymoclean kit (R1011).

3.1.5 Capping and 2′-O-methylation

1. Dilute 20 μg of RNA in 26 μL RNase-free H_2O in a PCR tube
2. Prepare a 4 mM stock of SAM (supplied with Vaccinia kit) in RNase-free H_2O from the supplied 32 mM stock

   Note: SAM is labile and should not be freeze-thawed repeatedly. If reusing SAM supplied in a kit, aliquot on first thaw to avoid freeze-thaws. Degraded SAM will interfere with proper mRNA capping and reduce translation efficiency.
3. Assemble a master mix according to the following recipe

Reagent	Per 40 μL reaction
10 × NEB capping buffer	4 μL
10 mM GTP	2 μL
4 mM SAM	2 μL
NEB Vaccinia capping enzyme	2 μL
NEB 2′-O-Me transferase	2 μL
RNasin	2 μL

4. Add 14 μL of the master mix to each PCR tube from step 1

5. Incubate at 37 °C for 60 min
6. Purify over a Zymo Clean and Concentrator 25 column following the small RNA removal protocol and elute in 30 μL RNase-free H_2O
7. **Optional:** Run a second glyoxal gel as in step 3.1.4 to verify your RNAs are the correct size

3.2 *In vitro* translation assay

The RNA reporters transcribed in Step 3.1 can be used in an *in vitro* translation assay. In this assay, the transcribed reporters are translated for 1 h in whole-cell lysate. The luciferase produced is then measured (see Fig. 1B). The whole-cell lysate, while not perfectly reproducing the biology of a living cell, presents the opportunity to easily manipulate the reaction by adding or removing chemical compounds, proteins, or other factors. In addition, the optimized *in vitro* reaction provides a short timeframe (which reduces unwanted secondary effects) and a high translation signal.

3.2.1 *Materials*

Lysis buffer

10 mM Hepes, pH 7.6
10 mM KOAc
0.5 mM MgOAc2
5 mM DTT
Complete EDTA-free protease inhibitor (1 tablet per 10 mL buffer)
(make in RNase-free water and keep on ice)

Note: The buffer can be stored at 4 °C, but the protease inhibitor must be added fresh.

3.2.2 *Making the lysate*

1. Grow cells in 15 cm plates, making 5–10 plates at a time

   Note: Do not let cells reach high confluency levels since this inhibits translation. Actively replicating cells are best suited for this protocol.
2. Treat your cells to obtain cells with the desired helicase perturbation and the appropriate control cells (as discussed in Section 2.4).

Note: For example, cells expressing WT or mutant helicase can be used to create a WT and mutant lysates. Alternatively, cells expressing normal levels and knock-out for the helicase of interest can be used to obtain lysates. See Section 2.5.

3. Trypsinize the cells by adding 4 mL trypsin to each plate and stop by adding 4 mL complete media and move into a conical tube
4. Spin 1 min at 1000 × *g*, 4 °C to pellet
5. Wash the pellet with PBS and transfer to a 1.5 mL microcentrifuge tube
6. Spin down the pellet and remove the PBS
7. Add equal volume of Lysis Buffer to the pellet

   Note: The volume of Lysis Buffer added needs to match the pellet approximately 1:1. To obtain this, check the volume of the pellet at the bottom of the 1.5 mL tube by eye and add the equal amount of Lysis Buffer. To ensure all lysates are made the same way, you can weigh all the pellets first. Add the Lysis Buffer to the first tube, and then adjust the amount of Lysis Buffer added to the other tubes in proportion to the weight.
8. Dissolve the pellet by pipetting up and down, do not vortex
9. Incubate the tubes on ice for 45 min. This induces hypotonic swelling in the cells
10. Homogenize the cells. This is done by aspirating slowly the lysate into a 1 mL syringe with 26G (subcutaneous) needle. Then, push the lysate out quickly with the needle against the side of the tube. Minimize bubbles. Repeat 13 times

    Note: Squirting the cells through the needle and against the tube helps completely break down the cell membranes, which were previously made fragile by the hypotonic swelling.
11. Check under light microscope for disruption with trypan blue (blue nuclei signify lysis). Disrupt until ~95% is done
12. Centrifuge the tubes at 14,000 × *g* for 1 min, 4 °C to remove cell debris and nuclei
13. Collect the supernatant. These are the lysates for the assay

    Note: any antibody depletion or other biochemical manipulations to the lysate can be performed at this step.
14. Aliquot and flash freeze the lysate in liquid nitrogen. Put at −80 °C

    Note: An aliquot of the lysate (diluted into lysate buffer 10:1) can be run on a western blot.

    Pause point: The lysates can be stored at −80 °C.

3.2.3 Running the assay

1. Each RNA assay will be run as three technical replicates; the number of total reactions will be the number of RNAs times three

   Note: If you do more than eight total replicates, it is easiest to run the assay in a 96-well PCR plate, and then image it on a 96-well opaque flat-bottom plate.
2. For each RNA, mix in a tube:

Reagent	1 replicate
100 mM ATP	0.042 μL
10 mM GTP	0.105 μL
1 M Creatine Phosphate	0.105 μL
7.5 U/μL Creatine phosphokinase	0.03 μL
1 M Hepes pH 7.5	0.05 μL
0.1 M DTT	0.1 μL
40 mM MgOAc	0.25 μL
2 mM KOAc	0.25 μL
1 mM amino acids	0.04 μL
Spermidine (0.02295 M)	0.056 μL
Murine RNAse inhibitor	0.125 μL
Lysate	2.5 μL
(Usually 100-200 ng) RNA	1.348 μL

Note: The concentration of MgOAc and KOAc may require optimization based on the control 5′UTR and/or the cell type used. It is recommended to test ranges of 25–150 mM for the MgOAc and ranges 1–3 mM for the KOAc.

Note: To minimize variability, we recommend making a 3X master mix for each RNA construct, and then splitting into three replicates.

Note: The concentration of RNA needed to obtain enough signal will vary by 5′UTR construct since different 5′UTR have different translational activity. Optimization might be required for each construct. Initial values can range from 20 to 200 ng total RNA per replicate.

3. Incubate for 1 hour at 30 °C (PCR Block).
4. Transfer the content of the 96-well PCR plate into a 96-well opaque flat-bottom plate

   Note: Make sure the plate is opaque and compatible with luciferase luminescence measurement.
5. Image the luciferase signal

   Note: Luciferase measurement protocols will vary based on the kit and apparatus used. Example: For the Renilla Luciferase Assay System (Promega) measured on a GloMax Explorer luminometer (Promega) Steps: (1) Transfer each reaction to a well in a 96-well plate, (2) add lysis buffer to each well, (3) leave 15 min on rocker and (4) place in machine and follow machine measurement protocol.

3.3 *In cell* translation assay

The RNA reporters transcribed in Section 3.1 can also be used for an *in cell* assay. In this assay, transcribed reporters are transfected into living cells that produce luciferase protein for 24 h, which is then measured (see Fig. 1C). The longer incubation time for the translation step compared to the *in vitro* assay in step 3.2 may leave more time for non-translation related confounding effects to take place. On the other hand, it provides a more physiologically relevant system since it takes place in living cells instead of lysate.

3.3.1 *Grow cells and transfect*

1. Grow cells in a sterile 96-well opaque flat-bottom plate

   Critical: This plate will be used for the whole experiment, including growing cells, transfecting them with reporter RNAs, and measuring the luciferase signal. Make sure the plate is compatible with the luminometer (in particular, the plate must be opaque to prevent signal leaking between wells).

   Note: Since the plate is opaque, it is not be possible to observe the cells under a microscope during the experiment. To obviate this issue, we recommend plating a "mirror" 96-well transparent culture plate in parallel with the same number of cells and transfecting them in the same manner as in the experimental plate. These cells will be able to be observed and will mirror the conditions of the cells used for the experiment. Alternatively, cells can be grown and transfected in a transparent culture plate. Cells will then be lysed, and the lysate moved to the opaque plate for luciferase measurement.

2. Plate cells at ~80% confluency

   Note: Confluency and plating timeline may need to be optimized based on cell type.
3. After 48 hours, transfect cells with a mixture of 10 ng *in vitro* transcribed firefly luciferase (Fluc) RNA with a minimal 5′UTR and between 50 and 200 ng renilla luciferase RNA (Rluc) downstream of selected 5′ UTRs using the TransIT mRNA Transfection Kit (Mirus Bio) per manufacturer's instructions
4. Lyse cells 24 hours post transfection
5. Measure luciferase values with the Dual-Luciferase Reporter Assay System (Promega) on a GloMax Explorer luminometer (Promega) or similar instrument, following manufacturer's instructions

4. Quantification and statistical analysis

4.1 For the *in vitro* translation assay (Section 3.2)

1. The data obtained from the plate reader will consist of a luminescence value for each replicate

Note: These values are dimensionless (see Table 1).

Table 1 Quantification and statistical analysis.

	Lysate - control						Lysate – experimental condition 1					
	Reporter: control			Reporter 1			Reporter: control			Reporter 1		
Replicate:	1	2	3	1	2	3	1	2	3	1	2	3
Raw values	225000	230000	250000	380000	400000	450000	470000	505000	550000	320000	440000	450000
	average is 235000						*average is 508333*					
Step 1	*divide by 235000*			*divide by 235000*			*divide by 508333*			*divide by 508333*		
	0.957	0.979	1.064	1.617	1.702	1.915	0.925	0.993	1.082	0.630	0.866	0.885
	average is 1			*average is 1.7446*								
Step 2	*divide by 1*			*divide by1.7446*			*divide by 1*			*divide by 1.7446*		
Final values	0.957	0.979	1.064	0.927	0.976	1.098	0.925	0.993	1.082	0.361	0.496	0.507
	average is 1			*average is 1*			*average is 1*			*average is 0.4547*		

The raw data obtained from the plate reader will consist of numerical luciferase luminescence value for each technical replicate. The averages that need to be calculated for step 1 are shown. (Step 1) The raw value of each replicate is divided it by the average value of the replicates obtained from the same lysate as with the control RNA reporter. This normalizes each replicate's value to the baseline translational activity of its lysate. The averages that need to be calculated for step 2 are shown. (Step 2) The intermediate resulting value for each replicate was divided by the average value of all replicates obtained with the same RNA reporter for the control lysate. This step will return the signal obtained for each reporter relative to the same reporter in the control lysate.

2. Inspect the luminescence values to ensure there is sufficient signal to analyze the results. Typical successful experiments have ranged from luminescence values in the 10^3 to 10^8 range. Observe if there are large differences between neighboring wells – note that luminescence will bleed from high signal wells into lower signal wells. Multiple order-of-magnitude signal differences should be avoided between neighboring wells
3. Divide the Renilla luminescence value of each replicate by the average value of the replicates obtained from the same lysates as that well and the control RNA reporter. This step normalizes the values of all RNA reporters to that of the control RNA reporter translated with the same lysate. This returns values for each reporter and condition that are normalized to the control and helps account for differences in translational activity of each lysate (see step 1 in Table 1).
4. Secondly, divide the resulting value for each replicate by the average value of all replicates for the same RNA reporter for the control lysate. This step will return the relative signal obtained for each reporter with that of the same reporter in the control lysate

Note: The values for all reporters in the control lysates will be equal to 1, while the values for the reporters with the experimental lysates will represent level of translation relative to the control (see step 2 in Table 1).

5. Following the above steps, the average value of all replicates of mRNA reporters in control lysates are normalized to 1. The impact of the perturbation of an RNA helicase on translation of a given mRNA is the difference between the experimental lysate *versus* the control

Note: The values for the control mRNA reporters will average to 1 across all lysates, since they are the normalization factor. This reflects the in-built assumption of this assay that the control reporter is insensitive to perturbations of the helicase. This assumption can be tested using an orthogonal luciferase, but there may be perturbations that impact all translation.

6. Statistical analysis can be run for each reporter to compare the values of the replicates in each lysate relative to the other lysates. For the example provided, statistical significance between conditions was calculated with an unpaired two-tailed t-test
7. Finally, the data can be visualized, for example using a box plot with data points (see Fig. 2).

4.2 For the *in cell* translation assay (Section 3.3)

1. Normalize the Renilla luciferase signal in each well by dividing it by the Firefly luciferase signal from the same well. This is done to control for cell number, transfection efficiency, and basal translation levels of each well

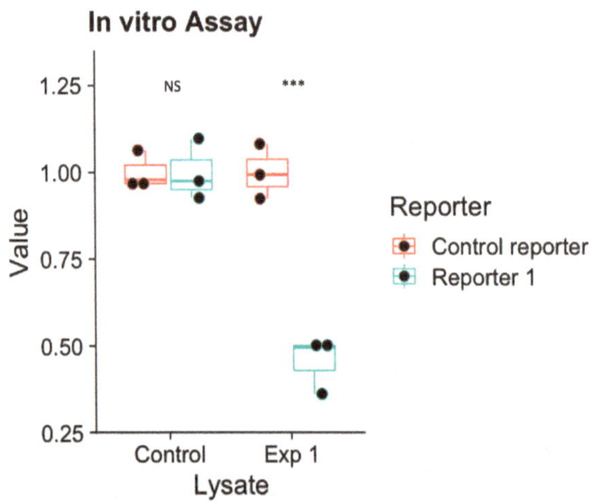

Fig. 2 Data visualization for the translation assay. The final values data obtained from normalization in Table 1 can be plotted. Both the control and experimental reporters will give final normalized values of 1 with the control lysate since they are the values on which the normalization is calculated. This is also true for the value of the control reporter in the experimental lysate. This graph shows that the translation of the experimental reporter goes down in the lysate experimental lysate compared to the control lysate. The dots represent the three technical replicates. Statistical significance was calculated with an unpaired two-tailed *t*-test (***$P < 0.005$).

2. The resulting values obtained for each well are then divided by the average of the values of the wells with control cells
3. The average value of all replicates of the mRNA reporters transfected in the control cells will be equal to 1. The average values of the replicates of a given mRNA reporter transfected in the treated cell will reflect the translational level relative to the control

Note: The values for the control mRNA reporters will average to 1 across all cells, since they are the normalization factor. This reflects the in-built assumption of this assay that the control reporter is insensitive to perturbations of the helicase.

4. Statistical analysis can be run for each reporter to compare the values of the replicates in each cell treatment relative to control and other treatments. For the example, statistical significance between conditions was calculated with an unpaired two-tailed *t*-test
5. Finally, the data can be visualized, for example using a bar graph as for the *in vitro* data (see Fig. 2).

5. Expected outcomes

The luciferase luminescence values will change both depending on the intrinsic translation level of different reporter constructs and the perturbations to RNA helicases. Reporters with decreased luciferase levels upon helicase perturbation depend on the helicase of interest for efficient translation, while reporters with increased luciferase levels are inhibited by the helicase of interest. Some helicases that operate during translation initiation like eIF4A will likely be overall promoters of translation (Linder & Jankowsky, 2011; Sen, Zhou, Ingolia, & Hinnebusch, 2015), although they might inhibit translation of a subset of transcripts (Calviello et al., 2021). It is important to note that reporters might behave differently to knockdown *vs* missense mutations of helicases (Calviello et al., 2021; Lennox et al., 2020).

6. Advantages and limitations

6.1 Advantages

The mRNAs reporters, as constructed, isolate a specific part of the mRNA on which the effect of the helicase can be determined. In the case of the reporters described here, they differ solely in their 5′UTR, hence the impact that is measured on translation can be attributed to translation initiation. The reporters in this protocol are transcribed, capped, and polyadenylated mRNA *in vitro*. This allows isolating the effect of the helicase on translation alone, without having to account for potential effects during DNA or pre-mRNA processing.

6.1.1 The short 1 hour translation timeframe of *in vitro* assay allows for tight control and focuses the assay predominantly on translation. The time constraint limits the possibility of secondary effects due to proteome changes and RNA-processing steps in which the helicase might be involved. The protocol, which has been optimized to increase translational output, allows for increased signal. Additionally, conducting the assay in a whole-cell lysate makes it amenable to manipulations that might be impossible in a cell system. This includes the addition of drugs, purified proteins, and reagents that are hard to deliver in cells or the removal of factors from the lysate with IP or similar techniques. Biochemical manipulation of the lysate can be performed, such as immunodepletion of factors, addition of recombinant proteins, or addition of compounds that are not cell permeable.

6.1.2 Compared the *in vitro* assay, the *in cell* assay offers more physiological relevance since translation occurs within living cells. For example, our prior work identified a difference in *CCNE1* 5′UTR translation between *in vitro* and in cell experiments, which could be caused by cell cycling (Calviello et al., 2021).

6.2 Limitations

There are situations where reporters do not recapitulate native regulation (Mitschka & Mayr, 2021). Another factor to consider is that the reporters contain a limited amount of the mRNA of interest (in this case the 5′UTRs), which might exclude other important regulatory features and miss important biological information. Additionally, given the overexpression of the reporters and/or the helicase, the stoichiometry of factors in the assay might differ from physiological conditions. Finally, given how data normalization is conducted, is necessary to find control 5′UTRs that are not regulated by helicase of interest, which may not always be possible.

6.2.1 The *in vitro* assay is more removed from the biology of the cells, since it lacks important features of the cells such as the cell wall, organelles, and the cell cycle. Similarly, the whole-lysate reaction is composed of diluted cytoplasm and additional reagents, providing for an environment that may differ in composition and stoichiometry from physiological conditions. Finally, there can be substantial variability in specific activity between batches of lysate, although the normalization to the control reporter is designed to minimize this noise.

6.2.2 The timeframe of *in cell* assay is significantly longer than for the *in vitro* assay, which may lead to a greater opportunity for secondary effect downstream of the perturbation of the helicase to act as confounding factors. Additionally, there is a greater possibility that the helicase might affect the mRNA reporters by non-translational means, such as through mRNA nuclear shuttling or mRNA decay. This assay tends to give lower signal compared to the *in vitro* assay, given the fact that lacks the additional reagents that optimize and boost its translational output. Additionally, given the inefficiency of transfection, it requires higher amounts of mRNA.

7. Optimization and troubleshooting

7.1.1 Poor or no signal:

- *In vitro* transcription is a delicate procedure and improperly transcribed mRNA, the transcription of an incorrect isoform, or insufficient levels of reporter may result in poor or no signal. Ensure to run a glyoxal gel to control the size, integrity, and amount of product. Ensure that all reagents are free from RNase which will degrade the reporters
- Another possibility is low signal whole-cell lysate which is not translationally active. This might result from the lysate having been collected when the cells were too confluent (as this inhibits translation), the frozen lysate being too old or having been freeze-thawed too many times, or the use of cell lines that are not optimal for translation

7.1.2 The experimental signal matches the control

- A lack of differences between the measurement obtained by the experimental and the control conditions may be due to the control 5′UTR (or control mRNA element) being sensitive to the perturbation of the helicase to the same degree as the experimental 5′UTR. To solve this problem, select a different control 5′UTR that is likely to be the least sensitive to regulation by the helicase of interest. Another solution is to employ a firefly reporter as internal control (as described in step 3.3). It may be useful to employ an internal ribosomal entry site (IRES) to drive the translation of the internal control firefly reporter, as many IRESs require different and often fewer translation factors and may not be sensitive to the helicase of interest (Pisarev, Shirokikh, & Hellen, 2005).

7.1.3 The *in vitro* data does not match the *in cell* data

- The data collected with the *in vitro* assay (step 3.2) may not match up with that collected with the *in cell* assay (step 3.3), even if the same reporters and same experimental conditions are employed. This might be due to non-translation related processes (such as mRNA export or decay) or other secondary effects of the helicase. These may be occurring in the *in cell* setup but may not have the time to take place in the short time span of *in vitro* assay. Additionally, it may be caused by the involvement of biological features of the cell, such as organelles or the cell cycle, that are lost in the whole-cell lysate setup. Finally, the amount of mRNA per reaction/cell, the confluency and translational potential of cells, and other variables might be different between the two assays

8. Alternative methods/procedures

In addition to the translational assays here presented, there are several alternative methods to measure the translational effect of a given helicase:

8.1 Ribosome profiling is a powerful technique that can provide genome wide and transcript specific information on the translational effect of RNA helicases (Calviello et al., 2021; Ingolia, Hussmann, & Weissman, 2019). This technique maps all footprints of actively translating ribosomes onto the transcriptome, allowing to detect changes in translation due to the perturbation of a helicase (Guenther et al., 2018).

8.2 Other techniques that measure global translation include polysome profiling, 35S methionine, FUNCAT, or puromycin staining (Aryanpur et al., 2019; Dieterich et al., 2010; Goodman, Pierre, & Hornberger, 2012; Guenther et al., 2018; Murat et al., 2018). Combined with the appropriate experimental setup, these methods can provide insight in the effects of helicases on the translatome

8.3 For the impact of helicases on the translation of specific proteins, methods such as fluorescence-based reporters in cells, western blotting, or direct measurement of protein levels can be employed. The use of translation inhibitors can test whether the changes observed are translationally mediated

8.4 Rabbit reticulocyte lysate and wheat germ extract are well established and highly efficient *in vitro* eukaryotic protein synthesis systems that can be employed to translate reporters in a cell-free context (Chong, 2014). While these systems may have less physiological relevance for the investigation of human helicases compared to the protocols described in Section 3, they may be more robust and more amenable for certain reporters and experimental setups. These systems also generally have higher specific activity

9. Summary and conclusion

As the first step of protein synthesis, translation initiation is fundamental and tightly regulated process in every cells. Helicases play an important role by modulating and interacting with the 5′UTR and its many structural features. In particular, several helicases unwind complex secondary structures, such as hairpins and g-quadruplexes, or modulate uORF selection allowing for proper regulation of ribosome scanning. Gaining insight of how helicases affect translation will unlock a more complete understanding of their role as regulators of the proteome both in physiological and diseased settings.

This chapter explores helpful ways to investigate the effect of helicases on mRNA translation initiation. Starting with *in vitro* transcribed, capped, and polyadenylated luciferase 5′UTR reporters allows for assays that restrict probing to the aspect of translation itself, removing confounding effects such as DNA and RNA processing. Two assays are described in this chapter: an *in vitro* approach, with employs the mRNA reporters in a cell-free whole-cell lysate system, and an *in cell* approach, in which the reporters are transfected into living cells. Either method can be used to investigate how translation initiation at 5′UTR is affected by RNA helicases. While the *in vitro* method offers tighter controls and fewer confounding effects, the *in cell* approach can offer a more physiologically relevant setup.

Although the protocol described here measures the effect of helicases on translation initiation by using 5′UTR reporters, it is adaptable to further use. Other mRNA features, such as the 3′UTR, individual regulatory elements and structures, ORF features, or even point mutations in the mRNA can be used as reporters in the same fashion. Similarly, this protocol can be used to investigate not only the translational role of helicases, but of any protein or factor with translational implication, including chemicals compounds.

Acknowledgments

The authors wish to thank Malvika Tejura and Bao Thai for their contributions. This work was supported by the National Institutes of Health DP2GM132932 and R01NS120667 (to S.N.F.) and the California Tobacco-Related Disease Research Grants Program 27KT-0003 (to S.N.F.) and T30DT1004 (to K.C.W.). S.N.F. is a Pew Scholar in the Biomedical Sciences, supported by The Pew Charitable Trusts.

References

Acker, M. G., Kolitz, S. E., Mitchell, S. F., Nanda, J. S., & Lorsch, J. R. (2007). Reconstitution of yeast translation initiation. *Methods in Enzymology*, *430*, 111–145. https://doi.org/10.1016/S0076-6879(07)30006-2.

Aitken, C. E., Beznosková, P., Vlčkova, V., Chiu, W.-L., Zhou, F., Valášek, L. S., et al. (2016). Eukaryotic translation initiation factor 3 plays distinct roles at the mRNA entry and exit channels of the ribosomal preinitiation complex. *eLife*, *5*, e20934. https://doi.org/10.7554/eLife.20934.

Arrowsmith, C. H., Audia, J. E., Austin, C., Baell, J., Bennett, J., Blagg, J., et al. (2015). The promise and peril of chemical probes. *Nature Chemical Biology*, *11*(8), 536–541. https://doi.org/10.1038/nchembio.1867.

Aryanpur, P. P., Renner, D. M., Rodela, E., Mittelmeier, T. M., Byrd, A., & Bolger, T. A. (2019). The DEAD-box RNA helicase Ded1 has a role in the translational response to TORC1 inhibition. *Molecular Biology of the Cell, 30*(17), 2171–2184. https://doi.org/10.1091/mbc.E18-11-0702.

Bai, Y., Zhou, K., & Doudna, J. A. (2013). Hepatitis C virus 3'UTR regulates viral translation through direct interactions with the host translation machinery. *Nucleic Acids Research, 41*(16), 7861–7874. https://doi.org/10.1093/nar/gkt543.

Cai, W., Xiong Chen, Z., Rane, G., Satendra Singh, S., Choo, Z., Wang, C., et al. (2017). Wanted DEAD/H or alive: Helicases winding up in cancers. *JNCI: Journal of the National Cancer Institute, 109*(6), djw278. https://doi.org/10.1093/jnci/djw278.

Calviello, L., Venkataramanan, S., Rogowski, K. J., Wyler, E., Wilkins, K., Tejura, M., et al. (2021). DDX3 depletion represses translation of mRNAs with complex 5' UTRs. *Nucleic Acids Research, 49*(9), 5336–5350. https://doi.org/10.1093/nar/gkab287.

Calvo, S. E., Pagliarini, D. J., & Mootha, V. K. (2009). Upstream open reading frames cause widespread reduction of protein expression and are polymorphic among humans. *Proceedings of the National Academy of Sciences, 106*(18), 7507–7512. https://doi.org/10.1073/pnas.0810916106.

Chong, S. (2014). Overview of cell-free protein synthesis: Historic landmarks, commercial systems, and expanding applications. *Current Protocols in Molecular Biology/Edited by Frederick M. Ausubel...[et al.], 108*, 16.30.1–16.30.11. https://doi.org/10.1002/0471142727.mb1630s108.

Dieterich, D. C., Hodas, J. J. L., Gouzer, G., Shadrin, I. Y., Ngo, J. T., Triller, A., et al. (2010). In situ visualization and dynamics of newly synthesized proteins in rat hippocampal neurons. *Nature Neuroscience, 13*(7), 897–905. https://doi.org/10.1038/nn.2580.

Fairman-Williams, M. E., Guenther, U.-P., & Jankowsky, E. (2010). SF1 and SF2 helicases: Family matters. *Current Opinion in Structural Biology, 20*(3), 313–324. https://doi.org/10.1016/j.sbi.2010.03.011.

Floor, S. N., Barkovich, K. J., Condon, K. J., Shokat, K. M., & Doudna, J. A. (2016). Analog sensitive chemical inhibition of the DEAD-box protein DDX3. *Protein Science: A Publication of the Protein Society, 25*(3), 638–649. https://doi.org/10.1002/pro.2857.

Floor, S. N., & Doudna, J. A. (2016). Tunable protein synthesis by transcript isoforms in human cells. *eLife, 5*, e10921. https://doi.org/10.7554/eLife.10921.

Fuller-Pace, F. V. (2013). DEAD box RNA helicase functions in cancer. *RNA Biology, 10*(1), 121–132. https://doi.org/10.4161/rna.23312.

Goodman, C. A., Pierre, P., & Hornberger, T. A. (2012). Imaging of protein synthesis with puromycin. *Proceedings of the National Academy of Sciences of the United States of America, 109*(17), E989. https://doi.org/10.1073/pnas.1202000109.

Guenther, U.-P., Weinberg, D. E., Zubradt, M. M., Tedeschi, F. A., Stawicki, B. N., Zagore, L. L., et al. (2018). The helicase Ded1p controls use of near-cognate translation initiation codons in 5' UTRs. *Nature, 559*(7712), 130–134. https://doi.org/10.1038/s41586-018-0258-0.

Hao, P., Yu, J., Ward, R., Liu, Y., Hao, Q., An, S., et al. (2020). Eukaryotic translation initiation factors as promising targets in cancer therapy. *Cell Communication and Signaling: CCS, 18*(1), 175. https://doi.org/10.1186/s12964-020-00607-9.

Hinnebusch, A. G., Ivanov, I. P., & Sonenberg, N. (2016). Translational control by 5′-untranslated regions of eukaryotic mRNAs. *Science (New York, N.Y.), 352*(6292), 1413–1416. https://doi.org/10.1126/science.aad9868.

Ingolia, N. T., Hussmann, J. A., & Weissman, J. S. (2019). Ribosome profiling: Global views of translation. *Cold Spring Harbor Perspectives in Biology, 11*(5), a032698. https://doi.org/10.1101/cshperspect.a032698.

Iwasaki, S., Floor, S. N., & Ingolia, N. T. (2016). Rocaglates convert DEAD-box protein eIF4A into a sequence-selective translational repressor. *Nature, 534*(7608), 558–561. https://doi.org/10.1038/nature17978.

Kozak, M. (1990). Downstream secondary structure facilitates recognition of initiator codons by eukaryotic ribosomes. *Proceedings of the National Academy of Sciences*, *87*(21), 8301–8305. https://doi.org/10.1073/pnas.87.21.8301.
Kozak, M. (2002). Pushing the limits of the scanning mechanism for initiation of translation. *Gene*, *299*(1), 1–34. https://doi.org/10.1016/S0378-1119(02)01056-9.
Lee, A. S. Y., Kranzusch, P. J., & Cate, J. H. D. (2015). EIF3 targets cell-proliferation messenger RNAs for translational activation or repression. *Nature*, *522*(7554), 111–114. https://doi.org/10.1038/nature14267.
Lennox, A. L., Hoye, M. L., Jiang, R., Johnson-Kerner, B. L., Suit, L. A., Venkataramanan, S., et al. (2020). Pathogenic DDX3X mutations impair RNA metabolism and neurogenesis during fetal cortical development. *Neuron*, *106*(3), 404–420.e8. https://doi.org/10.1016/j.neuron.2020.01.042.
Leppek, K., Das, R., & Barna, M. (2018). Functional 5' UTR mRNA structures in eukaryotic translation regulation and how to find them. *Nature Reviews. Molecular Cell Biology*, *19*(3), 158–174. https://doi.org/10.1038/nrm.2017.103.
Linder, P., & Jankowsky, E. (2011). From unwinding to clamping—The DEAD box RNA helicase family. *Nature Reviews Molecular Cell Biology*, *12*(8), 505–516. https://doi.org/10.1038/nrm3154.
Mitschka, S., & Mayr, C. (2021). Endogenous p53 expression in human and mouse is not regulated by its 3′UTR. *eLife*, *10*, e65700. https://doi.org/10.7554/eLife.65700.
Murat, P., Marsico, G., Herdy, B., Ghanbarian, A., Portella, G., & Balasubramanian, S. (2018). RNA G-quadruplexes at upstream open reading frames cause DHX36- and DHX9-dependent translation of human mRNAs. *Genome Biology*, *19*(1), 229. https://doi.org/10.1186/s13059-018-1602-2.
Naineni, S. K., Itoua Maïga, R., Cencic, R., Putnam, A. A., Amador, L. A., Rodriguez, A. D., et al. (2020). A comparative study of small molecules targeting eIF4A. *RNA*, *26*(5), 541–549. https://doi.org/10.1261/rna.072884.119.
Nashchekin, D., Zhao, J., Visa, N., & Daneholt, B. (2006). A novel Ded1-like RNA helicase interacts with the Y-box protein ctYB-1 in nuclear mRNP particles and in polysomes. *The Journal of Biological Chemistry*, *281*(20), 14263–14272. https://doi.org/10.1074/jbc.M600262200.
Parsyan, A., Shahbazian, D., Martineau, Y., Petroulakis, E., Alain, T., Larsson, O., et al. (2009). The helicase protein DHX29 promotes translation initiation, cell proliferation, and tumorigenesis. *Proceedings of the National Academy of Sciences*, *106*(52), 22217–22222. https://doi.org/10.1073/pnas.0909773106.
Pestova, T. V., & Kolupaeva, V. G. (2002). The roles of individual eukaryotic translation initiation factors in ribosomal scanning and initiation codon selection. *Genes & Development*, *16*(22), 2906–2922. https://doi.org/10.1101/gad.1020902.
Pisarev, A. V., Shirokikh, N. E., & Hellen, C. U. T. (2005). Translation initiation by factor-independent binding of eukaryotic ribosomes to internal ribosomal entry sites. *Comptes Rendus Biologies*, *328*(7), 589–605. https://doi.org/10.1016/j.crvi.2005.02.004.
Pisareva, V. P., Pisarev, A. V., Komar, A. A., Hellen, C. U. T., & Pestova, T. V. (2008). Translation initiation on mammalian mRNAs with structured 5'UTRs requires DExH-box protein DHX29. *Cell*, *135*(7), 1237–1250. https://doi.org/10.1016/j.cell.2008.10.037.
Rio, D. C. (2013). Expression and purification of active recombinant T7 RNA polymerase from E. coli. *Cold Spring Harbor Protocols*, *2013*(11). https://doi.org/10.1101/pdb.prot078527. pdb.prot078527.
Rio, D. C. (2015). Denaturation and electrophoresis of RNA with Glyoxal. *Cold Spring Harbor Protocols*, *2015*(2). https://doi.org/10.1101/pdb.prot081000. pdb.prot081000.
Rio, D. C. (Ed.). (2011). *RNA: A laboratory manual* Cold Spring Harbor Laboratory Press.
Sauer, M., Juranek, S. A., Marks, J., De Magis, A., Kazemier, H. G., Hilbig, D., et al. (2019). DHX36 prevents the accumulation of translationally inactive mRNAs with

G4-structures in untranslated regions. *Nature Communications*, *10*(1), 2421. https://doi.org/10.1038/s41467-019-10432-5.

Sen, N. D., Zhou, F., Ingolia, N. T., & Hinnebusch, A. G. (2015). Genome-wide analysis of translational efficiency reveals distinct but overlapping functions of yeast DEAD-box RNA helicases Ded1 and eIF4A. *Genome Research*, *25*(8), 1196–1205. https://doi.org/10.1101/gr.191601.115.

Sharma, D., & Jankowsky, E. (2014). The Ded1/DDX3 subfamily of DEAD-box RNA helicases. *Critical Reviews in Biochemistry and Molecular Biology*, *49*(4), 343–360. https://doi.org/10.3109/10409238.2014.931339.

Soto-Rifo, R., Rubilar, P. S., Limousin, T., de Breyne, S., Décimo, D., & Ohlmann, T. (2012). DEAD-box protein DDX3 associates with eIF4F to promote translation of selected mRNAs. *The EMBO Journal*, *31*(18), 3745–3756. https://doi.org/10.1038/emboj.2012.220.

Sweeney, T. R., Dhote, V., Guca, E., Hellen, C. U. T., Hashem, Y., & Pestova, T. V. (2021). Functional role and ribosomal position of the unique N-terminal region of DHX29, a factor required for initiation on structured mammalian mRNAs. *Nucleic Acids Research*, *49*(22), 12955–12969. https://doi.org/10.1093/nar/gkab1192.

van Voss, M. R. H., van Diest, P. J., & Raman, V. (2017). Targeting RNA helicases in cancer: The translation trap. *Biochimica et Biophysica Acta*, *1868*(2), 510–520. https://doi.org/10.1016/j.bbcan.2017.09.006.

Venkataramanan, S., Gadek, M., Calviello, L., Wilkins, K., & Floor, S. N. (2021). DDX3X and DDX3Y are redundant in protein synthesis. *RNA*, *27*(12), 1577–1588. https://doi.org/10.1261/rna.078926.121.

CHAPTER EIGHT

Bulk phase biochemistry of PIF1 and RecQ4 family helicases

Prasangi Rajapaksha, Robert H. Simmons III, Spencer J. Gray, David J. Sun, Phoebe Nguyen, David G. Nickens, and Matthew L. Bochman*

Molecular & Cellular Biochemistry Department, Indiana University, Bloomington, IN, United States

*Corresponding author: e-mail address: bochman@iu.edu

Contents

Abstract

DNA helicases are involved in nearly all facets of genome integrity, and in humans, mutations in helicase-encoding genes are often linked to diseases of genomic instability. Two highly studied and evolutionarily conserved helicase families are the PIF1 and RecQ helicases. Enzymes in these families have known roles in DNA replication, recombination, and repair, as well as telomere maintenance, DNA recombination, and transcription. Although genetics, structural biology, and a variety of other techniques have been used to study these helicases, ensemble analyses of their basic biochemical activities such as DNA binding, ATP hydrolysis, and DNA unwinding have made significant contributions to our understanding of their physiological roles. Here, we present general methods to generate recombinant proteins from both helicase families, as well as standard biochemical assays to investigate their activities on DNA.

Methods in Enzymology, Volume 673
ISSN 0076-6879
https://doi.org/10.1016/bs.mie.2022.03.031

1. Introduction

DNA helicases can be grouped into families and superfamilies based on their conserved sequence motifs, oligomeric state, and polarity of action on DNA (Berger, 2008). For instance, PIF1 family helicases all contain a 21-amino acid signature sequence (Bochman, Judge, & Zakian, 2011; Bochman, Sabouri, & Zakian, 2010) between their conserved helicase motifs II and III (Hall & Matson, 1999), are monomeric in solution, and unwind DNA in a 5′-3′ direction, making them members of Superfamily IB (SF-IB). In contrast, as members of SF-IIA, RecQ family helicases all contain a RecQ C-terminal domain (RQC) adjacent to their central helicase domain, are monomeric in solution, and display 3′-5′ polarity (Berger, 2008). Although these basic features hold true for all of the studied family members, species-specific differences often occur in the form of additional domains. For example, one of the human RecQ helicases, WRN, includes a 3′-5′ exonuclease domain that isn't found in yeast or bacterial RecQs (Bernstein, Gangloff, & Rothstein, 2010), and eukaryotic PIF1 helicases often contain N-and C-terminal domains flanking their conserved helicase domain that are not found in bacterial PIF1s (Andis, Sausen, Alladin, & Bochman, 2018; Bochman et al., 2011, 2010). Based on this protein-to-protein variability, members of both the PIF1 and RecQ helicase families have evolved to perform a variety of distinct functions *in vivo*.

The first PIF1 helicase, known simply as Pif1 or ScPif1, was identified 40 years ago in *Saccharomyces cerevisiae*, where it is involved in mitochondrial DNA replication and recombination (Bochman et al., 2010; Foury & Dyck, 1985). Since then, an additional family member called Rrm3 was discovered in yeast, and various functions for the *S. cerevisiae* PIF1 helicases have been described, such as telomere maintenance, nuclear DNA replication, DNA break repair, and G-quadruplex structure resolution (reviewed in Muellner & Schmidt, 2020). Additionally, PIF1 helicases have been identified and studied in other organisms, such as Pfh1 from the fission yeast *Schizosaccharomyces pombe*, multiple family members in the kinetoplastid parasite *Trypanosoma brucei*, the cancer-linked human PIF1, and the modular PIF1 helicases from thermostable bacteria (Andis et al., 2018; Bochman et al., 2010; Muellner & Schmidt, 2020). Such studies have shed light on the physical and genetic interactomes of these helicases (Stundon & Zakian, 2015; Zybailov et al., 2015), their regulation by post-translational modifications (Ononye, Sausen, Balakrishnan, & Bochman, 2020; Ononye, Sausen, Bochman, & Balakrishnan, 2021; Rossi, Ajazi, Carotenuto, Foiani, & Giannattasio, 2015; Vasianovich, Harrington, & Makovets, 2014), and the contributions of their unique non-helicase domains (Andis et al., 2018; Bessler & Zakian, 2004; Nickens, Sausen, & Bochman, 2019).

The RecQ family helicases share a similar lengthy and impressive history as the focus of many researchers (Chu & Hickson, 2009). Originally identified in *Escherichia coli*, the RecQs are now known to be highly conserved throughout evolution, with more complex organisms generally encoding an increasing number of family members. For instance, simple eukaryotes like *S. cerevisiae* contain two RecQ helicases (Hrq1 and Sgs1), while the human genome encodes five (RECQL1, BLM, WRN, RECQL4, and RECQL5) (Bochman, 2014). In this case, Hrq1 and Sgs1 are functionally homologous to RECQL4 and BLM, respectively. Similar to the PIF1 helicases, RecQ helicases function in a variety of processes *in vivo*, from DNA replication and transcription to DNA recombination and repair (Bachrati & Hickson, 2008; Bernstein et al., 2010; Bohr, 2008; Chu & Hickson, 2009; Gupta & Schmidt, 2020).

We have recently focused on the RecQ4 subfamily of RecQ helicases because they are the only RecQs found in all three domains of life and play important but often undefined roles in genome integrity. Additionally, mutations in human RECQL4 cause three distinct diseases (reviewed in Rogers, Simmons, Fluhler Thornburg, Buehler, & Bochman, 2020). RecQ4 subfamily helicases also genetically, physically, and functionally interact with other RecQs and PIF1 family helicases (Bochman, Paeschke, Chan, & Zakian, 2014; Nickens, Rogers, & Bochman, 2018; Rogers, Sanders, et al., 2020; Sanders, Nguyen, Rogers, & Bochman, 2020; Simmons, Rogers, & Bochman, 2021). In *S. cerevisiae*, these interactions impact telomerase regulation and the cellular responses to various forms of DNA damage.

Here, we provide detailed methods for the construction of DNA substrates, the purification of recombinant proteins, and biochemical assays to measure the activities of PIF1 family and RecQ4 subfamily helicases *in vitro*. These bulk phase assays can be used to test hypotheses about the functions of the helicases based on *in vivo* experiments. Conversely, they can act as a starting point to learn about the basic activities of the helicases and generate testable hypotheses for cell-based investigations. Indeed, we have recently used this type of give and take to begin to unravel the mechanism of DNA inter-strand crosslink (ICL) repair in *S. cerevisiae*, which involves the Hrq1 helicase (Rogers, Lee, et al., 2020).

2. Before you begin

Our best advice to those interested in performing the types of biochemical assays described below, especially for those who are new to protein purification and *in vitro* assays in general, is to review the Kornberg Commandments (Kornberg, 2003). Commandments IV (Do not waste clean thinking on dirty enzymes) and V (Do not waste clean enzymes on dirty substrates) are particularly germane here. A beautiful DNA substrate preparation will be ruined by a helicase stock that is contaminated with an expression host nuclease, and likewise, a crystallography-worthy helicase preparation can only tell you so much if your DNA substrate is not pristine. In other words, biochemistry is often slow and front-loaded by the necessity to troubleshoot protein and substrate purification problems, but rushing ahead with so-so reagents often just wastes more time.

In our experience, protein over-expression and purification is usually the limiting step in completing these investigations. Many of the PIF1 and RecQ helicases, particularly those found in eukaryotes, are large proteins and/or contain sizeable domains that are predicted to be natively disordered. These qualities make over-expression difficult in *E. coli*, for instance, but baculovirus-infected insect cell culture tends to perform much better. Although the latter requires more time and more expensive growth media, the time and money are worth it when a single 1-L insect cell preparation yields orders of magnitude more protein than tens of liters of *E. coli* cells over-expressing your protein (Nickens & Bochman, 2021).

Finally, it should be noted that there are limitations to the ensemble assays described below. Most are endpoint experiments and do not yield kinetics data. Even those that can be performed in a time course fashion are still limited by how quickly time points can be taken. Further, bulk phase assays can lack sufficient resolution to yield detailed mechanistic data, a situation that we encountered when investigating ICL repair by Hrq1 (Rogers, Lee, et al., 2020). In such cases, it is advisable to consider single-molecule approaches that enable real-time data collection and high-resolution observation of the reaction being investigated. Regardless, the methods described herein are an excellent starting point that can guide the establishment of proper single-molecule experiments. They are also quite versatile—being amenable to order-of-addition investigations, reagent titrations vs time course comparisons, and supplementation with various reagents like small molecule inhibitors—with only minimal and straightforward modifications of the protocols necessary.

3. Materials and equipment

3.1 General equipment

- Vertical polyacrylamide gel electrophoresis equipment (CBS Scientific Vertical Mini-gel System)
- Standing incubators: 27, 30, and 37 °C
- Shaking incubators capable of heating (≥37 °C) and cooling (≤18 °C)
- Centrifuges (micro, table-top, and super-speed) capable of cooling to 4 °C
- Sonicator or alternative method for bacterial cell lysis
- Medium-pressure liquid chromatography instrument (AKTA Pure)
- Liquid nitrogen Dewar
- LI-COR Odyssey DLx
- pH meter

3.1.1 Strains

- *E. coli* strain Rosetta 2(DE3)pLysS (F^- *ompT hsdS*$_B$(r_B^- m_B^-) *gal dcm* (DE3) pLysSRARE2 (Cam^R))
- *Spodoptera frugiperda* cell line Sf9

3.1.2 Reagents and chemicals

- Protease inhibitor mix: 600 nM leupeptin, 2 μM pepstatin A, 2 mM benzamidine, and 1 mM phenylmethanesulfonyl fluoride (PMSF). Commercially available protease inhibitor mixes are also appropriate
- NaCl
- Glycerol
- Lysozyme
- β-Mercaptoethanol
- HEPES
- EDTA
- Tris
- HCl
- NaOH
- NaOAc
- $MgCl_2$
- Triton-X100
- NP-40 or NP-40 substitute
- $Mg(OAc)_2$
- $Zn(OAc)_2$

- NH_4OAc
- Dithiothreitol (DTT)
- Tween-20
- SDS
- Proteinase K
- 100 mM ATP (pH 7.0): dissolve 1.2 g ATP in 15 mL Milli-Q H_2O, adjust the pH to 7.0 with 0.1 N NaOH, bring the volume to 20 mL with H_2O, make 1-mL aliquots, and store them at −80 °C
- 10 mM phosphoenolpyruvic acid (PEP): dissolve 20.6 mg PEP in 10 mL Milli-Q H_2O, make 500-μL aliquots, and store them at −80 °C
- 10 mM NADH: dissolve 43mg NADH in 10 mL Milli-Q H_2O, make 400-μL aliquots, and store them at −80 °C
- SDS-Proteinase K: Dissolve Proteinase K to 4 mg/mL in 50 mM Tris-HCl (pH 8.0) and 3 mM $CaCl_2$. Make small aliquots, and store them at −80 °C

3.1.3 Homemade buffers

- Pif1 lysis buffer: 50 mM sodium phosphate buffer (pH 7.5), 300 mM NaCl, 5 mM β-mercaptoethanol, 10% (w/v) glycerol, protease inhibitor mixture, and 0.5 mg/mL lysozyme
- Pif1 IEX buffer: 50 mM sodium phosphate buffer (pH 6.8), 10% (w/v) glycerol, 0.5 mm EDTA (pH 8.0), and 4 mM β-mercaptoethanol
- Pif1 storage buffer: 25 mM HEPES (pH 7.5), 150 mM NaCl, 2 mM β-mercaptoethanol, 0.1 mM EDTA (pH 8.0), and 30% (w/v) glycerol
- L buffer: 40 mM Tris-HCl (pH 8.0), 500 mM NaOAc, 1.1 mM $MgCl_2$, 0.1 mM EDTA, 0.1% (w/v) Triton X-100, 0.2% (w/v) NP-40, 10% (w/v) glycerol, and 0.1 mM PMSF
- 2 × resuspension buffer: 100 mM Na-HEPES (pH 8.0), 10% (w/v) glycerol, 300 mM NaOAc (pH 8.0), 10 mM $Mg(OAc)_2$, and 0.1% Tween-20
- Gel filtration buffer: 25 mM Na-HEPES (pH 7.5), 5% glycerol, 300 mM NaCl, 5 mM $MgCl_2$, and 0.05% Tween-20
- Hrq1 storage buffer: 25 mM Na-HEPES (pH 8.0), 30% (w/v) glycerol, 300 mM NaOAc (pH 7.6), 25 mM NaCl, 5 mM $Mg(OAc)_2$, 1 mM DTT and 0.1% (v/v) Tween-20
- 2 × annealing buffer: 40 mM Tris–HCl (pH 8.0), 8% glycerol, 0.2 mM EDTA, 80 μg/mL BSA, 20 mM DTT, and 20 mM $Mg(OAc)_2$
- 1 × ATPase buffer: 25 mM Na-HEPES (pH 8.0), 5% (w/v) glycerol, 50 mM NaOAc (pH 7.5), 150 mM NaCl, and 0.01% (v/v) NP-40

- 2 × binding buffer: 50 mM Na-HEPES (pH 8.0), 10% (w/v) glycerol, 100 mM NaOAc (pH 7.6), 300 μM NaCl, 15 mM $Mg(OAc)_2$, and 0.02% Tween-20
- 6 × nonfluorescent loading buffer: 50 mM Tris (pH 8.0), 25% (w/v) glycerol, and 0.025% Orange G
- 1 × TG running buffer: 25 mM Tris and 185 mM glycine (pH 8.8)
- 1 × TBE running buffer: 90 mM Tris base, 90 mM boric acid, and 2 mM EDTA (pH 8.0)

3.1.4 Oligonucleotides

We exclusively purchase our modified and unmodified oligonucleotides from Integrated DNA Technologies (IDT). The price per nucleotide at the smallest production scale (≤60 nt) is competitive and sufficient for hundreds of reactions for most of the substrates that we typically use *in vitro*. In most cases, standard desalting is the only purification required, though when using substrates in gels with single-nucleotide resolution (e.g., Nickens et al., 2018, 2019; Rogers, Lee, et al., 2020), PAGE purification is necessary. Further, IDT offers a wide variety of 5′, 3′, and internal modifications, including the near-infrared (NIR) fluorophores IRDye® 700 and 800 suggested below.

4. Step-by-step method details

4.1 Recombinant protein over-expression and purification

4.1.1 PIF1 family helicases

PIF1 family helicases have been investigated from a variety of organisms (Bochman et al., 2010; Muellner & Schmidt, 2020), and we recently published a detailed protocol (Nickens & Bochman, 2021) for the over-expression and purification of *S. cerevisiae* Pif1 and mutants thereof (Nickens et al., 2018, 2019; Ononye, Sausen, et al., 2020). Here, we include a protocol for the generation of recombinant PIF1 helicases from bacteria (Andis et al., 2018), which is also amenable to *S. cerevisiae* Pif1.

1. Transform a PIF1 expression plasmid such as pMB427 (Andis et al., 2018) into Rosetta 2(DE3) pLysS (Novagen) *E. coli* cells. Select for transformants on LB medium supplemented with 100 μg/mL ampicillin and 34 μg/mL chloramphenicol at 37 °C.
2. Inoculate one or more 5-mL cultures of LB medium supplemented with the same antibiotics with a match head-sized swab of cells from

the transformation plate, and grow for ~6 h at 30 °C with aeration. Only fresh transformants should be used as we have observed little-to-no expression from transformants stored on plates for extended periods of time or from glycerol stocks of expression strains carrying the pSUMO-Pif1 plasmid.

3. Back-dilute these starter cultures 1:100 into ZYP-5052 autoinduction medium (Studier, 2005) containing 1 × trace metals mix, 100 μg/mL ampicillin, and 34 μg/mL chloramphenicol, and incubate at 22 °C with agitation to OD_{600} >3 (15–18 h). Unless otherwise stated, all subsequent steps should be performed at 4 °C or on ice using instruments and equipment that is likewise chilled.
4. Harvest the cells by centrifugation at 5500 × *g* for 10 min, decant the growth medium, weigh the pellet(s), and freeze the cells at −80 °C prior to lysis.
5. Resuspend the cells by gently stirring in 4 mL/g cell pellet 1 × resuspension buffer supplemented with protease inhibitor mix and 20 μg/mL DNase I.
6. Lyse cells using a high-pressure cell cracker operated at >1000 psi, clarify the lysate by centrifugation at 33,000 × *g* for 30 min, and filter the lysate through a 0.22-μm membrane.
7. Apply the filtrate to a 1-mL Strep-Tactin Sepharose (IBA) column pre-equilibrated in resuspension buffer using an ÄKTA Pure (Cytiva) chromatography system.
8. Wash with 20 mL 1 × resuspension buffer, 10 mL 1 × resuspension buffer supplemented with 5 mM ATP, and 10 mL of 1 × resuspension buffer.
9. Elute protein with 15 mL 1 × resuspension buffer supplemented with 2.5 mM desthiobiotin (IBA).
10. Identify the peak fractions by SDS-PAGE and Coomassie staining analysis of 10-μL aliquots of each elution fraction, and pool the peak fractions.
11. Concentrate the protein with an Amicon Ultra-4 30 K centrifugal filter, and load it onto a HiPrep 16/60 Sephacryl S-200 HR column (GE Healthcare Life Sciences) pre-equilibrated in Gel Filtration Buffer.
12. Elute the protein with 1.5 column volumes of Gel Filtration Buffer, and analyze the fractions by SDS-PAGE as above.
13. Pool peak fractions, snap-freeze them with liquid nitrogen, and store at −80 °C.

4.1.2 RecQ4 family helicases

Recombinant *S. cerevisiae* Hrq1 was originally generated using *E. coli* (Bochman et al., 2014), but baculovirus-infected insect cell culture has more recently been adopted for its greatly increased yields of protein (Rogers & Bochman, 2017; Rogers et al., 2017). Furthermore, the catalytically inactive Hrq1-K318A mutant and various truncation mutants (e.g., Hrq1ΔN (Rogers, Lee, et al., 2020)) can be over-expressed and purified likewise, which was recently reviewed (Nickens & Bochman, 2021). Here, we focus on methods to over-express and purify the *Mycobacterium smegmatis* Hrq1 homolog called SftH (Rogers, Lee, et al., 2020).

1. Transform the expression strain of *E. coli* with plasmid pET28a-MsSftH (Rogers, Lee, et al., 2020). We have achieved excellent over-expression with Rosetta 2(DE3) pLysS cells. Select for transformants on LB medium supplemented with the appropriate antibiotics, e.g., 50 μg/mL kanamycin and 34 μg/mL chloramphenicol for the above-suggested strains, at 37 °C.
2. Inoculate 10-mL of LB medium supplemented with the same antibiotics with a match head-sized swab of cells from the transformation plate, and grow at 30 °C with aeration for 6–8 h.
3. Inoculate 1 L of ZY-5052 autoinduction medium (Studier, 2005) supplemented with antibiotics with the 10-mL culture, and incubate overnight at 22 °C with aeration. Unless otherwise stated, all subsequent steps should be performed at 4 °C or on ice using instruments and equipment that is likewise chilled.
4. Harvest the cells by pelleting in a centrifuge at 5500 × *g* for 10 min, decant the growth medium, weigh the pellet(s), store at −80 °C for ≥1 h.
5. Resuspend the cells by gently stirring 2 mL 1 × resuspension buffer per gram of cell weight, supplemented with protease inhibitor mix and 10 μg/mL DNase I.
6. Lyse cells, and clarify the lysate by centrifugation at 33,000 × *g* for 30 min. We have used a Fisher Scientific FB505 sonicator, a Microfluidics microfluidizer, an Avestin EmulsiFlex-C3 emulsifier, and a high-pressure cell cracker to lyse cells with comparable results, so any lysis method that does not introduce excessive amounts of heat is likely to be acceptable.
7. Apply the supernatant to a 1-mL His60 column, and wash with 10 mL resuspension buffer.

8. Next, wash with 20 mL 1 × resuspension buffer supplemented with 5 mM ATP, followed by 10 mL of 1 × resuspension buffer supplemented with 10 mM imidazole.
9. Elute SftH with 1 × resuspension buffer supplemented with 250 mM imidazole.
10. Identify the peak SftH-containing fractions by SDS-PAGE and Coomassie staining analysis of 10-μL aliquots of each elution fraction, pool the peak fractions, and dialyze them against a 20-fold excess of Hrq1 storage buffer.
11. Concentrate the SftH with Amicon Ultra-4 centrifugal filters with a 30,000 cutoff, and store at −80 °C in Hrq1 storage buffer.

4.2 DNA substrate preparation

PIF1 and RecQ helicases are known to interact with a diverse array of nucleic acid substrates *in vivo* and *in vitro* (Bernstein et al., 2010; Bochman et al., 2010), and many of these structures mimicking physiological substrates can be prepared with commercially available oligonucleotides. Here, we describe a simple method to anneal ssDNAs into more complex structures and provide the sequences that can be used to create a variety of model substrates for the biochemical assays listed below.

1. Combine equimolar amounts of the oligonucleotides to be annealed in 1 × Annealing Buffer.
2. Incubate overnight at 37 °C.
3. Assess the annealing by running a small portion of the reaction on an 8% 19:1 acrylamide:bisacrylamide native gel in 1 × TG Running Buffer and comparing to the migration of the unannealed labeled oligonucleotide. The gel percentage may need to be optimized to discriminate between the migration of free labeled ssDNA and annealed ssDNA.

If unannealed labeled oligonucleotide is present at >5% of the total, repeat the annealing process using a slight molar excess of the unlabeled oligonucleotide to drive the incorporation of more of the labeled strand into the desired substrate. The excess amount added should be kept to a minimum to avoid unwanted complications in biochemical assays such as binding to the remaining unlabeled ssDNA instead of the experimental substrate. We have had great luck with the above isothermal annealing method for a host of different substrates (Table 1). However, a more common method is to incubate the annealing reaction in a 95 °C heat block for 5 min, turn off the power, and let the block slowly cool to room temperature over several

Table 1 Common DNA substrates used in PIF1 and RecQ experiments.

Name	Sequence (5′-3′)	Notes
Poly(dT)5	TTTTT	This series of substrates can be used to gauge the binding site size of helicases using gel shift and, indirectly, ATPase assays (Nickens & Bochman, 2022; Rogers et al., 2017). Similar poly-oligonucleotides can be synthesized using any of the other dNTPs and/or with greater length depending on experimental needs
Poly(dT)10	TTTTTTTTTT	
Poly(dT)15	TTTTTTTTTTTTTTT	
Poly(dT)20	TTTTTTTTTTTTTTTTTTTT	
Poly(dT)25	TTTTTTTTTTTTTTTTTTTTTTTTT	
Poly(dT)30	TTTTTTTTTTTTTTTTTTTTTTTTTTTTTT	
MB733	ACCGTTGTGCAACTGAGTGGACAACGTGTCACTC ACATAGCGTTC	Anneal to generate a DNA fork with 25-nt random-sequence ssDNA arms and a 20-bp duplex
MB734	GAACGCTATGTGAGTGACACCAACAGGTGA GTCAACGTGTTGCCA	
MB1057	CTACTCTAACTCCGACCGCTTGCATGCCTGCAGG	MB1057 and MB1059 form a fork in annealing assays (Rogers & Bochman, 2017). MB1058 can additionally be added in strand exchange assays to convert the fork to blunt dsDNA upon exchange
MB1059	ATCCTCTCTAGAGTCAAGGTCGGAGTTAGAGTAG	
MB1058	CCTGCAGGCATGCAAGCGGTCGGAGTTAGAGTAG	
MB583	GACGCTGCCGAATTCTGGCTTGCTAGGACA TCTTTGCCCACGTTGACCC	These four oligonucleotides can be annealed to create a blunt-ended Holliday junction-like substrate
MB584	TGGGTCAACGTGGGCAAAGATGTCCTAGCA ATGTAATCGTCTATGACGTT	
MB585	CAACGTCATAGACGATTACATTGCTAGGAC ATGCTGTCTAGAGACTATCGA	
MB586	ATCGATAGTCTCTAGACAGCATGTCCTAGC AAGCCAGAATTCGGCAGCGT	

Continued

Table 1 Common DNA substrates used in PIF1 and RecQ experiments.—cont'd

Name	Sequence (5′-3′)	Notes
MB808	GGTGAACCTGCAGGTGGGCGGCTGCTCATC GTAGGTTAGTTGGTAGAATTCGGCAGCGTC	Can be annealed to form a ssDNA bubble flanked by blunt dsDNA arms (Rogers et al., 2017)
MB811	GACGCTGCCGAATTCTAGGTTGATTGGATG CTACTCGTCGGCGCCACCTGCAGGTTCACC	
MB1162	GGGTGAACCTGCAGGTGGGCGGCTGCTCAT CGTAGGTTAGTGGTAGAATTCGGCAGCGTC	Can be annealed to create a three-stranded D-loop (Guy & Bolt, 2005)
MB1163	GACGCTGCCGAATTCTACCATGCCTTGCTA GGACATCTTTGCCCACCTGCAGGTTCACCC	
MB1165	TAAGAGCAAGATGTTCTATAAAAGATGTCC TAGCAAGGCA	

hours or, often more conveniently, overnight. In the absence of a heat block, a thermocycler can be utilized in the same way.

4.3 DNA binding assay

Many methods, such as nitrocellulose filter binding (Bochman & Schwacha, 2007) and biolayer interferometry (Ononye, Sausen, et al., 2020), can be used to measure helicase-DNA binding, but here, we describe a simple gel mobility shift assay. Classically, such assays utilized radiolabeled DNA substrates, but we have recently transitioned to using substrates labeled with NIR fluorophores such as the IRDye® 700 and 800 modifications available from IDT and others. These dyes can be detected with comparable sensitivity to ^{32}P but do not require shielding and specialized work areas like radioisotopes and are not subject to the rapid photobleaching in ambient light of the more traditional fluorophores used to label DNA (e.g., fluorescein).

1. Prepare an $n + 10\%$ volume of DNA master mix on ice, where n is the number of reactions to be performed, including a no-protein control. Each reaction requires 4 nM DNA substrate, 5 μL 2 × binding buffer, and Milli-Q H_2O to 10 μL.
2. On ice, dilute the helicase with 1 × binding buffer to the concentrations to be tested. Generally, a range from 1 to 100 nM is a good place to start.
3. Aliquot 10 μL of DNA master mix into each tube on ice, add 10 μL of 1 × binding buffer (no-protein control) or helicase dilution to each tube, and mix thoroughly by pipetting up and down.
4. Incubate the reactions (e.g., at 30 °C for *S. cerevisiae* proteins or 37 °C for human proteins) for 30 min.
5. Add 4 μL of 6 × nonfluorescent loading buffer to each reaction, and mix thoroughly.
6. Load 20 μL of each reaction onto an 8% 19:1 acrylamide:bisacrylamide native gel in 1 × TG running buffer, and electrophorese at 100 V for 30–45 min.
7. The gels can be imaged with a laser scanning device such as a LI-COR Odyssey DLx and DNA binding quantified with the associated software or freeware such as ImageJ (https://imagej.nih.gov/ij/).
8. The percent of DNA bound is calculated by dividing the signal of the shifted DNA by the total DNA signal in the lane and multiplying by 100:

$$DNA\ bound = \frac{Shifted\ DNA}{(Shifted + unshifted\ DNA)} \times 100\%$$

9. DNA binding curves are often hyperbolic for one-site binding or sigmoidal for cooperative binding. These curves can be fitted with software such as GraphPad Prism.

4.4 ATPase assay

This NADH-coupled assay (Fig. 1) (Sausen, Rogers, & Bochman, 2019) has been useful to measure basal and DNA-stimulated ATPase activities of PIF1 (Nickens & Bochman, 2022) and RecQ helicase (Rogers et al., 2017). It can be performed one reaction at a time and monitored spectrophotometrically in standard-sized cuvettes, but here, we present a high-throughput version that can be used to follow up to 96 reactions per run. Although further decreases in reaction size to use less reagents and an increase in the throughput are possible using 384-well plates, pipetting error becomes such a confounding issue that many technical replicates per condition are necessary to average out the noise in the data, negating the throughput and reagent-saving gains. However, the use of liquid handling robotics may be able to overcome this obstacle should it be necessary.

1. Prepare an $n + 10\%$ volume of ATPase reaction master mix on ice, where n is the number of reactions to be performed, including no-protein controls. A single reaction receives 6 μL 100 mM ATP (pH 7.0), 0.6 μL 1 M $MgCl_2$, 6 μL 10 mM PEP, 4.6 μL 10 mM NADH, 1.2 μL pyruvate kinase (PK, 10 U/mL; diluted 1:5 with 1 × ATPase Buffer), 1.28 μL lactate dehydrogenase (LDH, 16 U/mL; diluted 1:10 with 1 × ATPase Buffer), and 40.12 μL 1 × ATPase buffer.

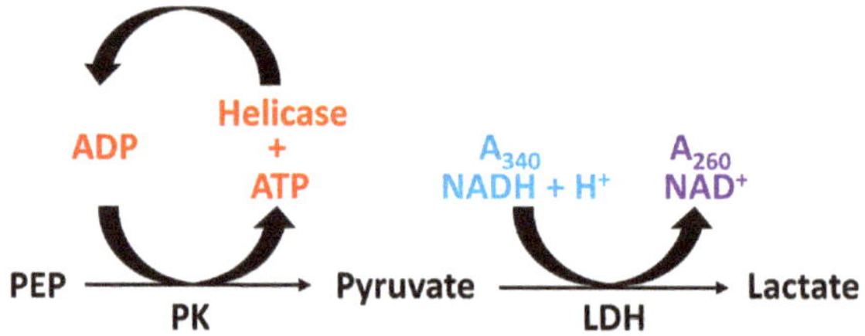

Fig. 1 NADH-coupled ATPase assay. In this assay, a helicase (+/− DNA, not shown) binds and hydrolyzes ATP into ADP. To maintain the ATP at a constant concentration, phosphoenol pyruvate (PEP) and pyruvate kinase (PK) are used as an ATP regeneration system, recycling ADP into ATP and generating pyruvate. This reaction is coupled to lactate production by the action of lactate dehydrogenase (LDH) on pyruvate in the presence of NADH. The conversion of NADH to NAD^+ is monitored by recording the decrease in absorbance at 340 nm (A_{340}). One ATP is hydrolyzed for every NADH oxidized.

2. In a separate tube on ice, dilute the helicase of interest to 10 nM in 1 × ATPase buffer. Make this as a protein master mix with enough volume for 60 μL per well.
3. Chill a 96-well plate on ice.
4. Pipet 60 μL of the protein master mix into each well of the 96-well plate to be monitored. Similarly, add 60 μL of 1 × ATPase Buffer to the no-protein control well(s).
5. If assessing the effects of DNA on the ATP hydrolysis rate of the helicase, add 1.2 μL of 1 μM DNA substrate (ssDNA, dsDNA, fork, etc.) to each well or an equal volume of buffer to no-DNA wells. Some autohydrolysis of ATP also occurs, so an ATP-only control lacking added helicase is also advisable.
6. Add 60 μL of ATPase reaction master mix to each well, and mix by pipetting up and down. A multichannel pipette can be used to simultaneously mix many wells at once.
7. Wipe off the bottom of the wells to remove ice and moisture, place the plate in the plate reader (a BioTek Synergy H1 or similar instrument), and monitor absorbance at 340 nm at the desired temperature every 10 min for 1 h. Only the linear portion of the data is used to calculate the rate of hydrolysis (i.e., the slope of the line) in μM/min. For very fast ATPases, more frequent measurements may be needed to capture the linear portion of the reaction, or a lower reaction temperature can be used to slow the rate of hydrolysis.
8. Absorbance readings are converted to ATP turnover (μM) based on the NADH concentration. It is assumed that 1 μM NADH oxidized is proportional to 1 μM ATP hydrolyzed.

NADH and PEP are consumed in a 1:1 ratio during the reaction to produce a change in A_{340}. By doubling their concentrations, the linear range of the reaction can be doubled, which is useful for fast reactions. If a reaction ends (i.e., runs out of NADH) in <10 min, it may be approaching the maximal rate of the PK and LDH enzymes, and therefore, they may become rate limiting. In this case, the reaction needs to be adjusted to increase the time domain. If the reaction cannot be slowed, the PK and LDH enzyme concentrations can be increased to compensate. If NADH and PEP are doubled, then reactions ending in 20 min may be approaching the rate limit of PK and LDH, and the same adjustments need to be made.

When adding ssDNA to the reactions, secondary structures may cause confounding results. To minimize these effects, the ssDNA oligonucleotides should be diluted to their working concentration(s), boiled, and snap-cooled prior to addition to the ATPase assays.

4.5 Helicase assay

As with DNA binding, there are many ways to monitor DNA unwinding (e.g., Ononye, Sausen, et al., 2020; Zhou, Zhang, Bochman, Zakian, & Ha, 2014), but we again present a simple gel-based method (Rogers, Sausen, & Bochman, 2019) using NIR-labeled substrates for its simplicity. All manipulations should be performed on ice unless otherwise stated.

1. Dilute recombinant helicase from freezer stocks to fivefold the desired final concentration using 1 × binding buffer. For Pif1, concentrations of ≤100 nM are generally appropriate, but for less active enzymes like Hrq1, up to 500 nM may be required to reach maximal activity.
2. Mix 2 μL of helicase with 1 μL 10 nM DNA substrate, 0.5 μL 100 mM ATP, 1 μL H_2O, and 5 μL 2 × binding buffer. For proteins that exhibit strong annealing activity, such as human RECQL4 and *S. cerevisiae* Pif1, add 10 nM cold trap DNA (i.e., a molar excess of unlabeled ssDNA equivalent to the labeled strand in your substrate). Two reactions should omit protein: one for use as a negative control to measure spontaneous DNA dissociation during the reaction; and the other for a boiled control (Fig. 2; see below).
3. Incubate the reactions (e.g., at 30 °C for *S. cerevisiae* proteins or 37 °C for human proteins) for 30 min, and then place them on ice.
4. To each, add 5 μL of 3 × nonfluorescent loading buffer supplemented with 400 μg/mL SDS-Proteinase K, and incubate the reactions at 37 °C for 10 min to release protein from the DNA.
5. Heat the boiled control to 95 °C for 5 min, and snap-cool on ice to use as a reference for the migration of fully unwound substrate.
6. Load 10 μL of the reactions on an 8% 19:1 acrylamide:bisacrylamide gel using 1 × TBE running buffer, and run at 10 V/cm until the dye front reaches the bottom of the gel.
7. The gels can be imaged as described above for DNA binding gel shifts. The percent of DNA unwound can be calculated by dividing the ssDNA signal by the total signal in each lane, and the amount of ssDNA in the no-protein control should be subtracted from each of the reactions:

$$DNA\ unwound = \left(\left(\frac{ssDNA}{ssDNA + dsDNA}\right) - (spontaneousssDNA)\right) \times 100\%$$

8. DNA unwinding curves are usually hyperbolic and can be fitted with software such as GraphPad Prism.

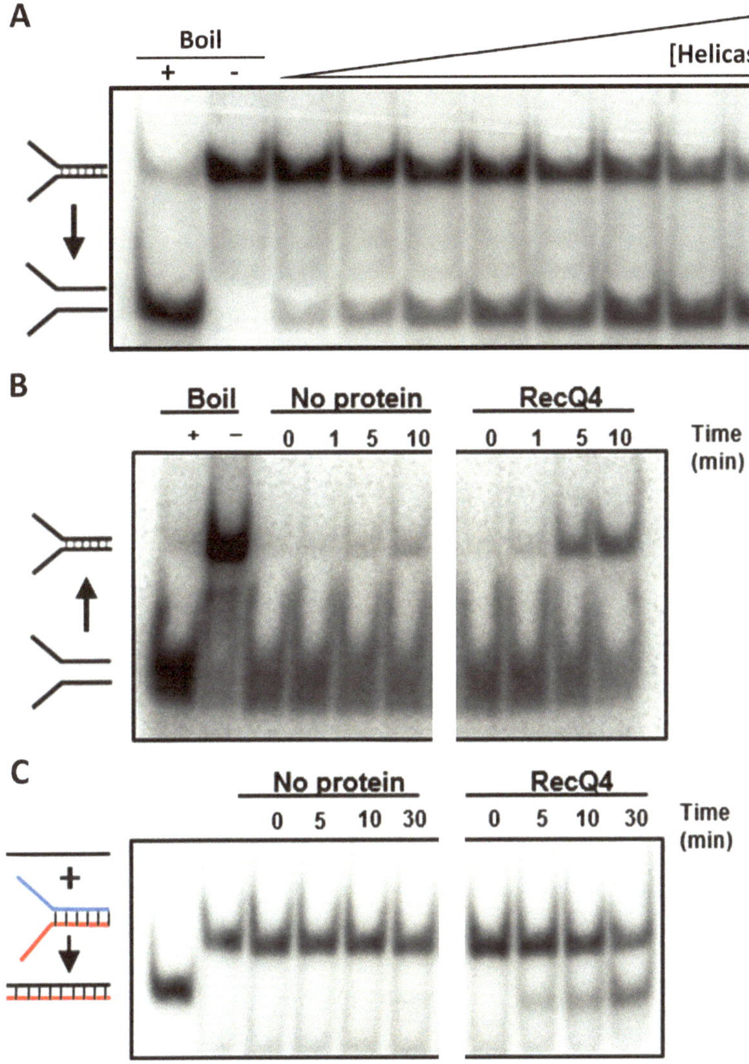

Fig. 2 Example helicase, DNA annealing, and stand exchange assays. (A) Conversion of the fork substrate into ssDNA by unwinding the duplex portion of the fork substrate is catalyzed by the addition of a DNA helicase in a concentration-dependent manner (1–100 nM). The boiled control (+) shows the position at which the ssDNA should migrate through the gel, and the unboiled control (–) shows where the fork substrate migrates ($t = 30$ min endpoints for each control). The fork substrate shown was created by annealing oligonucleotides MB1057 and MB1058 (Table 1). (B) Annealing of two partially complementary ssDNAs (MB1057 and MB1058) into a DNA fork by the RECQL4 (100 nM; RecQ4) helicase. Here, the annealed product (i.e., the fork) was boiled (+) to indicate the position where the unannealed ssDNAs migrate through the gel, and the unboiled (–) control indicates the migration position of the annealed fork. (C) Strand exchange catalyzed by 50 nM RECQL4. In this assay, the unlabeled partially complementary oligonucleotide (blue, MB1058) that forms half of the fork substrate

(Continued)

4.6 DNA annealing assay

DNA annealing is the opposite of helicase activity, that is, the annealing of two ssDNAs into a dsDNA substrate. Many PIF1 and RecQ family helicases display this activity (Lu & Davis, 2021; Ramanagoudr-Bhojappa, Byrd, Dahl, & Raney, 2014), though there are some species-specific differences among homologs, such as Hrq1 and RECQL4 (Rogers & Bochman, 2017). Regardless, the assay is quite similar to the helicase assay described above, but it omits ATP to avoid the confounding effects of the helicase cycling between annealing and unwinding modes.

1. On ice, dilute helicase protein from freezer stocks to fivefold the desired final concentration (10–100 nM) using 1 × binding buffer.
2. Prepare the complementary ssDNAs by heat-denaturing a helicase assay substrate such as a fork for 10 min at 95 °C and snap-cooling on ice.
3. Mix 2 μL helicase, 1 μL of 10 μM complementary ssDNAs, 2 μL Milli-Q H_2O, and 5 μL 2 × binding buffer. Omit protein from one reaction to use as a negative control for spontaneous annealing. Another reaction should likewise contain no protein but an equimolar amount of annealed substrate (i.e., that was not heat-denatured in step 2) as a control for where the fully annealed substrate should migrate.
4. Incubate the reactions (e.g., at 30 °C for *S. cerevisiae* proteins or 37 °C for human proteins) for 30 min.
5. Stop the reactions by adding 5 μL of 3 × nonfluorescent loading buffer supplemented with 400 μg/mL SDS-Proteinase K to each, and incubate the reactions at 37 °C for 10 min to release protein from the DNA.
6. The gels can be imaged as described above for DNA binding gel shifts. The percent of annealing can be calculated by dividing the annealed signal by the total signal in each lane, and the amount of spontaneously annealed product in the no-protein control should be subtracted from each of the reactions:

$$Annealed\ DNA = \left(\left(\frac{dsDNA}{dsDNA + ssDNA}\right) - (spontaneous\ dsDNA)\right) \times 100\%$$

Fig. 2—Cont'd in (A) and (B) can be exchanged for a completely complementary unlabeled oligonucleotide (black, MB1059) to produce blunt dsDNA, which migrates more quickly in the gel than the fork. *The images in (B) and (C) are adapted from Rogers, C. M., & Bochman, M. L. (2017). Saccharomyces cerevisiae Hrq1 helicase activity is affected by the sequence but not the length of single-stranded DNA.* Biochemical and Biophysical Research Communications, 486*(4), 1116–1121. doi:10.1016/j.bbrc.2017.04.003.*

4.7 Strand exchange assay

Some helicases that display annealing activity are additionally capable of strand exchange (Keller et al., 2014), i.e., the trading of one strand of a dsDNA duplex for a complementary ssDNA, similar to what occurs during strand exchange for homologous recombination (Daley, Kwon, Niu, & Sung, 2013). In general, substrates for this assay must be designed such that the annealed DNA structure has a different migration pattern than the strand exchange product on a native gel. In this example, it is the conversion of a fork substrate to a blunt dsDNA substrate. Alternatively, two different NIR labels can be used to discriminate between the original substrate and the strand exchange product.

1. On ice, dilute helicase protein from freezer stocks to fivefold the desired final concentration using 1 × binding buffer.
2. Mix 2 μL helicase, 5 μL 2 × Binding Buffer, 1 μL Milli-Q H_2O, 1 μL of 10 nM substrate (e.g., a DNA fork), and 1 μL of 10 nM complementary ssDNA. Omit protein from one reaction to use as a negative control for spontaneous strand exchange. Another reaction should likewise contain no protein but an equimolar amount of dsDNA substrate as a control for where the fully strand exchanged substrate should migrate.
3. Incubate the reactions (e.g., at 30 °C for *S. cerevisiae* proteins or 37 °C for human proteins) for 30 min.
4. Stop the reactions by adding 5 μL of 3 × nonfluorescent loading buffer supplemented with 400 μg/mL SDS-Proteinase K to each and incubate the reactions at 37 °C for 10 min to release protein from the DNA.
5. The gels can be imaged as described above for DNA binding gel shifts. The percent of strand exchange can be calculated by dividing the dsDNA signal by the total signal in each lane:

$$Strand\ exchange = \frac{Blunt\ dsDNA}{Blunt\ dsDNA = Fork\ DNA} \times 100\%$$

References

Andis, N. M., Sausen, C. W., Alladin, A., & Bochman, M. L. (2018). The WYL domain of the PIF1 helicase from the thermophilic bacterium Thermotoga elfii is an accessory single-stranded DNA binding module. *Biochemistry*, *57*(7), 1108–1118. https://doi.org/10.1021/acs.biochem.7b01233.

Bachrati, C. Z., & Hickson, I. D. (2008). RecQ helicases: Guardian angels of the DNA replication fork. *Chromosoma*, *117*, 219–233. Retrieved from http://www.ncbi.nlm.nih.gov/entrez/query.fcgi?cmd=Retrieve&db=PubMed&dopt=Citation&list_uids=18188578.

Berger, J. M. (2008). SnapShot: Nucleic acid helicases and translocases. *Cell, 134*. https://doi.org/10.1016/j.cell.2008.08.027. 888-888.e881.

Bernstein, K. A., Gangloff, S., & Rothstein, R. (2010). The RecQ DNA helicases in DNA repair. *Annual Review of Genetics, 44*, 393–417. https://doi.org/10.1146/annurev-genet-102209-163602.

Bessler, J. B., & Zakian, V. A. (2004). The amino terminus of the *Saccharomyces cerevisiae* DNA helicase Rrm3p modulates protein function altering replication and checkpoint activity. *Genetics, 168*, 1205–1218.

Bochman, M. L. (2014). Roles of DNA helicases in the maintenance of genome integrity. *Molecular & Cellular Oncology, 1*(3), e963429. https://doi.org/10.4161/23723548.2014.963429.

Bochman, M. L., Judge, C. P., & Zakian, V. A. (2011). The Pif1 family in prokaryotes: What are our helicases doing in your bacteria? *Molecular Biology of the Cell, 22*(12), 1955–1959. https://doi.org/10.1091/mbc.E11-01-0045.

Bochman, M. L., Paeschke, K., Chan, A., & Zakian, V. A. (2014). Hrq1, a homolog of the human RecQ4 helicase, acts catalytically and structurally to promote genome integrity. *Cell Reports, 6*(2), 346–356. https://doi.org/10.1016/j.celrep.2013.12.037.

Bochman, M. L., Sabouri, N., & Zakian, V. A. (2010). Unwinding the functions of the Pif1 family helicases. *DNA Repair (Amst), 9*, 237–249. https://doi.org/10.1016/j.dnarep.2010.01.008.

Bochman, M. L., & Schwacha, A. (2007). Differences in the single-stranded DNA binding activities of MCM2-7 and MCM467: MCM2 and MCM5 define a slow ATP-dependent step. *The Journal of Biological Chemistry, 282*, 33795–337804. https://doi.org/10.1074/jbc.M703824200.

Bohr, V. A. (2008). Rising from the RecQ-age: The role of human RecQ helicases in genome maintenance. *Trends in Biochemical Sciences, 33*, 609–620. https://doi.org/10.1016/j.tibs.2008.09.003.

Chu, W. K., & Hickson, I. D. (2009). RecQ helicases: Multifunctional genome caretakers. *Nature Reviews. Cancer, 9*, 644–654. https://doi.org/10.1038/nrc2682.

Daley, J. M., Kwon, Y., Niu, H., & Sung, P. (2013). Investigations of homologous recombination pathways and their regulation. *The Yale Journal of Biology and Medicine, 86*(4), 453–461. Retrieved from https://www.ncbi.nlm.nih.gov/pubmed/24348209.

Foury, F., & Dyck, E. V. (1985). A PIF-dependent recombinogenic signal in the mitochondrial DNA of yeast. *The EMBO Journal, 4*, 3525–3530. Retrieved from http://www.ncbi.nlm.nih.gov/entrez/query.fcgi?cmd=Retrieve&db=PubMed&dopt=Citation&list_uids=16453651.

Gupta, S. V., & Schmidt, K. H. (2020). Maintenance of yeast genome integrity by RecQ family DNA helicases. *Genes (Basel), 11*(2), 205. https://doi.org/10.3390/genes11020205.

Guy, C. P., & Bolt, E. L. (2005). Archaeal Hel308 helicase targets replication forks in vivo and in vitro and unwinds lagging strands. *Nucleic Acids Research, 33*(11), 3678–3690. https://doi.org/10.1093/nar/gki685.

Hall, M. C., & Matson, S. W. (1999). Helicase motifs: The engine that powers DNA unwinding. *Molecular Microbiology, 34*, 867–877. Retrieved from http://www.ncbi.nlm.nih.gov/htbin-post/Entrez/query?db=m&form=6&dopt=r&uid=0010594814.

Keller, H., Kiosze, K., Sachsenweger, J., Haumann, S., Ohlenschlager, O., Nuutinen, T., et al. (2014). The intrinsically disordered amino-terminal region of human RecQL4: Multiple DNA-binding domains confer annealing, strand exchange and G4 DNA binding. *Nucleic Acids Research, 42*(20), 12614–12627. https://doi.org/10.1093/nar/gku993.

Kornberg, A. (2003). Ten commandments of enzymology, amended. *Trends in Biochemical Sciences, 28*(10), 515–517. https://doi.org/10.1016/j.tibs.2003.08.007.

Lu, H., & Davis, A. J. (2021). Human RecQ helicases in DNA double-Strand break repair. *Frontiers in Cell and Development Biology*, *9*, 640755. https://doi.org/10.3389/fcell.2021.640755.

Muellner, J., & Schmidt, K. H. (2020). Yeast genome maintenance by the multifunctional PIF1 DNA helicase family. *Genes (Basel)*, *11*(2), 224. https://doi.org/10.3390/genes11020224.

Nickens, D. G., & Bochman, M. L. (2021). Characterization of the telomerase modulating activities of yeast DNA helicases. *Methods in Enzymology*, *661*, 327–342. https://doi.org/10.1016/bs.mie.2021.08.005.

Nickens, D. G., & Bochman, M. L. (2022). Pif1 activity is modulated by DNA sequence and structure. *Biochemistry*, *61*(1), 10–20. https://doi.org/10.1021/acs.biochem.1c00614.

Nickens, D. G., Rogers, C. M., & Bochman, M. L. (2018). The Saccharomyces cerevisiae Hrq1 and Pif1 DNA helicases synergistically modulate telomerase activity in vitro. *The Journal of Biological Chemistry*, *293*(37), 14481–14496. https://doi.org/10.1074/jbc.RA118.004092.

Nickens, D. G., Sausen, C. W., & Bochman, M. L. (2019). The biochemical activities of the Saccharomyces cerevisiae Pif1 helicase are regulated by its N-terminal domain. *Genes (Basel)*, *10*(6), 411. https://doi.org/10.3390/genes10060411.

Ononye, O. E., Sausen, C. W., Balakrishnan, L., & Bochman, M. L. (2020). Lysine acetylation regulates the activity of nuclear Pif1. *The Journal of Biological Chemistry*, *295*(46), 15482–15497. https://doi.org/10.1074/jbc.RA120.015164. Epub 2020 Sep 2.

Ononye, O. E., Sausen, C. W., Bochman, M. L., & Balakrishnan, L. (2021). Dynamic regulation of Pif1 acetylation is crucial to the maintenance of genome stability. *Current Genetics*, *67*(1), 85–92. https://doi.org/10.1007/s00294-020-01116-5. Epub 2020 Oct 20.

Ramanagoudr-Bhojappa, R., Byrd, A. K., Dahl, C., & Raney, K. D. (2014). Yeast Pif1 accelerates annealing of complementary DNA strands. *Biochemistry*, *53*(48), 7659–7669. https://doi.org/10.1021/bi500746v.

Rogers, C. M., & Bochman, M. L. (2017). Saccharomyces cerevisiae Hrq1 helicase activity is affected by the sequence but not the length of single-stranded DNA. *Biochemical and Biophysical Research Communications*, *486*(4), 1116–1121. https://doi.org/10.1016/j.bbrc.2017.04.003.

Rogers, C. M., Lee, C. Y., Parkins, S., Buehler, N. J., Wenzel, S., Martinez-Marquez, F., et al. (2020). The yeast Hrq1 helicase stimulates Pso2 translesion nuclease activity and thereby promotes DNA interstrand crosslink repair. *The Journal of Biological Chemistry*, *295*(27), 8945–8957. https://doi.org/10.1074/jbc.RA120.013626.

Rogers, C. M., Sanders, E., Nguyen, P. A., Smith-Kinnaman, W., Mosley, A. L., & Bochman, M. L. (2020). The genetic and physical Interactomes of the Saccharomyces cerevisiae Hrq1 helicase. *G3 (Bethesda)*, *10*(12), 4347–4357. https://doi.org/10.1534/g3.120.401864.

Rogers, C. M., Sausen, C. W., & Bochman, M. L. (2019). Gel-based assays for measuring DNA unwinding, annealing, and strand exchange. *Methods in Molecular Biology*, *1999*, 255–264. https://doi.org/10.1007/978-1-4939-9500-4_16.

Rogers, C. M., Simmons, R. H., III, Fluhler Thornburg, G. E., Buehler, N. J., & Bochman, M. L. (2020). Fanconi anemia-independent DNA inter-strand crosslink repair in eukaryotes. *Progress in Biophysics and Molecular Biology*, *158*, 33–46. https://doi.org/10.1016/j.pbiomolbio.2020.08.005. Epub 2020 Aug 30.

Rogers, C. M., Wang, J. C., Noguchi, H., Imasaki, T., Takagi, Y., & Bochman, M. L. (2017). Yeast Hrq1 shares structural and functional homology with the disease-linked human RecQ4 helicase. *Nucleic Acids Research*, *45*(9), 5217–5230. https://doi.org/10.1093/nar/gkx151.

Rossi, S. E., Ajazi, A., Carotenuto, W., Foiani, M., & Giannattasio, M. (2015). Rad53-mediated regulation of Rrm3 and Pif1 DNA helicases contributes to prevention of aberrant fork transitions under replication stress. *Cell Reports*, *13*(1), 80–92. https://doi.org/10.1016/j.celrep.2015.08.073.

Sanders, E., Nguyen, P. A., Rogers, C. M., & Bochman, M. L. (2020). Comprehensive synthetic genetic array analysis of alleles that interact with mutation of the Saccharomyces cerevisiae RecQ helicases Hrq1 and Sgs1. *G3 (Bethesda)*, *10*(12), 4359–4368. https://doi.org/10.1534/g3.120.401709.

Sausen, C. W., Rogers, C. M., & Bochman, M. L. (2019). Thin-layer chromatography and real-time coupled assays to measure ATP hydrolysis. *Methods in Molecular Biology*, *1999*, 245–253. https://doi.org/10.1007/978-1-4939-9500-4_15.

Simmons, R. H., 3rd, Rogers, C. M., & Bochman, M. L. (2021). A deep dive into the RecQ interactome: Something old and something new. *Current Genetics*, *67*(5), 761–767. https://doi.org/10.1007/s00294-021-01190-3.

Studier, F. W. (2005). Protein production by auto-induction in high density shaking cultures. *Protein Expression and Purification*, *41*(1), 207–234. Retrieved from http://www.ncbi.nlm.nih.gov/pubmed/15915565.

Stundon, J. L., & Zakian, V. A. (2015). Identification of Saccharomyces cerevisiae genes whose deletion causes synthetic effects in cells with reduced levels of the nuclear Pif1 DNA helicase. *G3 (Bethesda)*, *5*(12), 2913–2918. https://doi.org/10.1534/g3.115.021139.

Vasianovich, Y., Harrington, L. A., & Makovets, S. (2014). Break-induced replication requires DNA damage-induced phosphorylation of Pif1 and leads to telomere lengthening. *PLoS Genetics*, *10*(10), e1004679. https://doi.org/10.1371/journal.pgen.1004679.

Zhou, R., Zhang, J., Bochman, M. L., Zakian, V. A., & Ha, T. (2014). Periodic DNA patrolling underlies diverse functions of Pif1 on R-loops and G-rich DNA. *eLife*, *3*, e02190. https://doi.org/10.7554/eLife.02190.

Zybailov, B., Gokulan, K., Wiese, J., Ramanagoudr-Bhojappa, R., Byrd, A. K., Glazko, G., et al. (2015). Analysis of protein-protein interaction Interface between yeast mitochondrial proteins Rim1 and Pif1 using chemical cross-linking mass spectrometry. *The Journal of Proteomics & Bioinformatics*, *8*(11), 243–252. https://doi.org/10.4172/jpb.1000376.

CHAPTER NINE

Biochemical and single-molecule techniques to study accessory helicase resolution of R-loop proteins at stalled replication forks

Alisa E. Shaw[a], Sreeya Kairamkonda[a], Harshad Ghodke[b,c], and Grant D. Schauer[a,*]

[a]Department of Biochemistry and Molecular Biology, Colorado State University, Fort Collins, CO, United States
[b]Molecular Horizons and School of Chemistry and Molecular Bioscience, University of Wollongong, Wollongong, NSW, Australia
[c]Illawarra Health & Medical Research Institute, Wollongong, NSW, Australia
*Corresponding author: e-mail address: grant.schauer@colostate.edu

Contents

Methods in Enzymology, Volume 673
ISSN 0076-6879
https://doi.org/10.1016/bs.mie.2022.03.012

Abstract

R-loop proteins present a stable and robust blockade to the progression of a DNA replication fork during S-phase. The consequences of this block can include mutagenesis and other irreversible chromosomal catastrophes, causing genomic instability and disease. As such, further investigation into the molecular mechanisms underlying R-loop protein resolution is warranted. The critical role of non-replicative accessory helicases in R-loop protein resolution has increasingly come into light in recent years. Such helicases include the Pif1-family, monomeric helicases that have been studied in many different contexts and that have been ascribed to a multitude of separable protective functions in the cell. In this chapter, we present protocols to study R-loop protein resolution by Pif1 helicase at stalled replication forks using purified proteins, both at the biochemical and single-molecule level. Our system uses recombinant proteins expressed in *Saccharomyces cerevisiae* but could apply to practically any organism of interest due to the high interspecies homology of the proteins involved in DNA replication. The methods we outline are extensible to many systems and should be applicable to studying R-loop clearance by any Superfamily (SF) 1B helicase. These techniques will further enable mechanistic research on these critical but understudied components of the genomic maintenance program.

1. Introduction

High fidelity DNA replication requires the parental genomic template to be free of obstacles that may impede progression of the replication machinery. In eukaryotes, DNA replication is coordinated by the replisome, a multienzymatic machine comprising CMG helicase (Cdc45, Mcm2-7 core helicase, and GINS), the leading strand polymerase (Pol ε), the lagging strand polymerase (Pol δ), and the primase-polymerase (Pol α), among other associated factors like Mcm10, PCNA (Proliferating Cell Nuclear Antigen) sliding clamps, the RFC (Replication Factor C) clamp loader, and single-stranded binding protein RPA (Replication protein A) (Bai et al., 2017; Bell & Labib, 2016; Burgers & Kunkel, 2017; Kornberg & Baker, 2005; O'Donnell, Langston, & Stillman, 2013). CMG helicase fully encircles the leading strand during active replication and progresses in a 3'–5' direction, excluding the lagging strand (Bochman & Schwacha, 2008; Douglas, Ali, Costa, & Diffley, 2018; Georgescu et al., 2017; Ilves, Petojevic, Pesavento, & Botchan, 2010; Lõoke, Maloney, & Bell, 2017; Meagher, Epling, & Enemark, 2019; Vijayraghavan & Schwacha, 2012).

Roadblocks on parental DNA pose a threat to genomic integrity. These blocks can include RNA-DNA hybrids with displaced single-stranded (ss) DNA (a.k.a. R-loops), DNA secondary structures such as G-quadruplexes, interstrand crosslinks from chemical agents and chemotherapeutics, actively transcribing RNA polymerases, and other protein complexes (Brüning, Howard, & McGlynn, 2014; Gómez-González & Aguilera, 2019; Perera, Behrmann, Hoang, Griffin, & Trakselis, 2019; Rickman & Smogorzewska, 2019; Semlow & Walter, 2021). Of particular interest are R-loop protein blocks such as active transcription complexes, as they are endogenous to the cell yet present robust barriers to replication that are often directly mutagenetic (Azvolinsky, Giresi, Lieb, & Zakian, 2009; Crossley, Bocek, & Cimprich, 2019; García-Muse & Aguilera, 2019; Helmrich, Ballarino, & Tora, 2011; Kemiha, Poli, Lin, Lengronne, & Pasero, 2021; Lalonde, Trauner, Werner, & Hamperl, 2021; Saponaro, 2022; St Germain, Zhao, & Barlow, 2021). Replication fork pausing at R-loops can arise from topological stress from positive supercoiling ahead of the fork, or from a simple physical blockade of replisome translocation along its template (Shyian & Shore, 2021). In addition to basal transcription by RNA polymerase II (RNAPII), highly transcribed elements such as tRNA and rRNA genes have been shown to cause replication fork pausing and damage in yeast (Alzu et al., 2012; Azvolinsky et al., 2009; Prado & Aguilera, 2005). Transcription complexes on the leading strand are particularly problematic in S-phase, where they may cause direct "head-on" collisions with actively elongating replisomes, although co-directional collisions are often deleterious as well (Hamperl, Bocek, Saldivar, Swigut, & Cimprich, 2017; Hamperl & Cimprich, 2014; Helmrich et al., 2011; Kumar, Batra, Griffith, & Remus, 2021). Indeed, R-loops have been demonstrated to be directly associated with human carcinogenesis (Kotsantis et al., 2016; Macheret & Halazonetis, 2018; Stork et al., 2016). In humans and other eukaryotes, replication initiation and termination is intimately coordinated with transcription to avoid deleterious head-on collisions with R-loop blocks (Candelli, Gros, & Libri, 2018; Chen et al., 2019; Dellino et al., 2013; Felipe-Abrio, Lafuente-Barquero, García-Rubio, & Aguilera, 2015; Hamperl et al., 2017; Macheret & Halazonetis, 2018; Saponaro, 2022). Still, many harmful R-loop protein blocks associated with cellular metabolism remain in place during S-phase.

Various accessory helicases enable the replication fork to bypass these problematic blockades, some of which translocate in a 5'–3' direction opposite from the replisome (Gao & Yang, 2020; Hizume & Araki, 2019; Su et al., 2019). The PIF1 helicase family, part of the helicase Superfamily (SF) 1B, is conserved from bacteria to mammals and consists of multifunctional monomeric accessory helicases (Bessler, Torres, & Zakian, 2001; Byrd & Raney, 2017; Muellner & Schmidt, 2020). In humans, only one PIF1 family helicase,

Pif1, is expressed, while *S. cerevisiae* encode two: Pif1 and Rrm3. This chapter will primarily focus on the *S. cerevisiae* Pif1 accessory helicase. Pif1 can unwind double-stranded (ds) DNA, but demonstrates better proficiency at unwinding RNA:DNA hybrids (Boulé & Zakian, 2007; Chib, Byrd, & Raney, 2016; Mohammad, Wallgren, & Sabouri, 2018; Ramanagoudr-Bhojappa et al., 2013), and it also has the ability to strip proteins off of DNA (Koc, Singh, Stodola, Burgers, & Galletto, 2016; Ramanagoudr-Bhojappa et al., 2013). It has been studied under a wide variety of contexts and has been shown to have many roles in the nucleus, including Okazaki fragment processing (Osmundson, Kumar, Yeung, & Smith, 2017; Pike, Burgers, Campbell, & Bambara, 2009; Sparks, Burgers, & Galletto, 2020), replication through G-quadruplexes and telomeres (Dahan et al., 2018; Geronimo & Zakian, 2016; Kumar et al., 2021; Nickens & Bochman, 2022; Sparks, Singh, Burgers, & Galletto, 2019), break-induced repair (Buzovetsky et al., 2017; Li et al., 2021; Wilson et al., 2013), regulation of the ribosomal DNA barrier (Ivessa, Zhou, & Zakian, 2000), and removal of replication barriers (Koc et al., 2016; Kumar et al., 2021; Schauer et al., 2020; Tran et al., 2017), among others. Interestingly, Pif1 has been demonstrated to make physical contacts with replisome components (Buzovetsky et al., 2017; McDonald et al., 2016; Wilson et al., 2013). Given the role of similar accessory SF1B helicases in bacteria such as Rep and UvrD that travel opposite to the replicative helicase in removing protein blocks ahead of the replication fork in bacteria (Brüning et al., 2014), it is perhaps unsurprising that Pif1 may also play a similar vital role in stripping R-loop protein blocks alongside the replication fork. The goal of this chapter is to enable further studies into this critical activity.

We previously showed that dCas9 (nuclease deactivated Cas9) is a programmable, potent block to viral, bacterial, and eukaryotic replication forks (Whinn et al., 2019; see Fig. 1A), and this block has since been shown to cause replication fork stalling and mutagenesis in yeast (Doi et al., 2021; Laughery, Mayes, Pedroza, & Wyrick, 2019). Recently, we demonstrated that Pif1 helicase can liberate replisomes stalled at dCas9 R-loop blocks, regardless of block orientation (Schauer et al., 2020). As a general proxy for R-loop proteins (e.g., RNAPII), the biochemical assays we discuss in this chapter use dCas9 associated with an R-loop that has 20 nucleotide (nt) complementarity to the template. dCas9 blocks will stably block replisome progression for at least an hour, providing a convenient platform to study replication fork pausing at R-loop proteins. The primary output of the assay, DNA strand extension, is a proxy for Pif1 activity, i.e., rescued

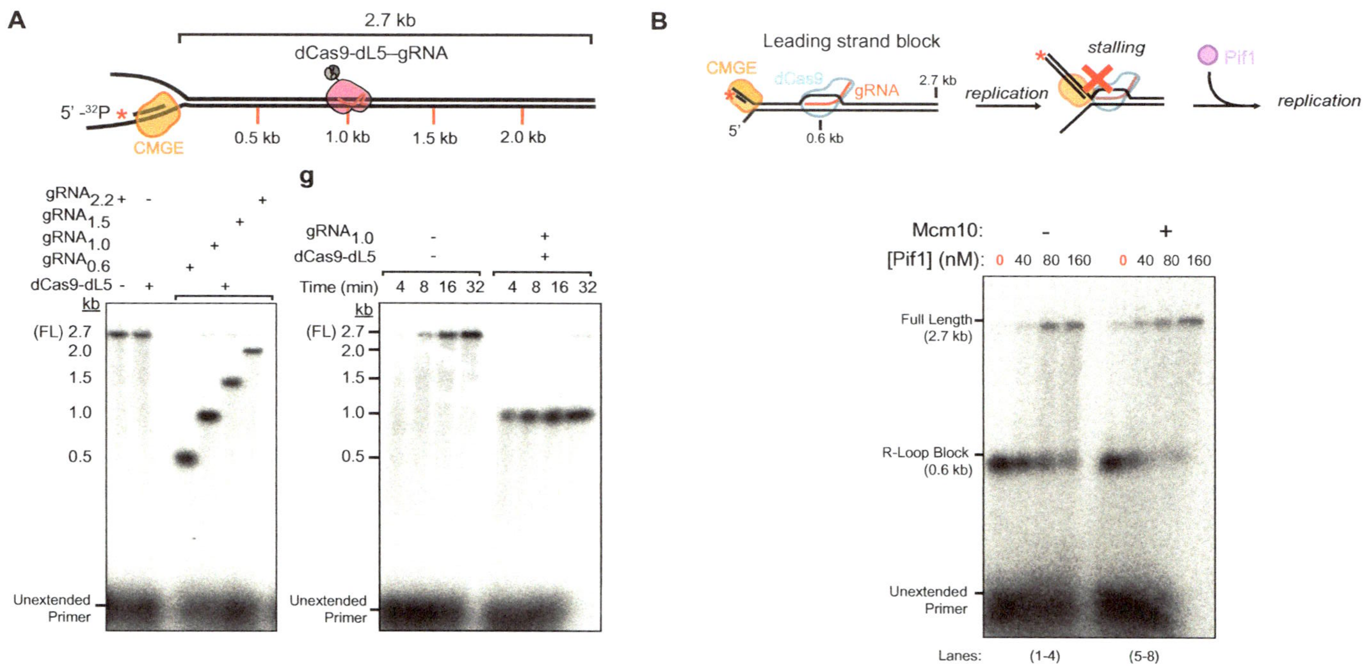

Fig. 1 Using dCas9 as a model for replication stalling and liberation at R-loops. (A) Top: Illustration of the primed forked DNA and dCas9 R-loop block. The ^{32}P-labeling position of the primer is denoted as a red asterisk. The CMG–Pol ε complex is labeled as CMGE (RFC, PCNA, Mcm10, and RPA left out for clarity). Left: dCas9-dL5 is a strong, programmable block to replication *in vitro*. The cgRNAs used to specifically target the template are indicated. cgRNA0.6 and cgRNA2.2 block the leading strand; cgRNA1.0 and cgRNA1.5 block the lagging strand. All reactions were stopped at 16 min. Right: Time course of eukaryotic replication in the presence or absence of dCas9-dL5 and cgRNA1.0. Reactions were stopped at 4, 8, 16 and 32 min as indicated. (B) Pif1 facilitates replisome bypass of a leading-strand dCas9 block. Top: Illustration of the primed forked DNA and dCas9 R-loop block. Bottom: Alkaline agarose gel analysis of the products. Where indicated, replication reactions were carried out with Mcm10 present in the initial incubation. *Panel A was modified from Whinn, K. S., Kaur, G., Lewis, J. S., Schauer, G. D., Mueller, S. H., Jergic, S., et al. (2019). Nuclease dead Cas9 is a programmable roadblock for DNA replication.* Scientific Reports, *9(1), 13292. doi:10.1038/s41598-019-49837-z; Panel B was modified from Schauer, G. D., Spenkelink, L. M., Lewis, J. S., Yurieva, O., Mueller, S. H., Oijen, A. M. van, et al. (2020). Replisome bypass of a protein-based R-loop block by Pif1.* Proceedings of the National Academy of Sciences of the United States of America, *117(48), 30354–30361. doi:10.1073/pnas.2020189117.*

fork progression after R-loop protein clearance (Fig. 1B). Using single-molecule Total Internal Reflection Fluorescence Microscopy (TIRFM), we also tracked the fate of the displaced dCas9 as well as its guide RNA (gRNA) and found that Pif1 works processively to evict the dCas9 protein and the gRNA simultaneously (Fig. 3B). In this chapter, we will discuss how to reconstitute helicase-directed block clearance at stalled replication forks at both the biochemical and single-molecule level. Although the methods we present specifically utilize *S. cerevisiae* replication proteins, they should also be generalizable among similar proteins (for example, one may wish to study other SFB1 helicases including Rrm3 or Sen1), and across all species.

2. Using dCas9 as a model R-loop protein block to replication

2.1 Expression and purification of dCas9Cys

Nuclease-deactivated Cas9 (dCas9) is easily expressed in *E. coli* and purified by a simple Ni-NTA affinity column and SP column (cation exchange) cleanup. This protocol assumes dCas9Cys has been modified to incorporate a single surface cysteine for maleimide labeling (see Section 2.2, Site-specific fluorophore labeling of dCas9Cys, for discussion). Neither the amino acid modifications nor the fluorophore alters the potent blocking activity of dCas9; therefore dCas9Cys can be used safely in both biochemical and single-molecule assays.

2.1.1 Materials and reagents

- Lysogeny Broth (LB) Ampicillin (Amp) plates and media (100 μg/mL Amp).
- dCas9Cys expression plasmid. 10xHis-MBP-TEV-S. pyogenes dCas9 M1C D10A C80S H840A C574S (heretofore called dCas9Cys) was provided to Addgene by the Doudna lab (O'Connell et al., 2014) Addgene Cat # 60815).
- DH5α *E.coli* cells (New England Biolabs # C2987H).
- Rosetta 2(DE3) *E.coli* overexpression cells (Sigma #71400).
- Thymine (Sigma #T0376).
- 1 M Isopropyl-β-D-thiogalactoside (IPTG; Sigma #16758).
- 1 M polyethylenimine (PEI; Sigma #408727).
- Ammonium sulfate (Sigma #A4418).
- Lysis buffer (20 mM Tris–HCl pH 7.6, 0.1 mM EDTA, 1 mM DTT, 150 mM NaCl, 5% (vol/vol) glycerol).

- 1 M imidazole (Sigma #15513).
- HisTrap buffer: 50 mM Tris–HCl pH 7.6, 0.5 mM EDTA, 1 mM DTT, 20 mM imidazole, 500 mM NaCl, 5% (vol/vol) glycerol.
- HiTrap SP buffer: 20 mM HEPES pH 7.6, 0.5 mM EDTA, 2 mM DTT, 150 mM NaCl, 5% (vol/vol) glycerol.
- 2X EDTA-free proteinase inhibitor tablets (Roche #5056489001).
- 0.2 M phenylmethylsulfonyl fluoride (PMSF; Sigma #11359061001) in isopropanol.
- 1 mL HisTrap Ni-NTA column (Cytiva).
- 1 mL HiTrap SP cation exchange column (Cytiva).
- TEV protease (Sigma #T4455).
- TEV buffer: 20 mM HEPES pH 7.6, 0.5 mM EDTA, 1 mM DTT, 150 mM NaCl, 10% (vol/vol) glycerol.

2.1.2 Overexpression of dCas9Cys

- Transform Rosetta cells with the dCas9Cys plasmid per manufacturer protocol, plate the outgrowth on LB-Amp plates, and grow for ~16 hours (h) at 37 °C.
- Grow two single colonies in 2 x 1 L of LB-Amp supplemented with thymine (25 μg/mL) for ~16 h at 37 °C overnight, shaking at 250 rpm, monitoring growth by A_{600}.
- When the culture reaches $A_{600} = 0.8$, reduce the temperature to 16 °C using an ice water bath for roughly 5 min. Return to the shaker and add 0.5 mM IPTG. Continue shaking at 16 °C for 16 h.
- Chill on ice. Harvest cells by centrifugation at 5000 × *g* in a swinging bucket rotor. Discard supernatant and leave cells in –80 °C (flash freeze in liquid N_2 optional) until ready for purification.

2.1.3 Purification of dCas9Cys

Perform the following steps in a cold room and/or using a chilled centrifuge and rotor (~6 °C) with 0.2 μM filtered, degassed, buffers:

- Resuspend cells in Lysis Buffer and quickly add 2 x protease inhibitor tablets and 0.7 mM PMSF.
- To lyse cells, pass twice through a French press at 12,000 psi. Alternative: Use a probe sonicator for lysis. Sonicate on ice at 60% output for seven cycles of 15s on, 45s off.
- Remove cell debris by centrifugation at 48,000 × *g* for 1 h (see note below).

- Carefully aspirate the supernatant into a fresh tube, avoiding the interface between pellet and supernatant.
- Add 0.3% (vol/vol) PEI and vigorously stir. After 1 h, centrifuge for another hour at 48,000 × *g* to remove the white DNA precipitate.
- Homogenize the remaining pellet in Lysis Buffer supplemented with 350 mM NaCl for 15 by mechanical resuspension with a spatula followed by stirring with a magnetic stir bar at 200 rpm for 15 min.
- Centrifuge at 48,000 × *g* for 1 h. Collect supernatant (~50–100 mL).
- Precipitate proteins from supernatant by adding 0.4 mg/mL solid ammonium sulfate and stirring for 60 min.
- Centrifuge at 48,000 × *g* for 30 min.
- Discard supernatant and dissolve precipitate in 30 mL of HisTrap buffer.
- Load onto a HisTrap column equilibrated with HisTrap buffer (see note below).
- Elute dCas9Cys with a linear gradient from 20 to 300 mM imidazole over 20 column volumes (CV), collecting 0.5 mL fractions.
- Pool peak fractions and dialyze against 2 L of TEV buffer for at least 3 h. Collect dialysate.
- Add 1,000 U TEV protease (Sigma) for 16 h to cleave the 10xHis-MBP tag.
- Reverse purify on a HisTrap column to remove uncleaved 10xHis-MBP-TEV-dCas9Cys by loading this fraction onto a HisTrap column preequilibrated with HisTrap buffer and collecting the flow-through.
- Pool fractions and dialyze overnight against 2 L of HiTrap SP buffer.
- Load onto a HiTrap SP column equilibrated with HiTrap SP buffer at 0.5 mL/min.
- Elute with a linear gradient from 150 mM to 1 M NaCl over 20 CV in HiTrap buffer.
- Collect peak fraction. Dialyze against 2 L of storage buffer (50 mM HEPES pH 7.6, 1 mM EDTA, 3 mM DTT, 300 mM NaCl, 50% (vol/vol) glycerol).
- Aliquot peak fractions into 20 μL multi-use aliquots into 0.6 mL microcentrifuge tubes, flash freeze in liquid nitrogen, and store at −80 °C. Aliquots may be thawed and flash frozen three times before discarding.

Note: Centrifuge speed for steps that note 48,000 × *g* can safely be raised as much as physically possible with your rotor of choice, often with benefit in yield. Conversely, it is possible to spin as slow as ~24,000 × *g* accompanied by doubling spin times and careful recovery/removal of the supernatant due to less efficient separation at the pellet: supernatant interface.

Note: In general, we do not recommend using more than 1 mM DTT or any other reducing agent with a Ni-NTA column, as it will reduce the Ni^{2+} on the resin. Although manufacturer instructions often claim that up to 2 mM DTT is tolerated by these columns, we have found a loss in yield above 1 mM. You can therefore safely opt to bring your buffers to 0.5 mM DTT, with no discernable effect on the outcome of the prep. If for some reason you decide to try any other tweaks to the buffer, we recommend to stay $\geq$100 mM NaCl to stabilize dCas9Cys, and to avoid chelators like EDTA, as they will chelate the nickel and drastically reduce binding affinity.

2.2 Site-specific fluorophore labeling of dCas9Cys

In our original report on dCas9 blocks, we used dCas9-dL5, which is a fusion of dCas9 with a photostable fluoromodule called dL5, which is activated upon binding of malachite green ester (Whinn et al., 2019). This construct allows long observation times under continuous fluorescence excitation (unpublished observations) and allowed us to perform single-particle tracking in live cells. In our subsequent *in vitro* work (Schauer et al., 2020), we found that site-specific maleimide-dye labeling was sufficient to observe fluorescence over an extended period of time, particularly when sequentially imaging many regions in serial (*i.e.*, cycling between flow cell regions during acquisition), as this minimizes photobleaching associated with continuous laser excitation (see Section 5.4, Imaging of Single-Molecule Reactions). Furthermore, site-specific fluorescence labeling is the superior alternative to nonspecific fluorescence labeling for single-molecule experiments, which are predicated on the observation of individual fluorophores. We therefore recommend using dCas9Cys for the *in vitro* experiments we outline here. However, if you plan to perform live-cell imaging of R-loop blocks (not discussed here), we recommend generating the dCas9-dL5 fusion as described in Whinn et al. (2019).

2.2.1 Materials and reagents

- Alexa Fluor™ C_5-maleimide (*e.g.*, Thermo Fisher #A10254) or equivalent dye-maleimide conjugate.
- Reducing buffer (50 mM HEPES pH 7.6, 500 mM NaCl, 1 mM EDTA, 70% (wt/vol) ammonium sulfate, and 10% (vol/vol) glycerol.
- Ammonium sulfate.
- Superdex G-25 resin (Cytiva # 17003301).
- Filtration buffer (50 mM HEPES pH 7.6, 3 mM DTT, 1 mM EDTA, 300 mM NaCl, 10% (vol/vol) glycerol).

- 1.5 x 10 cm glass column (*e.g.*, BioRad Econo-Column® #7374151).
- Storage buffer (50 mM Tris–HCl pH 7.6, 1 mM EDTA, 500 mM NaCl, 25% (vol/vol) glycerol).

2.2.2 Method of site-specific fluorophore labeling of dCas9Cys

The following methods were adapted from Kim et al. (2008). We used Alexa Fluor 488 (AF488) in Schauer et al. (2020), but any color maleimide dye can be substituted. Unless otherwise noted, perform the following steps in a cold room and/or using a chilled centrifuge and rotor (~6 °C) with 0.2 μM filtered, degassed, buffers:

- Reduce dCas9Cys (ideally ≥1 mg; obtained from Section 2.1, Expression and Purification of dCasy9Cys) by adding Reducing Buffer supplemented with 3 mM TCEP, pH 7.6 for 1 h at 6 °C with gentle rotation.
- Perform the following steps three times:
  (1) Centrifuge at 21,000 × *g* for 15 min. Carefully remove and discard supernatant.
  (2) Wash the precipitate with Reducing Buffer.
- dCas9Cys is now fully reduced and in a buffer absent of reducing agent. Perform the following steps in the dark with gentle rotation in a microcentrifuge tube backfilled with inert gas (N_2 or Ar)
- Label with fluorophore using a 10-fold molar excess of maleimide-dye in 600 μL labeling buffer for 3 h at 23 °C.
- Continue labeling for 16 h in the cold room (6 °C).
- Quench the reaction with 3 mM TCEP for 1 h at 6 °C.
- Equilibrate G-25 resin to Room Temperature (RT). Pour in 1.5 x 10 cm glass column and equilibrate with 3 CV of Filtration Buffer by gravity flow.
- Carefully apply your sample to the column bed. Immediately after your sample has fully penetrated the resin, apply 20 mL Filtration Buffer.
- Collect (with a fraction collector or manually), 7-drop (~250 μL) fractions. You should clearly see fluorescently labeled dCas9 migrating faster than the free dye. If you prefer to use an FPLC to automate this process, an alternative to this method would be to use a comparable FPLC column, such as a Superose 6 10/300 GL (Cytiva #29-0915-96).
- Pool peak fractions and dialyze against 2 L Storage Buffer for 16 h.
- Estimate labeling efficiency by comparing UV/Vis spectrophotometry, comparing the protein concentration determined by A260 vs the dye concentration determined by $c = AX/(l \times \varepsilon)$ where x = your dye's peak absorbance wavelength and ε is the published molar extinction coefficient of the dye. Other protein quantification methods such as the Bradford reagent are good alternatives to A260 measure your dCas9Cys

amount when its concentration is low. See Thermo Scientific Tech Tip #31 for more information: https://tools.thermofisher.com/content/sfs/brochures/TR0031-Calc-FP-ratios.pdf.

- *Note*: You may also wish to visualize labeling on your SDS-PAGE gel to determine the efficiency of separation of dye. Run the loading dye front about 3/4 of the way down on a 12.5% Tris-Glycine SDS-Gel. Before staining with Coomassie, scan the gel at the appropriate excitation wavelength for your dye on a multi-wavelength scanner (e.g. Typhoon 9500). Use standard densitometry routines in the FIJI image analysis software (Schindelin et al., 2012) to integrate the total signal of your protein vs. that of the free dye to estimate the percentage of dye incorporated.
- Aliquot into 20 μL multi-use aliquots into 0.6 mL microcentrifuge tubes. Aliquots may be thawed and flash frozen three times before discarding.

2.3 Designing and synthesizing sgRNA for targeting regions of template DNA

dCas9 requires gRNA for site-specific DNA template targeting. In *S. pyogenes*, dCas9 uses two RNAs as gRNA: tracRNA and cRNA. We use two methods of making gRNA for dCas9. In the method discussed in this chapter, we use a kit to generate single guide RNA (sgRNA), a convenient linear combination tracRNA and cRNA in a single RNA that can be quickly made with recombinant DNA and an *in vitro* sgRNA transcription kit.

Note: Another method to obtain gRNA is to simply order synthetic tracRNA and cRNA—this route is more expensive but is particularly useful for using fluorescently modified RNA for single-molecule assays. However, we have observed salt intolerance with this method, requiring low ionic strength for efficient targeting activity and stability. This effect was not observed once we began to use sgRNA.

Designing gRNA is simple: find a protospacer adjacent motif (PAM) site (typically 5'-NGG) near a region of interest on your replication template (see Section 4, Biochemical reconstitution of Pif1 clearance of R-loop stalled replication forks). Your targeting region will be the 20 bases upstream of the PAM site. As CRISPR/Cas9 is extremely efficient even when targeting whole yeast genomes in nuclei, it is that much more efficient on smaller and simpler plasmid-based linear templates. Based on our experience we would consider scenarios with poor targeting and/or off-targeting to be highly unlikely. You may still wish to check your targeting sequence for such effects using various web-based target selection tools like CHOPCHOP (Labun et al., 2019) that predict these and other useful parameters for CRISPR-based strategies.

2.3.1 Materials and reagents

- sgRNA synthesis kit (New England Biolabs EnGen #E3322S or equivalent)
- ∼55 nt synthetic sgRNA DNA template DNA (IDT or equivalent)

2.3.2 Method of preparing sgRNA in vitro

For producing sgRNA, we recommend following the manufacturer sgRNA synthesis protocol without modification. For the EnGen kit, this entails ordering the target-specific synthetic DNA required to make sgRNA: Using the manufacturer guidelines, a T7 promoter sequence is added to the 5'- end of 20 nt targeting region and a special overlap sequence is added to the 3'-end of the targeting region. In a one-step reaction, the overlap sequence tails of the ssDNA strands anneal in an antiparallel manner, T7 DNA polymerase fills in the gaps, and T7 RNA polymerase uses the resultant ∼110 bp DNA as a 2-copy template for sgRNA synthesis. Cost-saving alternatives to using this kit include making a gRNA expression template yourself by PCR and transcribing it with T7 RNA Polymerase as detailed in Zhang, Gao, Wang, & Zhao (2017). The sgRNA is subsequently cleaned up by silica-based spin column designed for RNA (e.g., New England Biolabs #T2040), and stored at −80 °C until required. The resultant sgRNA can be freeze/thawed dozens of times without discernable loss in targeting efficiency.

3. Purification of Pif1 helicase

As the generation of protein overexpression strains and general protein preparation is beyond the scope of this article, we recommend referring to the detailed chapter in *Methods in Enzymology* on expression of recombinant protein complexes in yeast from the O'Donnell lab (Mayle & O'Donnell, 2021). The protocol we present here assumes you have performed the steps outlined in that chapter to make an overexpression strain of your accessory helicase of interest, and that you have prepared a glycerol stock for safe keeping in the ultracold freezer. An alternative to integrating a GAL-inducible Pif1 overexpression cassette into the yeast genome is to simply clone Pif1 into a yeast episomal overexpression plasmid, or YEP. Typically, these plasmids contain an inducible GAL1/10 promoter for overexpression of your gene of interest and a fragment from the 2 μm circle that contains the elements necessary to for autonomous high copy replication when transformed into yeast cells. Both methods—strain-integrated overexpression cassettes and YEPs—require auxotrophic selection of a nutritional marker present in the cassette by growing in minimal media lacking your nutritional marker of interest. While nutritional selection can be performed on overexpression strains at the plating stage only and carried forward with rich media for the remainder of the prep, a drawback of YEPs is that auxotrophic selection must be maintained throughout the entire growth, resulting in longer media prep times and higher reagent costs. Thus, we recommend integrating your overexpression cassette.

Although purification of Pif1 is a straightforward two-column preparation, we have yet to identify conditions that result in what we consider to be high yield (roughly ≥2 mg per 12 L culture), however, a single preparation is usually sufficient for dozens of experiments. Though we have no experience with other methods of purification and thus cannot attest to relative yields, if your experiments require large amounts of Pif1, it may be worth trying to express and purify this helicase (or other accessory helicases) in *E. coli*, as described in Deegan, Baxter, Ortiz Bazán, Yeeles, & Labib (2019), as bacterial expression can often result in higher yields. Of course, Pif1 or your helicase of interest will not bear post-translational modifications with this method, so it may not be appropriate for certain accompanying lines of investigation (*e.g.*, protein-protein interaction studies). It may also be worth exploring the use of other affinity tags that can help solubility and yield, such as maltose binding protein (MBP).

3.1 Materials and reagents

- 3x-Flag-Pif1 overexpression strain (TRP1:GAL1,10-3xFlag-Pif1, ade2-1 ura3-1 his3-11,15 trp1-1 leu2-3,112 can1-100 bar1Δ MAT a pep4::KANMX6).
- Synthetic drop-out mix without yeast nitrogen base, adenine rich minus tryptophan, without Yeast Nitrogen Base (YNB; US Biological Life Sciences #D9531).
- YNB without amino acids (US Biological Life Sciences Y2025).
- Agar (Fisher # BP1423-500).
- Peptone (Fisher # DF0118-07-2).
- Yeast extract (Fisher #AAJ23547A1).
- Dextrose (Sigma #G7021-10KG).
- Glycerol (Fisher # BP229-4).
- Galactose, low endotoxin (US Biological Life Sciences #G1030).
- Polyvinylpyrrolidone (Sigma # PVP40-100G).

- Proteinase inhibitor cocktail for yeast extracts (Sigma #P8125).
- CryoMill (SPEX Sample Prep).
- Anti-flag M2 resin (Sigma #A2220).
- 3xFLAG peptide (Glpbio # GP10149).
- Dialysis buffer (0.1 M NaCl, 25 mM HEPES pH 7.6, 5% (vol/vol) glycerol, 1 mM DTT).
- 1 mL Mono S 5/50 cation exchange column (Cytiva #17516801).

3.2 Method of overexpression and purification of Pif1 accessory helicase

- Prepare dropout plates: Make a solution of 2% dextrose, 0.7% YNB, 0.2% synthetic drop-out mix. pH to 5.2 with NaOH (up to 5.8 is OK). Add 0.2% agar and autoclave. Pour ~25 mL per plate when warm to touch.
  - o Note: Occasionally the amino acids in the drop-out mix can burn off in the autoclave. You may wish to add more than necessary or add the dropout powder shortly after autoclaving.
- Streak 3x-Flag-Pif1 overexpression strain onto minimal media plates and grow for 16–24 h at 30 °C.
- Make Yeast Peptone Dextrose (YPD): Add 20 g peptone, 10 g yeast extract, 20 g of dextrose to 1 L H_2O. Autoclave, then cool to RT.
- Make 12x 1 L of Yeast Peptone Glycerol (YPG). Add 20 g peptone, 10 g yeast extract, and 20 g Galactose to each 1 L in a 2 L baffled flask.
- Grow a few colonies for 16 h in 5 mL YPD at 30 °C. Add saturated culture to remaining 1 L of YPD. Grow until OD_{600} = 0.6–1.2, ensuring your inoculant never reaches stationary phase.
- Split YPD culture into 12 YPG flasks and grow at 30 °C.
  - o Note: Depending on the doubling rate of your strain (which you should calculate yourself, but which should fall around 1.5 h for a "healthy" strain) we find that starting YPG cultures around OD_{600} = 0.01 allows them to get to an induction-ready density by morning, *e.g.* in 16 h. In the case of a starter culture at OD_{600} = 0.8, that would mean adding 12.5 mL to each flask. We have found that yeast make a better transition from glucose to glycerol when adding >10 mL and <25 mL of starter YPD culture, as a way to supplement the culture with trace amounts of glucose that becomes fully metabolized while making the carbon source switch.
- When the YPG cultures have reached OD_{600}, induce overexpression by adding 20 g galactose.

- Harvest cells by centrifugation at 1,500 × *g* in a chilled (6 °C) swinging-bucket rotor for 15 min. Pour off supernatant and re-use centrifuge bottles to pellet the remainder of YPG cultures if necessary.
- With a serological pipette, resuspend yeast with 5–10 mL of 20 mM HEPES pH 7.6, 1.2% polyvinylpyrrolidone (optional detergent that aids resuspension), and 1 mL protease inhibitor cocktail per 20 g yeast cells.
- Using the pipette, freeze your biomass dropwise in a vat of liquid nitrogen. Cells can be stored for several weeks in an ultracold freezer without reduction in yield.
- Lyse cells in cryomill set to 15 min precool followed by 15 cycles of 2 min run time and 2 min cool time at 12 cycles per second (CPS).

Perform the following steps in a cold room and/or using a chilled centrifuge and rotor (~6 °C) with 0.2 mM filtered, degassed, buffers:

- Clarify lysate by centrifugation at 48,000 × *g* in an SS-34 rotor or equivalent for 1 h at 4 °C. Transfer to new tube, avoiding pellet: supernatant interface, and centrifuge for an additional 1 h at 4 °C.
- Apply the supernatant to a 1 mL anti-FLAG M2 column equilibrated with buffer A (350 mM potassium glutamate, 25 mM HEPES, pH 7.5, 10% glycerol, 1 mM DTT).
- Wash with 30 column volumes (CV) of Buffer A.
- Elute protein in Buffer A supplemented with 200 mg/mL 3xFLAG peptide. Add 0.75 CV of elution buffer six times, pausing 30 min between each elution step. Collect each fraction in a new 0.5 mL tube.
- Pool elutions and dialyze eluate against 2 L of dialysis buffer.
- Load onto a Mono S column using a 10 mL syringe or superloop.
- Elute with a 0.1–0.8 M NaCl gradient over 20 CV in elution buffer (25 mL HEPES, 5% (vol/vol) glycerol, 1 mM DTT).
- Aliquot peak fractions into 20 μL multi-use aliquots into 0.6 mL microcentrifuge tubes, flash freeze in liquid nitrogen, and store at −80 °C. Aliquots may be thawed and flash frozen three times before discarding.

4. Biochemical reconstitution of Pif1 clearance of R-loop stalled replication forks

One important aspect of our biochemical results is that Pif1 is unable to strip off R-loop proteins without the replisome present (Schauer et al., 2020). Thus, the primary readout of R-loop clearance activity by accessory helicases arises from a DNA replication assay: the replisome will pause indefinitely at dCas9 until rescued by an accessory helicase. For a more general overview of replication strand-extension assays, including detailed protocols, see Schauer, Finkelstein, & O'Donnell (2017). As the protocol described herein is designed to work with replisomes from any species, we assume that the reader has access to a reconstituted replication system of their own, the construction of which is well beyond the scope of this chapter. If you indeed want to reconstitute *S. cerevisiae* replication, we refer you to the detailed chapter on expression of recombinant protein complexes in yeast from the O'Donnell lab (Mayle & O'Donnell, 2021) or in the chapter "Optimizing CMG helicase and CMG-dependent replication assays by designing DNA fork substrates and choosing nucleotide analogues for helicase preloading" by Yao and O'Donnell. The protein complexes we use herein can be reproducibly purified following the detailed protocols outlined in their respective references. This includes CMG helicase and Pol ε (Georgescu et al., 2014), RFC (Finkelstein, Antony, Hingorani, & O'Donnell, 2003), PCNA (Yao et al., 2003), Mcm10 (Lewis et al., 2017), MTC (Langston et al., 2017), Pol δ and Pol α, (Georgescu et al., 2015), and RPA (Henricksen, Umbricht, & Wold, 1994). Note that we often approach studies with a "leading-strand only" replisome, as it is fully functional in leading strand replication without the lagging strand machinery (Lewis et al., 2017, 2020; Schauer & O'Donnell, 2017; Wasserman, Schauer, O'Donnell, & Liu, 2019). This approach additionally helps eliminate variables arising from adding additional polymerases Pol α and Pol δ. For instance, Pol α might reprime ahead of the R-loop (Guilliam & Yeeles, 2021). Also, with this approach, only the complex between CMG and Pol ε is able unwind dsDNA and continue replication, as Pol ε does not possess strand displacement capability on its own like Pol δ, which for this reason must be added carefully. Thus, we know for sure that all DNA polymerization visualized on the gel was driven by CMG helicase. However, if you decide to exclude any part of the replisome in your studies, it is critical to perform control experiments with the full set of replisome proteins as we have in these references, to demonstrate that your primary observations do not change by altering the replisome composition.

4.1 Materials and reagents

- dCas9Cys (made in Section 2.1) and your target-specific gRNA (made in Section 2.3).
- Stocks of replication proteins: CMG, Pol ε, RFC, PCNA, Mcm10, MTC, Pol δ, Pol α, and RPA.
- 3kb_160:C2* substrate. Preparation of the linear forked substrate 3kb_160 has been described in detail elsewhere (Schauer et al., 2017). The process involves linearizing the 2.8 kb pUC19 plasmid with BsaI-HF

(New England Biolabs), using T7 ligase (New England Biolabs #M0318L) to ligate a synthetic, non-complementary fork with a site on the leading strand complementary to a 37 nt primer ("C2"), and removing excess fork with size exclusion chromatography (Sepharose 4B; Cytiva #17012001). Adjust the fork design as necessary for your replication system of interest. Since the plasmid is linearized in the AmpR cassette, you can also choose a wide variety of plasmids as templates while retaining the same fork overhang for sticky-end ligation. The primer is labeled with T4 polynucleotide kinase (PNK; New England Biolabs #M0201L) by ATP-γ-^{32}P (Perkin Elmer #BLU002Z250UC). Excess radioactive ATP is removed with a G-25 SEC column (Cytiva #27532501), and the primer is annealed to the 3kb_160 fork to yield a substrate for eukaryotic strand extension, 3kb_160:C2*.
- 10X MK (0.1 M MgOAc and 0.5 M KOAc).
- 5X TDBG (125 mM Tris-acetate pH 7.5, 25% (vol/vol) glycerol, 200 μg/mL BSA, 15 mM TCEP, 10 mM DTT).
- Recirculating, temperature-controlled water bath with tube rack.
- dA/dC mix (10 mM dATP and 10 mM dCTP in 10 mM Tris–HCl pH 7.5 and 1 mM EDTA).
- dT/dG mix (10 mM dTTP and 10 mM dGTP in 10 mM Tris–HCl pH 7.5 and 1 mM EDTA).
- Stop solution (1% Sodium dodecyl sulfate (SDS), 40 mM EDTA, 10% (vol/vol) glycerol, 0.02% bromophenol blue, 0.02% xylene cyanol).
- Alkaline running buffer (30 mM NaOH and 1.125 mM EDTA).
- LE agarose (GoldBio #A-201-100).
- 15 x 20 cm horizontal alkaline agarose apparatus (Biorad #1704483).
- C-fold paper towels.
- 3MM Whatman paper (Sigma #WHA3030917; 46 x 57 cm).
- DE81 DEAE cellulose paper or equivalent (e.g., Hybond-N+, Cytiva # RPN303B).
- Phosphorimager screen (Cytiva).
- Screen exposure cassette (Cytiva).
- Multiwavelength scanner (e.g., Typhoon FLA 9500, Cytiva).

4.2 Method of reconstituting Pif1 rescue of replication fork stalling at R-loop block

Typically, 20 or 25 μL volumes are sufficient for these reactions. If you are running a time course, add more volume (e.g., 1 more reaction) than you need. The reaction is separated into three phases (see Fig. 2A): (i) helicase

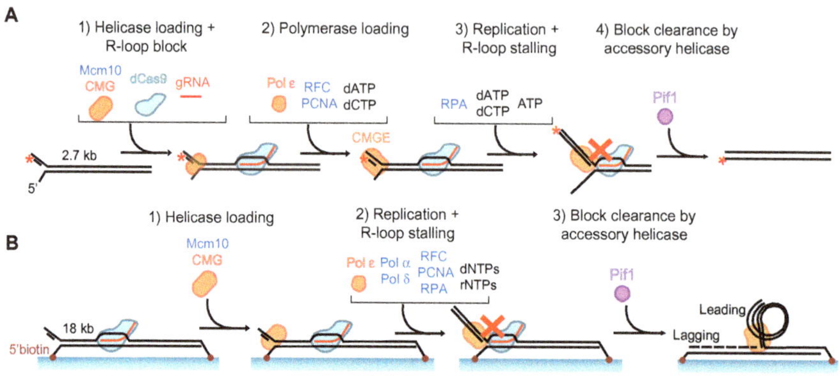

Fig. 2 Schematic of replication assay staging in biochemical and single-molecule experiments. The order-of-addition of assay components (fully described in text) is shown schematically for (A) biochemical experiments described in Section 4 and (B) single-molecule experiments described in Section 5. Components noted in blue font are omitted from the schematic for clarity.

loading onto the substrate. We separate this phase because loading requires threading onto the leading strand and is relatively inefficient. Mcm10 helps helicase loading to ssDNA (Wasserman et al., 2019). ATP (250 μM, which is low concentration relative to the 5 mM required for replication) is added to help the CMG translocate to the fork junction without unwinding the substrate. As an alternative, you may wish to use AMP-PNP, which was recently shown to aid CMG loading without causing unwinding (Yao, Zhang, Yurieva, & O'Donnell, 2022). Small volumes help raise the effective concentration of CMG before it is further diluted by the next phases. For related reasons, we add dCas9/gRNA at this step too. (ii) Polymerase-clamp loading, where polymerases and clamp use two dNTPs to allow "idling" (repeated cycles of nucleotide incorporation and exonuclease proofreading) of Pol ε to prevent primer degradation and also to allow Pol ε to polymerize up to the helicase, forming the stable "CMGE" complex (Langston et al., 2014). Finally, (iii) replication initiation by adding the full complement of 4 dNTPs, ATP (at the 5 mM concentration required for replication-driven unwinding by CMG), and RPA. Keep each phase roughly equal volume to aid mixing by pipette. As each phase is prepared separately before mixing, it is important to independently add salts and buffers (i.e., TDBG, MK, and any additional salts required) to each phase before adding proteins to prevent aggregation and/or precipitation that may occur when adding proteins directly to water at low ionic strength. When different among experimental conditions, be sure to correct for incoming salt of various proteins by adding the appropriate amount of salt to tubes that do not contain those proteins.

Note: Unless noted, the concentrations listed in this protocol are to be calculated for the *final* reaction volume (25 μL), not for the individual phases.

- Starting with ddH_2O, bring Mix 1 to 1X TDBG and 1X MK. Load 1 nM of 3kb_160:C2* substrate with 60 nM Mcm10, 30 nM CMG, 20 nM dCas9Cys and 60 nM gRNA in the presence of 0.25 mM ATP (calculated for this phase only).
- Starting with ddH_2O, bring Mix 2 to 1X TDBG and 1X MK. Add 5 nM RFC, 25 nM PCNA, 20 nM Pol ε, and 120 μM dA/dC.
- Starting with ddH_2O, bring Mix 3 to 1X TDBG and 1X MK. Add 5 mM ATP, 120 μM dT/dG, and 600 nM RPA.
- Split tube 1 into individual reactions. Example: For four time course reactions at three time points each, split Tube 1 into 4x8 μL tubes. Incubate for 10' at 30 °C.
- Add 8 μL of Mix 2 to each reaction, mixing by pipette. Incubate for 5' at 30 °C.
- Add 9 μL of Mix 3 to each reaction, mixing by pipette.
- Allow replisome to stall at R-loop block. Anywhere from 10 to 30 min is sufficient.
- Rescue stalled fork with helicase by doping in 80 nM of Pif1 and mixing well by pipette. Replication should continue to full length (2.8 kb) within ~10 min.
- During any point of this assay, collect a time point of interest: Withdraw 20 μL of the final reaction and add directly to a microcentrifuge tube preloaded with 8 μL of stop buffer. The SDS and EDTA in the buffer will instantly stop the replication reaction.
- Prepare alkaline agarose gel: Microwave 200 mL 1.2% LE Agarose, allowing to boil for 5 s. When warm to touch (60 °C or lower), bring to 30 mM NaOH and 1.25 mM EDTA while stirring. Pour gel with 20-well comb.
- Load reactions into alkaline agarose gel, loading less volume than each time point, e.g., 20 μL. Run at a constant 35 V for 16 h. Note: we do not recommend going higher than 40 V, as the gel can overheat and cause significant diffusion of the DNA throughout the gel.
- Compress and dry the gel: Remove the gel, place on saran wrap, overlay with a sheet of DE81 paper followed by a sheet of 3MM paper and a stack

of C-fold paper towels. Place stacks of napkins on top, cover with something flat, and add something very heavy (*e.g.*, a lead brick) on top. Allow to dry for 6–12 h.

- Remove all paper except the DE81. Enshroud gel in saran wrap, making sure the bottom face contains no air bubbles. Using an exposure cassette, expose the bottom face of the compressed gel to a phosphorimager screen. Expose for at least 24 h.
- Read gel on a multimode scanner in Storage Phosphor mode, using 16-bit grayscale and 250 μm pixel size. Adjust image contrast as necessary.

4.3 Analysis of biochemical assay

Analysis of replication products depends on the experimental design. It can range from qualitative observation to quantitative measurement. For instance, you may want to calculate the percentage of R-loop blocked replication forks went to full length after adding your helicase. For band analysis, FIJI has built-in Gel Analysis routines which can help you make comparative densitometric measurements on your bands (i.e., you can integrate and compare band intensities). For anything more complex we would recommend using ImageQuant TL (Cytiva) if you have access to it. Perform at least three replicates of every biochemical experiment, averaging any bulk property (e.g., replication speed) derived from the data along with the associated standard error of measurement (SEM) or standard deviation (SD).

5. Single-molecule observation of Pif1 clearance of R-loop stalled replication forks

Recapitulating observations from bulk experiments at the single-molecule level can be fraught with challenges. However, we have had good success with single-molecule replication and related events, perhaps owing to the robustness of the replication complexes involved. In a collaborative effort between the O'Donnell lab and the van Oijen lab, we published the first single-molecule report on eukaryotic DNA replication, using a bead-tethered assay to directly track the dynamic effect of the Mrc1-Tof1-Csm3 (MTC) complex on replication rates (Lewis et al., 2017). We then moved to single-molecule fluorescence, a more information rich technique. In a separate collaboration between the O'Donnell and Liu labs, we reconstituted replication on DNA stretched between two beads with a dual-optical trap, while monitoring stretching force as well as fluorescence of several DNA replication components, demonstrating that CMG harbors a ssDNA gate that enables transitions between ss and dsDNA (Wasserman et al., 2019). In a follow-up collaboration with the van Oijen lab, we established a versatile technique using a long linear forked DNA replication template double-tethered to a flow cell to simultaneously monitor DNA replication fork movement and the presence of fluorescently labeled replication complexes (see also the chapter "Production of long linear DNA substrates with site-specific chemical lesions for single-molecule replisome studies" by Kaur et al.), where we elucidated the dynamic composition of the replisome with fluorescently labeled replication complexes (Lewis et al., 2020). Our recent report tracking replication stalling and rescue at R-loop blocks also employed this dual-tethered template method, monitoring replication progression while also monitoring the fate of dCas9 and, separately, of the gRNA (Schauer et al., 2020). In this section, we primarily discuss the methods developed in that report.

The single-molecule assay relies on a dual-tethered forked DNA stretched along the XY plane of a microscope for visualization. The replisome loads onto the fork at one end and progresses towards the other—the readout is a fluorescence spot (Sytox-stained DNA) that moves in a line while becoming increasingly larger and more intense (see Fig. 3A). Replication of this long (~18 kb) template takes around 30 min, so we discuss how to cycle between several regions during the observation time. As a general observation, once the replisome has loaded onto the DNA substrate, it demonstrates replication activity. However it should be noted that efficiency of replisome loading onto forked substrates is relatively low, so we tend to observe roughly only 3% of visible DNA substrates become active complexes in this and in other experiments mentioned above. However, the power of single-molecule techniques lies in the ability to observe rare and/or heterogenous events, so it is trivial to parse the data to contain only demonstrably active complexes.

5.1 TIRF microscope considerations

The single-molecule fluorescence experiments from Whinn et al. (2019) and Schauer et al. (2020) were performed by the van Oijen lab, which uses a commercial objective-based TIRF setup based on a Nikon Ti2E inverted fluorescence microscope modified with a stage to mount flow cells, collecting fluorescence emission with an Electron Multiplying Charged Coupled Device (EMCCD) camera (Evolve 512, Photometrics). The Schauer lab uses a prism-based TIRF setup, using scientific complementary metal-oxide Semiconductor (sCMOS) cameras (ORCA Fusion, Hamamatsu). Our setup has a few minor advantages due to increased signal-to-noise (S/N) resulting from the TIRF optics geometry, larger observation area, and reduced cost.

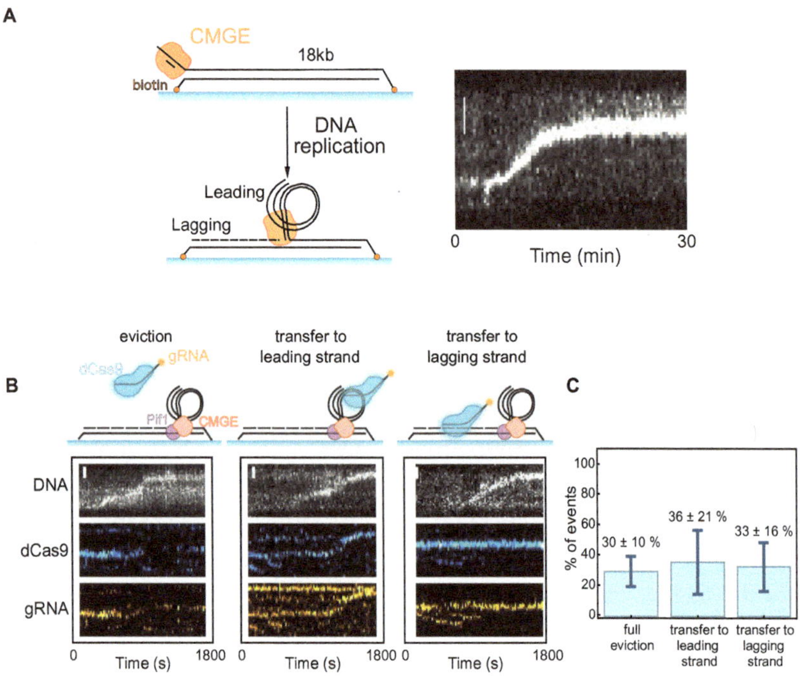

Fig. 3 Single-molecule assay of Pif1-dependent replisome bypass of a dCas9 blocked site. (A) Schematic representation of the single-molecule assay (described in Section 5) and a representative kymograph. Leading-strand DNA (stained with SYTOX Orange) accumulates along with a progressing replication fork. (B) Single-molecule analysis of the fate of the dCas9 and R-loop components of the block. Schematic illustrations and representative kymographs showing the DNA (Top, gray), fluorescently labeled dCas9 (Middle, blue), and fluorescently labeled gRNA (Bottom, orange). Three possible outcomes are observed: The block is evicted (Left), the block is transferred to the leading strand (Center), or the block is transferred to the lagging strand (Right). (C) Quantification of the observed frequency for the different behaviors outline in (A). The error bars represent the SEM from three separate experiments. *Figures were adapted from Schauer, G. D., Spenkelink, L. M., Lewis, J. S., Yurieva, O., Mueller, S. H., Oijen, A. M. van, et al. (2020). Replisome bypass of a protein-based R-loop block by Pif1.* Proceedings of the National Academy of Sciences of the United States of America, 117*(48), 30354–30361. doi:10.1073/pnas.2020189117.*

The disadvantages of our setup include the requirement to assemble and maintain a setup with even more custom-built parts (Gibbs, Kaur, Megalathan, Sapkota, & Dhakal, 2018; Martin-Fernandez, Tynan, & Webb, 2013) and the need for relatively high laser excitation power to achieve S/N comparable to or better than EMCCD (Juette et al., 2016). Also, prism-based setups require the use of fused-silica (quartz) slides, reducing excitation light scatter and thus background noise. Objective-based TIRF setups can also benefit from quartz coverslips, however, quartz is expensive, and regular borosilicate coverslips can also be used with only moderate detriment to S/N. Thus, we would recommend a commercial objective based TIRF microscope for most laboratories. Each setup will be different: for instance, you may need to design or modify your microscope stage to hold your flow cell of choice (see Section 5.2), or you may decide to use different lasers and/or cameras based on other needs in your laboratory. Choice of dyes and commonly used excitation wavelengths can vary, requiring custom filter sets to efficiently separate fluorescence emission. These filters must be matched to the unique spectral properties of the combination of your favorite dyes. For instance, our lab uses three compact continuous wave (CW) lasers: 488 nM/100 mW, 532 nm/80 mW, and 640 nm/75 mW (Coherent OBIS) to excite a 488/555/655 nm (green, red, far-red) fluorescence dye set. Via an image splitter (Cairn), emission collected through the objective is passed and/or reflected through a 560 nM longpass (LP) dichroic mirror (Chroma T560lpxr-UF1) and a 640 nM LP dichroic (Chroma T640lpxr-UF2), into three channels. The green and red dye emissions are filtered through cleanup filters (Chroma ET525/50 m and 585/65 m, respectively) and finally all images are projected onto the focal plane of three cameras. We recommend using an online tool like the Chroma Spectra Viewer to design your filter set (https://www.chroma.com/spectra-viewer). Since each setup can vary significantly, this section should be treated as a flexible set of recommendations rather than strict guidelines.

5.2 Considerations for TIRF microscope flow chambers

The single-molecule experiments we describe require the use of a flow cell to stretch and double-tether DNA templates onto the surface. Luckily, compared to some experiments which require high flow rates, the slow dual-tethered DNA deposition flow rates (described in Section 5.2) are moderate and not particularly demanding in terms of flow cell integrity. The options for constructing flow chambers are ever-increasing and thus cannot be fully discussed here. In our R-loop block studies, we poured and cured polydimethylsiloxane (PDMS; Slygard^{TM} 184 Elastomer Kit) on a mold, which was constructed through soft lithography on a silicon wafer, to produce a flow-cell lid with 0.5 mm x 1 cm (~1 μL) channels. To construct the flow chambers we placed the lid on top of 24 x 24 mm coverslips (Marienfield) treated with polyethylene glycol (PEG) and polyethylene glycol-biotin (bio-PEG) as previously detailed (Joo & Ha, 2012). The PDMS-based flow cell is placed in a custom-made holder, which sandwiches the flow cell and allows mounting to the microscope stage. The design and assembly of these flow cells has been previously described in detail (Geertsema, Duderstadt, & van Oijen, 2015). Conveniently, premade flow cells of all varieties designed to couple to coverslips can also be purchased (*e.g.*, from Darwin Microfluidics and Microfluidic Chipshop). Many of these systems incorporate multiple flow channels and may directly either incorporate common microfluidics connections (*e.g.*, mini-Luer ports from Darwin Microfluidics) or can be coupled to threaded ports (*e.g.* NanoPort Assemblies, IDEX Health & Science LLC). In lieu of PDMS flow chamber construction, simple PEGylated ~20 μL glass flow chambers can be made in-house, as previously described in detail elsewhere, e.g., Joo & Ha (2012), and are easily coupled to threaded ports assemblies with Epoxy (*e.g.* 5-min Epoxy, Devcon). For controlling flow, any programmable stepper-motor syringe pump will do provided it can precisely reach low flow rates (*e.g.*, Harvard Apparatus Infuse/Withdraw PHD). Precision glass syringes (*e.g.* Hamilton Gas Tight) are recommended. Flow cells most commonly use 2 ports (inlet and outlet). Using chemically resistant polyethylene tubing such as PE-60 (0.76 mm inner diameter and 1.22 mm outer diameter; Walker Scientific), we suggest withdrawing solution from a reservoir (*e.g.*, a microcentrifuge tube) into the inlet, through the observation chamber, and out of the outlet towards the syringe, as described in Geertsema et al. (2015). We recommend using an injection valve (IDEX 7725) to introduce small-volume (~20 μL) samples into the flow cell without disrupting laminar flow.

5.3 Reconstitution of replication fork stalling at R-loop block and roadblock removal by Pif1

Single-molecule reconstitution of replication fork stalling at R-loops and subsequent clearance by Pif1 is for all intents and purposes like the biochemical experiments outlined in Section 4.1, with a few minor changes where noted. First, although both substrates use the same synthetic fork, the DNA substrate for single-molecule experiments is longer to aid spatial resolution. To visualize double-tethered DNA, we used a template that we successfully used to monitor replication progression and/or blocking in previous studies (Lewis et al., 2020; Whinn et al., 2019). The template is an 18 kb DNA linear substrate ligated to a biotinylated synthetic replication fork on one end, and another biotinylated handle on the other end (see Figs. 2B and 3A). Linear template construction involves bulk supercoiled plasmid preparation, restriction cutting and linearization, ligation of the fork and handle, and purification to rid of excess fork. Since you may want to change the template and/or fork, we refer you to a detailed protocol (Mueller, Spenkelink, van Oijen, & Lewis, 2020) and also to the chapter "Production of long linear DNA substrates with site-specific chemical lesions for single-molecule replisome studies" by Kaur et al. Another difference is that we relied on Pol α to generate primers in the single-molecule assay, whereas we use a radio-labeled synthetic primer in the biochemical assays. Other changes include the addition of Sytox Orange to visualize the DNA and chloroquine to help extend the DNA during the tethering phase. Also, whereas in the biochemical experiments we had dCas9/gRNA present during replication (*e.g.*, Fig. 1), for the single-molecule experiments, we opted to first target dCas9/gRNA to the template and then purify the template to rid of free fluorescent dCas9/gRNA, which would otherwise contribute to background. Also, unlike the biochemical experiments, we used two phases for the single-molecule experiments (Fig. 2B): The first phase is CMG loading, while the second phase is adding the remainder of the replication proteins along with all the nucleotides necessary for replication. Note that because we are only briefly imaging each region well within the photobleaching lifetimes of the dyes (see Section 5.4), we do not require a photoprotective cocktail often used with single-molecule fluorescence experiments to mitigate photobleaching. However, adding oxygen scavengers is beneficial for long time scale experiments or imaging with high temporal resolution—if you decide to use one for your purposes, we suggest using the protocatechuic acid/protocatechuate dehydrogenase system developed by the Puglisi lab as it is pH stable, unlike the glucose oxidase system (Aitken, Marshall, & Puglisi, 2008).

5.3.1 Materials and reagents

- Long (>15 kb), linear, nick-free, forked DNA, biotinylated on the 3' end of the fork lagging strand, the 3' end of the leading strand of the distal end.
- Dye-labeled dCas9Cys (made in Section 2.1) and your target-specific gRNA (made in Section 2.3).
- Stocks of replication proteins: CMG, Pol ε, RFC, PCNA, Mcm10, MTC, Pol δ, Pol α, and RPA.
- 10X MK (0.1 M MgOAc and 0.5 M KOAc).
- 5X TDBG (125 mM Tris-Acetate pH 7.5, 25% (vol/vol) glycerol, 200 μg/mL BSA, 15 mM TCEP, 10 mM DTT).
- 5 M NaCl.
- Empty spin columns (Fisher # 89868).

- Sepharose 4B resin (Cytiva #17012001).
- Blocking buffer (50 mM Tris–HCl, pH 7.6, 50 mM KCl, 2% (vol/vol) Tween-20).
- Heparin (Fisher # BP252450).
- Streptavidin (Sigma # S4762-1MG).
- Chloroquine (Sigma #C6628).
- Inverted TIRF Microscope with 100X oil-immersion Apo TIRF objective (N.A. 1.49, Nikon) (see Section 5.2).
- Flow cell (see Section 5.3).
- Microscope flow chamber heater (Okolab).
- Nucleotides (dATP, dGTP, dCTP, dTTP, ATP, GTP, CTP, UTP).
- Sytox Orange (Fisher #S11368).

5.3.2 Methods

- Incubate DNA template with 450 nM gRNA (IDT) and 240 nM dye-labeled dCas9 (Section 2.1) in the presence of 100 μg/mL heparin for 2 h at RT.
- Bring Sepharose 4B to RT. Pour in 0.8 mL in empty spin column. Equilibrate column with 5 CV replication buffer (1X MK and 1X TDBG) supplemented with 300 mM NaCl.
- Allow column to drain by gravity, quickly adding DNA sample onto center of column bed.
- Remove excess dCas9-gRNA and heparin by collecting DNA in a microcentrifuge tube with a 1 min spin.
- Treat the flow cell with blocking buffer for 10 min.
- Treat the flow cell with 0.1 mg/mL streptavidin for 10 min.
- Flow ~20 μL of 20 pM double-biotin DNA substrate into the chamber in the presence of 200 μM chloroquine at 17 μL/min.
    - o Note: double tethering is repeatable enough that we generally do not visualize the double tethered DNA with Sytox Orange at this stage.
- Add 60 nM Mcm10, 30 nM CMG in the presence of 0.4 mM ATP in 20 mL of 1X MK and 1X TDBG. Allow CMG to load for 10 min at 31.2 °C.
- Add 60 nM Mcm10, 20 nM Pol ε, 20 nM Pol δ, 20 nM Pol α, 20 nM PCNA, 200 nM RPA, and 30 nM MTC in 20 μL of 1X MK and 1X TDBG replication buffer supplemented with 5 mM ATP, 125 μM dCTP, dGTP, dATP, and dTTP, and 250 μM CTP, GTP, ATP, and UTP, and 150 nM Sytox Orange.

- Turn on your fluorescence excitation laser(s) (in our experiments, we used 568 nm for Sytox Orange, 647-nm for AlexaFluor(AF)647-labeled gRNA, and 488-nm for AF488-dCas9), setting the power to an estimated 80 mW/cm^2. Always check the laser power independently at the back focal plane of the objective.
- Begin recording single-molecule replication movies. If you are using multiple excitation sources, alternate excitation with frames. Record DNA replication movies from several flow chamber regions (see Section 5.4).
- In reactions with Pif1 add 100 nM Pif1 after 15 min of replication and continue monitoring.

5.4 Imaging and analysis of single-molecule reactions

To obtain the maximum information from a single, irreversible flow chamber reaction, it is best to capture timelapse images from multiple XY locations. During typical acquisition of a replication movie, we typically cycle between nine fields of view, taking an image every 10 s. Below we offer tips for using NIS-Elements acquisition software to accomplish this, but a similar routine should be achievable in other acquisition suites as well:

In NIS-Elements, activate the "XY" tab in the ND Acquisition window. In this tab, define multiple XYZ locations. The number of fields to acquire may be limited by a combination of exposure time and the desired speed of the timelapse interval. Check "use PFS" (Perfect Focus System) to also save and use a PFS offset. The PFS is a hardware-based system for maintaining focus during timelapse imaging, in which focal drift from thermal or other forces can occur. Update the saved XYZ points and/or PFS offsets to the current location by clicking the → or ← buttons, or revisit the locations by double clicking the point name. When multi-point capture is activated during timelapse acquisition (the "time" tab in the ND Acquisition window), acquisition will occur at each selected XY location in sequence. In an example with three XY points, if the "time" tab is checked and set for a "no delay" interval, images will be acquired of XY1, XY2, XY3, XY1, XY2, XY3, etc., as fast as possible. NIS-Elements can also be used to deploy multiple OCs (Optical Configurations) in sequence by checking the "λ" tab in the ND Acquisition window. Building on the above example, activate this option and check the OCs (previously defined and named within the software) for "green" (488 nm excitation/510 nm emission) and "red" (532 nm excitation/573 nm emission) to acquire in sequence: XY1 green,

XY1 red, XY2 green, XY2 red, XY3 green, XY3 red, XY1 green, XY1 red, XY2 green, XY2 red, XY3 green, XY3 red, etc. Depending on your microscope/software setup, you may need to acquire more excitation/emission channels than necessary. To save a subset of channels in NIS-Elements, select the desired channels with CTRL + left click (red outline), then right click for the "set selected channels as custom" option. In the bottom tabs on the resulting image, right click "custom" then select "extract selected channels." This will generate a new image that can be saved separately.

5.5 Analysis of single-molecule replication events

Data analysis can be performed in FIJI and/or MATLAB with fully documented code available online: https://github.com/LMSpenkelink/ and https://github.com/SingleMolecule. Many of these and similar packages fit and track the centroids of fluorescence foci and provide the user with an output of foci location at every frame, which can be used in a variety of ways. For instance, to obtain the product length at every frame for every replication event, you can track the position and the integrated intensity of a leading-strand spot over time. Discard foci that do not move unidirectionally over the course of the experiment while growing in intensity, as well as events where the tethered lagging strand differs from the location of the original dsDNA template. Tracking either fluorescent dCas9 or gRNA (or both) you can also measure the average position of the dCas9–gRNA relative to the length of the DNA substrate. It is important to perform at least three technical replicates of every single-molecule experiment, reporting the mean and standard deviation of "bulk" properties (*e.g.*, average replication speed) reconstructed from numerous molecules.

One useful way to visualize replication forks moving on flow-stretched DNA is to present the timelapse acquisition as a kymograph, which displays the intensity of a predefined linear path over time (see Fig. 3A and B for examples). The software suite mentioned above can generate kymographs from movies. For a quick analysis after acquisition, NIS-Elements software also has a simple kymograph functionality: Open the timelapse image and in the selected channel (or overlay image of multiple channels), right click then select "create kymograph by line" and draw a line on your image. The line can be a click-to-click straight or segmented line, or hold down the left mouse button to draw a continuous non-linear line. Double click to terminate the line, then immediate right click again to create the new kymograph image, which can be saved in the desired image format. This creates a kymograph from a line that is static throughout all timepoints, so if the region you want to graph over time is moving too much (shifting or drifting), a more powerful kymograph creation software will be needed which can take into account a dynamic linear path. There are also numerous kymograph building plugins for FIJI. Two examples are dynamic reslice (https://imagej.net/plugins/dynamic-reslice) or dynamic kymograph (https://imagej.net/plugins/dynamic-kymograph) (Zhou, Liu, Ju, & Dixit, 2020).

6. Overview and conclusion

Replication fork stalling at R-loop blocks is a critical and persistent source of damage in the cell. It is important to understand the mechanisms that accessory helicases use to regulate R-loop roadblocks, protecting genomic integrity during this relentless source of replication stress. We present a general toolkit for understanding the clearance of R-loop blocks at stalled replication forks further. The biochemical methods provide a clear picture of the events at replication forks that are stalled at—and rescued from—R-loop blocks. The single-molecule methods offer a glimpse into the presence, composition, and dynamics of each player during fork rescue. We present these complementary techniques in hopes that researchers can use these tools to provide new mechanistic glimpses into this and other related mechanisms of R-loop clearance at a stalled replication fork.

Acknowledgments

A.E.S., S.K., and G.D.S. are supported by the National Institutes of Health (NIH) [R00GM126143]. H.G. is supported by the NIH [RM1GM130450] and Australian Research Council [DP210100365] (to Antoine van Oijen, University of Wollongong). We acknowledge Dr. Lisanne Spenkelink for critical feedback.

References

Aitken, C. E., Marshall, R. A., & Puglisi, J. D. (2008). An oxygen scavenging system for improvement of dye stability in single-molecule fluorescence experiments. *Biophysical Journal*, *94*(5), 1826–1835. https://doi.org/10.1529/biophysj.107.117689.

Alzu, A., Bermejo, R., Begnis, M., Lucca, C., Piccini, D., Carotenuto, W., et al. (2012). Senataxin associates with replication forks to protect fork integrity across RNA-polymerase-II-transcribed genes. *Cell*, *151*(4), 835–846. https://doi.org/10.1016/j.cell.2012.09.041.

Azvolinsky, A., Giresi, P. G., Lieb, J. D., & Zakian, V. A. (2009). Highly transcribed RNA polymerase II genes are impediments to replication fork progression in Saccharomyces cerevisiae. *Molecular Cell*, *34*(6), 722–734. https://doi.org/10.1016/j.molcel.2009.05.022.

Bai, L., Yuan, Z., Sun, J., Georgescu, R., O'Donnell, M. E., & Li, H. (2017). Architecture of the Saccharomyces cerevisiae replisome. *Advances in Experimental Medicine and Biology, 1042*, 207–228. https://doi.org/10.1007/978-981-10-6955-0_10.

Bell, S. P., & Labib, K. (2016). Chromosome duplication in Saccharomyces cerevisiae. *Genetics, 203*(3), 1027–1067. https://doi.org/10.1534/genetics.115.186452.

Bessler, J. B., Torres, J. Z., & Zakian, V. A. (2001). The Pif1p subfamily of helicases: Region-specific DNA helicases? *Trends in Cell Biology, 11*(2), 60–65. https://doi.org/10.1016/s0962-8924(00)01877-8.

Bochman, M. L., & Schwacha, A. (2008). The Mcm2-7 complex has in vitro helicase activity. *Molecular Cell, 31*(2), 287–293. https://doi.org/10.1016/j.molcel.2008.05.020.

Boulé, J.-B., & Zakian, V. (2007). The yeast Pif1p DNA helicase preferentially unwinds RNA DNA substrates. *Nucleic Acids Research, 35*(17). https://doi.org/10.1093/nar/gkm613.

Brüning, J.-G., Howard, J. L., & McGlynn, P. (2014). Accessory replicative helicases and the replication of protein-bound DNA. *Journal of Molecular Biology, 426*(24), 3917–3928. https://doi.org/10.1016/j.jmb.2014.10.001.

Burgers, P. M. J., & Kunkel, T. A. (2017). Eukaryotic DNA replication fork. *Annual Review of Biochemistry, 86*(1), 417–438. https://doi.org/10.1146/annurev-biochem-061516-044709.

Buzovetsky, O., Kwon, Y., Pham, N. T., Kim, C., Ira, G., Sung, P., et al. (2017). Role of the Pif1-PCNA complex in Pol δ-dependent strand displacement DNA synthesis and break-induced replication. *Cell Reports, 21*(7), 1707–1714. https://doi.org/10.1016/j.celrep.2017.10.079.

Byrd, A. K., & Raney, K. D. (2017). Structure and function of Pif1 helicase. *Biochemical Society Transactions, 45*(5), 1159–1171. https://doi.org/10.1042/BST20170096.

Candelli, T., Gros, J., & Libri, D. (2018). Pervasive transcription fine-tunes replication origin activity. *eLife*, 7, e40802. https://doi.org/10.7554/eLife.40802.

Chen, Y.-H., Keegan, S., Kahli, M., Tonzi, P., Fenyö, D., Huang, T. T., et al. (2019). Transcription shapes DNA replication initiation and termination in human cells. *Nature Structural & Molecular Biology, 26*(1), 67–77. https://doi.org/10.1038/s41594-018-0171-0.

Chib, S., Byrd, A. K., & Raney, K. D. (2016). Yeast helicase Pif1 unwinds RNA:DNA hybrids with higher processivity than DNA:DNA duplexes. *Journal of Biological Chemistry, 291*(11), 5889–5901. https://doi.org/10.1074/jbc.M115.688648.

Crossley, M. P., Bocek, M., & Cimprich, K. A. (2019). R-loops as cellular regulators and genomic threats. *Molecular Cell, 73*(3), 398–411. https://doi.org/10.1016/j.molcel.2019.01.024.

Dahan, D., Tsirkas, I., Dovrat, D., Sparks, M. A., Singh, S. P., Galletto, R., et al. (2018). Pif1 is essential for efficient replisome progression through lagging strand G-quadruplex DNA secondary structures. *Nucleic Acids Research, 46*(22), 11847–11857. https://doi.org/10.1093/nar/gky1065.

Deegan, T. D., Baxter, J., Ortiz Bazán, M.Á., Yeeles, J. T. P., & Labib, K. P. M. (2019). Pif1-family helicases support fork convergence during DNA replication termination in eukaryotes. *Molecular Cell.* https://doi.org/10.1016/j.molcel.2019.01.040.

Dellino, G. I., Cittaro, D., Piccioni, R., Luzi, L., Banfi, S., Segalla, S., et al. (2013). Genome-wide mapping of human DNA-replication origins: Levels of transcription at ORC1 sites regulate origin selection and replication timing. *Genome Research, 23*(1), 1–11. https://doi.org/10.1101/gr.142331.112.

Doi, G., Okada, S., Yasukawa, T., Sugiyama, Y., Bala, S., Miyazaki, S., et al. (2021). Catalytically inactive Cas9 impairs DNA replication fork progression to induce focal genomic instability. *Nucleic Acids Research, 49*(2), 954–968. https://doi.org/10.1093/nar/gkaa1241.

Douglas, M. E., Ali, F. A., Costa, A., & Diffley, J. F. X. (2018). The mechanism of eukaryotic CMG helicase activation. *Nature*, *555*(7695), 265–268. https://doi.org/10.1038/nature25787.

Felipe-Abrio, I., Lafuente-Barquero, J., García-Rubio, M. L., & Aguilera, A. (2015). RNA polymerase II contributes to preventing transcription-mediated replication fork stalls. *The EMBO Journal*, *34*(2), 236–250. https://doi.org/10.15252/embj.201488544.

Finkelstein, J., Antony, E., Hingorani, M. M., & O'Donnell, M. (2003). Overproduction and analysis of eukaryotic multiprotein complexes in Escherichia coli using a dual-vector strategy. *Analytical Biochemistry*, *319*(1), 78–87. https://doi.org/10.1016/S0003-2697(03)00273-2.

Gao, Y., & Yang, W. (2020). Different mechanisms for translocation by monomeric and hexameric helicases. *Current Opinion in Structural Biology*, *61*, 25–32. https://doi.org/10.1016/j.sbi.2019.10.003.

García-Muse, T., & Aguilera, A. (2019). R loops: From physiological to pathological roles. *Cell*, *179*(3), 604–618. https://doi.org/10.1016/j.cell.2019.08.055.

Geertsema, H. J., Duderstadt, K. E., & van Oijen, A. M. (2015). Single-molecule observation of prokaryotic DNA replication. *Methods in Molecular Biology (Clifton, N.J.)*, *1300*, 219–238. https://doi.org/10.1007/978-1-4939-2596-4_14.

Georgescu, R. E., Langston, L., Yao, N. Y., Yurieva, O., Zhang, D., Finkelstein, J., et al. (2014). Mechanism of asymmetric polymerase assembly at the eukaryotic replication fork. *Nature Structural & Molecular Biology*, *21*(8), 664–670. https://doi.org/10.1038/nsmb.2851.

Georgescu, R. E., Schauer, G. D., Yao, N. Y., Langston, L. D., Yurieva, O., Zhang, D., et al. (2015). Reconstitution of a eukaryotic replisome reveals suppression mechanisms that define leading/lagging strand operation. *eLife*, *4*, e04988. https://doi.org/10.7554/eLife.04988.

Georgescu, R., Yuan, Z., Bai, L., de Luna Almeida Santos, R., Sun, J., Zhang, D., et al. (2017). Structure of eukaryotic CMG helicase at a replication fork and implications to replisome architecture and origin initiation. *Proceedings of the National Academy of Sciences of the United States of America*, *114*(5), E697–E706. https://doi.org/10.1073/pnas.1620500114.

Geronimo, C. L., & Zakian, V. A. (2016). Getting it done at the ends: Pif1 family DNA helicases and telomeres. *DNA Repair*, *44*, 151–158. https://doi.org/10.1016/j.dnarep.2016.05.021.

Gibbs, D. R., Kaur, A., Megalathan, A., Sapkota, K., & Dhakal, S. (2018). Build your own microscope: Step-by-step guide for building a prism-based TIRF microscope. *Methods and Protocols*, *1*(4), 40. https://doi.org/10.3390/mps1040040.

Gómez-González, B., & Aguilera, A. (2019). Transcription-mediated replication hindrance: A major driver of genome instability. *Genes & Development*, *33*(15–16), 1008–1026. https://doi.org/10.1101/gad.324517.119.

Guilliam, T. A., & Yeeles, J. T. (2021). The eukaryotic replisome tolerates leading-strand base damage by replicase switching. *The EMBO Journal*, *40*(5), e107037. https://doi.org/10.15252/embj.2020107037.

Hamperl, S., Bocek, M. J., Saldivar, J. C., Swigut, T., & Cimprich, K. A. (2017). Transcription-replication conflict orientation modulates R-loop levels and activates distinct DNA damage responses. *Cell*, *170*(4), 774–786.e19. https://doi.org/10.1016/j.cell.2017.07.043.

Hamperl, S., & Cimprich, K. A. (2014). The contribution of co-transcriptional RNA:DNA hybrid structures to DNA damage and genome instability. *DNA Repair*, *19*, 84–94. https://doi.org/10.1016/j.dnarep.2014.03.023.

Helmrich, A., Ballarino, M., & Tora, L. (2011). Collisions between replication and transcription complexes cause common fragile site instability at the longest human genes. *Molecular Cell*, *44*(6), 966–977. https://doi.org/10.1016/j.molcel.2011.10.013.

Henricksen, L. A., Umbricht, C. B., & Wold, M. S. (1994). Recombinant replication protein A: Expression, complex formation, and functional characterization. *The Journal of Biological Chemistry*, *269*(15), 11121–11132.

Hizume, K., & Araki, H. (2019). Replication fork pausing at protein barriers on chromosomes. *FEBS Letters*, *593*(13), 1449–1458. https://doi.org/10.1002/1873-3468.13481.

Ilves, I., Petojevic, T., Pesavento, J. J., & Botchan, M. R. (2010). Activation of the MCM2-7 helicase by association with Cdc45 and GINS proteins. *Molecular Cell*, *37*(2), 247–258. https://doi.org/10.1016/j.molcel.2009.12.030.

Ivessa, A. S., Zhou, J. Q., & Zakian, V. A. (2000). The Saccharomyces Pif1p DNA helicase and the highly related Rrm3p have opposite effects on replication fork progression in ribosomal DNA. *Cell*, *100*(4), 479–489. https://doi.org/10.1016/s0092-8674(00)80683-2.

Joo, C., & Ha, T. (2012). Preparing sample chambers for single-molecule FRET. *Cold Spring Harbor Protocols*, *2012*(10). https://doi.org/10.1101/pdb.prot071530. pdb.prot071530-pdb.prot071530.

Juette, M. F., Terry, D. S., Wasserman, M. R., Altman, R. B., Zhou, Z., Zhao, H., et al. (2016). Single-molecule imaging of non-equilibrium molecular ensembles on the millisecond timescale. *Nature Methods*, *13*(4), 341–344. https://doi.org/10.1038/nmeth.3769.

Kemiha, S., Poli, J., Lin, Y.-L., Lengronne, A., & Pasero, P. (2021). Toxic R-loops: Cause or consequence of replication stress? *DNA Repair*, *107*, 103199. https://doi.org/10.1016/j.dnarep.2021.103199.

Kim, Y., Ho, S. O., Gassman, N. R., Korlann, Y., Landorf, E. V., Collart, F. R., et al. (2008). Efficient site-specific labeling of proteins via cysteines. *Bioconjugate Chemistry*, *19*(3), 786–791. https://doi.org/10.1021/bc7002499.

Koc, K. N., Singh, S. P., Stodola, J. L., Burgers, P. M., & Galletto, R. (2016). Pif1 removes a Rap1-dependent barrier to the strand displacement activity of DNA polymerase δ. *Nucleic Acids Research*, *44*(8), 3811–3819. https://doi.org/10.1093/nar/gkw181.

Kornberg, A., & Baker, T. A. (2005). *DNA Replication*. University Science Books.

Kotsantis, P., Silva, L. M., Irmscher, S., Jones, R. M., Folkes, L., Gromak, N., et al. (2016). Increased global transcription activity as a mechanism of replication stress in cancer. *Nature Communications*, 7(1), 13087. https://doi.org/10.1038/ncomms13087.

Kumar, C., Batra, S., Griffith, J. D., & Remus, D. (2021). The interplay of RNA:DNA hybrid structure and G-quadruplexes determines the outcome of R-loop-replisome collisions. *eLife*, *10*, e72286. https://doi.org/10.7554/eLife.72286.

Labun, K., Montague, T. G., Krause, M., Torres Cleuren, Y. N., Tjeldnes, H., & Valen, E. (2019). CHOPCHOP v3: Expanding the CRISPR web toolbox beyond genome editing. *Nucleic Acids Research*, *47*(W1), W171–W174. https://doi.org/10.1093/nar/gkz365.

Lalonde, M., Trauner, M., Werner, M., & Hamperl, S. (2021). Consequences and resolution of transcription–Replication conflicts. *Life*, *11*(7), 637. https://doi.org/10.3390/life11070637.

Langston, L. D., Mayle, R., Schauer, G. D., Yurieva, O., Zhang, D., Yao, N. Y., et al. (2017). Mcm10 promotes rapid isomerization of CMG-DNA for replisome bypass of lagging strand DNA blocks. *eLife*, *6*. https://doi.org/10.7554/eLife.29118.

Langston, L. D., Zhang, D., Yurieva, O., Georgescu, R. E., Finkelstein, J., Yao, N. Y., et al. (2014). CMG helicase and DNA polymerase ε form a functional 15-subunit holoenzyme for eukaryotic leading-strand DNA replication. *Proceedings of the National Academy of Sciences of the United States of America*, *111*(43), 15390–15395. https://doi.org/10.1073/pnas.1418334111.

Laughery, M. F., Mayes, H. C., Pedroza, I. K., & Wyrick, J. J. (2019). R-loop formation by dCas9 is mutagenic in Saccharomyces cerevisiae. *Nucleic Acids Research*, *47*(5), 2389–2401. https://doi.org/10.1093/nar/gky1278.

Lewis, J. S., Spenkelink, L. M., Schauer, G. D., Hill, F. R., Georgescu, R. E., O'Donnell, M. E., et al. (2017). Single-molecule visualization of *Saccharomyces cerevisiae* leading-strand synthesis reveals dynamic interaction between MTC and the replisome. *Proceedings of the National Academy of Sciences of the United States of America, 114*(40), 10630–10635. https://doi.org/10.1073/pnas.1711291114.

Lewis, J. S., Spenkelink, L. M., Schauer, G. D., Yurieva, O., Mueller, S. H., Natarajan, V., et al. (2020). Tunability of DNA polymerase stability during eukaryotic DNA replication. *Molecular Cell,* 77(1), 17–25.e5. https://doi.org/10.1016/j.molcel.2019.10.005.

Li, S., Wang, H., Jehi, S., Li, J., Liu, S., Wang, Z., et al. (2021). PIF1 helicase promotes break-induced replication in mammalian cells. *The EMBO Journal, 40*(8), e104509. https://doi.org/10.15252/embj.2020104509.

Lõoke, M., Maloney, M. F., & Bell, S. P. (2017). Mcm10 regulates DNA replication elongation by stimulating the CMG replicative helicase. *Genes & Development, 31*(3), 291–305. https://doi.org/10.1101/gad.291336.116.

Macheret, M., & Halazonetis, T. D. (2018). Intragenic origins due to short G1 phases underlie oncogene-induced DNA replication stress. *Nature, 555*(7694), 112–116. https://doi.org/10.1038/nature25507.

Martin-Fernandez, M., Tynan, C., & Webb, S. (2013). A 'pocket guide' to total internal reflection fluorescence. *Journal of Microscopy, 252*(1), 16–22. https://doi.org/10.1111/jmi.12070.

Mayle, R., & O'Donnell, M. (2021). Chapter One—Expression of recombinant multi-protein complexes in Saccharomyces cerevisiae. In W. B. O'Dell, & Z. Kelman (Eds.), *Vol. 660. Methods in enzymology* (pp. 3–20). Academic Press. https://doi.org/10.1016/bs.mie.2021.05.016.

McDonald, K. R., Guise, A. J., Pourbozorgi-Langroudi, P., Cristea, I. M., Zakian, V. A., Capra, J. A., et al. (2016). Pfh1 Is an accessory replicative helicase that interacts with the replisome to facilitate fork progression and preserve genome integrity. *PLoS Genetics, 12*(9). https://doi.org/10.1371/journal.pgen.1006238.

Meagher, M., Epling, L. B., & Enemark, E. J. (2019). DNA translocation mechanism of the MCM complex and implications for replication initiation. *Nature Communications, 10*(1), 3117. https://doi.org/10.1038/s41467-019-11074-3.

Mohammad, J. B., Wallgren, M., & Sabouri, N. (2018). The Pif1 signature motif of Pfh1 is necessary for both protein displacement and helicase unwinding activities, but is dispensable for strand-annealing activity. *Nucleic Acids Research, 46*(16), 8516–8531. https://doi.org/10.1093/nar/gky654.

Mueller, S. H., Spenkelink, L. M., van Oijen, A. M., & Lewis, J. S. (2020). Design of customizable long linear DNA substrates with controlled end modifications for single-molecule studies. *Analytical Biochemistry, 592,* 113541. https://doi.org/10.1016/j.ab.2019.113541.

Muellner, J., & Schmidt, K. H. (2020). Yeast genome maintenance by the multifunctional PIF1 DNA helicase family. *Genes, 11*(2), 224. https://doi.org/10.3390/genes11020224.

Nickens, D. G., & Bochman, M. L. (2022). Pif1 activity is modulated by dna sequence and structure. *Biochemistry, 61*(1), 10–20. https://doi.org/10.1021/acs.biochem.1c00614.

O'Connell, M. R., Oakes, B. L., Sternberg, S. H., East-Seletsky, A., Kaplan, M., & Doudna, J. A. (2014). Programmable RNA recognition and cleavage by CRISPR/Cas9. *Nature, 516*(7530), 263–266. https://doi.org/10.1038/nature13769.

O'Donnell, M., Langston, L., & Stillman, B. (2013). Principles and concepts of DNA replication in bacteria, archaea, and eukarya. *Cold Spring Harbor Perspectives in Biology, 5*(7), a010108. https://doi.org/10.1101/cshperspect.a010108.

Osmundson, J. S., Kumar, J., Yeung, R., & Smith, D. J. (2017). Pif1-family helicases cooperatively suppress widespread replication-fork arrest at tRNA genes. *Nature Structural & Molecular Biology, 24*(2), 162–170. https://doi.org/10.1038/nsmb.3342.

Perera, H. M., Behrmann, M. S., Hoang, J. M., Griffin, W. C., & Trakselis, M. A. (2019). Contacts and context that regulate DNA helicase unwinding and replisome progression. *The Enzymes*, *45*, 183–223. https://doi.org/10.1016/bs.enz.2019.08.001.

Pike, J. E., Burgers, P. M. J., Campbell, J. L., & Bambara, R. A. (2009). Pif1 helicase lengthens some Okazaki fragment flaps necessitating Dna2 nuclease/helicase action in the two-nuclease processing pathway. *The Journal of Biological Chemistry*, *284*(37), 25170–25180. https://doi.org/10.1074/jbc.M109.023325.

Prado, F., & Aguilera, A. (2005). Impairment of replication fork progression mediates RNA polII transcription-associated recombination. *The EMBO Journal*, *24*(6), 1267–1276. https://doi.org/10.1038/sj.emboj.7600602.

Ramanagoudr-Bhojappa, R., Chib, S., Byrd, A. K., Aarattuthodiyil, S., Pandey, M., Patel, S. S., et al. (2013). Yeast Pif1 helicase exhibits a one-base-pair stepping mechanism for unwinding duplex DNA. *The Journal of Biological Chemistry*, *288*(22), 16185–16195. https://doi.org/10.1074/jbc.M113.470013.

Rickman, K., & Smogorzewska, A. (2019). Advances in understanding DNA processing and protection at stalled replication forks. *The Journal of Cell Biology*, *218*(4), 1096–1107. https://doi.org/10.1083/jcb.201809012.

Saponaro, M. (2022). Transcription-replication coordination. *Life (Basel, Switzerland)*, *12*(1), 108. https://doi.org/10.3390/life12010108.

Schauer, G., Finkelstein, J., & O'Donnell, M. (2017). In vitro assays for eukaryotic leading/lagging strand DNA replication. *Bio-Protocol*, 7(18). https://doi.org/10.21769/BioProtoc.2548.

Schauer, G. D., & O'Donnell, M. E. (2017). Quality control mechanisms exclude incorrect polymerases from the eukaryotic replication fork. *Proceedings of the National Academy of Sciences of the United States of America*. https://doi.org/10.1073/pnas.1619748114.

Schauer, G. D., Spenkelink, L. M., Lewis, J. S., Yurieva, O., Mueller, S. H., Oijen, A. M.v., et al. (2020). Replisome bypass of a protein-based R-loop block by Pif1. *Proceedings of the National Academy of Sciences of the United States of America*, *117*(48), 30354–30361. https://doi.org/10.1073/pnas.2020189117.

Schindelin, J., Arganda-Carreras, I., Frise, E., Kaynig, V., Longair, M., Pietzsch, T., et al. (2012). Fiji: An open-source platform for biological-image analysis. *Nature Methods*, *9*(7), 676–682. https://doi.org/10.1038/nmeth.2019.

Semlow, D. R., & Walter, J. C. (2021). Mechanisms of vertebrate DNA interstrand cross-link repair. *Annual Review of Biochemistry*, *90*, 107–135. https://doi.org/10.1146/annurev-biochem-080320-112510.

Shyian, M., & Shore, D. (2021). Approaching protein barriers: Emerging mechanisms of replication pausing in eukaryotes. *Frontiers in Cell and Developmental Biology*, *9*, 672510. https://doi.org/10.3389/fcell.2021.672510.

Sparks, M. A., Burgers, P. M., & Galletto, R. (2020). Pif1, RPA and FEN1 modulate the ability of DNA polymerase δ to overcome protein barriers during DNA synthesis. *Journal of Biological Chemistry*. https://doi.org/10.1074/jbc.RA120.015699. jbc.RA120.015699.

Sparks, M. A., Singh, S. P., Burgers, P. M., & Galletto, R. (2019). Complementary roles of Pif1 helicase and single stranded DNA binding proteins in stimulating DNA replication through G-quadruplexes. *Nucleic Acids Research*, *47*(16), 8595–8605. https://doi.org/10.1093/nar/gkz608.

St Germain, C., Zhao, H., & Barlow, J. H. (2021). Transcription-replication collisions—A series of unfortunate events. *Biomolecules*, *11*(8), 1249. https://doi.org/10.3390/biom11081249.

Stork, C. T., Bocek, M., Crossley, M. P., Sollier, J., Sanz, L. A., Chédin, F., et al. (2016). Co-transcriptional R-loops are the main cause of estrogen-induced DNA damage. *eLife*, *5*, e17548. https://doi.org/10.7554/eLife.17548.

Su, N., Byrd, A. K., Bharath, S. R., Yang, O., Jia, Y., Tang, X., et al. (2019). Structural basis for DNA unwinding at forked dsDNA by two coordinating Pif1 helicases. *Nature Communications*, *10*. https://doi.org/10.1038/s41467-019-13414-9.

Tran, P. L. T., Pohl, T. J., Chen, C.-F., Chan, A., Pott, S., & Zakian, V. A. (2017). PIF1 family DNA helicases suppress R-loop mediated genome instability at tRNA genes. *Nature Communications*, *8*, 15025. https://doi.org/10.1038/ncomms15025.

Vijayraghavan, S., & Schwacha, A. (2012). The eukaryotic Mcm2-7 replicative helicase. *Sub-Cellular Biochemistry*, *62*, 113–134. https://doi.org/10.1007/978-94-007-4572-8_7.

Wasserman, M. R., Schauer, G. D., O'Donnell, M. E., & Liu, S. (2019). Replication fork activation is enabled by a single-stranded DNA gate in CMG helicase. *Cell*, *178*(3), 600–611.e16. https://doi.org/10.1016/j.cell.2019.06.032.

Whinn, K. S., Kaur, G., Lewis, J. S., Schauer, G. D., Mueller, S. H., Jergic, S., et al. (2019). Nuclease dead Cas9 is a programmable roadblock for DNA replication. *Scientific Reports*, *9*(1), 13292. https://doi.org/10.1038/s41598-019-49837-z.

Wilson, M. A., Kwon, Y., Xu, Y., Chung, W.-H., Chi, P., Niu, H., et al. (2013). Pif1 helicase and Polδ promote recombination-coupled DNA synthesis via bubble migration. *Nature*, *502*(7471), 393–396. https://doi.org/10.1038/nature12585.

Yao, N., Coryell, L., Zhang, D., Georgescu, R. E., Finkelstein, J., Coman, M. M., et al. (2003). Replication factor C clamp loader subunit arrangement within the circular pentamer and its attachment points to proliferating cell nuclear antigen. *Journal of Biological Chemistry*, *278*(50), 50744–50753. https://doi.org/10.1074/jbc.M309206200.

Yao, N. Y., Zhang, D., Yurieva, O., & O'Donnell, M. E. (2022). CMG helicase can use ATPγS to unwind DNA: Implications for the rate-limiting step in the reaction mechanism. *Proceedings of the National Academy of Sciences of the United States of America*, *119*(4). https://doi.org/10.1073/pnas.2119580119.

Zhang, T., Gao, Y., Wang, R., & Zhao, Y. (2017). Production of guide RNAs in vitro and in vivo for CRISPR using ribozymes and RNA polymerase II promoters. *Bio-Protocol*, 7(4), e2148. https://doi.org/10.21769/BioProtoc.2148.

Zhou, R., Liu, H., Ju, T., & Dixit, R. (2020). Chapter 16—Quantifying the polymerization dynamics of plant cortical microtubules using kymograph analysis. In C. T. Anderson, E. S. Haswell, & R. Dixit (Eds.), *Vol. 160. Methods in cell biology* (pp. 281–293). Academic Press. https://doi.org/10.1016/bs.mcb.2020.04.006.

CHAPTER TEN

Transcriptional regulation by a RecQ helicase

Subrata Debnath[a], Xing Lu[a], and Sudha Sharma[a,b,*]

[a]Department of Biochemistry and Molecular Biology, College of Medicine, Howard University, Washington, DC, United States
[b]National Human Genome Center, College of Medicine, Howard University, Washington, DC, United States
*Corresponding author: e-mail address: sudha.sharma@howard.edu

Contents

Abstract

RecQ helicases participate in a variety of DNA metabolic processes through their multiple biochemical activities. In vitro characterization and cellular studies have suggested that RECQ1 (also known as RECQL or RECQL1) performs its diverse functions through specific interactions with DNA and protein partners. We have taken an unbiased approach to determine the contribution of RECQ1 in genome maintenance and as a putative susceptibility factor in breast cancer. Here, we provide methodology to map the genome-wide binding sites of RECQ1 together with the profiling of RECQ1-dependent transcriptome to investigate its role in gene regulation. The described approach will be helpful to develop a mechanistic framework for elucidating critical functions of RECQ1 and other RecQ homologs in distinct chromatin and biological contexts.

Methods in Enzymology, Volume 673
ISSN 0076-6879
https://doi.org/10.1016/bs.mie.2022.03.057

1. Introduction

The RecQ helicases, represented by five known homologs in humans, constitute a group of highly conserved proteins essential for maintaining genomic stability in all living organisms (Croteau, Popuri, Opresko, & Bohr, 2014). Susceptibility to develop breast cancer is significantly increased in individuals with germline mutations in *RECQ1* (also known as *RECQL* or *RECQL1*), a gene encoding the most abundant RecQ protein essential for genomic integrity in humans (Cybulski et al., 2015; Debnath & Sharma, 2020; Sami & Sharma, 2013; Sharma et al., 2007; Sharma & Brosh, 2007; Sun et al., 2015). Moreover, biallelic mutations in *RECQ1* were recently linked to a previously unidentified genomic instability disease named as RECON syndrome (Abu-Libdeh et al., 2022). It remains to be seen whether RECON syndrome patients will display cancer predisposition, but these discoveries confirm the vital role of RECQ1 in genome maintenance and cancer prevention. Indeed, emerging evidence suggests that RECQ1 expression maybe correlated with progression, therapeutic response, and survival in cancer patients (Arora et al., 2017; Mendoza-Maldonado et al., 2011; Sharma, 2014; van Kan, Burns, & Helsby, 2021; Viziteu et al., 2017). These studies emphasize the need to investigate the molecular function(s) of RECQ1 at a genome-wide level using unbiased approaches.

RECQ1, a ubiquitously expressed and predominantly nuclear protein, shares a highly conserved centrally located helicase domain that couple nucleotide hydrolysis to DNA unwinding and the conserved RecQ-specific C-terminal (RQC) domain that defines the RecQ family helicases (Sami & Sharma, 2013; Sharma, Doherty, & Brosh, 2006). As is true for most RecQ helicases, recombinant RECQ1 is a DNA-dependent ATPase, binds to single stranded and double stranded DNA, unwinds DNA duplex in 3′-5′ direction and promotes strand annealing (Sharma et al., 2005). RECQ1 directly employs multitude of its catalytic activities and via collaborating protein partners mediates restart of arrested replication and repair a variety of DNA damage incurred by genotoxin exposure (Banerjee, Sommers, Huang, Seidman, & Brosh, 2015; Berti et al., 2013; Bugreev, Brosh, & Mazin, 2008; Cui et al., 2003; Doherty et al., 2005; Parvathaneni, Stortchevoi, Sommers, Brosh, & Sharma, 2013; Popuri et al., 2008, 2014; Popuri, Croteau, Brosh, & Bohr, 2012; Sami et al., 2015; Sharma, Phatak, Stortchevoi, Jasin, & Larocque, 2012; Woodrick et al., 2017).

Mechanistic basis of how a genetic or functional loss of RECQ1 leads to genomic instability or cancer remains unclear.

As modulators of chromatin dynamics, RecQ DNA helicases function in genomic stabilization and repair by binding to specific regions, such as origins of replication, replication forks, G-quadruplex structures, R-loops and telomeres (Brosh, 2013; Lu, Parvathaneni, Hara, Lal, & Sharma, 2013; Opresko, 2008; Sharma, 2011; Tan, Wang, Phoon, Yang, & Lan, 2020). We reasoned that human RECQ1 could be a multifunctional protein due to its high abundance and specific interactions with chromatin. This is supported by the fact that RECQ1 homologs in Neurospora (Cogoni & Macino, 1999) and Rat (Lau et al., 2006) have important gene regulatory functions. Furthermore, knockdown of *RECQ1* in breast cancer cells significantly changes the expression of genes associated with tumorigenesis (Li et al., 2014). One of the potential mechanisms by which RECQ1 may modulate gene expression is by binding to G-rich sequences predicted to form secondary DNA structures in their target gene promoters (Li et al., 2014). In addition, the DNA binding and unwinding activity of RECQ1 may contribute to transcriptional regulation in cooperation with specific transcription factors and accessory proteins. The first step to investigate these mechanistic scenarios is identifying regions of the genome bound by RECQ1.

2. Experimental approach and applications

ChIP-seq combines chromatin immunoprecipitation (ChIP) with high-throughput next generation sequencing to identify the genome-wide binding sites of DNA-associated proteins (Park, 2009). The combination of ChIP-seq and transcriptome analysis is a powerful but underutilized approach to investigate genomic functions of RecQ helicases (Höllbacher, Balázs, Heinig, & Uhlenhaut, 2020; Jiang & Mortazavi, 2018). The data from ChIP-seq of RecQ helicases can strengthen information generated from the differential gene expression data in their deficiency (Gray, Vallur, Eddy, & Maizels, 2014; Johnson, Cao, Ryvkin, Wang, & Johnson, 2010; Nguyen et al., 2014; Tang et al., 2015). For example, cellular deficiency of BLM, another RecQ homolog, leads to differential expression of genes containing G-quadruplexes (Johnson et al., 2010; Nguyen et al., 2014). Comparison of genome-wide binding sites of RecQ proteins with publicly available ChIP-seq data for chromatin modulators could identify new roles of these proteins as modulators of RecQ binding. The approach described here can be used for investigating RecQ helicases in general and to (i) identify their target genes, (ii) determine the genomic sequences or structural motifs they bind preferentially, (iii) discover the chromatin features (activating or repressive) of their binding, and (iv) unravel their cooperating factors in transcriptional regulatory networks.

We developed a robust RECQ1 ChIP-seq protocol in MCF7 cells to investigate the molecular mechanism of RECQ1 in breast cancer susceptibility and therapeutic response in estrogen receptor (ER) positive tumors (Fig. 1A). In parallel, we determined global changes in gene expression that occur upon RECQ1 depletion by performing RNA-seq following *RECQ1* knockdown in these ER-positive cells (Fig. 1B). ChIP-seq provided a map of the global binding sites of RECQ1, whereas RNA-seq provided transcriptome-wide gene expression changes upon depletion of RECQ1 in ER-positive cells. Integrating these ChIP-seq and RNA-seq data allowed us to discover RECQ1 helicase as a novel regulator of the master regulatory transcription factor Estrogen Receptor alpha (ERα) elucidating a mechanism for RECQ1 target gene regulation in breast cancer cells (Lu et al., 2021) (Fig. 2).

Detailed protocols for both RECQ1 ChIP-seq and RNA-seq in *RECQ1* knockdown are presented below. We believe that such an unbiased, genome-wide approach may uncover new roles of RECQ1 and advance mechanistic understanding of RECQ1 in human disease and cancer. In principle, such approach can also be applied to investigate diverse functions of other human RecQ helicases.

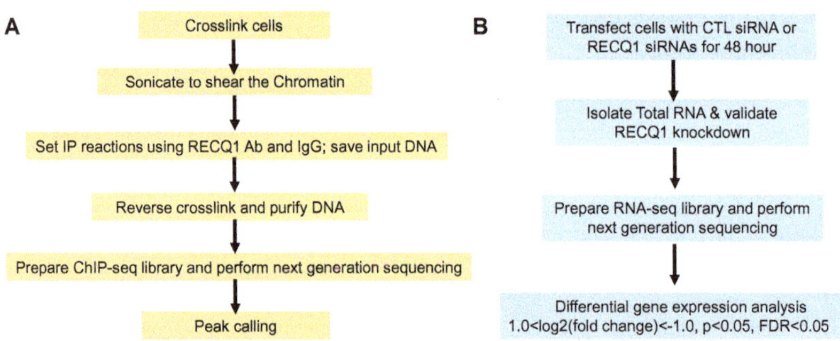

Fig. 1 Schematic of experimental workflow for unbiased investigation of genome-wide roles of RECQ1 (and in principle, other RecQ helicases) in transcriptional regulation. Methodology for ChIP-seq (A) and RNA-seq (B) are outlined in this chapter.

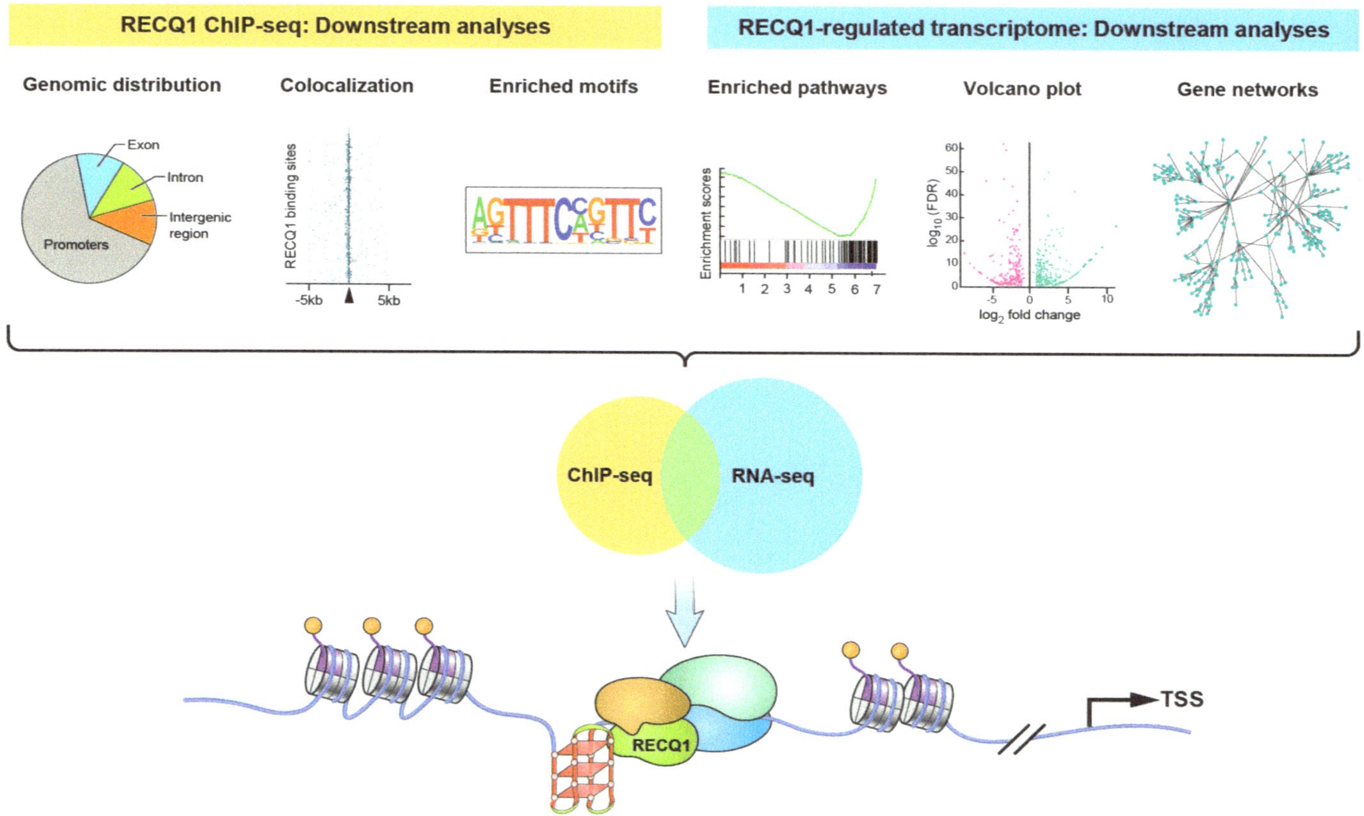

Fig. 2 Elucidating gene regulatory mechanisms of RECQ1 by integrating ChIP-seq and RNA-seq data. Determination of the genomic landscape of RECQ1 binding and integrating that with the transcriptomic changes in RECQ1-deficient cells allows identification of genome-wide regions where RECQ1 functions, in cooperation with other transcription factors and co-factors to regulate transcriptional outcomes. RecQ helicases likely employ their catalytic activities and structure specific DNA binding in regulating gene expression. We recently employed this approach to uncover a new role of RECQ1 in regulation of estrogen response pathway. TSS denotes transcriptional start site.

3. Material and equipment

3.1 Reagents and buffers

3.1.1 Cell culture, siRNA transfection, western blotting, and qPCR

- Human breast cancer cell lines MCF7 (ATCC, HTB-22) and T47D (ATCC, HTB-133). Both cell lines are Estrogen and Progesterone receptor positive
- Dulbecco's Modified Eagle Medium (DMEM, Thermo Fisher, 12430112)
- Opti-MEM (Thermo Fisher, 11058021)
- Fetal Bovine Serum (FBS, Thermo Fisher, 26140-079)
- Penicillin–Streptomycin (Sigma-Aldrich, P4333)
- Trypsin–EDTA (Thermo Fisher, 25200056)
- Mycoplasma PCR Detection Kit (Sigma-Aldrich, MP0035)
- On-Target Plus SMARTpool siRNAs against *RECQ1* (Dharmacon, J-013597-07)
- Non-Targeting (Negative control) siRNA, (Dharmacon, D-001206-13-05)
- Lipofectamine™RNAiMAX Reagent (Thermo Fisher, 13778)
- PierceTM RIPA Buffer (Thermo Fisher, 89901)
- Halt™ Protease Inhibitor Cocktail (Thermo Fisher, 78429)
- Pierce™ BCA Protein Assay Kit (Thermo Fisher, 23227)
- 4x Laemmli Sample Buffer (BIO-RAD, 1610747)
- PVDF Transfer Membrane (Thermo Fisher, 88518)
- Anti-RecQ1 Antibody (Bethyl Laboratories, A300-450A)
- Anti-GAPDH Antibody (Cell Signaling Technology, 2118)
- Anti-Rabbit secondary antibody conjugated to HRP (Vector Laboratories, PI-1000-1)
- SuperSignal West Pico PLUS Chemiluminescent Substrate (Thermo Fisher, 34580)
- TRIzol™ Reagent (Thermo Fisher, 15596018)
- iScript Reverse Transcription Supermix (BIO-RAD, 1708841)
- iTaq Universe SYBR Green Supermix ((BIO-RAD, 1725121)
- PCR Primers are listed in Tables 1 and 2

3.1.2 ChIP-seq

- 37% Formaldehyde Solution (Sigma-Aldrich, 252549)
- Protease Inhibitor Cocktail (Sigma-Aldrich, P8340)

Table 1 Primer sequence for RT-qPCR validation of RNA-seq of RECQ1 knockdown in MCF7 cells.

Gene symbol	NCBI accession number	Primer sequence	Product size
RECQ1	XM_005253462.4	F: CAATGGCTGGAAAGGAGGTA R: AATGGGCAAATGACGAGTGT	112 bp
ESR1	NM_001122742.1	F: AGGTGCCCTACTACCTGGAG R: ACACTGCACAGTAGCGAGTC	183 bp
FOXA1	XM_017021246.1	F: ATGAAACCAGCGACTGGAAC R: GGTGTTCATGGTCATGTAGGTG	128 bp
TFF1	NM_003225.2	F: TCCCCTGGTGCTTCTATCCT R: GGACTAATCACCGTGCTGGG	125 bp
GATA3	XM_005252443.4	F: CAGACCACCACAACCACACTCT R: GGATGCCTTCCTTCTTCATAGTCA	124 bp
Internal controls in RT-qPCR			
GAPDH	NM_001256799.3	F: GTCAAGGCTGAGAACGGGAA R: AAATGAGCCCCAGCCTTCTC	158 bp
SDHA	NM_001294332.1	F: GTCTGCGCAGGGACTGG R: AGCAGATGCCCTCTTGTTCC	179 bp

Table 2 Primers for validation of RECQ1 ChIP-seq targets by qPCR in MCF7 cells.

Gene symbol		NCBI accession number	Primer sequence	Product size
ESR1	Enh2	NC_000006.12	F: CAGTCTTGGCCCTTTACCCA R: GCTCTTAAGGGTTCCTGGTGG	74 bp
	ProA		F: CACATAAGGCAGCACATTAGAGAAA R: GAAGACTGGGCTTAAAATAAACGC	72 bp
	+ 284		F: AAAGGGGAGAAGAGCTGAGG R: GAGGTGGCAGATAGCAGAGG	70 bp
NCOA3		NC_000020.11	F: AACAGCCTCGTGGCTTCC R: GCGTTTCTGGTATGCCAAGC	116 bp
MYC		NC_000008.11	F: CCACTCTCCCTGGGACTCTT R: CGCATCCTTGTCCTGTGAGT	167 bp
E2F3		NC_000006.12	F: CTTTGCCCCACCCCCTAAAA R: GGCCGCTACCTCCTTACTTC	163 bp
BAZ1A		NC_000014.9	F: AAAGCTCCTCCCAACAGCTC R: CTTCTCCCGCTGCCACTG	134 bp
Positive control in RECQ1-ChIP-qPCR validation				
Origin of replication	B48	NC_000019.10	F: CTCCACCCCCAAGGAAAAAG R: GGCAGGGTCCCATGCA	77 bp
	B13		F: CCCCAGGGAGTAGGTTGTGA R: TGTTATTTGAGAAAAGCCCAA	100 bp
Negative control in RECQ1-ChIP-qPCR validation				
GAPDH		NC_000012.12	F: CGGGATTGTCTGCCCTAATTAT R: GCACGGAAGGTCACGATGT	66 bp

- Rabbit IgG (Bethyl Laboratories, P120-101)
- Protein A Agarose/Salmon Sperm DNA (Millipore-Sigma, 16-157)
- RNase A, 20 mg/mL (Thermo Fisher, 12091021)
- Proteinase K, 20 mg/mL (Thermo Fisher, 25530049)
- 2.5 M Glycine Solution (Fisher Scientific, C43755)
- QIAquick PCR Purification Kit (QIAGEN, 28104)
- Cell Lysis Buffer: 10 mM Tris (pH 8.0), 1 mM EDTA, 0.5 mM EGTA, 0.5% SDS and 1 mM DTT supplemented with 1x Protease Inhibitor Cocktail and 1 mM PMSF immediately before use.
- Sonication/IP Buffer: 50 mM HEPES pH 8.0, 140 mM NaCl, 1 mM EDTA, 1% Triton X-100, 0.1% Na-deoxycholate, 0.01% SDS supplemented with 1x Protease Inhibitor and 1 mM PMSF immediately before use.
- Wash Buffers
  - Low Salt Buffer: 20 mM Tris–HCl pH 8.0, 150 mM NaCl, 2 mM EDTA, 0.1% SDS, and 0.1% Triton X-100
  - High Salt Buffer: 20 mM Tris–HCl pH 8.0, 500 mM NaCl, 2 mM EDTA, 0.1% SDS, and 0.1% Triton X-100
  - IP Wash Buffer: 500 mM LiCl, 100 mM Tris–HCl pH 9.0, 1% deoxycholic acid, and 1% NP-40
- Elution Buffer: 50 mM Tris–HCl pH 8.0, 1 mM EDTA, 50 mM $NaHCO_3$, and 1% SDS

3.1.3 RNA-seq

- RNeasy Plus Mini Kit (QIAGEN, 74134)
- TruSeq RNA Library Prep Kit (Illumina, RS-122-2001)

3.2 Equipment

- Nanodrop spectrophotometer, or equivalent
- Criterion™ Vertical Electrophoresis Cell (BIO-RAD) or equivalent
- Mini Trans-Blot Electrophoretic Transfer Cell (BIO-RAD) or equivalent
- T100 Thermal Cycler (BIO-RAD) or equivalent
- CFX96 Touch Real-Time PCR Detection System (BIO-RAD) or equivalent
- Bioanalyzer (Agilent)
- Diagenode Bioruptor or Misonix Ultrasonic Liquid Processor
- Centrifuge/Microcentrifuge (Refrigerated)
- Rotating wheel/platform
- Vortex mixer

4. Methods

4.1 ChIP-seq

To characterize genome-wide binding patterns of RECQ1 in breast cancer cell lines, we optimized a ChIP-seq protocol and successfully applied in MCF7 cells. This protocol includes procedures to isolate nuclei, sonicate chromatin, IP RECQ1-DNA complex, and recover RECQ1 ChIP-enriched DNA fragments (see Note 1).

4.1.1 Cell culture and crosslinking

1. Culture human ERα-positive MCF7 breast cancer cell lines in DMEM supplemented with 10% FBS and 100 U/mL penicillin–streptomycin in a humidified incubator maintaining 5% CO_2 and 37°C (see Note 2).
2. For ChIP-seq experiments, plate sufficient cells in complete growth medium to attain 70–80% confluency 24 h after plating, which corresponds to $\sim 1 \times 10^7$ MCF7 cells in a 150 mm plate. Since a typical ChIP-seq experiment is performed from two to three replicates, it is suggested to use $\sim 3 \times 10^7$ cells in total by culturing the cells in three 150 mm plates.
3. To crosslink the chromatin, slowly add a volume of 37% formaldehyde equal to 1/36th of the volume of 20 mL medium (550 μL of 37% formaldehyde, 1% final concentration) to the cells in the plate, shake gently at room temperature for 15 min.
4. Quench the crosslinking by adding 1.1 mL of 2.5 M glycine (125 mM final concentration) to the crosslinked cells, mix gently and incubate at room temperature for 5 min.
5. Collect cells from the plate using a cell scraper and transfer the cell suspension to a 50 mL falcon tube on ice.
6. Spin the cells down at 1000 × *g* for 5 min at 4°C and carefully remove supernatant.
7. Resuspend the cells with 10 mL ice-cold 1X PBS, spin cells at 1000 × *g* for 5 min at 4°C and discard the supernatant.
8. Resuspend the cells for a second time with 10 mL ice-cold 1X PBS, spin cells at 1000 × *g* for 5 min at 4°C and discard the supernatant.
   - At this point the cell pellets may be flash frozen using dry ice-ethanol or liquid nitrogen bath and stored at −80°C.
9. Resuspend the cell pellet in 5 mL ice-cold cell lysis buffer containing 1 × Protease Inhibitor (one protease inhibitor tablet per 10 mL cell lysis

buffer). Pipet up and down to mix. Incubate on ice for 15 min, vortex the cell suspension briefly every 5 min.

- (Optional) Following incubation, homogenize the cell suspension 10 times using a Dounce homogenizer to facilitate the release of the nuclei.

10. Centrifuge the nuclei suspension at 1000 × *g* at 4°C for 5 min to pellet the nuclei.

4.1.2 Sonication to shear DNA

1. Resuspend the nuclei pellet in 650 μL of ice-cold Sonication/IP buffer supplemented with 1x Protease Inhibitor Cocktail and 1 mM PMSF (immediately before use), transfer the contents to a 2 mL microcentrifuge tube and start the sonication process to shear the chromatin to 150–500 bp DNA fragments.

- At this step, the nuclei pellet in Sonication/IP buffer can be immediately flash frozen and stored at −80°C, if needed.

2. We used Bioruptor at 9 strength output (30 cycles, 30 s on plus 60 s off) to generate the fragmented chromatin. During the sonication, make sure that samples are always ice cold (see Note 3).

3. After sonication, spin the lysate for 10 min at 4°C to remove insoluble material. Transfer the cleared supernatant to a new 2 mL microfuge tube.

4. Keep 50 μL of sonicated chromatin aside on ice to be used as input control. Measure the DNA concentration.

4.1.3 Preparation of input DNA and measurement of the size and concentration of sheared chromatin

1. Add 2 μL RNase A (0.5 mg/mL) to 50 μL of sonicated chromatin. Incubate at 37°C for 30 min to remove RNA.

2. Add 5 μL Proteinase K (20 mg/mL), mix and incubate at 65°C for 4–5 h (or overnight) to reverse the cross-links.

3. Purify DNA using either a PCR purification kit following the manufacturer's protocol or phenol: chloroform followed by ethanol precipitation in the presence of 10 μL glycogen (5 mg/mL). Resuspend the DNA in 50 μL nuclease free water. The samples can now be stored at −20°C.

4. Determine the DNA concentration using a NanoDrop spectrophotometer. The concentration and range of DNA size will be more accurate when measured using an Agilent Bioanalyzer.

5. To analyze the DNA size, transfer 500 ng of purified input DNA to a 250 μL PCR tube and add 1 μL 0.5 M NaCl. Adjust the final volume to 10 μL with nuclease free water if needed.
6. Heat the samples in a thermocycler at 65°C for 4–5 h (or overnight).
7. Add gel loading buffer and run the DNA on a 2% agarose gel. For ChIP-seq or ChIP-qPCR, majority of the DNA must appear as a smear between 150–500 bp.

4.1.4 Immunoprecipitation

1. Prepare 1.5 mL microfuge tubes each containing 10–30 μg of sheared crosslinked chromatin. Use the DNA concentration for sonicated chromatin sample (input) to calculate the volume to be used. 10–30 μg of chromatin is roughly equivalent to $1.5–4.5 \times 10^6$ cells.
2. To the sheared crosslinked chromatin, set up IP reactions in triplicate by adding 4 μg of anti-RECQ1 antibody per tube. Set up a parallel set of reactions with the same amount of rabbit IgG as a negative control. Adjust the final volume to 200 μL with nuclease free water if needed (see Notes 4–6).
3. Incubate overnight at 4°C with rotation.
4. Add 30 μL of Protein A beads coated with sonicated single stranded salmon sperm DNA to each IP reaction and incubate at 4°C for 3–4 h with rotation.
5. Centrifuge the Protein A beads at 2000 × *g* for 1 min and remove the supernatant.
6. Wash beads twice with 1 mL low salt buffer.
7. Wash beads twice with 1 mL high salt buffer.
8. Wash beads twice with 1 mL IP wash buffer.
9. Wash beads twice with 1 mL 1 × TE buffer.
   - For each wash, rotate at 4°C for 3 min and centrifuge at 2000 × *g* for 1 min, carefully removing the supernatant.
10. Elute IP DNA by adding 100 μL of Elution Buffer to the Protein A beads and incubate at room temperature for 15 min with rotation.
11. Centrifuge at 2000 × *g* for 1 min and transfer the supernatant to a fresh microfuge tube.

4.1.5 Reversal of cross-links and DNA purification using spin column

1. Add 5 μL Proteinase K (20 mg/mL) to the eluted IP DNA, heat at 55°C for 30 min, and then increase the temperature to 80°C for 3 h or overnight.

2. Allow the samples to cool to room temperature by leaving them at room temperature for 1 h.
3. Purify free DNA using a PCR purification kit or phenol: chloroform extraction followed by ethanol precipitation in the presence of 10 μL glycogen (5 mg/mL).
4. Measure DNA concentration using a NanoDrop spectrophotometer.
   - If DNA concentrations are too low to give reliable results using a spectrophotometer, Agilent Bioanalyzer should be used.
5. Purified DNA can be used to prepare the ChIP-enriched DNA library suitable for high-throughput sequencing immediately or stored at −20°C (see Note 7).

4.1.6 ChIP-seq library preparation, peak calling and data analysis

1. After purification of the ChIP DNA, prepare the libraries according to standard Illumina's protocol using the TruSeq ChIP Sample Preparation kit. The ChIP DNA is blunt-ended and phosphorylated. A single 'A' nucleotide is added to the 3′ ends of the fragments in preparation for ligation to an adapter that has a single-base 'T' overhang. The ligation products are purified and size selected. The size-selected DNA is then purified and PCR-amplified to enrich for fragments that have indexed adapters on both ends. The final purified product is then quantitated by qPCR before cluster generation and then sequenced in single-end 75-bp sequencing using a NextSeq 500 system (Illumina). The samples typically have ~30 to 64 million pass filter reads with a base call quality of above ~90% of bases with Q30 and above.
2. For RECQ1 ChIP-seq peak calling against IgG control, use the MACS2 algorithm from the Genomatix genome analyzer (https://www.genomatix.de/solutions/genomatix-genome-analyzer.html), with the following parameters: broad region calling on, band width = 300, qvalue cutoff = 1.00e-02 and model fold = [5, 50]. Use *BAMscale* (Pongor et al., 2020) to generate the normalized sequencing coverage for the RECQ1 ChIP-seq data.
3. Use the genome inspector program of the genomatix genome analyzer for colocalization analyses. Generate a colocalization heatmap using the *rtracklayer* (version1.48.0) (Lawrence, Gentleman, & Carey, 2009) and *ComplexHeatmap* (Gu, Eils, & Schlesner, 2016) (version 2.4.3) packages in R (version 4.0.2, script available at: https://github.com/ncbi/BAMscale) using the RECQ1 called peaks and the transcriptional start

regions (TSR) or transcription regulator(s) of interest by downloading the ChIP-seq data files from ENCODE.

4. For colocalization analyses, peaks are clustered using k-means clustering, setting the centers to 3, and using the H3K4me1 (enhancer mark) and H3K4me3 (promoter) histone marks for cluster identification. Percent colocalization is measured as the fraction of RECQ1 peaks within the 5-kb window of the TSRs (see Note 8).
   - RECQ1 ChIP-seq data is available from GEO under accession number GSE153286.
5. Pathway analysis of statistically significant and biologically relevant RECQ1-bound genomic sites can be performed by freely accessible web-tools such as Gene Set Enrichment analysis (GSEA), Database for Annotation, Visualization and Integrated Discovery (DAVID), Ingenuity Pathways Analysis (IPA).

4.1.7 Validation of ChIP-seq data

1. Select 5–10 genes for validation of ChIP-seq data by ChIP-qPCR. The selection of genes is based on the following criteria. (1) The genomic sequence of the target gene had at least one binding site of RECQ1 in the promoter and/or enhancer region, (2) the expression of the target gene changed significantly upon *RECQ1* knockdown as measured by RNA-seq from MCF7 cells transfected with *CTL* siRNA or *RECQ1* siRNAs (see RNA-seq section below) (see Note 7).
2. Calculate the fold enrichment of the targeted genomic sequences over IgG as: fold enrichment $= 2^{-(Ct_{IP} - Ct_{IgG})}$, where Ct_{IP} and Ct_{IgG} are mean threshold cycles of PCR done in triplicates on DNA samples immunoprecipitated with RECQ1 antibody and control IgG, respectively (see Note 9).

4.2 RNA-seq

To determine global changes in gene expression that occur upon RECQ1 depletion, we performed RNA-seq in biological triplicates following siRNA mediated *RECQ1* knockdown in MCF7 cells grown in regular complete medium (see Notes 1 and 2).

4.2.1 Reverse transfection of siRNA

1. To each well of a 6-well plate, add 500 μL Opti-MEM medium and then 3 μL of 20 μM solution of control siRNA or *RECQ1* siRNAs. Mix

gently. Add 5 μL Lipofectamine RNAiMAX to each well, mix and incubate at room temperature for 20 min.

2. Harvest MCF7 cells in complete growth medium and dilute cells in Opti-MEM medium to 1×10^5 per mL. Add 2.5 mL of this cell suspension to each well containing the diluted siRNAs and Lipofectamine RNAiMAX mixture. The final concentration of the siRNAs in each well will be 20 nM.
3. Incubate the cells for 48 h at 37°C in a CO_2 incubator and confirm depletion of RECQ1 by Western Blotting and RT-qPCR. This would give nearly 90% reduction in RECQ1 protein and mRNA levels in MCF7 cells (see Note 10).

4.2.2 Validation of knockdown by western blotting and RT-qPCR

4.2.2.1 Western blotting

1. After 48 h of transfection, prepare whole cell lysates of transfected cells using radioimmunoprecipitation assay (RIPA) buffer containing Halt protease inhibitor cocktail. Measure protein concentration using bicinchoninic acid (BCA) protein quantification kit.
2. Mix 50 μg of the whole cell lysate with Laemmli Buffer supplemented with β-mercaptoethanol and boil at 95°C for 5 min followed by sodium dodecyl sulfate-polyacrylamide gel electrophoresis (SDS-PAGE). Transfer the proteins to a PVDF membrane.
3. After the transfer, block the PVDF membrane with 5% milk in PBS containing 0.1% Tween-20 for 1 h at room temperature on a shaking platform.
4. Incubate the membrane with primary antibodies, anti-RECQ1 (1:1000 dilution) and anti-GAPDH (1:1000 dilution) overnight at 4°C, followed by incubation with secondary antibody conjugated to HRP (1:2500 dilution) for 1 h at room temperature. Develop the immunoblots using SuperSignal West Pico PLUS Chemiluminescent Substrate Western kit.

4.2.2.2 RT-qPCR

1. Harvest MCF7 cells 48 h after transfection with siRNAs as mentioned above. Isolate total RNA using the TRIzol reagent following the manufacturer's instructions. Dissolve the RNA in 20 μL nuclease-free water and measure RNA concentration by Nanodrop spectrophotometer.
2. Use 0.5 μg of total RNA isolated above for reverse-transcription to cDNA using the iScript Reverse Transcription Supermix following the manufacturer's instructions. In a 200 μL PCR tube, add 4 μL

Supermix solution, 0.5 μg total RNA and nuclease free water such that the final volume is 20 μL. Gently mix by pipetting and spin down. Incubate the tube in the Thermal Cycler at 25°C for 6 min and at 46°C for 50 min.

3. Using this cDNA, perform real-time quantitative PCR in triplicate. A 10 μL PCR reaction included 5 μL of iTaq Universal SYBR green Supermix, 3 μL of cDNA, 1 μL of forward and reverse primer mixture (200 nM final concentration for each primer) and 1 μL nuclease free water. Run the reactions at 95°C for 30 s, followed by 40 cycles of 94°C for 10 s and 60°C for 15 s on the real-time PCR system. Normalize the relative transcript levels to the housekeeping gene *GAPDH* and measure differential expression using the 2–ΔΔCT method. The housekeeping gene *SDHA* can be used as an additional negative control for the RT-qPCR experiments.

4.2.3 mRNA-seq

1. Transfect ~1×10^7 MCF7 cells with *CTL* siRNA or *RECQ1* siRNAs as described above.
2. Isolate total RNA 48 h after transfection using the RNeasy Plus Mini Kit following the manufacturer's instructions. We use 350 μL RLT Plus Buffer (component of RNeasy Plus Mini Kit) for lysing up to 5×10^6 cells. RNeasy Plus Mini Kit would yield 10–35 μg total RNA from 1×10^6 cells depending on cell type. For MCF7 cells, we usually get about 15 μg total RNA from one well of a confluent 6-well plate.
3. Check RNA integrity with a Bioanalyzer. For RNA-seq, make sure that the samples have an RNA integrity number (RIN) of ~9.5–10.
4. Pool the mRNA-seq samples and perform paired-end sequencing on HiSeq using Illumina TruSeq mRNA Prep Kit. Typically, the samples should have ~80 to 100 million pass filter reads with a base call quality of >90% of bases with Q30 and above.
5. Before alignment of sequences to the reference genome (Human—hg19), sample reads should be trimmed to remove adapters and low-quality bases using Trimmomatic software and the annotated transcripts using STAR. The average mapping rate of all samples was ~95% and the unique alignment was above 89%. Using Picard software calculate the mapping statistics. Typically, the samples should have very low (<1%) ribosomal RNA reads, the percent coding bases ~65%, percent UTR bases ~30%, and mRNA bases ~94% for all the samples. Measure

the library complexity in terms of unique fragments in the mapped reads using Picard's MarkDuplicate utility. Non-duplicate reads should be ~65–70%.

6. Calculate read count per gene by HTSeq under the annotation of Gencode and normalize by size factor implemented in the DESeq2 package. Regularized-logarithm transformation (rlog) values of gene expression should be used to perform hierarchical clustering and principal component analysis.
7. To assess differential gene expression between different conditions (e.g., constructs vs. mocks), use a generalized linear model within DESeq2 that incorporates information from counts and uses negative binomial distribution with fitted mean and a gene-specific dispersion parameter. DESeq2 used Wald statistics for significance testing and the Benjamini-Hochberg adjustment for multiple corrections.
   - The RNA-seq data upon *RECQ1* knockdown in MCF7 cells is available in GEO under accession number GSE152323.
8. Select genes significantly downregulated or upregulated ($0.5 < \log_2$ fold change < -0.5; $P < 0.05$) upon knockdown of *RECQ1*. Using these statistical cutoffs, perform Gene set enrichment analyses (GSEA) to identify pathways regulated by RECQ1 (see Note 11).
9. Select 5–10 up-regulated and down-regulated genes identified in RNA-seq and validate gene expression changes by RT-qPCR (see Note 12).

5. Notes

1. Before the experimental set-up, it is important to consider which cells to perform RNA-seq or ChIP-seq. We used MCF7 cells for our study since we wanted to explore genome-wide role of RECQ1 in ERα-positive breast cancer and its association with ERα.
2. Do not use cells beyond the 20–25 passages for the experiments. Routinely test cells using the Mycoplasma PCR Detection Kit to ensure that the cells are Mycoplasma-free.
3. Efficient sonication is a critical step in ChIP experiments. Extensive care and optimization will be required for achieving sonication conditions for each cell type, sample volume and the kind of sonicator used. For example, when using Misonix Ultrasonic Liquid Processor

(XL-2000) probe sonicator to shear MCF7 cells prepared as above, optimal fragmentation is achieved with 20 cycles, each cycle consisting of 30 s on and 30 s off at 30% power output.

4. Another critical factor for a successful ChIP or ChIP-seq experiment is the quality of antibody. The RECQ1 antibody that we used is highly specific, has high IP efficiency, and is demonstrated to be effective in ChIP. The amount of antibody used per reaction will need to be optimized. Before setting up a ChIP-seq experiment, determine antibody specificity and IP efficiency. If the genomic binding sites for the target protein is known, ChIP-qPCR can be used to compare the various available antibodies against the target protein. For example, enrichment of replication origins (Lu et al., 2013)(primer sequence included in Table 2), telomeric sequences (Sami et al., 2015), or G4-containing genes (such as *MYC*) (Li et al., 2014) over the non-specific background may be tested in the ChIP material by qPCR to determine the specificity and efficiency of antibodies targeting human RecQ proteins. In addition, ChIP efficiency of an antibody can be determined by ChIP followed by Western blotting. Compare the amount of target protein pulled down with the input in each ChIP reaction to determine the most efficient antibody.
5. We recommend using a positive control for ChIP reactions. For example, in our RECQ1 ChIP-seq experiments, we included triplicate ChIP reactions with anti-ERα antibody as a positive control for ChIP because ERα targets are known. Comparison of our ERα ChIP-seq data with previously published ERα-ChIP-seq datasets from MCF7 cells showed ~75–95% colocalization validating the specificity of our experimental protocol for ChIP-seq. Other possible controls include well-established transcription factors such as p53 (if wild-type p53 is expressed and functional in the cell line used) or Histone H3 tri-methyl K4 (H3K4me3) and tri-methyl K9 (H3K9me3) when studying active and inactive genes, respectively.
6. ChIP reactions with control IgG antibody is a good control that gives an estimate of the background by pulling down irrelevant non-specific target and serves as negative control. Alternatively, input DNA purified from cells that are cross-linked, fragmented, but without adding any antibody for enrichment can be used as a negative control. Finally, the use of isogenic knockout cells for the target protein can serve as an additional control in which ChIP reactions can be performed from the knockout cells using the antibody against the target protein.

7. Fold enrichment in ChIP-seq over background is determined using a ChIP reaction such as an IgG control (as described) or input. Before proceeding with the ChIP samples for sequencing, we recommend testing a few regions of interest (if known) by qPCR to confirm that the enrichment has worked. Using the criteria as mentioned, we selected the following protein-coding genes to validate RECQ1 ChIP-seq data: *ESR1, FOXA1, MYC, NCOA3, E2F3, BAZ1P.* The binding of RECQ1 to the promoter and/or enhancer of these genes was validated by ChIP-qPCR and RT-qPCR was used to confirm altered expression of these genes upon siRNA-mediated knockdown of *RECQ1* in breast cancer cells. Detailed list of primer sequence can be found in reference (Lu et al., 2021).
8. A wealth of high-quality ChIP-seq and RNA-seq data can be freely retrieved from ENCODE database to find novel insights into the diverse roles of RecQ and related helicases in gene regulation. For example, comparing our RECQ1 ChIP-seq data with ChIP-seq data for a number of transcription factors including our ERα ChIP-seq showed that RECQ1 genome-wide binding positively correlates with ERα, FOXA1, GATA3, TCF7L2 and ZNF217 (Lu et al., 2021).
9. Check all qPCR reactions by melt curve analyses and agarose gel electrophoresis to confirm the presence of a single specific product of the expected size.
10. Transfection efficiency will vary for different cell lines and therefore conditions (siRNA concentration, time after transfection, 2 rounds of transfections, etc.) for efficient gene knockdown may need to be optimized for each cell line. RNA-seq can also be performed in isogenic knockout and wildtype cells or following stable gene knockdown.
11. Pathway analysis of statistically significant and biologically relevant RECQ1-bound genomic sites as well as RECQ1-dependent differential gene expression can be performed using freely accessible web-tools such as Gene Set Enrichment analysis (GSEA), Database for Annotation, Visualization, and Integrated Discovery (DAVID), Ingenuity Pathways Analysis (IPA).
12. It is important to validate RNA-seq results in multiple cell lines to verify cell-type dependence of differential gene expression. We used T47D cells another ERα-positive breast cancer cell line to verify differential expression of specific genes whose expression changed upon depletion of RECQ1 in MCF7 cells. To validate, T47D cells

were reverse transfected with *CTL* siRNA or *RECQ1* siRNAs as described above followed by RT-qPCR for 5–10 RECQ1-regulated genes.

Acknowledgments

We thank Drs. Mirit Aladjem and Ashish Lal and members of their lab at NCI for support and discussion. We also thank NCI sequencing core facility for performing RNA-seq. This work was supported by the National Institute of General Medical Sciences of the National Institutes of Health (under Award Number SC1GM093999) and National Science Foundation (under Award Number NSF1832163).

References

Abu-Libdeh, B., Jhujh, S. S., Dhar, S., Sommers, J. A., Datta, A., Longo, G. M., et al. (2022). RECON syndrome is a genome instability disorder caused by mutations in the DNA helicase RECQL1. *The Journal of Clinical Investigation*, e147301.

Arora, A., Parvathaneni, S., Aleskandarany, M. A., Agarwal, D., Ali, R., Abdel-Fatah, T., et al. (2017). Clinicopathological and functional significance of RECQL1 helicase in sporadic breast cancers. *Molecular Cancer Therapeutics*, *16*(1), 239–250. https://doi.org/10.1158/1535-7163.MCT-16-0290.

Banerjee, T., Sommers, J. A., Huang, J., Seidman, M. M., & Brosh, R. M., Jr. (2015). Catalytic strand separation by RECQ1 is required for RPA-mediated response to replication stress. *Current Biology : CB*, *25*(21), 2830–2838. https://doi.org/10.1016/j.cub.2015.09.026.

Berti, M., Ray Chaudhuri, A., Thangavel, S., Gomathinayagam, S., Kenig, S., Vujanovic, M., et al. (2013). Human RECQ1 promotes restart of replication forks reversed by DNA topoisomerase I inhibition. *Nature Structural & Molecular Biology*, *20*(3), 347–354. https://doi.org/10.1038/nsmb.2501.

Brosh, R. M., Jr. (2013). DNA helicases involved in DNA repair and their roles in cancer. *Nature Reviews.Cancer*, *13*(8), 542–558. https://doi.org/10.1038/nrc3560.

Bugreev, D. V., Brosh, R. M., Jr., & Mazin, A. V. (2008). RECQ1 possesses DNA branch migration activity. *The Journal of Biological Chemistry*, *283*(29), 20231–20242. https://doi.org/10.1074/jbc.M801582200.

Cogoni, C., & Macino, G. (1999). Posttranscriptional gene silencing in neurospora by a RecQ DNA helicase. *Science (New York, N.Y.)*, *286*(5448), 2342–2344. 8089.

Croteau, D. L., Popuri, V., Opresko, P. L., & Bohr, V. A. (2014). Human RecQ helicases in DNA repair, recombination, and replication. *Annual Review of Biochemistry*, *83*, 519–552. https://doi.org/10.1146/annurev-biochem-060713-035428.

Cui, S., Klima, R., Ochem, A., Arosio, D., Falaschi, A., & Vindigni, A. (2003). Characterization of the DNA-unwinding activity of human RECQ1, a helicase specifically stimulated by human replication protein A. *The Journal of Biological Chemistry*, *278*(3), 1424–1432. https://doi.org/10.1074/jbc.M209407200.

Cybulski, C., Carrot-Zhang, J., Kluzniak, W., Rivera, B., Kashyap, A., Wokolorczyk, D., et al. (2015). Germline RECQL mutations are associated with breast cancer susceptibility. *Nature Genetics*, *47*(6), 643–646. https://doi.org/10.1038/ng.3284.

Debnath, S., & Sharma, S. (2020). RECQ1 helicase in genomic stability and cancer. *Genes*, *11*(6), E622. https://doi.org/10.3390/genes11060622.

Doherty, K. M., Sharma, S., Uzdilla, L. A., Wilson, T. M., Cui, S., Vindigni, A., et al. (2005). RECQ1 helicase interacts with human mismatch repair factors that regulate genetic recombination. *The Journal of Biological Chemistry, 280*(30), 28085–28094. doi: M500265200.

Gray, L. T., Vallur, A. C., Eddy, J., & Maizels, N. (2014). G quadruplexes are genomewide targets of transcriptional helicases XPB and XPD. *Nature Chemical Biology, 10*(4), 313–318. https://doi.org/10.1038/nchembio.1475.

Gu, Z., Eils, R., & Schlesner, M. (2016). Complex heatmaps reveal patterns and correlations in multidimensional genomic data. *Bioinformatics (Oxford, England), 32*(18), 2847–2849. https://doi.org/10.1093/bioinformatics/btw313.

Höllbacher, B., Balázs, K., Heinig, M., & Uhlenhaut, N. H. (2020). Seq-ing answers: Current data integration approaches to uncover mechanisms of transcriptional regulation. *Computational and Structural Biotechnology Journal, 18*, 1330–1341. https://doi.org/10.1016/j.csbj.2020.05.018.

Jiang, S., & Mortazavi, A. (2018). Integrating ChIP-seq with other functional genomics data. *Briefings in Functional Genomics, 17*(2), 104–115. https://doi.org/10.1093/bfgp/ely002.

Johnson, J. E., Cao, K., Ryvkin, P., Wang, L. S., & Johnson, F. B. (2010). Altered gene expression in the werner and bloom syndromes is associated with sequences having G-quadruplex forming potential. *Nucleic Acids Research, 38*(4), 1114–1122. https://doi.org/10.1093/nar/gkp1103.

Lau, N. C., Seto, A. G., Kim, J., Kuramochi-Miyagawa, S., Nakano, T., Bartel, D. P., et al. (2006). Characterization of the piRNA complex from rat testes. *Science (New York, N.Y.), 313*(5785), 363–367. doi:1130164.

Lawrence, M., Gentleman, R., & Carey, V. (2009). Rtracklayer: An R package for interfacing with genome browsers. *Bioinformatics (Oxford, England), 25*(14), 1841–1842. https://doi.org/10.1093/bioinformatics/btp328.

Li, X. L., Lu, X., Parvathaneni, S., Bilke, S., Zhang, H., Thangavel, S., et al. (2014). Identification of RECQ1-regulated transcriptome uncovers a role of RECQ1 in regulation of cancer cell migration and invasion. *Cell Cycle (Georgetown, Tex.), 13*(15), 2431–2445. https://doi.org/10.4161/cc.29419.

Lu, X., Parvathaneni, S., Hara, T., Lal, A., & Sharma, S. (2013). Replication stress induces specific enrichment of RECQ1 at common fragile sites FRA3B and FRA16D. *Molecular Cancer, 12*(1). https://doi.org/10.1186/1476-4598-12-29. 29-4598-12-29.

Lu, X., Redon, C. E., Tang, W., Parvathaneni, S., Bokhari, B., Debnath, S., et al. (2021). Genome-wide analysis unveils DNA helicase RECQ1 as a regulator of estrogen response pathway in breast cancer cells. *Molecular and Cellular Biology, 41*(4), e00515–e00520. https://doi.org/10.1128/MCB.00515-20.

Mendoza-Maldonado, R., Faoro, V., Bajpai, S., Berti, M., Odreman, F., Vindigni, M., et al. (2011). The human RECQ1 helicase is highly expressed in glioblastoma and plays an important role in tumor cell proliferation. *Molecular Cancer, 10*. https://doi.org/10.1186/1476-4598-10-83. 83-4598-10-83.

Nguyen, G. H., Tang, W., Robles, A. I., Beyer, R. P., Gray, L. T., Welsh, J. A., et al. (2014). Regulation of gene expression by the BLM helicase correlates with the presence of G-quadruplex DNA motifs. *Proceedings of the National Academy of Sciences of the United States of America, 111*(27), 9905–9910. https://doi.org/10.1073/pnas.1404807111.

Opresko, P. L. (2008). Telomere ResQue and preservation- -roles for the werner syndrome protein and other RecQ helicases. *Mechanisms of Ageing and Development, 129*(1–2), 79–90. doi:S0047-6374(07)00168-6.

Park, P. J. (2009). ChIP-seq: Advantages and challenges of a maturing technology. *Nature Reviews Genetics, 10*(10), 669–680. https://doi.org/10.1038/nrg2641.

Parvathaneni, S., Stortchevoi, A., Sommers, J. A., Brosh, R. M., Jr., & Sharma, S. (2013). Human RECQ1 interacts with Ku70/80 and modulates DNA end-joining of double-strand breaks. *PLoS One, 8*(5), e62481. https://doi.org/10.1371/journal.pone.0062481.

Pongor, L. S., Gross, J. M., Vera Alvarez, R., Murai, J., Jang, S. M., Zhang, H., et al. (2020). BAMscale: Quantification of next-generation sequencing peaks and generation of scaled coverage tracks. *Epigenetics & Chromatin, 13*(1). https://doi.org/10.1186/s13072-020-00343-x. 21-020-00343-x.

Popuri, V., Bachrati, C. Z., Muzzolini, L., Mosedale, G., Costantini, S., Giacomini, E., et al. (2008). The human RecQ helicases, BLM and RECQ1, display distinct DNA substrate specificities. *The Journal of Biological Chemistry, 283*(26), 17766–17776. https://doi.org/10.1074/jbc.M709749200.

Popuri, V., Croteau, D. L., Brosh, R. M., Jr., & Bohr, V. A. (2012). RECQ1 is required for cellular resistance to replication stress and catalyzes strand exchange on stalled replication fork structures. *Cell Cycle (Georgetown, Tex.), 11*(22), 4252–4265. https://doi.org/10.4161/cc.22581.

Popuri, V., Hsu, J., Khadka, P., Horvath, K., Liu, Y., Croteau, D. L., et al. (2014). Human RECQL1 participates in telomere maintenance. *Nucleic Acids Research, 42*(9), 5671–5688. https://doi.org/10.1093/nar/gku200.

Sami, F., Lu, X., Parvathaneni, S., Roy, R., Gary, R. K., & Sharma, S. (2015). RECQ1 interacts with FEN-1 and promotes binding of FEN-1 to telomeric chromatin. *The Biochemical Journal, 468*(2), 227–244. https://doi.org/10.1042/BJ20141021.

Sami, F., & Sharma, S. (2013). Probing genome maintenance functions of human RECQ1. *Computational and Structural Biotechnology Journal, 6*, e201303014. https://doi.org/10.5936/csbj.201303014.

Sharma, S. (2011). Non-B DNA secondary structures and their resolution by RecQ helicases. *Journal of Nucleic Acids, 2011*, 724215. https://doi.org/10.4061/2011/724215.

Sharma, S. (2014). An appraisal of RECQ1 expression in cancer progression. *Frontiers in Genetics, 5*, 426. https://doi.org/10.3389/fgene.2014.00426.

Sharma, S., & Brosh, R. M., Jr. (2007). Human RECQ1 is a DNA damage responsive protein required for genotoxic stress resistance and suppression of sister chromatid exchanges. *PLoS One, 2*(12), e1297. https://doi.org/10.1371/journal.pone.0001297.

Sharma, S., Doherty, K. M., & Brosh, R. M., Jr. (2006). Mechanisms of RecQ helicases in pathways of DNA metabolism and maintenance of genomic stability. *The Biochemical Journal, 398*(3), 319–337. doi:BJ20060450.

Sharma, S., Phatak, P., Stortchevoi, A., Jasin, M., & Larocque, J. R. (2012). RECQ1 plays a distinct role in cellular response to oxidative DNA damage. *DNA Repair, 11*(6), 537–549. https://doi.org/10.1016/j.dnarep.2012.04.003.

Sharma, S., Sommers, J. A., Choudhary, S., Faulkner, J. K., Cui, S., Andreoli, L., et al. (2005). Biochemical analysis of the DNA unwinding and strand annealing activities catalyzed by human RECQ1. *The Journal of Biological Chemistry, 280*(30), 28072–28084. doi:M500264200.

Sharma, S., Stumpo, D. J., Balajee, A. S., Bock, C. B., Lansdorp, P. M., Brosh, R. M., Jr., et al. (2007). RECQL, a member of the RecQ family of DNA helicases, suppresses chromosomal instability. *Molecular and Cellular Biology, 27*(5), 1784–1794. doi: MCB.01620-06.

Sun, J., Wang, Y., Xia, Y., Xu, Y., Ouyang, T., Li, J., et al. (2015). Mutations in RECQL gene are associated with predisposition to breast cancer. *PLoS Genetics, 11*(5), e1005228. https://doi.org/10.1371/journal.pgen.1005228.

Tan, J., Wang, X., Phoon, L., Yang, H., & Lan, L. (2020). Resolution of ROS-induced G-quadruplexes and R-loops at transcriptionally active sites is dependent on BLM helicase. *FEBS Letters, 594*(9), 1359–1367. https://doi.org/10.1002/1873-3468.13738.

Tang, W., Robles, A. I., Beyer, R. P., Gray, L. T., Nguyen, G. H., Oshima, J., et al. (2015). Werner syndrome helicase modulates G4 DNA-dependent transcription and opposes mechanistically distinct senescence-associated gene expression programs. *bioRxiv*. https://doi.org/10.1101/023739.

van Kan, M., Burns, K. E., & Helsby, N. A. (2021). A systematic review of inter-individual differences in the DNA repair processes involved in melphalan monoadduct repair in relation to treatment outcomes. *Cancer Chemotherapy and Pharmacology*, *88*(5), 755–769. https://doi.org/10.1007/s00280-021-04340-z.

Viziteu, E., Klein, B., Basbous, J., Lin, Y. L., Hirtz, C., Gourzones, C., et al. (2017). RECQ1 helicase is involved in replication stress survival and drug resistance in multiple myeloma. *Leukemia*, *31*(10), 2104–2113. https://doi.org/10.1038/leu.2017.54.

Woodrick, J., Gupta, S., Camacho, S., Parvathaneni, S., Choudhury, S., Cheema, A., et al. (2017). A new sub-pathway of long-patch base excision repair involving 5′ gap formation. *The EMBO Journal*, *36*(11), 1605–1622. https://doi.org/10.15252/embj.201694920.

CHAPTER ELEVEN

Analysis of the conformational space and dynamics of RNA helicases by single-molecule FRET in solution and on surfaces

Dagmar Klostermeier*
University of Muenster, Institute for Physical Chemistry, Muenster, Germany
*Corresponding author: e-mail address: dagmar.klostermeier@uni-muenster.de

Contents

Abstract

RNA helicases are a diverse group of enzymes that catalyze the unwinding of RNA duplex regions in an ATP-dependent reaction. Both the helicase itself and its RNA substrate undergo conformational changes during the reaction, which are amenable to Förster resonance energy transfer (FRET) studies. Single-molecule FRET studies in

Methods in Enzymology, Volume 673
ISSN 0076-6879
https://doi.org/10.1016/bs.mie.2022.03.043

solution by confocal microscopy and on surfaces by total internal reflection microscopy provide information on different conformers present, their fractional populations in equilibrium, and the rate constants of their inter-conversion. Collectively, the information gained can be integrated into a kinetic and thermodynamic framework that quantitatively describes the conformational dynamics of the helicase studied. FRET experiments also provide distance information to map and model the structures of individual conformational states. The integrated model provides a comprehensive description of the structure and dynamics of the helicase, which can be linked to its biological function.

Single-molecule FRET studies have tremendous potential to define the relationship between structure, function and dynamics of RNA helicases and to understand the mechanistic basis for their broad range of biological functions. The focus of this chapter is on providing guidance in the design of single-molecule FRET experiments and on the interpretation of the data obtained. Selected examples illustrate important considerations when analyzing single-molecule experiments, as well as their limitations and possible pitfalls.

Abbreviations

CCD	charge-coupled device
CMOS	complementary metal-oxide semiconductor
CDTH	cumulative dwell-time histogram
DTH	dwell-time histogram
EM-CCD	electron-multiplying charge-coupled device
FlAsH	fluorescent arsenical hairpin
FRET	Förster resonance energy transfer
HMM	hidden Markov modeling
PEG	polyethylene glycol
RBD	RNA-binding domain
RecA_N, _C	N- or C-terminal RecA-like domain
TIR(F)	total internal reflection (fluorescence)

1. Introduction

RNA helicases are a diverse group of enzymes that catalyze the unwinding of RNA duplex regions in an ATP-dependent reaction. Based on sequence and structural similarities as well as common features in their mode of action, RNA helicases are grouped into different helicase families (Fairman-Williams, Guenther, & Jankowsky, 2010; Gorbalenya & Koonin, 1993). Members of the same family share a common, family-specific core (Fairman-Williams et al., 2010) of approx. 400 amino acids, formed by two RecA-like domains (RecA_N, RecA_C) connected by a flexible linker. However, the structures of the core and the conserved motives present vary between different helicase families. In most RNA helicases, the helicase core is flanked by variable additional domains that differ between different families, and between members of the same family. In the following, the focus will be on DEAD-box helicases, the largest RNA helicase family, termed according to a conserved motif of the sequence D-E-A-D (Schmid & Linder, 1991, 1992). In DEAD-box proteins, the helicase core is flanked by a diverse set of domains that modulate its functions by affecting RNA and ATP binding, ATP hydrolysis, or RNA unwinding (reviewed in Donsbach & Klostermeier, 2021; Rudolph & Klostermeier, 2015). During the catalysis of ATP hydrolysis and RNA unwinding, the helicase core alternates between open and closed states (Aregger & Klostermeier, 2009; Sengoku, Nureki, Nakamura, Kobayashi, & Yokoyama, 2006; Theissen, Karow, Kohler, Gubaev, & Klostermeier, 2008) (Fig. 1A, B). The conformational cycle of the helicase core is intimately linked to strand separation. The RNA substrate also alters its conformation during the reaction—in the simplest case the duplex is converted into two single strands, in more physiological scenarios RNAs undergo large-scale conformational rearrangements (Cardo, Karunatilaka, Rueda, & Sigel, 2012; Steiner, Karunatilaka, Sigel, & Rueda, 2008). Changes in enzyme and RNA conformation involve changes in distance between reference points on the order of a few nanometer and are thus amenable to Förster resonance energy transfer (FRET) studies. In a helicase-centered approach, the helicase is labeled with donor and acceptor fluorophores to monitor conformational changes through changes in FRET between these dyes (Fig. 1C). In an RNA-centered approach, on the other hand, the donor and acceptor dyes are introduced into the RNA; and the effect of the helicase on RNA conformation is monitored (Fig. 1D). Such studies can identify different conformers with different FRET efficiencies and quantify their population. Furthermore, they provide information on the rate constants of inter-conversion of these states and help formulate kinetic models that can then be tested further. FRET experiments also provide distance information that can be used to map and model global structures.

This article gives an overview over single-molecule FRET studies of RNA helicases in solution using confocal microscopy, and on surface-immobilized helicases using total internal reflection microscopy. The focus is on providing guidance in the design of single-molecule FRET experiments and in the analysis of single-molecule data. Data acquisition depends strongly on the experimental platform used, such as the type of microscope and the software for immediate analysis of the raw data, and is not treated here. The examples presented were not chosen to comprehensively review

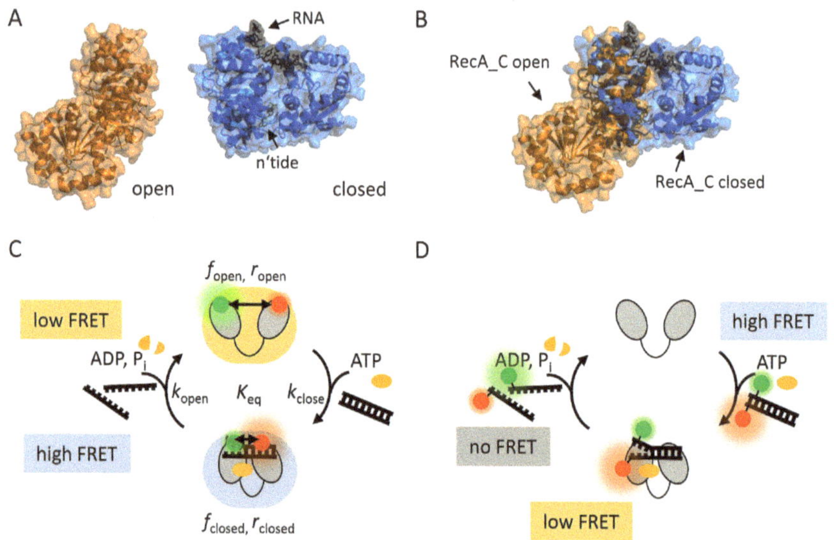

Fig. 1 Conformational changes in DEAD-box helicases and their RNA substrates, and information accessible by single-molecule FRET approaches. (A) The common DEAD-box helicase core consists of two RecA-like domains (RecA_N and RecA_C), that are flexibly linked. In the absence of ligands, the RecA-like domains are separated (open state; orange, PDB-ID 2hxy) (Andersen et al., 2006). When ATP and RNA bind, the core switches to the closed state (blue, PDB-ID 3ex7) (Nielsen et al., 2008), in which both domains interact with each other and with the RNA (gray) and the nucleotide (n'tide). (B) Superposition of open and closed states on the RecA_N domains. The different positions of RecA_C relative to RecA_N in the open and closed states are clearly visible. (C) Schematic of the conformational cycle of DEAD-box helicases. By attaching donor and acceptor fluorophore to RecA_N and RecA_C, the conformational cycle can be investigated in single-molecule FRET experiments. Single-molecule FRET experiments on freely diffusing helicases in solution by confocal microscopy provide information on different FRET states present (low FRET, high FRET), their relative populations (f_{open}, f_{close}), and the distances between the dyes in these states (r_{open}, r_{close}). From the relative populations, the equilibrium constant K_{eq} for closing can be calculated. A set of inter-dye distances can be used as restraints to define the relative position of the two RecA-like domains of the helicase and to generate structural models. Single-molecule experiments on surface-immobilized molecules provide the same type of information. In addition, they enable the long-term observation of individual molecules on a time scale of up to minutes, and the determination of rate constants of inter-conversion. (D) DEAD-box helicase activity from the perspective of the RNA. By introducing donor and acceptor fluorophores at suitable sites in the RNA, conformational changes of the RNA during its interaction with the helicase can be monitored. This is illustrated schematically using a simple RNA duplex substrate, but such studies become more insightful with larger, physiologically relevant helicase substrates. *Figure modified after Chakraborty, A., Krause, L., & Klostermeier, D. (2022). Determination of rate constants for conformational changes of RNA helicases by single-molecule FRET TIRF microscopy.* Methods, *https://doi.org/10.1016/j.ymeth.2022.03.004; reprinted with permission.*

the insights into the mechanism of helicases gained by single-molecule FRET studies but were specifically selected to illustrate the possibilities and potential pitfalls of this technique. For an overview of the mechanism of DEAD-box helicases, the readers are referred to a set of pertinent reviews (Donsbach & Klostermeier, 2021; Henn, Bradley, & De La Cruz, 2012; Jarmoskaite & Russell, 2014; Putnam & Jankowsky, 2013; Rudolph & Klostermeier, 2015).

2. Förster resonance energy transfer (FRET) as a molecular ruler

Förster resonance energy transfer is a distance-dependent energy transfer process between two fluorescent dyes, termed donor and acceptor (Fig. 2A). On excitation of the donor fluorophore, it is converted into an electronically excited state of higher energy. The change in charge distribution that accompanies the excitation process is described by the transition dipole moment. The dipole moment of the donor can interact in a non-radiative process through space with the transition dipole of a suitable acceptor. This interaction is distance-dependent, hence the efficiency of the energy transfer from donor to acceptor also depends on the distance separating these two fluorophores. Energy transfer leads to the excitation of the acceptor fluorophore, with concomitant non-radiative return of the donor fluorophore to the ground state. Eventually, the acceptor returns to the ground state and emits fluorescence. The competition of donor relaxation with fluorescence emission and non-radiative transfer of energy to the acceptor determines the energy transfer efficiency. A low FRET efficiency is characterized by a high donor and low acceptor fluorescence, a high FRET efficiency is associated with a low donor and high acceptor fluorescence. Likewise, changes in energy transfer lead to anti-correlated changes of donor and acceptor fluorescence; an increase in energy transfer is associated with a decrease in donor fluorescence and an increase in acceptor fluorescence, whereas a decrease in transfer efficiency is characterized by an increase in donor fluorescence and a decrease in acceptor fluorescence (Fig. 2B).

The distance dependence of the FRET efficiency can be exploited to determine distances between two positions in a molecule or in a complex of different molecules, and to monitor changes in these distances. In principle, such experiments can be done by measuring donor and acceptor intensities after excitation of the donor in a standard fluorescence

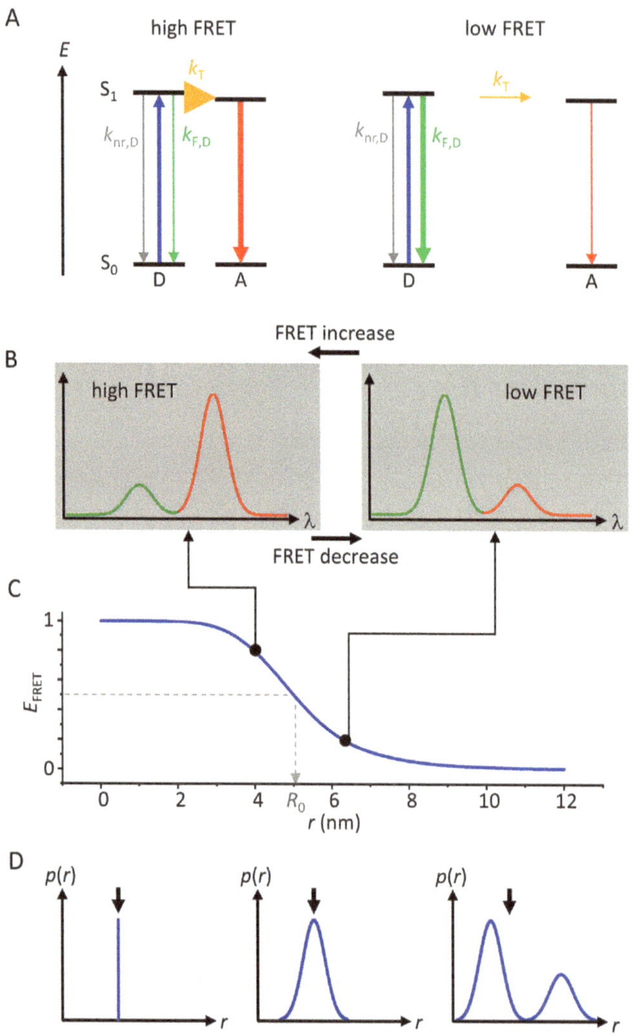

Fig. 2 Förster resonance energy transfer (FRET). (A) Jablonski diagram showing the electronic ground states (S_0) and excited states (S_1) for donor (D) and acceptor (A). Arrows indicate donor excitation (blue), relaxation of the excited state of the donor through non-radiative decay (gray, $k_{nr,D}$), relaxation through fluorescence emission (green, $k_{F,D}$) and energy transfer to the acceptor (orange, k_T), and acceptor emission associated with fluorescence (red). The rate constant for energy transfer is high when donor and acceptor dipoles are close-by (left, high-FRET state), the rate constant for energy transfer is low when they are far apart (right, low-FRET state). The widths of the arrows indicate the flux through the respective paths. (B) Fluorescence spectra showing donor and acceptor emission (green, red) for high-FRET states (left) and low-FRET states (right). (C) Distance dependence of the FRET efficiency E_{FRET} (see Eq. 2). Changes in FRET

spectro-meter, and by converting these intensities I_D and I_A, into FRET efficiencies E:

$$E = \frac{I_A}{I_D + I_A} \tag{1}$$

According to Förster's theory (Förster, 1948), the distance-dependence of the FRET efficiency E is given by

$$E = \frac{R_0^6}{R_0^6 + r^6} \tag{2}$$

(Fig. 2C). In this equation, R_0 is the Förster or characteristic transfer distance, and r is the actual distance between donor and acceptor dyes. The Förster distance depends on the photophysical properties of the dye pair used. Its value may change with the attachment of these dyes to the molecule of interest, with binding of a small-molecule ligand or other interaction partner, and with solution conditions. Hence, determination of the Förster distance under the particular experimental conditions is an absolute must if distance information is important. It is also important when the focus of the experiment lies on ligand-induced changes in distances: Changes in spectral properties of one or both dyes, and hence in R_0, on ligand binding may feign changes in distance, although all that changes is the Förster distance! Large-scale conformational rearrangements, involving distances between reference points in the range of 2–8 nm, are amenable to FRET studies using standard organic fluorophores with typical R_0 values around 4–6 nm (Fig. 2C).

In classic experiments on a large ensemble of molecules, the donor and acceptor intensities measured collect contributions over all molecules sampled. If the sample is homogenous with respect to the distance separating the

efficiencies are associated with anti-correlated changes in donor and acceptor emission: An increase in FRET (=decrease in distance *r*; see panel (C)) leads to an increase in acceptor emission and a decrease in donor emission. Conversely, a decrease in FRET (=increase in distance *r*) leads to a decrease in acceptor emission and an increase in donor emission. (D) Distance probability distributions *p*(*r*) (blue) and average distance determined in ensemble experiments (arrow). Left: The distance between two reference points within the molecule of interest is identical in all molecules in the ensemble; center: The distance fluctuates around a mean distance that is determined in ensemble experiments; right: The molecule of interest exists in two different conformations with distinct distance distributions. The arrow indicates the population-weighted average, which does not have a molecular meaning in this case.

two dyes, i.e., if all molecules have the same, defined conformation, then the distance computed from these values corresponds to the distance separating the dyes in each molecule (Fig. 2D). If the sample is heterogeneous, however, i.e., if two or more forms of the molecules co-exist that differ in the distance separating the two dyes, then these different species contribute differently to the intensities measured, and the distance calculated from these intensities is an average that depends on the distance in each species and the relative population of these species (Fig. 2D). In other words, such an average distance does not have a molecular meaning. In the worst-case scenario, the distance determined from such a measurement may not occur in a single molecule in the entire sample!

The situation is entirely different when FRET experiments are performed on a single-molecule level using fluorescence microscopy. Here, the donor and acceptor fluorescence intensities are measured for each molecule individually, separated from the rest of the ensemble either in time (confocal microscopy, see Section 4) or in space (total internal reflection fluorescence (TIRF) microscopy, see Section 5). At each timepoint of the measurement, a single molecule must be in a single, defined conformation. Therefore, the distance calculated from the donor and acceptor intensities of a single, FRET-labeled molecule corresponds to the distance between the two dyes in this particular conformation.

Large-scale conformational changes typically occur on timescales of milliseconds to seconds or even minutes. FRET experiments using fluorescence microscopy cover a wide range of timescales, from microseconds (confocal microscopy) to minutes (TIRF microscopy) and are thus suitable to follow conformational changes in real time.

3. Strategies in the experimental design of single-molecule FRET studies

Mapping conformational states and following conformational changes by (single-molecule) FRET requires intramolecular labeling of the helicase or the RNA of interest with donor and acceptor fluorophores in suitable positions. For this purpose, two functional groups need to be introduced at the desired sites.

3.1 Labeling proteins with donor and acceptor fluorophores

Labeling requires three principal choices to be made: the choice of the labeling chemistry, the choice of a suitable dye pair, and the choice of suitable sites for labeling.

3.1.1 Choice of labeling chemistry

Fluorophores can be introduced into proteins through two major avenues: by the reaction of amino acid side-chains with suitable reactive forms of fluorescent dyes, or through genetically encoded tags.

Many labeling strategies exploit the addition of the thiol group in the side-chain of cysteines to maleimide derivatives of fluorescent dyes, a reaction that is highly efficient for solvent-accessible cysteines. The disadvantage is that many proteins contain intrinsic cysteines that may or may not be located in suitable positions for the FRET experiment planned. Hence, the first step often is the replacement of solvent-accessible, intrinsic cysteines in the helicase of interest by site-directed mutagenesis. Although straightforward in principle, this process can become time- and resource-consuming if the protein of interest contains several cysteines that need to be removed. In the second step, cysteines need to be introduced by site-directed mutagenesis in suitable positions for labeling (see Andreou & Klostermeier, 2012 for a flow chart guiding through this procedure; see Section 3.1.3). A detailed protocol for statistical labeling of helicases containing two solvent-accessible cysteines with a mixture of donor and acceptor fluorophore maleimides is provided in Krause and Klostermeier (2021).

The extensive rounds of mutagenesis that may be required to generate a FRET construct through labeling of cysteines can be avoided by incorporating non-natural amino acids into the protein of interest (Ryu & Schultz, 2006; Wang, Xie, & Schultz, 2006), followed by labeling through bio-orthogonal click reactions (Kolb, Finn, & Sharpless, 2001). Site-specific incorporation of non-natural amino acids is achieved by introducing a stop codon at the desired site and stop-codon suppression using tRNA-synthetase/tRNA pairs to incorporate the non-natural amino acid at this site (Wang, Brock, Herberich, & Schultz, 2001). Functional groups that can be introduced include keto groups (Brustad, Lemke, Schultz, & Deniz, 2008; Tyagi & Lemke, 2013; Wang, Zhang, Brock, & Schultz, 2003), azides (Agard, Baskin, Prescher, Lo, & Bertozzi, 2006), alkynes (Chin et al., 2002; Nguyen et al., 2009; Tyagi & Lemke, 2013), and many more (reviewed in Shandell, Tan, & Cornish, 2021). Click reactions, by definition, occur rapidly and with high yields, and provide a spectrum of reactions for fluorescent labeling (reviewed in Tyagi & Lemke, 2013). Depending on the type of reaction, however, labeling may require conditions that are not beneficial for protein activity and stability. For example, the copper-catalyzed cycloaddition of azides and alkynes (Wang, Chan, et al., 2003) requires the presence of Cu(I), which is generated in situ. Oxidative processes associated with this reaction system can have deleterious effects on proteins (Cheng, Uchida, & Kawakishi, 1992).

Copper-free alternatives are available but come at the expense of lower yields (Agard, Prescher, & Bertozzi, 2004; reviewed in Plass, Milles, Koehler, Schultz, & Lemke, 2011).

Using the same functional group for both sites (i.e., two cysteines, or two identical non-natural amino acids) leads to statistical labeling: The reaction products are a mixture of the desired donor/acceptor- and acceptor/donor-labeled species and molecules carrying two donor or two acceptor fluorophores. If labeling is incomplete, species carrying only one donor or only one acceptor moiety or no dye at all are also populated (Fig. 3A). Molecules carrying donor and acceptor give rise to a FRET peak, centered at the FRET efficiency determined by the inter-dye distance (Fig. 3A). Note that the photophysical properties of the dyes may depend on the labeling geometry (i.e., they may be different for donor/acceptor and acceptor/donor species). In this case, the two FRET species generated by statistical labeling may have different FRET efficiencies although the distance between the two dyes is identical, simply because the Förster distance is different (see Eq. 2)! In addition to the FRET peak, the molecules carrying only one type of dye also contribute to the distribution of FRET efficiencies observed (Fig. 3A): Molecules labeled with one or two donor moieties, but lacking the acceptor, will appear as a peak with a FRET efficiency of zero (only donor emission). If the fraction of molecules labeled with only the donor is large, then the concomitant large zero-peak may preclude the detection of molecules that carry both the donor and the acceptor but have a low FRET efficiency. Those molecules that carry only an acceptor dye will not be excited (and detected) efficiently. They may give rise to a small peak at a FRET efficiency of one (only acceptor emission).

Using two different functional groups enables the site-specific introduction of the donor fluorophore at one site, and of the acceptor fluorophore at the other site. In addition, the formation of molecules carrying two donor or two acceptor molecules is prevented, although incomplete labeling may still generate molecules carrying only the donor or only the acceptor (Fig. 3B). Site-specific labeling of proteins with donor- and acceptor fluorophores has been achieved by combining cysteine labeling with labeling of a non-natural amino acid (Brustad et al., 2008).

Site-specific introduction of the donor fluorophore at one site, and of the acceptor fluorophore at the other site, can also be achieved by employing labeling strategies that rely on native chemical ligation (Dawson, Muir, Clark-Lewis, & Kent, 1994) or expressed protein ligation (Muir, Sondhi, & Cole, 1998; reviewed in Blaschke, Silberstein, & Muir, 2000)

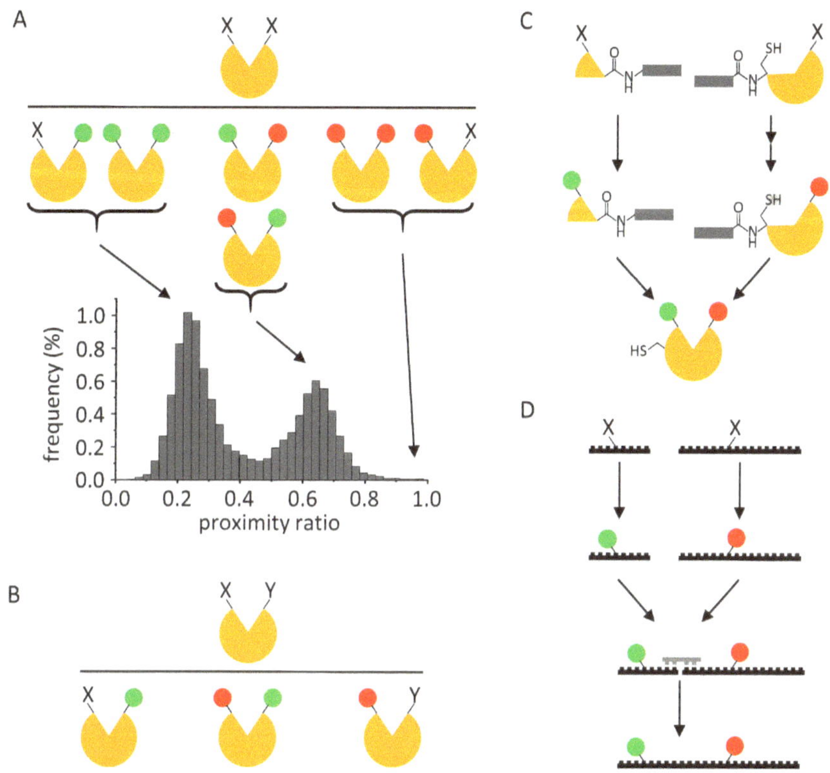

Fig. 3 Strategies to label helicase and RNA substrates. (A) Labeling the protein of interest (*orange*) through statistical modification of the same functional group X, e.g., cysteine side-chains, leads to a mixture of species carrying only the donor (*green*), only the acceptor (*red*), or both dyes in DA and AD configuration. If the labeling reactions are incomplete, some molecules remain singly labeled or unlabeled. The FRET histogram (bottom; uncorrected histogram showing the distribution of proximity ratios) then shows a low-FRET peak corresponding to the molecules labeled with only the donor, a peak representing the species carrying both dyes at intermediate FRET efficiencies, and possibly a small peak at high FRET efficiencies caused by molecules labeled with only the acceptor. (B) Labeling the protein of interest (*orange*) using two different functional groups X and Y, e.g., a cysteine side-chain and a non-natural amino acid, leads to a species carrying both dyes in a defined configuration (DA). If the labeling reactions are incomplete, some molecules remain singly labeled or unlabeled. (C) Labeling strategy using intein fusion proteins and expressed protein ligation. The N-terminal part of the protein of interest (*orange*) is produced as a fusion with a C-terminal intein (*gray*), the C-terminal part carries an intein at the N-terminus and a cysteine (SH) following the intein. The functional groups X in each fragment can be labeled separately. If X is a cysteine, then the cysteine following the N-terminal intein in the C-terminal part needs to be protected (indicated by the *gray sphere*). After labeling, the two protein fragments can be ligated to the full-length protein, carrying donor and acceptor at the desired sites in a unique configuration. (D) Labeling strategy for large RNA substrates using (splint) ligation. The RNA substrate can be split into two fragments carrying a functional group X, introduced through chemical synthesis, which are then separately labeled with the donor and the acceptor. The complete RNA substrate is generated by ligation of the two fragments, e.g., using splint ligation. Gray: DNA splint for splint ligation.

(Fig. 3C). While native chemical ligation links two chemically synthesized, labeled peptides with a peptide bond, expressed protein ligation applies the same strategy to two recombinantly produced protein fragments. The protein of interest is produced in two parts, in the form of intein fusions, with a cysteine or non-natural amino-acid for labeling in each fragment of the protein of interest (Karow, Theissen, & Klostermeier, 2007). Cleavage of the intein fusion proteins generates the N-terminal part of the protein of interest with a C-terminal thioester, and the C-terminal part with an N-terminal cysteine; these fragments are the reactive forms for ligation. The two parts of the protein of interest are then labeled with the two different dyes separately, either before or after cleavage of the fusion proteins. In the final step, the full-length protein is generated by ligation of the labeled fragments. The cysteine at the N-terminus of the C-terminal fragment, which acts as a nucleophile during ligation, may interfere with labeling approaches that rely on cysteine labeling. However, alternatives have been described where the N-terminal residue of the C-terminal fragment can be another residue (Dassa, Haviv, Amitai, & Pietrokovski, 2004), or where the N-terminal cysteine is protected during the labeling reaction, and subsequently deprotected for the ligation. Split inteins can also be used for such ligation approaches (Bachmann, Matern, Schutz, & Mootz, 2015; Brenzel, Kurpiers, & Mootz, 2006; Shi & Muir, 2005; Wu, Hu, & Liu, 1998). In this approach, a self-excising intein is split into two fragments that are inactive on their own but assemble into a functional intein on complex formation. One part of the protein of interest is fused to one intein fragment, and the other part to the second fragment. Both intein fusion proteins are then produced and labeled separately. On mixing, the intein fragments interact, leading to their excision and the covalent linkage of the two parts of the protein of interest by a natural peptide bond. Cysteine-free versions of split inteins, compatible with labeling through cysteines introduced by site-directed mutagenesis, have also been described (Bhagawati et al., 2019).

In principle, proteins can also be labeled by generating fusion proteins with fluorescent proteins. Because of the genetic encoding of these labels, this strategy comes with the advantage that it generates exclusively completely labeled species and can be applied for in vivo studies. However, fluorescent proteins are large entities: Green fluorescent protein has a molecular mass of 27 kDa, compared to the helicase core with approx. 46 kDa. Such a "heavy" modifier will most likely affect the dynamics and possibly also the conformational equilibrium of the molecule of interest. The same limitation applies to genetically encoded fluorogenic tags, such as the Halo-, SNAP-, or CLIP-tags (Gautier et al., 2008; Keppler et al., 2003; Los et al., 2008; Stagge, Mitronova, Belov, Wurm, & Jakobs, 2013) that also add an entity with a molecular mass of approx. 20–30 kDa to the protein of interest. Possible alternatives are the FlAsH tag, a short tag containing a tetra-cysteine motif that forms fluorescent complexes with arsenic-containing ligands (Luebke, 1998), and the five-amino-acid sortase tag that is recognized by the enzyme sortase (Popp & Ploegh, 2011). These considerably smaller tags can be introduced at the termini as well as internally, e.g., in loop regions. They will also have less influence on the conformational dynamics and equilibrium of the protein of interest.

The considerations above on the outcome of labeling reactions using different strategies are all valid for monomeric helicases. For dimeric enzymes, the situation is different, and additional considerations are necessary depending on the goal of the FRET experiment. If conformational changes across the dimer interface are of interest, then labeling of dimeric enzymes is simple: On introduction of only a single cysteine in each monomer, two reactive sites are available in the dimer for statistical labeling with donor and acceptor. As the local environment of these cysteines is identical in each monomer, donor/acceptor- and acceptor/donor-labeled species are indistinguishable and have the same Förster distance, hence a single conformation is represented as a single FRET peak. On the other hand, monitoring conformational changes within each subunit may become more challenging: Introducing two cysteines within one monomer leads to four cysteines in the dimer, and statistical labeling generates a mess of differently labeled species that give rise to complex FRET distributions that are difficult, if not impossible, to interpret (see Section 4.4.3). If the dimer dissociates readily, then it is possible to generate heterodimers with one subunit containing two cysteines, and a second subunit lacking cysteines, simply by mixing the cysteine-containing variant with a sufficiently large excess of the cysteine-free version. If the dimer is highly stable, then this approach may not be possible. In this case, it should be considered to coproduce the subunit lacking cysteines with a subunit containing two cysteines for labeling, and to purify heterodimers consisting of one cysteine-containing and one cysteine-free subunit, e.g., by tandem-affinity purification (Gubaev, Weidlich, & Klostermeier, 2016; Hartmann, Gubaev, & Klostermeier, 2017; Stelljes, Weidlich, Gubaev, & Klostermeier, 2018). Note that in this approach, naturally a lot of protein is lost during purification because co-production generates a statistical mixture of both homodimers and the desired heterodimer.

3.1.2 Choice of dye pair

The choice of dyes for a FRET experiment depends both on technical aspects and on the behavior of protein of interest. A number of dye series with absorption and emission spectra covering the entire visible spectral range are commercially available in reactive forms for cysteine modification, increasingly also for the modification of non-natural amino acids in click reactions. The photophysical properties of the dyes are important for several reasons: Naturally, the equipment, particularly the excitation lasers and filter sets for detection that are available, limits the set of dyes that can be used. The distance(s) of interest also matter(s), as the emission and absorption spectra of donor and acceptor determine the R_0 of the respective dye pair and hence the range of distances that can be determined reliably (see Fig. 2C). A high quantum yield is desired for sensitive detection. High photostability ensures extended observation times before the fluorophore bleaches and ceases to emit fluorescence, which is particularly important for single-molecule FRET experiments by TIRF microscopy (see Section 5). The choice of dyes may also be limited by the choice of the modification chemistry, as not all dyes are available with all types of functionalization. Finally, the behavior of the protein of interest when labeled with these dyes is an important factor, as dyes may affect its solubility and activity. Note that most organic dyes have low water-solubility and are therefore dissolved in DMSO and then diluted into the aqueous protein solution. DMSO can have deleterious effects on enzymatic activity, and its concentration should be kept a low as possible. In any case, the enzymatic activity of the protein of interest needs to be assessed after labeling to ensure that the experiments are performed with a functional helicase (see Section 3.1.4).

3.1.3 Choice of labeling sites

Labeling sites should be chosen carefully, based on three main criteria: the reactivity of the functional group at the site of choice, the integrity of the protein when this site is modified, and the specific experiment to be performed. The distance separating the labeling sites should be in the sensitive range around the Förster distance of the dye pair chosen (see Fig. 2C), and the expected changes should be associated with detectable changes in FRET efficiencies. Structural information on the protein of interest, or on individual domains, may serve as a guide in the assessment of side-chain accessibilities and in the choice of suitable sites for the specific experiment. In the absence of structural information, decisions on labeling sites have to be made by educated guesses, based on homology models and working hypotheses on structural changes occurring.

Reactivity. To achieve a high reactivity of the functional group chosen for labeling, labeling sites should be located on the surface of the protein of interest, accessible to solvent and solutes. In our experience, cysteine side-chains pointing into the solution are modified typically within a few minutes, whereas less accessible cysteines may not even be modified completely after extended incubation overnight. Long-term incubation does not help much because of the limited stability of maleimides in aqueous solution. If labeling yields are low, a different site needs to be chosen.

Integrity of the protein of interest. Side-chains used for labeling need to be in non-conserved regions of the protein of interest, outside conserved motifs. Residues in surface-exposed loops are particularly suited for labeling. Obviously, residues important for function and/or stability should not be touched. If multi-component systems are investigated, it needs to be tested whether labeling interferes with binding to interaction partners. Fluorescent dyes have different degrees of hydrophobicity and may affect protein solubility differently.

Suitability for the desired experiment. Within the limits set by reactivity and integrity of the protein of interest, labeling sites should be chosen based on the hypothesis of the experiment. For FRET experiments aimed at identifying different conformational states, without interest in the exact distances and possible distance changes involved, it may be appropriate to choose sites that give rise to extreme changes in FRET efficiencies, ideally between zero-FRET states and states with a FRET efficiency of one. In this regime, the two states can be distinguished unambiguously, but the FRET efficiencies are not sensitive to the inter-dye distance; only a lower or upper limit can be deduced for the low- and high-FRET states (see Fig. 2C). If absolute distances do matter, e.g., in experiments aimed at mapping global conformations, the sites should be chosen such that the expected FRET efficiencies be between 0.1 and 0.9 (see Fig. 2C) to ensure that the FRET efficiency measured is sensitive to the inter-dye distance. Here, a set of suitable sites needs to be selected that jointly enables monitoring both translational and rotational motion of domains (see Section 4.4.4).

In principle, labeling strategies used to identify conformational states in single-molecule FRET experiments by confocal microscopy can be transferred directly to TIRF experiments to monitor transitions between these states, provided that the same excitation wavelength and filter sets are available. This is particularly beneficial as measurements by confocal microscopy can be analyzed much faster, giving instant feedback on different FRET states populated, whereas the generation of FRET histograms from experiments done by TIRF microscopy is more time-consuming (see Sections 5.2 and 5.3). Even in cases that aim at the kinetic analysis of conformational changes, it is thus advisable to screen for suitable constructs that give promising results in solution studies, and then use these in follow-up kinetic studies by TIRF microscopy.

3.1.4 Labeling and enzymatic activity

Independent of the labeling strategy chosen, it is of vital importance to test for enzymatic activity of the labeled protein of interest. Enzyme activity may be reduced because of mutations introduced, the exposure to reagents during labeling, and the presence of the dyes. For RNA helicases, RNA-dependent ATPase activity or ATP-dependent RNA unwinding activity can be tested. While steady-state ATPase assays as a function of ATP and RNA concentrations will reveal the effect of labeling on interactions with ATP and RNA, helicase assays as a function of time and protein concentration will reveal if the enzyme has lost enzymatic activity and/or efficiency. It is important to note that in most cases, preparations of labeled protein contain some fraction of unlabeled protein. For efficient enzymes, this fraction might contribute substantially to the activity observed. Therefore, its fraction should be determined by measuring the labeling efficiency for both dyes. From these values, typically determined through absorption spectroscopy, all species present can be calculated (see Andreou & Klostermeier, 2012 for a detailed description). With (rather typical) labeling efficiencies of, e.g., 40% with donor and 70% with acceptor, a fraction of $(1\text{–}0.4)\cdot(1\text{–}0.7)=0.18$ or 18% of all molecules remain unlabeled. The corresponding concentration of the unlabeled enzyme should be used in parallel control reactions to appreciate its contribution to activity.

3.2 Labeling RNA substrates

To understand the mechanism of RNA helicases, it may be of interest to investigate conformational changes in the RNA substrate induced by the helicase. When these studies are performed with model RNA substrates, fluorescently labeled RNA oligonucleotides generated by chemical synthesis can be used to assemble the desired RNA structure. If the fluorophores do not withstand the solvents used in chemical synthesis, thiol or amine groups can be introduced during chemical synthesis and derivatized with appropriately functionalized dyes afterwards. However, many helicases act on large

RNA molecules that are not amenable to chemical synthesis. In these cases, RNAs have to be generated through in vitro transcription from suitable templates. Labels are then introduced after synthesis, although approaches to introduce fluorescent dyes during in vitro transcription have been described more recently (Chauvier et al., 2021). To investigate RNA remodeling of the group II intron by the DEAD-box helicase Mss116 by single-molecule FRET, short DNA oligonucleotides carrying the donor and acceptor dyes were hybridized to two loop regions of the RNA (Karunatilaka, Solem, Pyle, & Rueda, 2010; Steiner et al., 2008). Large, donor/acceptor labeled RNA molecules can also be assembled by ligation of shorter labeled fragments (Fig. 3D), either through splint ligation, using a DNA splint complementary to the ends of the fragments to be ligated (Moore & Query, 2000), or using direct RNA ligation (Hanspach, Trucks, & Hengesbach, 2019; see Akiyama & Stone, 2009 and Kershaw & O'Keefe, 2012 for a detailed description). Labeling through multi-step ligation of a combination of labeled and unlabeled fragments has also been described. However, it should be taken into account that the yields decline with each additional ligation step needed. Fluorogenic tags and fluorogenic aptamers are also available for labeling of RNA (Paige, Wu, & Jaffrey, 2011). While these modifications are very useful for RNA detection in vivo (reviewed in Ilgu et al., 2016; Schmidt, Gao, Little, Jalihal, & Walter, 2020), their considerable size limits their use for in vitro FRET studies of RNAs. Hanspach et al., (2019) and Schmidt et al. (2020) provide excellent overviews about fluorescent labeling of RNA.

3.3 Surface immobilization of proteins and RNA

For single-molecule experiments by TIRF microscopy, the donor/acceptor-labeled molecule needs to be immobilized on the functionalized surface of the microscopy slide. The most common interaction exploited for surface immobilization is the biotin-streptavidin interaction, using biotinylated molecules of interest and streptavidin-functionalized slides. A step-by-step protocol of slide preparation and functionalization is described in Krause and Klostermeier (2021). Biotinylated proteins can be produced recombinantly: A 14-amino-acid biotinylation tag (GLNDIFEAQ**K**IEWH; biotinylation site K highlighted in bold) is introduced at the N-terminus of the protein of interest (Beckett, Kovaleva, & Schatz, 1999; Harms, Andreou, Gubaev, & Klostermeier, 2014; Samatanga, Andreou, & Klostermeier, 2017), and the fusion protein is overproduced in *Escherichia coli* strains that also overproduce biotin ligase, grown in media supplemented with biotin. Thereby, the protein is biotinylated in vivo, and can be purified from crude extracts using standard chromatographic techniques. Other immobilization approaches employ His-tagged versions of the protein of interest in combination with nitrilo triacetate-functionalized surfaces (Lata & Piehler, 2005), or epitope/antibody interactions (Pal, Lesoine, Lieb, Novotny, & Knauf, 2005).

Immobilization of RNAs on streptavidin-coated surfaces can be achieved directly with biotinylated RNA, which can be generated by chemical synthesis or by in vitro transcription with biotinylated starter nucleotides. Alternatively, immobilization can be achieved through hybridization of an additional sequence (handle), introduced at the 5′- or 3′-end of the RNA, to a complementary, biotinylated DNA oligonucleotide immobilized on the surface (Karunatilaka et al., 2010).

4. Confocal microscopy on helicases in solution: Equilibrium distributions of conformers

4.1 Separating molecules in time: Confocal microscopy

In confocal microscopy (Minsky, 1988) (Fig. 4), the donor/acceptor-labeled molecules diffuse freely in solution. By focusing the excitation light tightly into this solution using a microscope objective, an observation volume, termed the confocal volume, is generated in the focal plane (Fig. 4A). Within this confocal volume, the excitation intensity is sufficiently high to excite the donor fluorophore. Fluorescence emitted from either donor or acceptor is then collected by the objective, separated into the spectral regions of donor and acceptor emission, and detected on point detectors, typically avalanche photodiodes.

The single molecules diffuse freely through the observation volume. They are observed as they pass by, one after the other, isolated in time. To ensure single-molecule conditions, the concentration of the donor/acceptor-labeled molecule must be sufficiently low, such that at any given time, there is at most one molecule in the observation volume. We typically use concentrations of the molecule of interest of 100–150 pM. This concentration refers to the concentration of the donor fluorophore, i.e., on the concentration of the detected species. If the total concentration of protein is used as a basis, then the unlabeled protein present will lead to an overestimation of the concentration of species that are excited and detected,

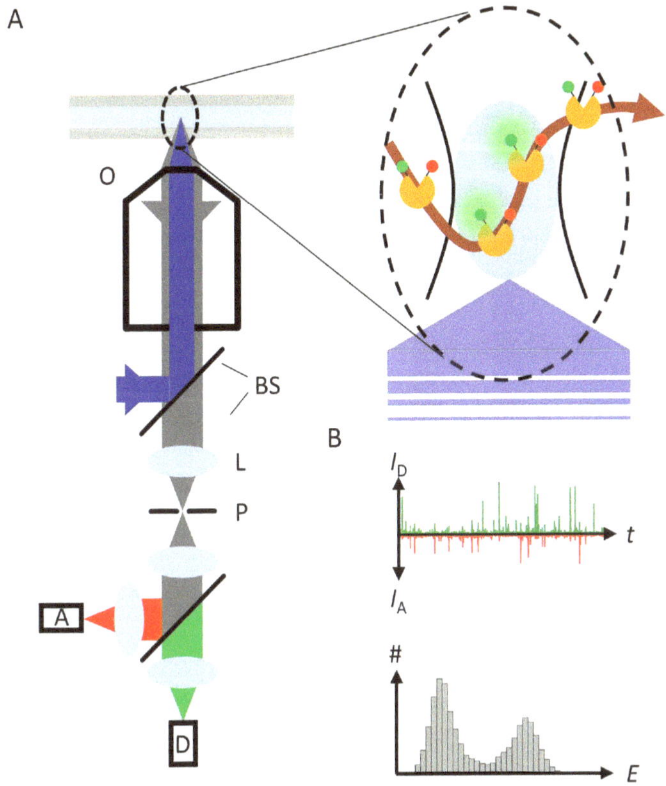

Fig. 4 Single-molecule experiments on molecules diffusing freely in solution by confocal microscopy. (A) Schematic depiction of a confocal microscope. The excitation light (blue) is coupled into an objective (O) and focused into a tiny volume within the sample, termed the confocal volume (blue oval in enlarged depiction on the right). Molecules diffusing though the confocal volume are excited and emit fluorescence (green halo); labeled molecules outside the confocal volume are not excited and do not emit fluorescence. Fluorescence (gray) emitted from molecules within the confocal volume is collected by the objective, focused into the aperture of a pinhole (P) by a lens (L), then split into donor and acceptor contributions (green, red) by a beam splitter (BS), and detected as a function of time using two separate point detectors (D, A). (B) Single-molecule experiments by confocal microscopy generate pairs of time traces of fluorescence (top), one for the donor fluorescence (green) and one for the acceptor fluorescence (red). After identification of fluorescence bursts, caused by single molecules traversing the confocal volume, FRET efficiencies can be calculated for each single-molecule event, and a FRET histogram (bottom) for all molecules detected one after the other can be generated.

to fewer single-molecule events per time, and hence to extended measurement times to interrogate a certain number of molecules.

Because of the limited fluorescence signal from a single emitter, it is particularly important to minimize background fluorescence from impurities. This requires not only very pure preparations of proteins and RNA, but also necessitates special consideration of the main component in the measurement, the buffer used. The specific composition of the buffers used for single-molecule experiments depends on the requirements of the helicase studied. Buffers for single-molecule FRET experiments by confocal microscopy are pre-treated with active charcoal overnight before the experiment to remove any fluorescent contaminants. Components associated with high background fluorescence, such as glycerol or detergents, should be avoided if possible. Buffers should be stored in glassware instead of plastic vials to prevent the leaking of softeners. A protocol for single-molecule measurements by confocal microscopy is given in Krause and Klostermeier (2021).

4.2 From fluorescence bursts to FRET histograms and conformational states

The raw data obtained from single-molecule FRET experiments by confocal microscopy are two fluorescence time traces, one for the donor fluorescence, one for the acceptor fluorescence (Fig. 4B). Single-molecule conditions mean that most of the time, no molecule at all is present in the confocal volume, and background counts due to shot noise of the detector, scattering and fluorescence impurities are recorded, collectively termed background. Only during very brief periods of time, a molecule enters the confocal volume, the donor is excited, and donor and/or acceptor emission give rise to a fluorescence burst, until the molecule diffuses out of the confocal volume, and the signal returns to background levels. The size of these bursts depends on the time the molecule spends in the confocal volume and the specific path it takes to traverse it.

To determine FRET efficiencies for each molecule observed and to collect all events in FRET histograms, the raw data need to be processed. Data processing is done differently in different laboratories, depending on the type of confocal microscope and software packages used. The description in the following therefore is limited to the general principle and to the main steps of data processing. First, the background counts need to be subtracted from the measured intensities, and single-molecule events need to be assigned. Single-molecule events are commonly identified by applying threshold criteria, either absolute ($> n$ photons, with n typically between 50 and 100) or relative ($> n$-fold above background). From the background-corrected donor and acceptor intensities, proximity ratios can be determined, either by calculating the ratio of I_A and I_D, or by calculating the fraction of I_A over the total intensity ($I_D + I_A$) (Eq. 1). Either way, a high proximity ratio corresponds to a high FRET efficiency and hence a small distance between donor and acceptor, while a low proximity ratio corresponds to a low FRET efficiency and a large inter-dye distance. The proximity ratio helps identify different conformational states and quantify their populations. If the inter-dye distance is important, for example to collect distance restraints for modeling, then the background-corrected intensities need to be corrected further for instrument effects, i.e., for leakage of the donor emission into the acceptor channel and vice versa, for different sensitivity of the system for donor and acceptor emission and the different quantum yields of these dyes, and for the direct excitation of the acceptor at the excitation wavelength used (see Section 4.4.1 for an example). The required correction parameters can be determined in control experiments, using protein labeled with only the donor, and with only the acceptor. A detailed description of the procedure to determine these correction parameters and to apply the corrections is given in Andreou and Klostermeier (2012), Hartmann, Weidlich, and Klostermeier (2016), and Krause and Klostermeier (2021).

The conversion of FRET efficiencies into inter-dye distances requires the knowledge of the Förster distance for the dye pair used. It is important to note that it is not sufficient to apply tabulated values determined in a different context! The Förster distance depends on the photophysical properties of the dyes used, and these in turn depend strongly on the molecular environment of the dye. As a consequence, R_0 needs to be determined for the specific construct under study, under conditions matching the experimental conditions as closely as possible (i.e., the same buffer, and the same ligands present, if possible). The procedure to determine R_0 for the specific construct used is described in detail in Andreou and Klostermeier (2012) and Hartmann et al. (2016). With the specific value for R_0, FRET efficiencies can then be converted into distances using Eq. (2), and the values obtained can be compared with structural information available for calibration or can be used to generate structural models through trilateration procedures (see Section 4.4.4 for an example).

4.3 Kinetic information from single-molecule FRET data by confocal microscopy

The duration of a typical fluorescence burst from a single molecule is limited to a few milliseconds, which is the time a single molecule needs to traverse the confocal volume. From the donor and acceptor photons detected within these bursts, kinetic information on the timescale of sub-micro- to milliseconds can be extracted. Such approaches, based on recurrence analyses (Hoffmann et al., 2011), correlation analyses (Nettels, Gopich, Hoffmann, & Schuler, 2007) or direct analyses of photon streams combined with hidden Markov modeling (HMM) (Pirchi et al., 2016), have been applied to characterize polypeptide chain dynamics. For further information, the interested reader is referred to the original references. Large-scale conformational changes are much slower and are thus not (or only very rarely) observed in the millisecond time window of single-molecule experiments by confocal microscopy.

4.4 Example data illustrating possibilities and potential pitfalls

Single-molecule FRET experiments in solution have defined the conformational states of DEAD-box helicases (Theissen et al., 2008), and have delineated the conformation of the different nucleotide states during the catalytic cycle (Aregger & Klostermeier, 2009; Theissen et al., 2008). They have revealed that binding of ATP and the RNA substrate to the helicase core triggers closing of the core, whereas phosphate release, coupled to dissociation of the second strand of the RNA duplex, is the trigger for re-opening and re-setting of the helicase core for subsequent cycles. Closing leads to distortion of the RNA duplex in preparation for strand separation (Mallam, Del Campo, Gilman, Sidote, & Lambowitz, 2012; Sengoku et al., 2006), and occurs before ATP hydrolysis, in-line with the observation that ATP binding is sufficient for duplex unwinding (Liu, Putnam, & Jankowsky, 2008). Single-molecule FRET experiments by confocal microscopy have also shown that accessory factors can stabilize the helicase core in half-open and closed states (Andreou & Klostermeier, 2014; Hilbert, Kebbel, Gubaev, & Klostermeier, 2011). Finally, distances determined by single-molecule FRET have been used as restraints to model the conformation of the helicase core, and to generate a structural model of a full-length helicase comprising a helicase core and an RNA-binding domain (RBD) (Karow & Klostermeier, 2010; Samatanga et al., 2017), based on a homology model of the helicase core and the crystal structure of the isolated RBD (Wang et al., 2006) (see Section 4.4.4.). These studies provided evidence for an allosteric activation of this and possibly other helicases by RNA binding to the RBD (Samatanga et al., 2017). For experimental details, the reader is referred to the original literature cited. In the following, selected examples are described to illustrate important considerations when analyzing these experiments, their limitations, and possible pitfalls.

4.4.1 Uncorrected and corrected histograms

The background-corrected intensities of donor and acceptor fluorescence still contain contributions from cross-talk of the donor emission into the acceptor channel and vice versa, and from direct excitation of the donor. They are also distorted because of different detection efficiencies of donor and acceptor fluorescence. Nevertheless, these intensities can be used without further corrections to calculate a proximity ratio, or apparent FRET efficiency, for each single-molecule event to generate histograms. This is illustrated exemplarily using data for the dimeric DEAD-box helicase Hera from *Thermus thermophilus*, labeled with donor and acceptor fluorophores in each RecA-like domain of the helicase core (Fig. 5A). The uncorrected histogram contains two peaks, indicating that two different species are present. The peak at a proximity ratio of 0.25 represents the species carrying only a donor, the peak at a proximity ratio of 0.65 is caused by donor/acceptor-labeled molecules. Using the correction parameters determined in control experiments, the raw intensities can be corrected, and the resulting FRET efficiencies can be collected in a histogram (Fig. 5A). Note that the peak for the donor/acceptor-labeled species is now centered at a FRET efficiency of 0.5, which can be converted into an inter-dye distance using Eq. (2). This distance obtained is the distance between the centers of the dyes, which can be compared to structural information available, or used for modeling of global conformations. The correction procedure shifts the donor-only peak to a FRET efficiency below zero. While the negative FRET efficiency has no physical meaning, this shift has the beneficial side-effect that low-FRET species can now easily be identified.

4.4.2 Quantifying the populations of species and extracting (apparent) equilibrium constants

Commonly, FRET histograms are well-described with Gaussian distributions. More sophisticated procedures that calculate the best fit to experimentally determined FRET histograms are treated in detail in Nir et al. (2006). By describing histograms with Gaussian distributions, the mean FRET efficiencies of the species detected, as well as the widths of these distributions are determined. Analysis of bimodal histograms showing two FRET states with the sum of two Gaussian distributions additionally retrieves the fractional population of the two species, represented by the area under the respective curve (Fig. 5B). Bimodal histograms can be obtained for different scenarios. In the simplest case, they arise when two states, A and B, inter-convert (Fig. 5C, model I). The equilibrium constant K_{eq} for this reaction scheme can be calculated by dividing the fractions (f_B/f_A) of the two species. Other possible

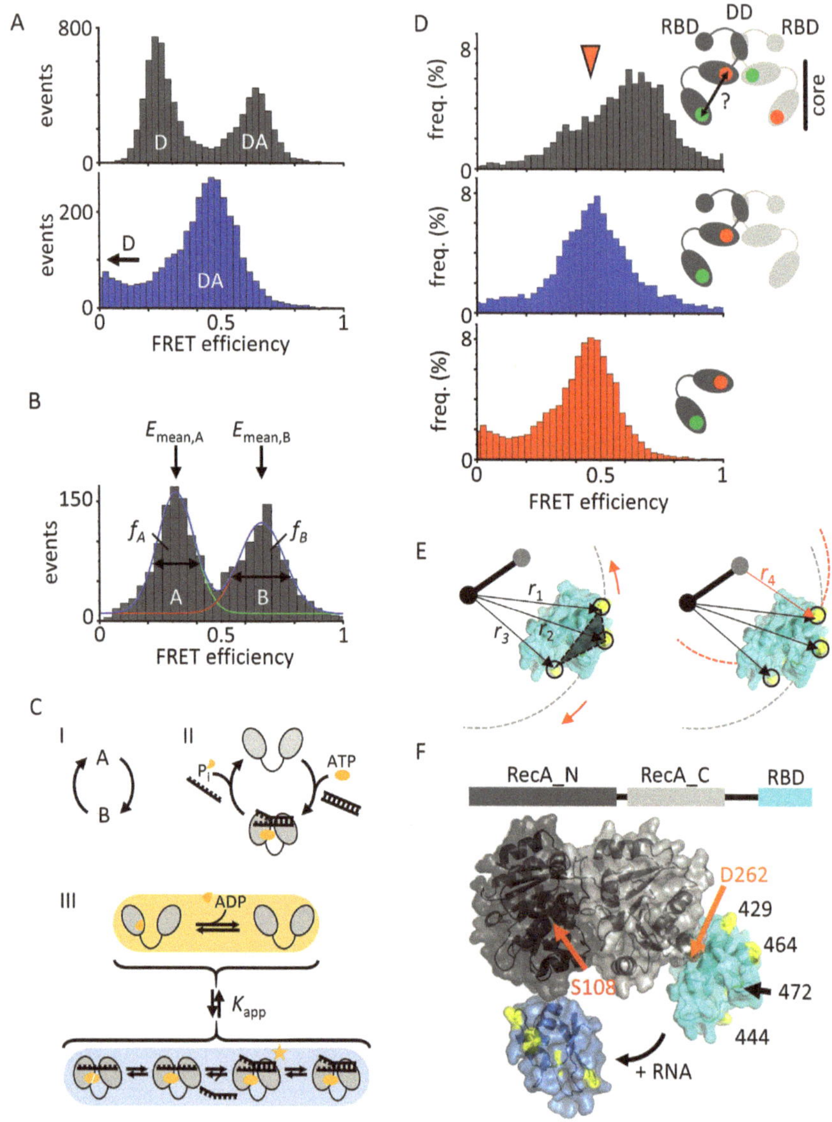

Fig. 5 Selected examples illustrating possibilities and pitfalls of single-molecule experiments in solution. (A) Uncorrected proximity-ratio- (top) and corrected (bottom) single-molecule FRET histogram of the isolated Hera helicase core, labeled with donor and acceptor in the two RecA-like domains. D: donor-only peak, DA: peak representing donor/acceptor-labeled molecules. Note the shift in the FRET efficiency of the donor/acceptor-labeled species from approx. 65% before correction to approx. 45% after correction for instrument non-idealities. The donor-only peak is shifted to negative values due to the correction procedure (P. Donsbach & D. Klostermeier, unpublished data). *(Continued)*

Fig. 5—Cont'd (B) Determination of mean FRET efficiencies and fractional populations from bimodal FRET histograms and formulation of working models. Top: By describing the FRET histogram with the sum of two Gaussian distributions (blue line; green, red: distribution for the individual FRET species A and B), the mean FRET efficiencies ($E_{FRET,A}$, $E_{FRET,B}$) of the two states observed (A,B), the width of these distributions (double arrows), and the fractional populations of the two FRET species (f_A, f_B, from the area under the respective Gaussian) are obtained (L. Krause & D. Klostermeier, unpublished data). (C) Models with different levels of detail that are consistent with a bimodal FRET histogram. I: Two states A and B with low and high FRET efficiencies inter-convert. The equilibrium constant K_{eq} can be calculated from the ratio of the fractional populations. II: In the context of a DEAD-box helicase, labeled with donor and acceptor dyes in the two RecA-like domains, the states A and B observed may reflect the open and closed states. The equilibrium constant for closing is the ratio f_B/f_A. III: Single-molecule FRET studies of the DEAD-box helicase core have shown that the ADP-bound and nucleotide-free states are in the low-FRET, open conformation (orange), whereas all other states in the catalytic cycle are in a high-FRET, closed state (blue). The different open states and the different closed states cannot be distinguished by FRET. In this case, the equilibrium constant calculated from the fractions observed is an apparent equilibrium constant, with the sum of the concentrations of all open states in the numerator, and the sum of the concentrations of all closed states in the denominator. (D) Single-molecule FRET histograms of different FRET constructs of the dimeric DEAD-box helicase Hera from *Thermus thermophilus* in solution, obtained from single-molecule FRET experiments by confocal microscopy. One protomer is shown in dark gray, the second in light gray. Core: helicase core formed by two RecA-like domains, RBD: RNA-binding domain, DD: dimerization domain. Top: FRET histogram for the dimer containing two cysteines in each protomer. Because of different species with different inter-dye distances, the histogram is very broad, and does not reflect the distance between the two positions within the helicase core (question mark). Middle: FRET histogram of heterodimeric Hera, in which one protomer (light gray) does not contain cysteines, the second (dark gray) contains a cysteine in each RecA-like domain of the core. The heterodimer was generated by mixing the homodimer in the top panel with a fivefold excess of unlabeled Hera dimer, and facilitating subunit exchange by incubation at 65 °C. The distribution observed in the FRET histogram has shifted, and the width is reduced. Bottom: FRET histogram for the isolated Hera core with a narrow, well-defined FRET peak (P. Donsbach & D. Klostermeier, unpublished data). (E) Multi-lateration principle to determine relative orientations of domains. The gray and black circles connected by a black axis symbolize two reference points on one part of the protein (e.g., on each RecA-like domain of the helicase core), the cyan structure is the domain whose position relative to the reference structure is to be determined (e.g., a flanking RNA-binding domain). The residues labeled on this domain are highlighted in yellow and with open circles. Left: The distances r_1–r_3 between the black reference position on one part and the three residues on the other part of the protein can be determined in single-molecule FRET experiments. They define the relative orientation of the plane of the triangle formed by the three positions to the reference point, but this triangle is free to rotate around the reference point (gray circle). Right: Introducing an additional distance between a second additional reference point (gray) enables the

(Continued)

models that give rise to bimodal FRET histograms include parallel or sequential reactions involving more than one state with low and/or high FRET efficiencies (see Fig. 9). In the context of DEAD-box helicases, the simple model of interconverting low- and high FRET states might be interpreted as the inter-conversion of open and closed states of the helicase core, coupled to duplex separation (Fig. 5C, model II). Dissection of the conformation of the helicase core throughout the catalytic cycle (Aregger & Klostermeier, 2009; Theissen et al., 2008) has shown that the helicase core is in the open state not only in the absence of nucleotide, but also in the ADP-bound state (Aregger & Klostermeier, 2009; Theissen et al., 2008). The closed state is populated in the ATP- and RNA-bound states, in an activated ATP- and RNA-bound states (Talavera & De La Cruz, 2005), and in the ADP·P_i- and RNA-bound state (Aregger & Klostermeier, 2009; Theissen et al., 2008). As the different open states and the different closed states have the same FRET efficiency, they cannot be distinguished in single-molecule FRET experiments. Therefore, the fraction f_A represents all low-FRET open states present (i.e., the sum of their concentrations), and the fraction f_B summarizes all high-FRET closed states (Fig. 5C, model III). The equilibrium constant calculated from these fractions is thus an apparent constant. Note that the interpretation of single-molecule FRET experiments in terms of the underlying reaction schemes is an iterative process: From a first set of experiments, working models are generated that are scrutinized in subsequent experiments. Often, additional information on the system from complementary biochemical or biophysical studies is key to refine these working models.

Fig. 5—Cont'd determination of the relative position of the cyan domain to the remainder of the protein (intersection of gray and red circles). (F) Distances from FRET histograms: modeling of the relative position of the RBD to the helicase core of YxiN. Using distance restraints from single-molecule FRET experiments of YxiN labeled either in the N-terminal RecA-like domain (dark gray) and a position in the RBD, or on the C-terminal RecA-like domain (light gray) and the RBD, the position of the RBD relative to the helicase core was modeled in the absence of RNA (light cyan) and in the presence of RNA (blue) using the FPS software (Kalinin et al., 2012). The RBD undergoes a large-scale movement involving a translation by 5.3 nm and a rotation by 37° on RNA binding. *Panel (C) Modified after Chakraborty, A., Krause, L., & Klostermeier, D. (2022). Determination of rate constants for conformational changes of RNA helicases by single-molecule FRET TIRF microscopy. Methods, https://doi.org/10.1016/j.ymeth.2022.03.004; reprinted with permission. Panel (F) Modified after Samatanga, B., Andreou, A. Z., & Klostermeier, D. (2017). Allosteric regulation of helicase core activities of the DEAD-box helicase YxiN by RNA binding to its RNA recognition motif.* Nucleic Acids Research, *45(4), 1994–2006; reprinted with permission.*

4.4.3 Intramolecular FRET experiments with dimeric enzymes

The *T. thermophilus* DEAD-box helicase Hera is a dimeric helicase, with two helicase cores (Fig. 5D). To monitor (changes in) the distance between the RecA-like domains, one cysteine was introduced into each of these domains, similar to the original approach to dissect the conformational changes of the monomeric helicase core during the catalytic cycle (Linden, Hartmann, & Klostermeier, 2008). However, for a dimeric enzyme, this approach leads to a species with four cysteines (Fig. 5D). Thus, statistical labeling of this construct generates a mixture of differently labeled species, with different distances between donor and acceptor fluorophores. All these species contribute to the FRET histogram, which, as a result, shows a broad distribution of FRET efficiencies ranging from 0.25 to 0.85. The Hera dimer is very stable, and subunit exchange only occurs at higher temperatures (Klostermeier & Rudolph, 2009). By incubating the labeled dimer with a five-fold molar excess of unlabeled wildtype protein, and allowing for subunit exchange at 65 °C, an appreciable fraction of heterodimer can be generated that consists of one unlabeled and one labeled subunit (Fig. 5D). The resulting FRET histogram is much cleaner, with a more symmetric and narrower peak centered at a FRET efficiency of 0.45. Notably, the histogram is still broader than the histogram for the labeled, monomeric Hera helicase core. This could be due to the dimeric species with both subunits labeled that are still present after subunit exchange. Cleaner results might be obtained by co-producing the cysteine variant and the cysteine-free wildtype protein, fused to different affinity tags, followed by tandem affinity purification of the heterodimer containing only two cysteines (see Gubaev et al., 2016; Hartmann et al., 2017; Stelljes et al., 2018 for the experimental design). Statistically, only half of the dimers are heterodimers, which means that, naturally, a large fraction of the protein of interest is "lost" during the purification.

4.4.4 Using distance restraints from FRET experiments to generate structural models

In single-molecule FRET experiments, distances between the dyes used can be determined that provide information on the structure of the protein. Distance determination absolutely requires the correction of the measured intensities for instrument non-idealities by applying a set of correction parameters (see Section 4.2). A high accuracy and high precision of the distances determined is key to arrive at well-defined structural models (Dimura et al., 2020; Hellenkamp et al., 2018; Kalinin et al., 2012).

The basic principle behind the generation of structural models relies on multi-lateration, i.e., determination of distances between a set of reference points to define their relative positions (Dimura et al., 2020; Muschielok et al., 2008; Muschielok & Michaelis, 2011). While for a limited number of distance restraints, the part to be mapped can be placed manually in principle, software available for automated analysis should be preferred for two reasons: First, the distances obtained contain contributions from the lengths of the linkers connecting the protein side-chains and the dyes and the resulting segmental flexibility of the dyes. The software packages available model the volume accessible for the dye, and take this contribution into account, leading to a higher accuracy of the structural model obtained. Second, the accuracy of the model obtained increases with the number of distance restraints used (Dimura et al., 2020). With an increasing number of distance restraints to be satisfied, manual placement of the unknown part becomes impossible.

Using a multi-lateration approach, a structural model for the *Bacillus subtilis* DEAD-box helicase YxiN, consisting of a helicase core and a C-terminal RBD, was generated (Samatanga et al., 2017). To this end, donor and acceptor fluorophores were introduced into the N-terminal RecA-domain and five different positions in the RBD, giving a set of five FRET constructs. With these constructs, single-molecule FRET experiments in solution were performed, FRET histograms were corrected for instrument effects, and the mean distances obtained were used to place the RBD relative to the label on the helicase core. This set of distances defines the possible positions of the RBD relative to the single labeled residue in the RecA_N. However, these experiments are not sufficient to place the RBD relative to the helicase core: The RBD can rotate around the label on the RecA_N on a sphere without violating the distance restraints (Fig. 5E)! This degree of freedom remains, no matter how many distances are determined between a single residue on RecA_N and positions in the RBD. To unambiguously place the RBD relative to the helicase core, a second position on the core (either in RecA_N, when the connection between RecA_N and Rec_C is known, or on RecA_C) has to be labeled, and its distance to a label on the RBD has to be determined (Fig. 5E). With this additional restraint, a structural model of YxiN was obtained with the program FPS (Dimura et al., 2020), both for the apo protein and for YxiN with RNA bound to the RBD (Fig. 5F). Comparison of these models reveals that RNA binding leads to a movement of the RBD from a position close to the RecA_C to a position close to the RecA_N (Fig. 5F). This movement of the RBD is linked to an allosteric activation of the helicase core (Samatanga et al., 2017).

5. Total internal reflection microscopy: Kinetics of conformational changes

In single-molecule FRET experiments by total internal reflection fluorescence (TIRF) microscopy (Ambrose, 1956) (Fig. 6), the donor/acceptor-labeled molecules are immobilized on the surface of the microscopy slide. The excitation light is coupled into the objective away from the optical axis. As a consequence, the excitation beam impinges onto the slide-sample interface at an angle above the critical angle, and becomes internally reflected, leading to the generation of an evanescent field at the surface. Surface-immobilized molecules, located within this evanescent field, are excited selectively (Fig. 6A). Fluorescent molecules in solution, on the other hand, further away from the slide surface, are not excited. Fluorescence emitted from donor and acceptor is collected by the objective, separated into the spectral regions of donor and acceptor emission, and imaged onto an area detector, typically a charge-coupled device camera (Fig. 6B).

The single molecules immobilized on the surface are separated in space. To ensure that each of these molecules is detected as a spot that is clearly separated from adjacent spots generated by neighboring molecules, the surface density of the donor/acceptor-labeled molecules must be sufficiently low. This density is determined by the functionalization density of the surface, which can be defined through the fraction of functionalized molecules in the mixture used for surface passivation and functionalization. For the immobilization of donor/acceptor-labeled proteins through the biotin-streptavidin interaction, we typically spike the polyethylene glycol (PEG)-solution used for surface passivation with 2% of biotin-functionalized PEG molecules. For experimental details on slide functionalization and flow cell fabrication for single-molecule experiments by TIRF microscopy, see Krause and Klostermeier (2021). To achieve extended observation times, photobleaching and blinking (Eggeling et al., 2006; Kong, Nir, Hamadani, & Weiss, 2007) of the fluorescent dyes is suppressed by supplementing buffers used with oxygen scavenging systems and triplet-state quenchers (Aitken, Marshall, & Puglisi, 2008; Rasnik, McKinney, & Ha, 2006; Swoboda et al., 2012). We use buffers containing 50 nM protocatechuate-3,4-dioxygenase, 2.5 mM protocatechuic acid, and 1 mM methyl viologen, 0.1% (*w*/*v*) BSA (Aitken et al., 2008) and 1 mM trolox (Rasnik et al., 2006). Practical guides on slide and sample preparation as well as sample handling during TIRF experiments are found in Roy, Hohng, and Ha (2008) and Krause and Klostermeier (2021).

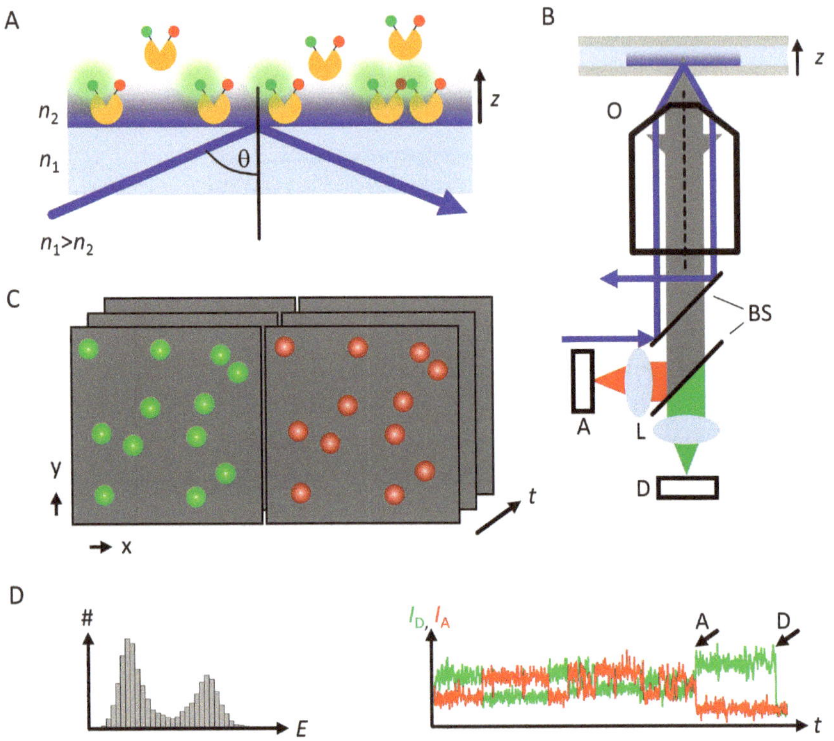

Fig. 6 Single-molecule experiments on surface-immobilized molecules by TIRF microscopy. (A) Total internal reflection: If an incident beam (blue) impinges on a glass-water interface at an angle θ above the critical angle θ_c, the light undergoes total internal reflection. An evanescent field on the distant side of the interface (blue gradient) enables the selective excitation of fluorophores close to the surface. The evanescent field decays exponentially with increasing distance *z* from the surface, and molecules in solution are not excited. *n*: refractive indices of glass (n_1) and water (n_2). (B) Schematic depiction of a total internal reflection microscope. The excitation light (blue) is coupled into an objective (O) to achieve total-internal-reflection conditions. The emitted light (gray) is collected by the objective, passes a beam splitter (BS), is separated into the spectral regions of two fluorophores used by a second beam splitter, and donor and acceptor fluorescence (green, red) are imaged onto different regions of an area detector (D, A). L: lens. (C) Single-molecule experiments by TIRF microscopy generate pairwise images of the molecules immobilized on the surface, one for the donor fluorescence (green) and one for the acceptor fluorescence (red). From these images, histograms of FRET efficiencies over all molecules detected can be calculated (D). Stacks of images, taken at different time-points, or videos, provide information on changes in donor and acceptor fluorescence for each molecule in the image as a function of time (E). (D) FRET histogram for all molecules detected. (E) FRET time trace for one single molecule.

5.1 Separating molecules in space: Total internal reflection microscopy

The raw data from single-molecule-FRET experiments by TIRF microscopy are pairwise series of images (videos), one for the donor fluorescence of each molecule, and one for the acceptor fluorescence (Fig. 6C). These image stacks enable the determination of FRET efficiencies for all molecules detected within the image from their donor and acceptor intensities, and the construction of FRET histograms (Fig. 6D). More importantly, from the series of images at different time points, donor and acceptor intensities can be extracted as a function of time for each single molecule detected (Fig. 6E). These donor and acceptor fluorescence time traces can then be converted into FRET time traces, which contain information on the kinetics of conformational changes occurring and enable the calculation of rate constants for these processes.

5.2 Molecules in space: From images to histograms and conformational distributions

The images generated by TIRF microscopy represent every single surface-immobilized molecule on the surface as a single, diffraction-limited spot. The brightness of these spots is a measure for the donor or acceptor intensity, respectively. These intensities can be calculated only after the images have been processed. Image processing consist of alignment of the donor and acceptor images, background subtraction, identification of donor and acceptor spots, and the calculation of donor and acceptor intensities for each molecule by integration over the entire spot. The details of image processing depend on the experimental platform used. For more details on the procedure used in our lab, see Chakraborty et al. (2022).

In principle, the FRET efficiencies of all molecules detected in one image pair can be collected in a FRET histogram to identify conformational states with different FRET efficiencies and their relative populations (Fig. 6D). In practice, the analysis of TIRF data is based on the entire series of images, (see Section 5.3), however, and the generation of FRET histograms is done after FRET time traces have been calculated.

While FRET histograms from confocal microscopy count each molecule detected only once, FRET histograms from TIRF microscopy collect FRET values of each molecule over the entire duration of the FRET time trace. Therefore, molecules that give rise to long time traces due to late photobleaching of the fluorophores are weighted more than those molecules where the fluorophores bleach early and the time traces are short. If possible, FRET histograms from TIRF experiments should be compared with FRET histograms obtained with the same FRET construct in solution by confocal microscopy to confirm that surface immobilization does not alter the conformational states present or the relative population of these states. FRET histograms from TIRF experiments thus serve as a useful internal control. The unique information, however, that cannot be obtained by confocal microscopy, is the information on changes in FRET efficiencies over time and hence on the rate constants of conformational changes occurring.

5.3 Molecules in time: From videos to rate constants and kinetic models

5.3.1 Generating donor and acceptor time traces

During image processing, fluorescence spots are located over the entire image stack of a video (see Chakraborty et al., 2022 for details). The coordinates of molecules detected in both donor and acceptor images are then used as a mask, and their donor/acceptor intensities are calculated by integration over the entire spot in each image of the entire stack. The resulting list of donor and acceptor intensities for each molecule as a function of time is visualized as a set of time traces of donor and acceptor intensities. These time traces are then inspected, either visually or increasingly by automated procedures, and time traces fulfilling pre-defined criteria are selected for further analysis. Selected time traces show anti-correlated donor and acceptor signals, indicating that intensity changes are caused by changes in FRET efficiencies, and show donor and acceptor bleaching in single steps, indicative of single molecules labeled with one donor and one acceptor dye. In addition, time traces should be inspected for "normal" intensities—intensities higher than the values typically detected may point to more than a single molecule present, e.g., oligomers or aggregates. Very short time traces and time traces showing no conformational change may also be excluded from further analysis. In most laboratories, the selection is done manually, which is very time-consuming. Despite pre-defined criteria, the selection may be subjective, and different users may select different sets of traces. With the advent of machine-learning methods, automated approaches have been developed that accelerate the selection process and minimize the user bias. AutoSiM uses a trained neural network for automated trace selection (Li, Zhang, Johnson-Buck, & Walter, 2020). Transfer learning enables rapid adaptation of the procedure to different data sets. Deep-FRET (Thomsen et al., 2020) takes automation one step further and covers the entire data processing pipeline from raw microscope images to FRET and dwell-time histograms, needing only a user-defined threshold as input. Thereby, the time needed for data analysis is reduced to 1% of the time required with human inspection and curation.

5.3.2 Calculation of FRET time traces and determination of dwell times

Once trace selection is completed, FRET time traces can be calculated. These FRET trajectories show spontaneous, rapid switches of the molecule of interest between different FRET states, separated by waiting times of varying durations, termed dwell times (Fig. 7, see also Fig. 8A). While the switching event itself is not resolved with the typical time resolution of TIRF experiments (typically 30–100 ms), the distribution of the waiting times observed contains the information on rate constants of the underlying conformational changes.

The first step in the analysis of FRET time traces is the assignment of the individual states detected. In simple systems, switching between two states with clearly distinct FRET efficiencies, the state of the molecule at each time point can be assigned by threshold criteria (Blanco & Walter, 2010; Chakraborty et al., 2022). For more complex systems, involving switches between several states and/or states with small differences in FRET efficiencies, this approach is not possible. In these cases, step-finding algorithms are needed (reviewed in Blanco & Walter, 2010). The most common procedure is the analysis of time traces by hidden Markov modeling (HMM) (Lee, 2009; McKinney, Joo, & Ha, 2006) (Fig. 7). HMM describes a FRET system as a set of different states that exhibit different FRET efficiencies. These states are connected by single-step processes that are governed by specific rate constants (Fig. 7A, C). The features of such a model are expressed in three matrices: One matrix collects the FRET efficiencies of each state, a second matrix collects the probabilities for conversion of one state in another state, and the third matrix defines the probabilities of finding the system in a particular state at the beginning of the experiment (Fig. 7A, C). These matrices are then used to identify the states present, and to determine the most probable path through these states, termed the Viterbi path. HMM analysis of single-molecule FRET time traces provides an idealized trajectory, showing only the mean FRET efficiency of the states populated and the switches between them (Fig. 7B). From the interval between two switching events, the waiting time or dwell time in a particular state can be calculated (see Fig. 8). The distribution of dwell times for single-step processes is exponential, and histograms of dwell times hence show exponential behavior.

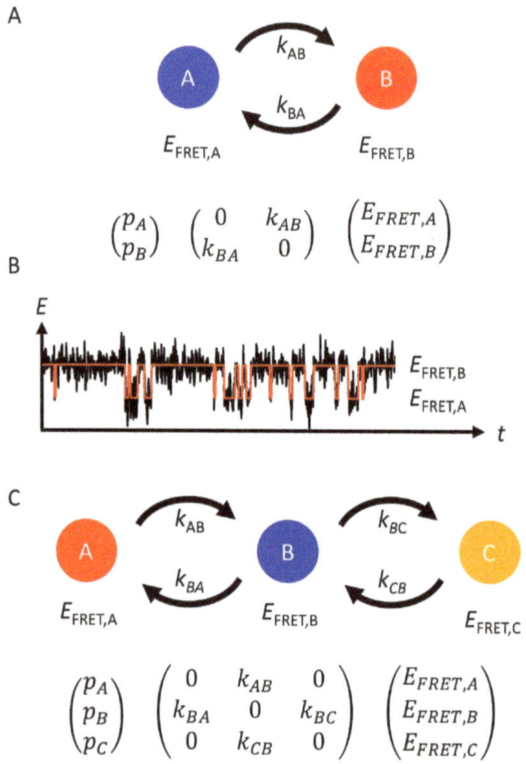

Fig. 7 Hidden Markov models to analyze kinetic data from single-molecule FRET experiments. (A) Kinetic model for hidden Markov analysis of a two-state system with two states, A and B, with FRET efficiencies $E_{FRET,A}$ and $E_{FRET,B}$, interconverting with rate constants k_{AB} and k_{BA}. For HMM analysis, the model is described by the initiation probability matrix, containing the starting probabilities for states A and B, p_A and p_B, the transition probability matrix containing the rate constants of inter-conversion, k_{AB} and k_{BA}, and the emission probability matrix, containing the emission values $E_{FRET,A}$ and $E_{FRET,B}$. (B) Experimental single-molecule FRET trace for a two-state system and Viterbi path (red), the most probable pathway through the two FRET states determined by HMM analysis. (C) Kinetic model for hidden Markov analysis of a three-state system with states, A, B, and C (top) and matrices for HMM analysis (bottom).

5.3.3 Dwell time histograms, cumulative dwell-time histograms and integrated rate laws

Dwell times can be collected in two types of histograms (Fig. 8). Dwell-time histograms (DTHs) collect the number of occurrences of a dwell time within a certain interval. Cumulative dwell-time histograms (CDTHs), on the other hand, collect the number of occurrences of a dwell time below a certain value (Fig. 8A). Understanding the relationship between dwell-time

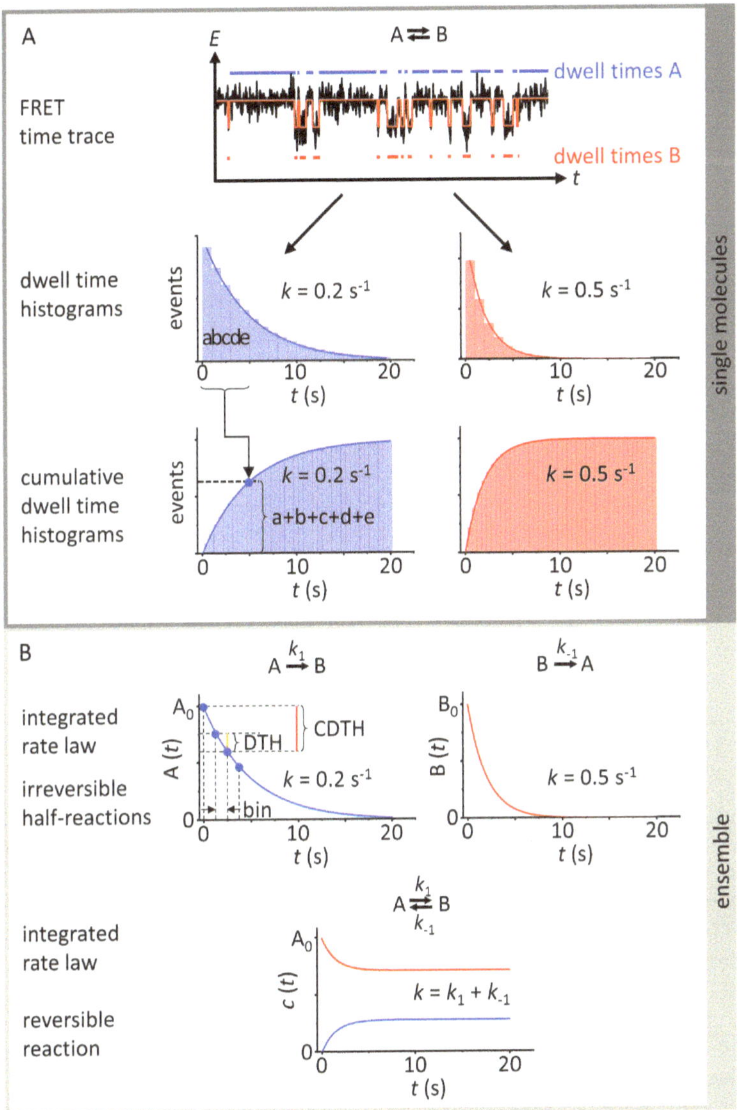

Fig. 8 Comparison of single-molecule and ensemble kinetics: Dwell-time histograms, cumulative dwell-time histograms, and integrated rate laws. (A) Determination of rate constants from single-molecule experiments. From FRET time traces obtained by TIRF microscopy (black), two FRET states A and B (blue, red) can be identified. From the idealized time trace (Viterbi path; red line), obtained by hidden Markov analysis with a model describing the interconversion of states A and B, their dwell times are determined. Dwell-time histograms (DTHs; blue and red bars) count the number with which a dwell time within a certain interval (bin) has been observed. The bin size used here is

(Continued)

and cumulative dwell-time histograms, and their relation to the integrated rate law describing the corresponding ensemble reaction, is key for the interpretation of these histograms.

5.3.3.1 Single-exponential dwell-time histograms

In a simple single-molecule FRET experiment, a large number of single molecules is observed, each switching multiple times between two different FRET states. Such a two-state FRET system is formally described by a reaction scheme in which a state A converts into state B by with a rate constant k_1 (forward reaction), and B converts into A with a rate constant k_{-1} (reverse reaction; Fig. 8B). The progress of the reaction, starting from a concentration of state A of A_0 at $t=0$, is described by integrated rate laws that express the concentration of states A and B as a function of time, in this case

$$A(t) = \frac{A_0}{k_1 + k_{-1}}\left(k_{-1} + k_1 e^{-(k_1+k_{-1})t}\right) \tag{3}$$

and

$$B(t) = \frac{A_0}{k_1 + k_{-1}} k_1 \left(1 - e^{-(k_1+k_{-1})t}\right) \tag{4}$$

When we characterize such a reaction in ensemble kinetic experiments, we typically start from defined initial conditions, e.g., a homogeneous ensemble

---

Fig. 8—Cont'd 1 s. From the exponential distribution, the rate constants k_1 and k_{-1} can be determined (blue and red lines). Cumulative dwell-time histograms (CDTHs) count the total number of dwell times up to a certain value and constitute the integrated form of dwell-time histograms. The bin size used here is 0.5 s. The relationship between dwell-time and cumulative dwell-time histograms is illustrated for the value at $t=5$ s: The value in the cumulative dwell-time (a+b+c+d+e) is the sum of the numbers in the first five bins of the dwell-time histogram (a, b, c, d, e). (B) Relationship between integrated rate laws for reversible reactions and the irreversible half-reactions, and dwell-time and cumulative dwell-time histograms. Bottom: integrated rate laws A(*t*), B(*t*) for the same reversible reaction analyzed on the single-molecule level in A, starting from A_0 at $t=0$ s. The rate constant of the forward reaction from A to B is 0.2 s^{-1}, the rate constant for the reverse reaction form B to A is 0.5 s^{-1}. Top: integrated rate laws for the constituent two half-reactions from A to B (left) and B to A (right). The values tabulated in dwell-time histograms (DTHs) and cumulated dwell-time histograms (CDTHs) are indicated in the left panel. The integrated rate law represents the molecules still present at the time *t*, whereas the CTDH represents the total number of molecules that have reacted at the time *t* (red). The CTH thus equals $A_0 - A(t)$. The DTH counts the number of molecules that have reacted since the last time point (orange) and hence is the first derivative of the CDTH. See text for details.

of molecules in state A with a starting concentration A_0, that, in a synchronized manner, reacts to state B until equilibrium is reached. Then, we monitor the concentration of A and/or B over time using a suitable probe and derive the underlying rate constant by describing the data obtained with Eqs. (3) and (4). In this setting, the forward and reverse reaction are coupled and the "apparent" rate constant for the approach to equilibrium is $k = k_1 + k_{-1}$.

On the other hand, in single-molecule experiments, we work with unsynchronized molecules and watch the dynamic behavior of the individual molecules over time. From the FRET time traces, we determine many dwell times for each of the two FRET states. These dwell times characterize the *decay* of the respective species, i.e., the reaction away from this state. The kinetic information, i.e., the rate constant for the underlying process, is encoded in the distribution of these dwell times. The dwell-time distributions are analyzed separately for each of the two states by collecting all dwell times measured for this state in a histogram. Thus, these histograms characterize only the forward or only the reverse reaction. In single-molecule experiments, we treat the underlying reversible reaction as two separate (half-)reactions: The conversion of state A into state B with k_1 and the conversion of state B into state A with k_{-1} (Fig. 8A, B).

The irreversible half-reactions for this two-state FRET system are described by very simple integrated rate laws in the form of a single-exponential decay:

$$A(t) = A_0 e^{-k_1 t} \tag{5}$$

or

$$B(t) = B_0 e^{-k_{-1} t} \tag{6}$$

To connect integrated rate laws with dwell-time histograms, it is helpful to note that the concentrations, $A(t)$ and $B(t)$, in these rate laws are directly proportional to the numbers of molecules present at the time t, at least for reactions without any change in volume. The integrated rate laws thus quantify the total number of molecules present at each time t, i.e., the molecules *have not* reacted (yet). In contrast, dwell-time histograms count the molecules that *have* reacted: Cumulative dwell-time histograms count all molecules that have reacted at time t, whereas dwell-time histograms count the molecules have reacted within a certain time interval (Fig. 8A, B).

Mathematically, the dwell-time histograms (DTHs) for the two states A and B correspond to the negative first derivative of the integrated rate laws in Eqs. (5) and (6), and follow single exponentials of the form:

$$DTH(A) = -\frac{dA}{dt} = k_1 A_0 e^{-k_1 t} \tag{7}$$

and

$$DTH(B) = -\frac{dB}{dt} = k_{-1} B_0 e^{-k_{-1} t} \tag{8}$$

The negative sign in Eqs. (7) and (8) converts the decrease in concentration into positive values, giving rise to an exponential decay (Fig. 8A). Conveniently, the calculation of the first derivative does not affect the rate constant of an exponential function. Therefore, we can determine the value for the forward and reverse rate constants of the reversible reaction, k_1 and k_{-1}, from single-molecule experiments by describing the dwell-time histograms for the forward and reverse half-reaction with single-exponential functions. The amplitude of the exponential function is not taken into account in this type of analysis.

Cumulative dwell-time histograms, on the other hand, counting *all* molecules that have reacted up to a certain time t (Fig. 8A), are the integrated form of the dwell-time histogram. Cumulative dwell-time histograms (CDTHs) for the inter-converting two-state FRET system are described by Eqs. (9) and (10):

$$CDTH(state\ A) = \int_0^t k_1 A_0 e^{-k_1 t} dt = A_0 \left(1 - e^{-k_1 t}\right) \tag{9}$$

and

$$CDTH(state\ B) = \int_0^t k_{-1} B_0 e^{-k_{-1} t} dt = B_0 \left(1 - e^{-k_1 t}\right) \tag{10}$$

Cumulative dwell-time histograms have the same amplitude as the corresponding integrated rate laws, which can be advantageous if amplitude information is important (see Section 5.4.2). Note that Eqs. (9) and (10) are not identical to the integrated rate laws (Eqs. 5 and 6): Instead of an exponential decay they are increasing exponential functions (Fig. 8A). The relationship between CDTHs and the integrated rate laws $A(t)$ and $B(t)$ is simply

$$CDTH(state\ A) = A_0 - A(t) \tag{11}$$

and

$$CDTH(state\ B) = B_0 - B(t) \tag{12}$$

To generate (cumulative) dwell-time histograms, the dwell times collected are binned into intervals with a width defined by the binning time or bin size t_{bin}. These intervals should be small enough to describe the underlying exponential function but large enough to have a reasonable number of events per bin. A histogram that represents the half-life of the species in three bins and extends to approx. three lifetimes enables reliable extraction of the underlying rate constant (see Section 5.4.1). As cumulative dwell-time histograms count dwell times of all molecules that have reacted up to a certain time, the number of events per bin is larger than for a dwell-time histogram with the same bin size and the same number of dwell times recorded. Hence, the intervals used for cumulative dwell-time histograms can be smaller, enabling the resolution of faster reactions (down to the time resolution of the measurement determined by the frame rate of the camera; see Sections 5.4.1 and 5.4.2).

5.3.3.2 Double-exponential dwell time histograms

In certain cases, dwell-time histograms are not single exponential, but have two or more phases. This can be the case for parallel or sequential reactions involving two kinetically different states with non-distinguishable FRET efficiencies.

Parallel reactions. An example for a parallel reaction giving rise to double-exponential dwell-time histograms is a reaction scheme in which state A can convert to state B_1 or to state B_2 with rate constants k_1 and k_2, and both states B_1 and B_2 re-convert to A with rate constants k_{-1} and k_{-2} (Fig. 9A). In this case, the dwell-time distribution for state A is single exponential, and the rate constant determined is the sum of the rate constants leading away from state A, i.e., $k_1 + k_2$. Now let us assume that B_1 and B_2 show the same FRET efficiency, meaning they are indistinguishable in single-molecule FRET experiments. Their individual dwell times will then be collected in a single dwell-time histogram, which shows double-exponential behavior due to the different rate constants of conversion of B_1 into A (with k_{-1}) compared to the conversion of B_2 into A (with k_{-2}). The rate constants k_{-1} and k_{-2} can be extracted by analyzing the dwell-time histogram with a double-exponential function. In addition to dwell times, single-molecule FRET histograms provide information on the relative populations of the two

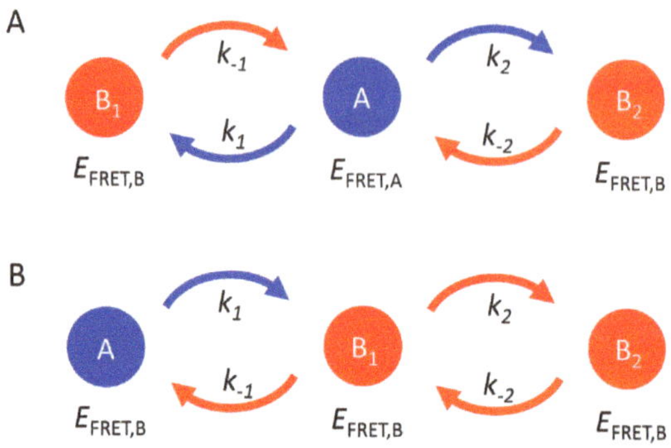

Fig. 9 Double-exponential dwell-time distributions for parallel and sequential reactions. (A) Parallel reaction of A (blue) to two kinetically different, but FRET-wise indistinguishable species B_1 and B_2 (red). Arrows are color-coded according to the state to whose dwell time they contribute: The dwell time of A is single-exponentially distributed, with a rate constant $k = k_1 + k_2$. The dwell-time distribution of state B (= B_1 and B_2) is double-exponential, $k_{obs,1}$ and $k_{obs,2}$ equal k_{-1} and k_{-2}. (B) Sequential reaction of A to B_2 with the intermediate B_1. Again, B_1 and B_2 are indistinguishable by FRET. Arrows are color-coded according to the state to whose dwell time they contribute: The dwell-time of A is single-exponentially distributed and controlled by k_1. The dwell-time distribution of state B (= B_1 and B_2) is double-exponential; $k_{obs,1}$ and $k_{obs,2}$ are determined by k_{-1}, k_2, and k_{-2}. See text for details.

FRET states, A and (B_1 + B_2), defining the apparent equilibrium constant K_{app} between these states. In combination with the relationship between K_{app} and the rate constants k_1, k_2, k_{-1}, and k_{-2}, we can ultimately calculate the microscopic rate constants k_1 and k_2 as well:

$$K_{app} = \frac{[B_1] + [B_2]}{[A]} = K_1 + K_2 = \frac{k_1}{k_{-1}} + \frac{k_2}{k_{-2}} \tag{13}$$

A step-by-step derivation of the equations to calculate all four rate constants can be found in Andreou, Harms, and Klostermeier (2019).

Sequential reactions. A second example for a reaction associated with double-exponential dwell-time histograms is a sequential reaction of the type A reacts to B_1 which is converted into B_2, where B_1 and B_2 have the same FRET efficiency, which is different from the FRET efficiency of A (Fig. 9B). The reaction is governed by four microscopic rate constants, the rate constants for the inter-conversion of A and B_1, k_1 and k_{-1}, and k_2 and k_{-2} for the conversion of B_1 and B_2. In the context of a single-molecule FRET experiment, this reaction scheme also appears as an equilibrium between the FRET state A on one side and the FRET state (B_1 + B_2) on the other side. Again, the dwell times of A show a single-exponential distribution, determined by the rate constant k_1. In contrast, the dwell-time histogram of (B_1 + B_2) is double-exponential as the dwell times of the kinetically distinct species B_1 and B_2 are collected in a single dwell-time histogram. The rate constants describing the two phases of this dwell-time histogram are controlled by k_1, k_2, and k_{+2}. Again, the four microscopic rate constants can be extracted by combining the rate constants determined from dwell-time histograms with the information on the equilibrium populations of the low and high FRET states observed: The rate constant k_1 is obtained directly from the analysis of the single-exponential dwell-time histogram of state A. The remaining three rate constants can be determined from the two dwell times observed for the second state (B_1 + B_2), $k_{obs,1}$ and $k_{obs,2}$. For this, we start with the relationships between the observed rate constants for this type of reaction and the microscopic rate constants in ensemble reactions (see Bernasconi, 1976 and Bagshaw, 2017 for more details):

$$k_{obs,1} + k_{obs,2} = k_1 + k_{-1} + k_2 + k_{-2} \tag{14}$$

and

$$k_{obs,1} \bullet k_{obs,2} = k_1 k_2 + k_1 k_{-2} + k_{-1} k_{-2} \tag{15}$$

and

$$K_{app} = \frac{[B_1] + [B_2]}{[A]} = \frac{K_1 \bullet K_2}{1 + K_2} \tag{16}$$

Note again that, in single-molecule experiments, the forward and reverse reactions are separated into individual half-reactions. As we only look at the decay of the states B_1 and B_2, we can set k_1 as zero in Eqs. (14) and (15) (but do have to carry it forward in Eq. (16) as it is relevant for the equilibrium constant determined from the FRET histogram!). Thus, we have three equations for three unknown parameters, k_1, k_2, and k_{-2}, which we can solve analytically.

5.3.3.3 Specific considerations in the analysis of double-exponential dwell-time histograms

Double-exponential dwell-time histograms come with specific challenges in data analysis (see Section 5.4.2 for examples). The first complication is associated with the binning process: The number of bins required to resolve two significantly different rate constants is substantially different. For example, for a reaction (phase) governed by a rate constant of $0.1\,s^{-1}$

($\tau = 1/k = 10\,s$), a bin size of approx. 3 s, with dwell times recorded up to 30 s, would be suitable. However, a second, faster phase with a rate constant of, e.g., $10\,s^{-1}$ ($\tau = 0.1\,s$), would require much smaller bins, otherwise most of the short dwell times representing this phase would be collected in the first bin and this phase would not be resolved (see Section 5.4.2 for an example). This problem can be overcome by analyzing dwell-time histograms with different bin sizes, particularly if a suspiciously high number of dwell times ends up in an over-proportionately large first bin. It can also be mitigated by constructing cumulative dwell-time histograms instead that contain a larger number of events per bin than dwell-time histograms (see Section 5.4.1). The second complication is related to the amplitude information of dwell-time histograms. The fact that dwell-time histograms constitute the first derivative of integrated rate laws becomes particularly important when the amplitudes of the two phases are analyzed. If the integrated rate law for a double-exponential reaction is

$$A(t) = xA_0e^{-k_1t} + (1-x)A_0e^{-k_2t} \tag{17}$$

with x describing the fraction of the reaction occurring with k_1, and $(1-x)$ the fraction reacting with k_2, then the negative first derivative describing the dwell-time histogram is

$$DTH(state\ A) = -\frac{dA}{dt} = xk_1A_0e^{-k_1t} + (1-x)k_2A_0e^{-k_2t} \tag{18}$$

As described before for single-exponential rate laws, the rate constants k_1 and k_2 can be determined directly from the dwell-time histogram. However, the amplitudes of the two phases in the integrated rate law and the dwell-time histogram are different: While the amplitudes of the rate law in Eq. (17) reflect the fraction of the two different states reacting with different rate constants directly, the amplitudes of the dwell-time histogram (Eq. 18) become weighted by the rate constants for the two processes. If k_1 and k_2 are similar, the resulting distortion of the amplitudes is small, but if the two processes occur on different time scales, then the effect of this weighting needs to be considered in an analysis of the amplitudes. Two illustrative examples addressing this issue are presented in more detail in Section 5.4.2.

5.3.4 Deriving kinetic models

While dwell-time histograms give rate constants for the reactions away from the state considered, and the assignment to individual processes is rather trivial for simple binary equilibria, it can become more complicated to define the underlying model and assign individual rate constants for more complex reaction schemes. Putting rate constants and the states observed into a biologically relevant context can be even more challenging.

In cases involving only few states, selecting a suitable model can be performed based on the number of (distinguishable) states observed in FRET histograms and time traces, and the connectivity can be determined from the number of rate constants obtained from dwell-time histograms of these states. For example, a scenario in which a low and high-FRET state inter-convert, and dwell times of both states are single-exponentially distributed, is consistent with a two-state kinetic system and two rate constants of inter-conversion, k_1, and k_{-1}. The conversion of two FRET states with a single-exponential distribution of dwell times for one state, but double-exponential behavior of the dwell-time histogram of the second state, may be rationalized by a three-state system with a low-FRET state and two high-FRET states, and a conversion of the low-FRET state into the two different high-FRET states in parallel reactions or in a sequential reaction. In both cases, the microscopic rate constants k_1, k_{-1}, k_2 and k_{-2} can be extracted from the rate constants governing the dwell-time histograms, provided that their orders of magnitude are not too different, and they are within the time resolution of TIRF experiments (see Section 5.4.1).

For more complex systems, the direct analysis is more challenging. In these cases, it may be helpful to use global data analysis procedures based on kinetic models instead of dwell-time analyses. One example is the program KinTek Explorer that enables the analysis of integrated single-molecule data, based on user-defined kinetic models (Johnson, 2009). Global analysis is not limited to single-molecule data sets, but kinetic transients from ensemble methods can also be included. The program SMACKS (Schmid, Gotz, & Hugel, 2016) uses a maximum-likelihood approach to retrieve all rate constants and their uncertainties from donor- and acceptor time traces. To do so, the routine uses a global kinetic model, includes every data point of the entire data set in the analysis, and allows for experimental variations of parameters between individual traces. Different models can be ranked according to the Bayesian information criterion to identify the simplest model that describes the data without over-fitting.

The model obtained is of course only a formal description of a kinetic system compatible with the data. Kinetic models are working models: They serve as a starting point to formulate hypotheses than can be tested in further experiments to lead to a more refined model. Ultimately, the formal kinetic model needs to be put into a biological context, i.e., the states present in the model need to be reconciled with functional and non-functional states of the molecule of interest, and different complexes containing a different set of constituents (see also Section 5.4.2).

5.4 Example data illustrating possibilities and potential pitfalls

FRET times traces of the donor/acceptor-labeled DEAD-box helicase eIF4A, obtained by TIRF microscopy, have unraveled the kinetics of conformational changes of opening and closing (Harms et al., 2014). They showed that eIF4A is predominantly in the open state in the absence of ligand and auxiliary factors (Harms et al., 2014). From the low-FRET open state, only very rare transitions to the high-FRET closed state are observed, i.e., closing is very slow. In the presence of ATP and RNA, eIF4A still remains in the open conformation, with only few short excursions to the closed state (Harms et al., 2014). This is in contrast to other DEAD-box helicases that predominantly occur in the high-FRET closed state when ATP and RNA are present (Aregger & Klostermeier, 2009; Del Campo & Lambowitz, 2009; Sengoku et al., 2006; Theissen et al., 2008), pointing to a lower stability of the closed state in eIF4A. In the presence of the auxiliary actor eIF4G, eIF4A is pre-aligned in a half-open state (Harms et al., 2014) with a different FRET efficiency, but remains rather static. Only when ATP and RNA are present, rapid switches between the half-open and a high-FRET state are observed. The dwell times of these states can be described by single exponentials, the rate constants from dwell-time histograms revealed that opening is still faster than closing, in agreement with the higher equilibrium population of the half-open state. When eIF4B is also present, the rate constant for opening of eIF4A is unaffected, but closing becomes faster, shifting the equilibrium toward the closed state (see Section 5.4.2). Similar experiments with different single- and double-stranded RNAs showed that the RNA substrate also modulates the kinetics of eIF4A conformational changes and regulates its activities (Andreou et al., 2019) (see also Section 5.4.2). TIRF microscopy on the surface-immobilized, donor/acceptor-labeled group II intron unraveled the conformational space and dynamics of this ribozyme (Steiner et al., 2008), and showed that the DEAD-box protein Mss116 mediates ribozyme folding by favoring dynamic sampling, and by stabilizing intermediates and transition states on the folding pathway (Karunatilaka et al., 2010). For experimental details and more information on the insights gained into the mechanism of RNA helicases in these experiments, the reader is referred to the original publications. In the following sections, selected examples are used to highlight important points that need to be considered in the analysis and interpretation of kinetic data from single-molecule FRET experiments by TIRF microscopy.

5.4.1 Generating dwell-time histograms: Choosing a suitable bin size

To generate dwell-time histograms, a suitable interval for binning of the dwell times measured must be chosen (Fig. 10). The bin size t_{bin} is a compromise between achieving the highest time resolution possible and obtaining a well-defined histogram with a sufficient number of events per bin. The highest resolution is achieved with a bin size corresponding to the time resolution of the experiment (typically 0.1 s). However, depending on the number of dwell times collected, such a small bin size may lead to a low number of events per bin, and rather rugged distributions (Fig. 10A) that may not be straight-forward to analyze. The minimum bin size necessary depends on the rate constant to be determined: A dwell-time histogram with three bins per lifetime τ (i.e., the inverse of the rate constant) is sufficient to define the underlying rate constant k. Hence, with the lower limit of the bin size t_{bin} of 0.1 s, reactions with lifetimes of $\tau > 0.3\,\mathrm{s}$ or $k < 3.3\,\mathrm{s}^{-1}$ can be determined reliably.

The total number of dwell times collected then needs to be sufficiently high to give a reasonable number of events per bin. For cumulative dwell-time histograms, the number of events per bin is higher because of the cumulation of events, and lower bin sizes can be used (Fig. 10A). The number of dwell times measured determines the quality of the dwell-time histogram obtained (Fig. 10B). The relationship is simple: the more, the better.

Apart from these general considerations, the choice of bin size becomes particularly important for the analysis of double-exponential dwell-time histograms. If the underlying rate constants k_1 and k_2 differ, then the optimal bin size to resolve each phase may be different: The appropriate bin size for the slow phase will not be sufficient to resolve the fast phase, and a suitable bin size to resolve the fast phase will distribute the events corresponding to the slow phase over too many bins. The fraction f of events in the first bin of a histogram with the bin size t_{bin} that correspond to a phase with the rate constant k can be calculated by dividing the number x of events in the first bin

$$x = \int_0^{t_{bin}} e^{-kt}\,dt \qquad (19)$$

by the total number of events

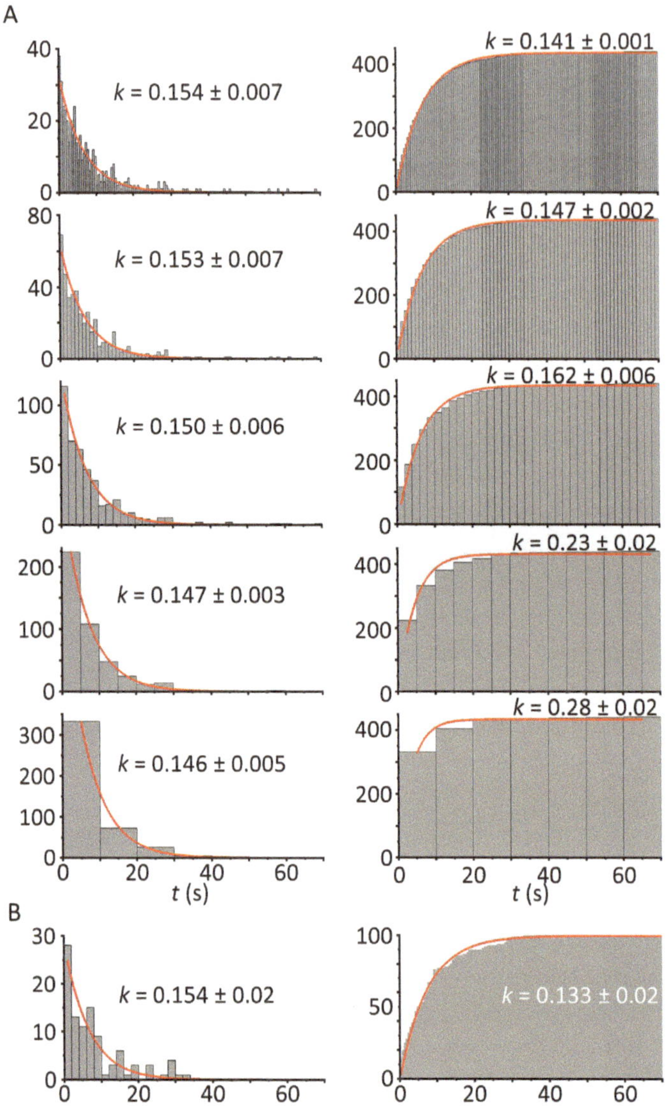

Fig. 10 Binning, dwell-time and cumulative dwell-time histograms. (A) Dwell-time histograms (left) and cumulative dwell-time histograms (right) for the open state of eIF4A in the presence of eIF4B, eIF4G, ATP, and a 14/10mer RNA (Andreou et al., 2019). The data set of 442 dwell times was processed with different bin sizes (from top to bottom: 0.5, 1, 2, 5, and 10 s). The red line is a single-exponential function fitted to the data. Numbers represent the rate constant retrieved from the fit and the fitting error. For the dwell-time histogram, a bin size approx. 1/3 of the lifetime is optimal (here: 2–5 s). For cumulative dwell-time histograms, smaller bin sizes, here down to 0.5 s, are possible because of the larger number of events per bin. With large binning intervals, the initial part of the exponential is not resolved. (B) Dwell-time and cumulative dwell-time histogram generated with only the first 100 events out of the 442 events recorded in the data set in (A). The bin sizes are 2 and 0.5 s, respectively. The quality of the histograms increases with increasing numbers of events analyzed. Errors are errors of the fit.

$$n = \int_0^{\infty} e^{-kt} dt \tag{20}$$

giving

$$f = \frac{x}{n} = \frac{\int_0^{t_{bin}} e^{-kt} dt}{\int_0^{\infty} e^{-kt} dt} = 1 - e^{-kt_{bin}} \tag{21}$$

For example, in a dwell-time histogram with 0.5 s bins, the dwell times for a reaction with $\tau > 1.5\,\text{s}$ or $k < 0.7\,\text{s}^{-1}$ are adequately distributed over different bins (Fig. 11A). For a fast phase with $k = 3.8\,\text{s}^{-1}$, on the other hand, 85% of all events end up in the first bin (Eq. 21, Fig. 11A; see Section 5.4.2 for examples). A cumulative dwell-time histogram with a bin size of 0.1 s, on the other hand, resolves both phases (Fig. 11A). Note that the amplitudes of the two phases in dwell-time histograms are weighted with the rate constant of the corresponding phase (Fig. 11A, B). If amplitudes are analyzed, e.g., to obtain the fractions of two species reacting with different rate constants, then this weighting needs to be taken into account. Cumulative dwell-time histograms directly provide this amplitude information (Fig. 11B; see Section 5.4.2 for examples).

5.4.2 Double-exponential dwell-time histograms

In the analysis of opening and closing of the DEAD-box helicase eIF4A, we encountered double-exponential dwell-time histograms in several cases. One example is the closing of eIF4A in the presence of eIF4B (Andreou et al., 2019; Andreou, Harms, & Klostermeier, 2016), a second example is the opening of eIF4A in the presence of RNAs containing a 10-bp duplex, flanked by a short single-stranded extension at their 5′-end (Andreou et al., 2019).

5.4.2.1 Opening and closing of eIF4A in the presence of different RNAs

eIF4A switches rapidly between the low-FRET, half-open and high-FRET, closed states in the presence of eIF4B, eIF4G, and RNA and ATP (Harms et al., 2014). With different model RNAs, the rate constants for opening and closing are different, indicating that the RNA substrate modulates the eIF4A helicase activity through a modulation of its conformational cycling (Harms et al., 2014). In the presence of single-stranded RNAs of different lengths, the dwell times of half-open and closed states show single-

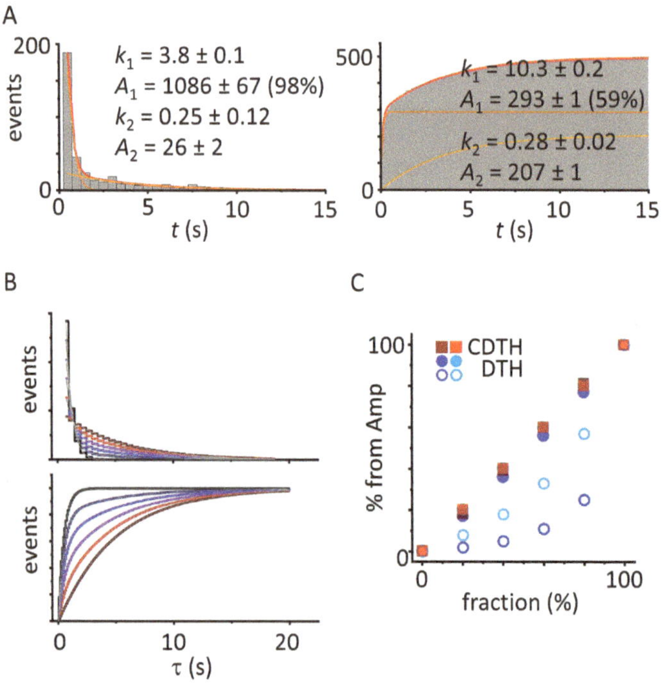

Fig. 11 Caveats in the analysis of double-exponential (C)DTHs. (A) Double-exponential dwell-time (left) and cumulative dwell-time histogram (right) for the open state of eIF4A in the presence of the auxiliary factors eIF4B, eIF4G, and of the substrates ATP and RNA (Andreou et al., 2019). Red: fit with a double-exponential function of the form $y = A_1 \cdot \exp.(-k_1 t) + A_2 \cdot \exp.(-k_2 t)$; light and dark orange: single exponentials contributing to the double-exponential fits. The relative amplitudes of the two phases are different because the dwell-time histogram is the first derivative of the cumulative dwell-time histogram, and its amplitudes are weighted with the rate constants (see main text). The correct amplitudes can be calculated as A_1/k_1 and $A_2/k2$ (see Eqs. 16 and 17), which gives a relative amplitude of 73% for the fast phase. Analysis of the cumulative dwell-time histogram yields a value of 59% for the relative amplitude of the fast phase. Note that the dwell-time histogram with the bin size of 0.5 s does not resolve the fast phase of the reaction: The vast majority of dwell times belonging to the fast phase are accumulated in the first bin. Total number of dwell times analyzed: $n = 501$. (B) Dwell-time histograms (left) and cumulative dwell-time histograms (right) for a biphasic reaction with $k_{obs,1} = 0.2\,s^{-1}$ and $k_{obs,2} = 2\,s^{-1}$ (simulated data), and different amplitudes of the fast and slow phases; dark red: 0%, red: 20%, purple: 40%, blue: 60%, dark blue: 80° and black: 100% fast phase. The total number of events simulated is $n = 100$. The data sets were analyzed with double-exponential functions to retrieve the two rate constants and the associated amplitudes (gray lines). DTHs were described with a standard equation of the form $y = A_1 \cdot \exp.(-k_1 t) + A_2 \cdot \exp.(-k_2 t)$, and with a modified version of this equation, of the form $y = k_1 \cdot A_1 \cdot \exp.(-k_1 t) + k_2 \cdot A_2 \cdot \exp.(-k_2 t)$ that takes the weighting of the

exponential distributions (Andreou et al., 2019). The rate constant for opening shows little dependence on the length of the RNA, while closing becomes faster with increasing RNA length. In the presence of double-stranded RNAs containing a 10-bp duplex flanked at the 5′-end by a short single-stranded region, however, the behavior of eIF4A is different: Dwell-time histograms for the half-open state remain single exponential, but dwell times of the closed state now show a double-exponential distribution (Andreou et al., 2019) (Fig. 12A), i.e., opening is a biphasic reaction. Analysis of the dwell-time histogram for the closed state with the smallest possible bin size (0.1 s, time resolution of the experiment; Fig. 12A) with a double-exponential function provides a first estimate of the rate constants of the two phases of $k_1 = 7.5\,\mathrm{s}^{-1}$ and $k_2 = 0.38\,\mathrm{s}^{-1}$, which can be used to decide on suitable bin sizes to further analyze this reaction. For the rapid phase, the bin size of 0.1 s is appropriate, for the slow phase, a bin size of 1 s would be more suitable. However, in a dwell-time histogram with a bin size of 1 s, essentially all dwell times belonging to the fast phase (99.9%, see Eq. 21) are sorted into the first bin, and this phase is not resolved. The large difference in rate constants in this case makes it possible to separate the analysis of the two phases: The rate constant for the fast phase can be determined from a dwell-time histogram with a bin size of 0.1 s, by restricting the analysis to the time window of $t < 1\,\mathrm{s}$ (Fig. 12A). The rate constant for the second phase, on the other hand, can be determined from a separate dwell-time histogram with a bin size of 1 s, now restricting the analysis to the time $t > 1\,\mathrm{s}$. This analysis gives the rate constants $k_1 = 6.4\,\mathrm{s}^{-1}$ and $k_2 = 0.29\,\mathrm{s}^{-1}$. Note that the fitted curve for the dwell-time histogram with a bin size of 0.1 s underestimates the data at the end of this time interval. This is because the events from the second phase start to contribute—from Eq. (21), we can estimate that approx. 30% of dwell times belonging to the

amplitudes by the rate constants into account; both sets of curves superimpose perfectly. CTHs were analyzed with the equation $y = A_1 \cdot \exp.(-k_1 t) + (n\text{-}A_1) \cdot \exp.(-k_2 t)$. All analyses recover the rate constants of $0.2\,\mathrm{s}^{-1}$ and $2\,\mathrm{s}^{-1}$. (C) Fractions of the amplitude for the slow phase, retrieved by analysis of the simulated data in B (reaction 1; blue, dark red), and of a data set for the same reaction with $k_{obs,1} = 0.4\,\mathrm{s}^{-1}$ and $k_{obs,2} = 1.2\,\mathrm{s}^{-1}$ (reaction 2; light blue, red). Squares: amplitudes derived by analysis of cumulative dwell time histograms. Open circles: amplitudes retrieved by analysis of dwell-time histograms with the standard double-exponential function, and closed circles: amplitudes derived with the modified equation that takes the weighting of the amplitudes by the rate constants into account. Note that in both cases, ignoring the weighting (open circles) leads to a tremendous underestimation of the amplitude of the slow phase that is more pronounced for reaction 1 where the fast phase is 10-fold faster than the slow phase, compared to reaction 2 where the fast phase is only threefold faster.

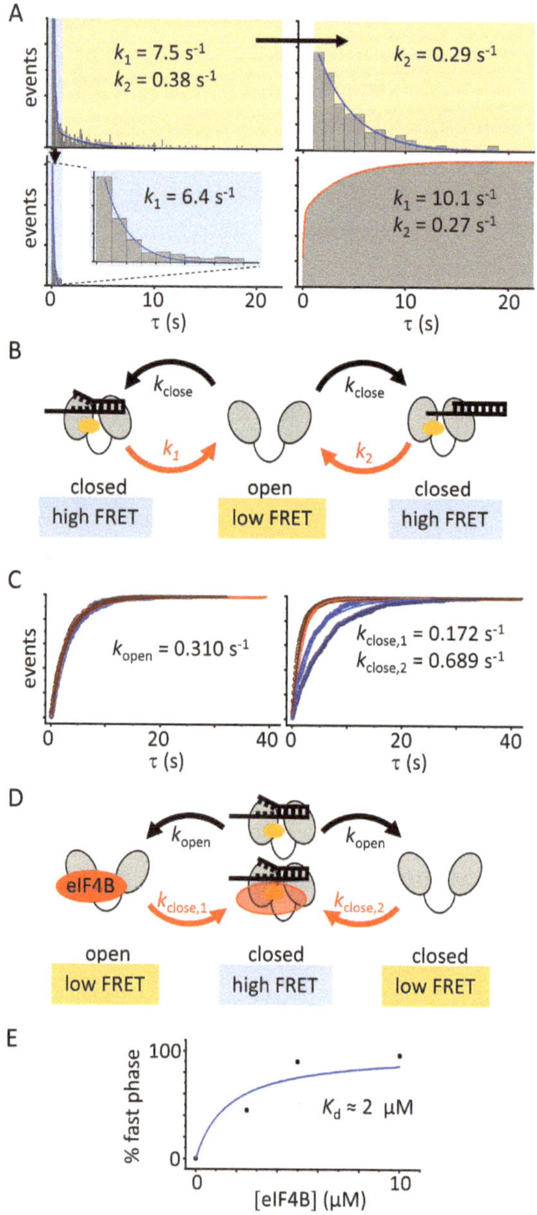

Fig. 12 Examples for double-exponential (C)DTHs. (A) Dwell-time analysis ($n = 500$) for the closed state of eIF4A in the presence of the auxiliary factors eIF4B, eIF4G, and of the substrates ATP and a 14/10mer RNA. The dwell-time histogram with the shortest bin size possible (0.1 s, integration time of the experiment; top left panel) can be described with a double-exponential function (blue), giving rate constants $k_1 = 7.5 \pm 0.5\,s^{-1}$ for the fast phase, $k_2 = 0.38 \pm 0.03\,s^{-1}$ for the slow phase. Thus, the fast phase would be best resolved with a bin size of 0.05 s, which is below the resolution of the experiment, hence,

slow phase are <1 s. On the other hand, the dwell time for the time window above 1 s is well-described by the fit, which is in agreement with only 0.1% of dwell times in this range corresponding to events from the fast phase. Finally, generation of a cumulative dwell-time histogram with a bin size of 0.1 s

---

the bin size of 0.1 s is the best choice. For the slow phase, binning with a bin size of 1 s is more suitable (top right panel). Analysis of dwell times <1 s (bottom left panel) gives a rate constant of $6.4 \pm 0.49\,s^{-1}$ for the fast phase; analysis of the dwell times >1 s with a bin size of 1 s gives a rate constant for the slow phase of $0.29 \pm 0.01\,s^{-1}$. Analysis of the cumulative dwell-time histograms directly yields the rate constants $k_1 = 10.1 \pm 0.2\,s^{-1}$ and $k_2 = 0.274 \pm 0.002\,s^{-1}$, and the relative amplitude of the slow phase of 41%. Note that a similar value for the contribution of the slow phase of 43% is determined from the dwell-time histogram when the weighting of the amplitudes by the rate constants is taken into account. Errors are errors of the fit. (B) Model consistent with the data in panel (A): In the presence of a 14/10mer RNA containing a single-stranded and a double-stranded region, eIF4A alternates between a half-open, low-FRET state and two different closed, high-FRET states. The dwell-time histogram for opening is biphasic, the two rate constants reflect opening of eIF4A interacting with the duplex region (left) and eIF4A interacting with the single-stranded region of the RNA (right); these different complexes return to the open state with different rate constants (red). (C) Cumulative dwell-time histograms for opening and closing of eIF4A in the presence of eIF4G, ATP and RNA and different concentrations of eIF4B. Dark blue: 0 μM eIF4B, blue: 2.5 μM, red: 5 μM, and dark red: 10 μM eIF4B. The different data sets contain different numbers of dwell times (374–948 events). They have been normalized to an amplitude of 100 to achieve equal weighting during global analysis. The histograms for the closed state (i.e., for opening; top panel) can be described by a single-exponential function; a global fit with the rate constant as a shared parameter shared over all data sets yields a rate constant of $k_{open} = 0.310 \pm 0.0009\,s^{-1}$. Opening of eIF4A is thus independent of eIF4B. In contrast, histograms of the dwell times of the half-open state (i.e., for closing; middle panel) follow a double-exponential function. A global fit with the rate constants as global parameters over all data sets gives rate constants $k_{close,1} = 0.172 \pm 0.001\,s^{-1}$ and $k_{close,2} = 0.689 \pm 0.007\,s^{-1}$. The amplitude of the fast phase increases with increasing concentrations of eIF4B (see panel (E)). (D) Model consistent with the data in panel (C), with one closed state of eIF4A and two kinetically different open states. The dwell-time histogram for closing is biphasic, the two rate constants reflect closing of eIF4A bound to eIF4B (left) and closing of eIF4A alone (right). In this scenario, the amplitude of the fast phase is a measure for the fraction of eIF4A bound to eIF4B (see panel (E)). Note that eIF4G is also present in these experiments, but not depicted for clarity. (E): Amplitudes of the fast phase for closing from data analysis in panel (C). A K_d value of approx. 2 μM can be estimated from the data, consistent with the range estimated from supernatant depletion experiments (Andreou et al., 2016). *Panels (A, C) Data from Andreou, A. Z., Harms, U., & Klostermeier, D. (2019). Single-stranded regions modulate conformational dynamics and ATPase activity of eIF4A to optimize 5′-UTR unwinding.* Nucleic Acids Research, *47(10), 5260–5275. Panels (B, D, E) Modified from Andreou, A. Z., Harms, U., & Klostermeier, D. (2019). Single-stranded regions modulate conformational dynamics and ATPase activity of eIF4A to optimize 5′-UTR unwinding. Nucleic Acids Research, 47(10), 5260–5275; reprinted with permission.*

and description by a double-exponential function gives rate constants of $k_1 = 10.1\,s^{-1}$ and $k_2 = 0.27\,s^{-1}$ (Fig. 12A; note that the rate constant of $10.1\,s^{1}$ is at the limit of the time resolution!). While the analysis based on dwell-time histograms provides the information on rate constants, cumulative dwell-time histograms would be preferred if amplitude information were to be extracted. Even when the rate constants have been determined, it remains unclear what their molecular origin is. A biphasic dwell-time histogram of the closed state implies that two kinetically different high-FRET states exist. For example, it is conceivable that the helicase interacts with the single-stranded region in one closed state, and the second state reflects the helicase interacting with the duplex region or with the single-stranded/double-stranded junction (Fig. 12B). Opening of eIF4A bound to single-stranded RNA is monophasic, with a rate constant of approx. $0.3\,s^{-1}$, comparable to the slow phase observed with duplexes flanked by short single-stranded regions (Andreou et al., 2019). Jointly, these data thus support a model in which eIF4A opens slowly when bound to single-stranded RNA, but rapidly when interacting with the duplex region (Fig. 12B). Other scenarios are possible, however, and subsequent experiments should be designed to be able to test working hypotheses, and to exclude ideally all but one model from those consistent with the data.

5.4.2.2 Opening and closing of eIF4A in the absence and presence of eIF4B

Addition of eIF4B to the eIF4A/4G complex in the presence of RNA and ATP leads to an increased population of the high-FRET, closed state of eIF4A, and to frequent conformational switching between the low- and high-FRET states (Harms et al., 2014). The effect of eIF4B on eIF4A can be titrated. With increasing concentrations of eIF4B, the fraction closed observed in single-molecule FRET histograms increases (Andreou et al., 2019). The dwell-time histograms for opening and closing show different responses to an increase in the concentration of eIF4B, however. The cumulative dwell-time histograms for the closed state are single-exponential, independent of the concentration of eIF4B, indicating that eIF4B does not affect opening of eIF4A ($k_{open} = 0.34\,s^{-1}$ from a global fit; Fig. 12C). In contrast, the cumulative dwell-time histograms for the open state are single-exponential in the absence of eIF4B, but become double-exponential with increasing concentrations of 4B (Fig. 12C). A global fit with the rate constants for the two phases as global parameters, gives the rate constants $k_{close,1} = 0.17\,s^{-1}$ and $k_{close,2} = 0.69\,s^{-1}$ (Fig. 12C). These rate constants are rather similar, facilitating their extraction from the same dwell-time or cumulative dwell-time histograms. Notably, the amplitudes of the two phases depend on the concentration of eIF4B. With increasing concentration of eIF4B, the amplitude of the slow phase decreases; inversely, the amplitude of the fast phase increases. At higher concentrations of eIF4B, the cumulative dwell-time histogram becomes single-exponential again, but now consists of only the faster phase. This behavior suggests that eIF4A/4G alone closes with the rate constant that is reflected in the slow phase, whereas closing is accelerated in the 4A/4B/4G complex, where it occurs with the rate constant of the faster phase (Fig. 12D). According to this model, the relative amplitude of the faster phase is a measure for the fraction of eIF4A bound to 4G and 4B. The amplitude information can be obtained from dwell-time histograms, considering the weighting by the rate constants of the two phases in the analysis, or directly from the amplitudes obtained by analysis of the cumulative dwell-time histogram with a double-exponential function. Plotting the amplitude as a function of the concentration of eIF4B thus gives a binding curve that can be analyzed for the K_d value for eIF4B dissociation from the eIF4A/4G complex (Fig. 12E). Again, the model presented here is only one possible scenario consistent with the data and with the system under investigation. Alternative models also need to be considered and tested in additional experiments.

6. Conclusions

Single-molecule FRET has provided valuable insight into the mechanism of RNA unwinding by RNA helicases in the past. FRET experiments in solution and on immobilized molecules provide both overlapping and complementary information on different conformers present, their fractional populations in equilibrium, and the rate constants of their interconversion. Single-molecule FRET is in fact the only method that makes it possible to monitor conformational changes of enzymes in real time. Collectively, the information gained can be integrated into a kinetic and thermodynamic framework that quantitatively describes the conformational dynamics of the helicase studied. Distances determined in FRET experiments complement the formal model with structural information on the different states detected. They define their global architecture, and, if collected in large numbers with sufficient accuracy or combined with insights from other structural methods, provide restraints for structural modeling. Once the states in this model have been aligned with functional states of the enzyme studied, the integrated model provides a comprehensive description of the structure and dynamics of the helicase, linked to its biological function.

Previous studies have addressed conformational changes of only few selected helicases, and even fewer helicase substrates. A lot remains to be learned from applications of this powerful method to other helicases, to physiologically relevant RNA substrates, to RNA/helicase complexes, and to larger assemblies of helicases with other factors. Although the sample preparation for single-molecule FRET experiments remains time-consuming and challenging, carefully designed constructs, paired with a clever choice of labeling strategies, make such studies feasible. With increasing complexity of the systems studied, data interpretation, the integration of data into kinetic models, and the connection to physiologically relevant species and the biological context become more challenging. Nevertheless, single-molecule FRET studies have tremendous potential to define the relationship between structure, function, and dynamics of RNA helicases and to understand the mechanistic basis for their broad range of biological functions.

Acknowledgments

I thank current and past members of the laboratory for their contributions.

Funding

Work in the author's laboratory was funded by the Deutsche Forschungsgemeinschaft (KL1153/7-1 and 7.2, to D.K.).

References

Agard, N. J., Baskin, J. M., Prescher, J. A., Lo, A., & Bertozzi, C. R. (2006). A comparative study of bioorthogonal reactions with azides. *ACS Chemical Biology*, *1*(10), 644–648.

Agard, N. J., Prescher, J. A., & Bertozzi, C. R. (2004). A strain-promoted [3 + 2] azide-alkyne cycloaddition for covalent modification of biomolecules in living systems. *Journal of the American Chemical Society*, *126*(46), 15046–15047.

Aitken, C. E., Marshall, R. A., & Puglisi, J. D. (2008). An oxygen scavenging system for improvement of dye stability in single-molecule fluorescence experiments. *Biophysical Journal*, *94*(5), 1826–1835.

Akiyama, B. M., & Stone, M. D. (2009). Assembly of complex RNAs by splinted ligation. *Methods in Enzymology*, *469*, 27–46.

Ambrose, E. J. (1956). A surface contact microscope for the study of cell movements. *Nature*, *178*(4543), 1194.

Andersen, C. B., Ballut, L., Johansen, J. S., Chamieh, H., Nielsen, K. H., Oliveira, C. L., et al. (2006). Structure of the exon junction core complex with a trapped DEAD-box ATPase bound to RNA. *Science*, *313*(5795), 1968–1972.

Andreou, A. Z., Harms, U., & Klostermeier, D. (2016). eIF4B stimulates eIF4A ATPase and unwinding activities by direct interaction through its 7-repeats region. *RNA Biology*, *14*(1), 113–123.

Andreou, A. Z., Harms, U., & Klostermeier, D. (2019). Single-stranded regions modulate conformational dynamics and ATPase activity of eIF4A to optimize 5'-UTR unwinding. *Nucleic Acids Research*, *47*(10), 5260–5275.

Andreou, A. Z., & Klostermeier, D. (2012). Conformational changes of DEAD-box helicases monitored by single molecule fluorescence resonance energy transfer. *Methods in Enzymology*, *511*, 75–109.

Andreou, A. Z., & Klostermeier, D. (2014). eIF4B and eIF4G jointly stimulate eIF4A ATPase and unwinding activities by modulation of the eIF4A conformational cycle. *Journal of Molecular Biology*, *426*(1), 51–61.

Aregger, R., & Klostermeier, D. (2009). The DEAD-box helicase YxiN maintains a closed conformation during ATP hydrolysis. *Biochemistry*, *48*(45), 10679–10681.

Bachmann, A. L., Matern, J. C., Schutz, V., & Mootz, H. D. (2015). Chemical-tag labeling of proteins using fully recombinant split inteins. *Methods in Molecular Biology*, *1266*, 145–159.

Bagshaw, C. R. (2017). *Biomolecular kinetics—A step-by-step guide*. New York: CRC Press.

Beckett, D., Kovaleva, E., & Schatz, P. J. (1999). A minimal peptide substrate in biotin holoenzyme synthetase-catalyzed biotinylation. *Protein Science*, *8*(4), 921–929.

Bernasconi, C. F. (1976). *Relaxation kinetics*. New York: Academic Press.

Bhagawati, M., Terhorst, T. M. E., Fusser, F., Hoffmann, S., Pasch, T., Pietrokovski, S., et al. (2019). A mesophilic cysteine-less split intein for protein trans-splicing applications under oxidizing conditions. *Proceedings of the National Academy of Sciences of the United States of America*, *116*(44), 22164–22172.

Blanco, M., & Walter, N. G. (2010). Analysis of complex single-molecule FRET time trajectories. *Methods in Enzymology*, *472*, 153–178.

Blaschke, U. K., Silberstein, J., & Muir, T. W. (2000). Protein engineering by expressed protein ligation. *Methods in Enzymology*, *328*, 478–496.

Brenzel, S., Kurpiers, T., & Mootz, H. D. (2006). Engineering artificially split inteins for applications in protein chemistry: Biochemical characterization of the split Ssp DnaB intein and comparison to the split Sce VMA intein. *Biochemistry*, *45*(6), 1571–1578.

Brustad, E. M., Lemke, E. A., Schultz, P. G., & Deniz, A. A. (2008). A general and efficient method for the site-specific dual-labeling of proteins for single molecule fluorescence resonance energy transfer. *Journal of the American Chemical Society*, *130*(52), 17664–17665.

Cardo, L., Karunatilaka, K. S., Rueda, D., & Sigel, R. K. (2012). Single molecule FRET characterization of large ribozyme folding. *Methods in Molecular Biology*, *848*, 227–251.

Chakraborty, A., Krause, L., & Klostermeier, D. (2022). Determination of rate constants for conformational changes of RNA helicases by single-molecule FRET TIRF microscopy. *Methods*. https://doi.org/10.1016/j.ymeth.2022.03.00 (in press).

Chauvier, A., St-Pierre, P., Nadon, J. F., Hien, E. D. M., Perez-Gonzalez, C., Eschbach, S. H., et al. (2021). Monitoring RNA dynamics in native transcriptional complexes. *Proceedings of the National Academy of Sciences of the United States of America*, *118*(45), e2106564118. https://doi.org/10.1073/pnas.2106564118.

Cheng, R. Z., Uchida, K., & Kawakishi, S. (1992). Selective oxidation of histidine residues in proteins or peptides through the copper(II)-catalysed autoxidation of glucosone. *Biochemical Journal*, *285*(Pt. 2), 667–671.

Chin, J. W., Santoro, S. W., Martin, A. B., King, D. S., Wang, L., & Schultz, P. G. (2002). Addition of p-azido-L-phenylalanine to the genetic code of *Escherichia coli*. *Journal of the American Chemical Society*, *124*(31), 9026–9027.

Dassa, B., Haviv, H., Amitai, G., & Pietrokovski, S. (2004). Protein splicing and auto-cleavage of bacterial intein-like domains lacking a C'-flanking nucleophilic residue. *Journal of Biological Chemistry*, *279*(31), 32001–32007.

Dawson, P. E., Muir, T. W., Clark-Lewis, I., & Kent, S. B. (1994). Synthesis of proteins by native chemical ligation. *Science*, *266*(5186), 776–779.

Del Campo, M., & Lambowitz, A. M. (2009). Structure of the yeast DEAD-box protein Mss116p reveals two wedges that crimp RNA. *Molecular Cell*, *35*(5), 598–609.

Dimura, M., Peulen, T. O., Sanabria, H., Rodnin, D., Hemmen, K., Hanke, C. A., et al. (2020). Automated and optimally FRET-assisted structural modeling. *Nature Communications*, *11*(1), 5394.

Donsbach, P., & Klostermeier, D. (2021). Regulation of RNA helicase activity: Principles and examples. *Biological Chemistry*, *402*(5), 529–559.

Eggeling, C., Widengren, J., Brand, L., Schaffer, J., Felekyan, S., & Seidel, C. A. (2006). Analysis of photobleaching in single-molecule multicolor excitation and Forster resonance energy transfer measurements. *The Journal of Physical Chemistry A: Molecules, Spectroscopy, Kinetics, Environment, & General Theory*, *110*(9), 2979–2995.

Fairman-Williams, M. E., Guenther, U. P., & Jankowsky, E. (2010). SF1 and SF2 helicases: Family matters. *Current Opinion in Structural Biology*, *20*(3), 313–324.

Förster, T. (1948). Zwischenmolekulare Energiewanderung und Fluoreszenz. *Annalen der Physik*, *437*(1), 55–75.

Gautier, A., Juillerat, A., Heinis, C., Correa, I. R., Jr., Kindermann, M., Beaufils, F., et al. (2008). An engineered protein tag for multiprotein labeling in living cells. *Chemistry & Biology*, *15*(2), 128–136.

Gorbalenya, A. E., & Koonin, E. V. (1993). Helicases: Amino acid sequence comparisons and structure-function relationships. *Current Opinion in Structural Biology*, *3*(3), 419–429.

Gubaev, A., Weidlich, D., & Klostermeier, D. (2016). DNA gyrase with a single catalytic tyrosine can catalyze DNA supercoiling by a nicking-closing mechanism. *Nucleic Acids Research*, *44*(21), 10354–10366.

Hanspach, G., Trucks, S., & Hengesbach, M. (2019). Strategic labelling approaches for RNA single-molecule spectroscopy. *RNA Biology*, *16*(9), 1119–1132.

Harms, U., Andreou, A. Z., Gubaev, A., & Klostermeier, D. (2014). eIF4B, eIF4G and RNA regulate eIF4A activity in translation initiation by modulating the eIF4A conformational cycle. *Nucleic Acids Research*, *42*(12), 7911–7922.

Hartmann, S., Gubaev, A., & Klostermeier, D. (2017). Binding and hydrolysis of a single ATP is sufficient for N-gate closure and DNA supercoiling by gyrase. *Journal of Molecular Biology*, *429*(23), 3717–3729.

Hartmann, S., Weidlich, D., & Klostermeier, D. (2016). Single-molecule confocal FRET microscopy to dissect conformational changes in the catalytic cycle of DNA topoisomerases. *Methods in Enzymology*, *581*, 317–351.

Hellenkamp, B., Schmid, S., Doroshenko, O., Opanasyuk, O., Kuhnemuth, R., Rezaei Adariani, S., et al. (2018). Precision and accuracy of single-molecule FRET measurements—A multi-laboratory benchmark study. *Nature Methods*, *15*(9), 669–676.

Henn, A., Bradley, M. J., & De La Cruz, E. M. (2012). ATP utilization and RNA conformational rearrangement by DEAD-box proteins. *Annual Review of Biophysics*, *41*, 247–267.

Hilbert, M., Kebbel, F., Gubaev, A., & Klostermeier, D. (2011). eIF4G stimulates the activity of the DEAD box protein eIF4A by a conformational guidance mechanism. *Nucleic Acids Research*, *39*(6), 2260–2270.

Hoffmann, A., Nettels, D., Clark, J., Borgia, A., Radford, S. E., Clarke, J., et al. (2011). Quantifying heterogeneity and conformational dynamics from single molecule FRET of diffusing molecules: Recurrence analysis of single particles (RASP). *Physical Chemistry Chemical Physics*, *13*(5), 1857–1871.

Ilgu, M., Ray, J., Bendickson, L., Wang, T., Geraskin, I. M., Kraus, G. A., et al. (2016). Light-up and FRET aptamer reporters; evaluating their applications for imaging transcription in eukaryotic cells. *Methods*, *98*, 26–33.

Jarmoskaite, I., & Russell, R. (2014). RNA helicase proteins as chaperones and remodelers. *Annual Review of Biochemistry*, *83*, 697–725.

Johnson, K. A. (2009). Fitting enzyme kinetic data with KinTek global kinetic explorer. *Methods in Enzymology*, *467*, 601–626.

Kalinin, S., Peulen, T., Sindbert, S., Rothwell, P. J., Berger, S., Restle, T., et al. (2012). A toolkit and benchmark study for FRET-restrained high-precision structural modeling. *Nature Methods*, *9*(12), 1218–1225.

Karow, A. R., & Klostermeier, D. (2010). A structural model for the DEAD box helicase YxiN in solution: Localization of the RNA-binding domain. *Journal of Molecular Biology*, *402*(4), 629–637.

Karow, A. R., Theissen, B., & Klostermeier, D. (2007). Authentic interdomain communication in an RNA helicase reconstituted by expressed protein ligation of two helicase domains. *FEBS Journal*, *274*, 463–473.

Karunatilaka, K. S., Solem, A., Pyle, A. M., & Rueda, D. (2010). Single-molecule analysis of Mss116-mediated group II intron folding. *Nature*, *467*(7318), 935–939.

Keppler, A., Gendreizig, S., Gronemeyer, T., Pick, H., Vogel, H., & Johnsson, K. (2003). A general method for the covalent labeling of fusion proteins with small molecules *in vivo*. *Nature Biotechnology*, *21*(1), 86–89.

Kershaw, C. J., & O'Keefe, R. T. (2012). Splint ligation of RNA with T4 DNA ligase. *Methods in Molecular Biology*, *941*, 257–269.

Klostermeier, D., & Rudolph, M. G. (2009). A novel dimerization motif in the C-terminal domain of the *Thermus thermophilus* DEAD box helicase Hera confers substantial flexibility. *Nucleic Acids Research*, *37*, 421–430.

Kolb, H. C., Finn, M. G., & Sharpless, K. B. (2001). Click chemistry: Diverse chemical function from a few good reactions. *Angewandte Chemie (International Ed. in English)*, *40*(11), 2004–2021.

Kong, X., Nir, E., Hamadani, K., & Weiss, S. (2007). Photobleaching pathways in single-molecule FRET experiments. *Journal of the American Chemical Society*, *129*, 4643–4654.

Krause, L., & Klostermeier, D. (2021). Probing RNA helicase conformational changes by single-molecule FRET microscopy. *Methods in Molecular Biology*, *2209*, 119–132.

Lata, S., & Piehler, J. (2005). Stable and functional immobilization of histidine-tagged proteins via multivalent chelator headgroups on a molecular poly(ethylene glycol) brush. *Analytical Chemistry*, 77(4), 1096–1105.

Lee, T. H. (2009). Extracting kinetics information from single-molecule fluorescence resonance energy transfer data using hidden Markov models. *Journal of Physical Chemistry B*, *113*(33), 11535–11542.

Li, J., Zhang, L., Johnson-Buck, A., & Walter, N. G. (2020). Automatic classification and segmentation of single-molecule fluorescence time traces with deep learning. *Nature Communications*, *11*(1), 5833–5843.

Linden, M. H., Hartmann, R. K., & Klostermeier, D. (2008). The putative RNase P motif in the DEAD box helicase Hera is dispensable for efficient interaction with RNA and helicase activity. *Nucleic Acids Research*, *36*, 5800–5811.

Liu, F., Putnam, A., & Jankowsky, E. (2008). ATP hydrolysis is required for DEAD-box protein recycling but not for duplex unwinding. *Proceedings of the National Academy of Sciences of the United States of America*, *51*(105), 20209–20214.

Los, G. V., Encell, L. P., McDougall, M. G., Hartzell, D. D., Karassina, N., Zimprich, C., et al. (2008). HaloTag: A novel protein labeling technology for cell imaging and protein analysis. *ACS Chemical Biology*, *3*(6), 373–382.

Luebke, K. J. (1998). A FLASH of insight into cellular chemistry: Genetically encoded labels for protein visualization *in vivo*. *Chemistry & Biology*, *5*(12), R317–R322.

Mallam, A. L., Del Campo, M., Gilman, B., Sidote, D. J., & Lambowitz, A. M. (2012). Structural basis for RNA-duplex recognition and unwinding by the DEAD-box helicase Mss116p. *Nature*, *490*(7418), 121–125.

McKinney, S. A., Joo, C., & Ha, T. (2006). Analysis of single-molecule FRET trajectories using hidden Markov modeling. *Biophysical Journal*, *91*(5), 1941–1951.

Minsky, M. (1988). Memoir on inventing the confocal microscope. *Scanning, 10*, 128–138.
Moore, M. J., & Query, C. C. (2000). Joining of RNAs by splinted ligation. *Methods in Enzymology, 317*, 109–123.
Muir, T. W., Sondhi, D., & Cole, P. A. (1998). Expressed protein ligation: A general method for protein engineering. *Proceedings of the National Academy of Sciences of the United States of America, 95*(12), 6705–6710.
Muschielok, A., Andrecka, J., Jawhari, A., Bruckner, F., Cramer, P., & Michaelis, J. (2008). A nano-positioning system for macromolecular structural analysis. *Nature Methods, 5*(11), 965–971.
Muschielok, A., & Michaelis, J. (2011). Application of the nano-positioning system to the analysis of fluorescence resonance energy transfer networks. *Journal of Physical Chemistry B, 115*(41), 11927–11937.
Nettels, D., Gopich, I. V., Hoffmann, A., & Schuler, B. (2007). Ultrafast dynamics of protein collapse from single-molecule photon statistics. *Proceedings of the National Academy of Sciences of the United States of America, 104*(8), 2655–2660.
Nguyen, D. P., Lusic, H., Neumann, H., Kapadnis, P. B., Deiters, A., & Chin, J. W. (2009). Genetic encoding and labeling of aliphatic azides and alkynes in recombinant proteins via a pyrrolysyl-tRNA Synthetase/tRNA(CUA) pair and click chemistry. *Journal of the American Chemical Society, 131*(25), 8720–8721.
Nielsen, K. H., Chamieh, H., Andersen, C. B., Fredslund, F., Hamborg, K., Le Hir, H., et al. (2008). Mechanism of ATP turnover inhibition in the EJC. *RNA, 15*, 67–75.
Nir, E., Michalet, X., Hamadani, K. M., Laurence, T. A., Neuhauser, D., Kovchegov, Y., et al. (2006). Shot-noise limited single-molecule FRET histograms: Comparison between theory and experiments. *Journal of Physical ChemistryB, 110*(44), 22103–22124.
Paige, J. S., Wu, K. Y., & Jaffrey, S. R. (2011). RNA mimics of green fluorescent protein. *Science, 333*(6042), 642–646.
Pal, P., Lesoine, J. F., Lieb, M. A., Novotny, L., & Knauf, P. A. (2005). A novel immobilization method for single protein spFRET studies. *Biophysical Journal, 89*(2), L11–L13.
Pirchi, M., Tsukanov, R., Khamis, R., Tomov, T. E., Berger, Y., Khara, D. C., et al. (2016). Photon-by-photon hidden Markov model analysis for microsecond single-molecule FRET kinetics. *Journal of Physical ChemistryB, 120*(51), 13065–13075.
Plass, T., Milles, S., Koehler, C., Schultz, C., & Lemke, E. A. (2011). Genetically encoded copper-free click chemistry. *Angewandte Chemie (International Ed. in English), 50*(17), 3878–3881.
Popp, M. W., & Ploegh, H. L. (2011). Making and breaking peptide bonds: Protein engineering using sortase. *Angewandte Chemie (International Ed. in English), 50*(22), 5024–5032.
Putnam, A. A., & Jankowsky, E. (2013). DEAD-box helicases as integrators of RNA, nucleotide and protein binding. *Biochimica et Biophysica Acta, 1829*(8), 884–893.
Rasnik, I., McKinney, S. A., & Ha, T. (2006). Nonblinking and long-lasting single-molecule fluorescence imaging. *Nature Methods, 3*(11), 891–893.
Roy, R., Hohng, S., & Ha, T. (2008). A practical guide to single-molecule FRET. *Nature Methods, 5*(6), 507–516.
Rudolph, M. G., & Klostermeier, D. (2015). When core competence is not enough: Functional interplay of the DEAD-box helicase core with ancillary domains and auxiliary factors in RNA binding and unwinding. *Biological Chemistry, 396*(8), 849–865.
Ryu, Y., & Schultz, P. G. (2006). Efficient incorporation of unnatural amino acids into proteins in *Escherichia coli*. *Nature Methods, 3*(4), 263–265.
Samatanga, B., Andreou, A. Z., & Klostermeier, D. (2017). Allosteric regulation of helicase core activities of the DEAD-box helicase YxiN by RNA binding to its RNA recognition motif. *Nucleic Acids Research, 45*(4), 1994–2006.

Schmid, S., Gotz, M., & Hugel, T. (2016). Single-molecule analysis beyond dwell times: Demonstration and assessment in and out of equilibrium. *Biophysical Journal, 111*(7), 1375–1384.

Schmid, S. R., & Linder, P. (1991). Translation initiation factor 4A from *Saccharomyces cerevisiae*: Analysis of residues conserved in the D-E-a-D family of RNA helicases. *Molecular and Cellular Biology, 11*(7), 3463–3471.

Schmid, S. R., & Linder, P. (1992). D-E-A-D protein family of putative RNA helicases. *Molecular Microbiology, 6*(3), 283–291.

Schmidt, A., Gao, G., Little, S. R., Jalihal, A. P., & Walter, N. G. (2020). Following the messenger: Recent innovations in live cell single molecule fluorescence imaging. *Wiley Interdisciplinary Reviews: RNA, 11*, e1587.

Sengoku, T., Nureki, O., Nakamura, A., Kobayashi, S., & Yokoyama, S. (2006). Structural basis for RNA unwinding by the DEAD-box protein Drosophila vasa. *Cell, 125*(2), 287–300.

Shandell, M. A., Tan, Z., & Cornish, V. W. (2021). Genetic code expansion: A brief history and perspective. *Biochemistry, 60*(46), 3455–3469.

Shi, J., & Muir, T. W. (2005). Development of a tandem protein trans-splicing system based on native and engineered split inteins. *Journal of the American Chemical Society, 127*(17), 6198–6206.

Stagge, F., Mitronova, G. Y., Belov, V. N., Wurm, C. A., & Jakobs, S. (2013). SNAP-, CLIP- and halo-tag labelling of budding yeast cells. *PLoS One, 8*(10), e78745.

Steiner, M., Karunatilaka, K. S., Sigel, R. K., & Rueda, D. (2008). Single-molecule studies of group II intron ribozymes. *Proceedings of the National Academy of Sciences of the United States of America, 105*(37), 13853–13858.

Stelljes, J. T., Weidlich, D., Gubaev, A., & Klostermeier, D. (2018). Gyrase containing a single C-terminal domain catalyzes negative supercoiling of DNA by decreasing the linking number in steps of two. *Nucleic Acids Research, 46*(13), 6773–6784.

Swoboda, M., Henig, J., Cheng, H. M., Brugger, D., Haltrich, D., Plumere, N., et al. (2012). Enzymatic oxygen scavenging for photostability without pH drop in single-molecule experiments. *ACS Nano, 6*(7), 6364–6369.

Talavera, M. A., & De La Cruz, E. M. (2005). Equilibrium and kinetic analysis of nucleotide binding to the DEAD-box RNA helicase DbpA. *Biochemistry, 44*(3), 959–970.

Theissen, B., Karow, A. R., Kohler, J., Gubaev, A., & Klostermeier, D. (2008). Cooperative binding of ATP and RNA induces a closed conformation in a DEAD box RNA helicase. *Proceedings of the National Academy of Sciences of the United States of America, 105*(2), 548–553.

Thomsen, J., Sletfjerding, M. B., Jensen, S. B., Stella, S., Paul, B., Malle, M. G., et al. (2020). DeepFRET, a software for rapid and automated single-molecule FRET data classification using deep learning. *eLife, 9*, e60404.

Tyagi, S., & Lemke, E. A. (2013). Genetically encoded click chemistry for single-molecule FRET of proteins. *Methods in Cell Biology, 113*, 169–187.

Wang, L., Brock, A., Herberich, B., & Schultz, P. G. (2001). Expanding the genetic code of Escherichia coli. *Science, 292*(5516), 498–500.

Wang, Q., Chan, T. R., Hilgraf, R., Fokin, V. V., Sharpless, K. B., & Finn, M. G. (2003). Bioconjugation by copper(I)-catalyzed azide-alkyne [3 + 2] cycloaddition. *Journal of the American Chemical Society, 125*(11), 3192–3193.

Wang, S., Hu, Y., Overgaard, M. T., Karginov, F. V., Uhlenbeck, O. C., & McKay, D. B. (2006). The domain of the *Bacillus subtilis* DEAD-box helicase YxiN that is responsible for specific binding of 23S rRNA has an RNA recognition motif fold. *RNA, 12*, 959–967.

Wang, L., Xie, J., & Schultz, P. G. (2006). Expanding the genetic code. *Annual Review of Biophysics and Biomolecular Structure, 35*, 225–249.

Wang, L., Zhang, Z., Brock, A., & Schultz, P. G. (2003). Addition of the keto functional group to the genetic code of *Escherichia coli*. *Proceedings of the National Academy of Sciences of the United States of America*, *100*(1), 56–61.

Wu, H., Hu, Z., & Liu, X. Q. (1998). Protein trans-splicing by a split intein encoded in a split DnaE gene of Synechocystis sp. PCC6803. *Proceedings of the National Academy of Sciences of the United States of America*, *95*(16), 9226–9231.

CHAPTER TWELVE

Long DNA constructs to study helicases and nucleic acid translocases using optical tweezers

Clara Aicart-Ramos[a,*], Silvia Hormeno[a], Oliver J. Wilkinson[b], Mark S. Dillingham[b], and Fernando Moreno-Herrero[a,*]
[a]Department of Macromolecular Structures, Centro Nacional de Biotecnología, Consejo Superior de Investigaciones Científicas, Madrid, Spain
[b]DNA:Protein Interactions Unit, School of Biochemistry, University of Bristol, Bristol, United Kingdom
*Corresponding author: e-mail address: caicart@cnb.csic.es; fernando.moreno@cnb.csic.es

Contents

Methods in Enzymology, Volume 673
ISSN 0076-6879
https://doi.org/10.1016/bs.mie.2022.03.010

Abstract

Single molecule biophysics experiments for the study of DNA-protein interactions usually require production of a homogeneous population of long DNA molecules with controlled sequence content and/or internal tertiary structures. Traditionally, Lambda phage DNA has been used for this purpose, but it is difficult to customize. In this article, we provide a detailed and simple protocol for cloning large (~25 kbp) plasmids with bespoke sequence content, which can be used to generate custom DNA constructs for a range of single-molecule experiments. In particular, we focus on a procedure for making long single-stranded DNA (ssDNA) molecules, ssDNA-dsDNA hybrids and long DNA constructs with flaps, which are especially relevant for studying the activity of DNA helicases and translocases. Additionally, we describe how the modification of the free ends of such substrates can facilitate their binding to functionalized surfaces allowing immobilization and imaging using dual optical tweezers and confocal microscopy. Finally, we provide examples of how these DNA constructs have been applied to study the activity of human DNA helicase B (HELB). The techniques described herein are simple, versatile, adaptable, and accessible to any laboratory with access to standard molecular biology methods.

1. Introduction

A central component of any single-molecule experiment designed to understand the mechanism of action of proteins that interact with DNA is the DNA construct itself (Bocanegra, Plaza, Pulido, & Ibarra, 2021; Dulin, Lipfert, Moolman, & Dekker, 2013; Kong & Greene, 2021; Ordu, Lusser, & Dekker, 2016; Spinks, Spenkelink, Dixon, & van Oijen, 2021; Yodh, Schlierf, & Ha, 2010). Moreover, the production of dedicated DNA constructs has been critical to unveil their mechanical properties and how these modulate the activity of processing enzymes; information that is hardly achievable using bulk methods (Basu, Bobrovnikov, & Ha, 2021; Brahmachari & Marko, 2018; Bustamante, Bryant, & Smith, 2003; Camunas-Soler, Ribezzi-Crivellari, & Ritort, 2016; Marin-Gonzalez, Vilhena, Perez, & Moreno-Herrero, 2021). In the last 30 years, many DNA-centric single-molecule methods have been developed which have deepened our understanding of the behavior of biomolecules (Bustamante, Chemla, Liu, & Wang, 2021; Ha, Kaiser, Myong, Wu, & Xiao, 2022; Miller, Zhou,

Shepherd, Wollman, & Leake, 2018). The development of single molecule techniques has in part been possible thanks to the implementation of methods for fabrication of very long *à-la-carte* DNA molecules. This manuscript aims to describe a simple procedure for fabrication of long DNA constructs (tens of kbp) with controlled sequences and/or internal tertiary structures, as well as with the chemical modifications necessary for single molecule biophysical experiments.

There are multiple single-molecule techniques dedicated to the study of DNA-protein interactions. Some involve no external manipulation of the DNA and are based on high-resolution imaging (Dufrêne et al., 2017; Wilson & Costa, 2017; Zhao, Jiang, & Qi, 2017). Examples of these techniques include atomic force microscopy (AFM), electron microscopy or fluorescence-based methods (DNA curtains, for instance (Fazio, Visnapuu, Wind, & Greene, 2008)). Other methods use an external force to manipulate the DNA. Seminal work by Bustamante and colleagues used the optical force induced by a highly-focused laser beam on a plastic microsphere to characterize the elastic properties of individual double-stranded DNA (dsDNA) and single-stranded DNA (ssDNA) molecules (Smith, Cui, & Bustamante, 1996). Alternatively, an external force induced by permanent magnets on superparamagnetic beads has also been used to stretch and twist individual DNA molecules (Strick, Allemand, Bensimon, Bensimon, & Croquette, 1996; Strick, Allemand, Bensimon, & Croquette, 1998). The optical tweezers (OT) and magnetic tweezers (MT) techniques not only opened the field to study the nanomechanical properties of biological polymers such as nucleic acids (Bustamante et al., 2003; Smith et al., 1996; Smith, Finzi, & Bustamante, 1992; Strick et al., 1996, 1998; Wang, Yin, Landick, Gelles, & Block, 1997) and polypeptides (Bustamante, Alexander, MacIuba, & Kaiser, 2020; Kellermayer, Smith, Granzier, & Bustamante, 1997) but also contributed to the elucidation and demonstration of mechanistic models of protein function, such as the dynamics of different molecular motors (Block, Goldstein, & Schnapp, 1990; Svoboda, Schmidt, Schnapp, & Block, 1993) that play fundamental roles in different cellular processes such as DNA replication (Maier, Bensimon, & Croquette, 2000; Wuite, Smith, Young, Keller, & Bustamante, 2000), transcription (Fazal, Meng, Murakami, Kornberg, & Block, 2015), translation (Der Wen et al., 2008), recombination (Bell & Kowalczykowski, 2016; Brouwer et al., 2017) or compaction (Smith et al., 2001; Strick, Kawaguchi, & Hirano, 2004). Other methods that use an external force to manipulate DNA are based on the drag force applied by liquid flow (Bianco & Kowalczykowski, 1997; Doyle, Ladoux, & Viovy, 2000; Fazio et al., 2008; Perkins, Smith, & Chu, 1997; Perkins, Smith, Larson, & Chu, 1995;

Van Oijen et al., 2003), or exploit the manipulation of microspheres using acoustic waves produced in a confined environment (Sitters et al., 2015). In any case, what is common to all of these single-molecule methods is the pre-requisite for design and fabrication of a bespoke DNA construct.

The study of DNA-protein interactions using single-molecule methods typically requires the production of a homogeneous population of long DNA constructs of several tens of kbps. This is particularly important in methods that combine fluorescence because light diffraction limits resolution to a few hundreds of nm, i.e., around 1 kbp. Among these methods the combination of OT with confocal microscopy is powerful because it allows simultaneous and correlative manipulation and visualization of single-molecule interactions in real time with sub-picoNewton force resolution and sub-nanometer localization (Balaguer et al., 2021; Gutierrez-Escribano et al., 2019; Hashemi Shabestari, Meijering, Roos, Wuite, & Peterman, 2017; Wasserman, Schauer, O'Donnell, & Liu, 2019). Moreover, these long DNA molecules often need to include specific DNA sequences, i.e., a protein-binding site, and/or to incorporate localized tertiary structures, such as a ssDNA flap or a hairpin that acts as a loading-site for a protein (Brutzer, Luzzietti, Klaue, & Seidel, 2010; Hormeno et al., 2022; Levikova, Klaue, Seidel, & Cejka, 2013; Luzzietti et al., 2011; Wilkinson, Carrasco, Aicart-Ramos, Moreno-Herrero, & Dillingham, 2020). Finally, they also need to be labeled at their ends with chemical modifications (typically biotins and/or digoxigenins) to allow tight non-covalent binding to functionalized surfaces for manipulation. Both the length and specificity requirements of these substrates have created technical challenges that the single-molecule community have dedicated considerable effort towards overcoming.

The problem of length has traditionally been solved by employing commercially available bacteriophage DNA, typically from Lambda (48.5 kbp) (Smith et al., 1996). However, with the use of Lambda, the problem of the sequence specificity remains. Therefore, several strategies have been developed for the re-design of this kind of substrate. Some require restriction enzyme cleavage and ligation (Li et al., 2021), but this approach is limited to the few unique restriction sites present along the Lambda sequence. *In vitro* site-specific recombination methods have also been described (Bell, Plank, Dombrowski, & Kowalczykowski, 2012; Thorpe & Smith, 1998), but these involve complex molecular cloning steps and are limited to specific non-coding regions in the Lambda genome. Moreover, the specialized recombinase enzymes required need to be produced in-house as they are not commercially available. *In vivo* recombineering methods have also been described (Brown, De La Torre, & Finkelstein, 2017; Kim, De La Torre, Leal, & Finkelstein, 2017) but the purification of recombinant Lambda DNA from lysogens is required. In summary, although these methods do allow the generation of customized long DNA molecules, they are relatively complex and inaccessible to many laboratories.

The most common vehicle for the facile design, production and purification of specific DNA sequences are plasmids. These circular molecules are very easy to store and amplify using commercially available cloning strains and plasmid purification kits. In addition, because plasmids are copied within bacteria which possess robust DNA repair mechanisms, the quality of the resulting DNA is very high and far superior to that obtained in a PCR reaction (Eckert & Kunkel, 1991). Importantly, plasmids are an excellent choice for producing unique and/or customized DNA constructs by exploiting the availability of synthetic DNA fragments which can be inserted by conventional cloning methods or by Gibson Assembly (Mueller, Spenkelink, van Oijen, & Lewis, 2020; Sánchez et al., 2021; Spakman, King, Peterman, & Wuite, 2020; Van Loenhout, De Grunt, & Dekker, 2012). However, while plasmid-based approaches allow the production of long dsDNA constructs containing the sequences of interest, they do not address the generation of long ssDNA or ssDNA-dsDNA hybrids molecules which are especially useful for the study of DNA helicases and translocases. So far, the strategy followed to produce long ssDNA and ssDNA-dsDNA hybrid constructs has been based on Lambda DNA and is subject to the limitations above with respect to sequence design (Belan, Moore, et al., 2021; Belan, Barroso, et al., 2021; Brouwer et al., 2017; Candelli et al., 2013, 2014; King et al., 2013; Smith et al., 1996).

In this article we provide a detailed and simple protocol, accessible to any molecular biology laboratory, for the design and cloning of large plasmids (up to ~25 kbp) with sequences of interest. These plasmids serve to generate long customized DNA constructs for different single-molecule experiments. In particular, we describe a procedure to produce long ssDNA and ssDNA-dsDNA hybrid constructs. We also describe a procedure to fabricate a long DNA construct containing a ssDNA flap that is useful to study the activity of DNA helicases (Levikova et al., 2013; Wilkinson et al., 2020). These three long DNA constructs are ideal for determining polarity of movement (Gilhooly, Gwynn, & Dillingham, 2013) and kinetic parameters such as translocation velocity or pausing frequency and duration. Besides the cloning steps, we also describe how to modify the substrates for immobilization in OT set-ups and provide examples of their use for the study of a human DNA helicase.

2. Molecular design of large DNA plasmids

We describe below the procedure to produce large customized plasmids of over 17 kbp, containing particular regions of interest. These plasmids comply with two key requirements for single-molecule optical tweezers experiments, the long length and the capacity to easily modify the sequence and incorporate chemical modifications or tertiary structures at a specific region.

Large plasmids were obtained by ligation of three DNA pieces (C1, C2, and C3, Fig. 1). C1 and C2 fragments are fabricated by PCR using Lambda DNA as the template and including suitable restriction sites in the PCR primers. The third key piece (C3) contains the sequence of interest (e.g., several copies of a *parS* recognition sequence (Balaguer et al., 2021)), and is obtained either by plasmid digestion or by PCR amplification. Importantly, the C3 fragment also includes both an antibiotic resistance gene, to allow correctly-transformed bacteria to be selected using the appropriate antibiotic, and an origin of replication, to allow the autonomous replication of the final plasmid. The three fragments are ligated and DH5α competent cells transformed by regular heat-shock procedure. Following this strategy, we have cloned plasmids up to 25 kbp in length.

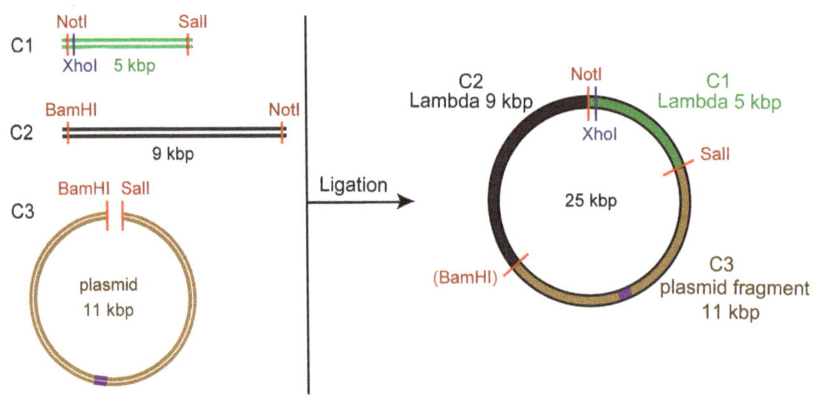

Fig. 1 Fabrication of large DNA plasmids. Scheme of the cloning of a large plasmid by ligation of C1, C2 and C3 fragments. C1 and C2 fragments are obtained by PCR-amplification of Lambda DNA, and C3 fragment is obtained by digestion or PCR amplification of a plasmid that contains a sequence of interest. In this example, the indicated restriction sites (except for BamHI) are unique sites.

This strategy is very versatile. One can use any plasmid of choice to produce the C3 piece and select different sections of Lambda DNA for PCR-amplification depending on the restriction sites present in them. Additionally, one can vary the restriction sites included in the primers employed to perform the PCRs depending on the ones that are present in the final fragments that need to be ligated. Here, it is good practice to select compatible restriction sites which, once cleaved, produce overhangs that can be ligated to overhangs obtained by different restriction enzymes (e.g., complementary SalI and XhoI, or NotI and PspOMI enzymes). This allows one to avoid the potential complications caused if one of the restriction sites is present in the original sequence. Also, note that although we have employed Lambda DNA to produce the C1 and C2 pieces by PCR, any other template DNA could be used.

2.1 Equipment

- Thin-walled 0.2-ml PCR tubes and 1.5-mL tubes (DNAse free and/or sterile).
- PCR thermocycler machine (Mastercycler nexus gradient, Eppendorf).
- Centrifuge (5424R, Eppendorf).
- Agarose gel electrophoresis chamber (Mini-Sub® Cell GT Cell, Bio-Rad) and electrophoresis power supply.
- Mini-Rocker Shaker (Mini-Rocker Shaker MR-1, BOECO).
- Blue light LED transilluminator (Transilluminator DR-46B, Clare Chemical).
- UV transilluminator (Gel Doc™ 2000, Bio-Rad).
- NanoDrop spectrophotometer (NanoDrop 1000, Thermo Scientific).
- Water bath (SUB Aqua Pro, Grant).
- Thermo block (PCH-1, Grant-bio).
- Shaking incubator (GFL Shaking Incubator 3033, GFL).
- Incubator (MIR-162, Sanyo).
- Drigalsky spreader.
- Bunsen burner (or a Butane/propane small gas bottle).

2.2 Buffers, strain and reagents

- PCR primers (Sigma-Aldrich).
- Plasmid DNA of interest.
- Lambda DNA (NEB).
- Restriction enzymes (NEB).

- Shrimp Alkaline phosphatase (rSAP, NEB).
- T4 DNA ligase (NEB).
- Nuclease-free water (Sigma).
- Phusion high-fidelity PCR kit (Thermo Fisher Scientific).
- GoTaq® PCR core system II kit (Promega).
- UltraPure agarose (Invitrogen).
- 10 × TBE buffer (Promega).
- 6 × Blue/Orange loading dye (Promega).
- 1 kb DNA ladder (NEB).
- 10,000 × SYBR® safe DNA gel stain (Invitrogen).
- QIAquick PCR purification kit (Qiagen).
- QIAquick gel extraction kit (Qiagen).
- QIAprep spin miniprep kit (Qiagen).
- Subcloning efficiency™ DH5α™ competent cells (F^- φ80*lac*ZΔM15 Δ(*lac*ZYA-*arg*F)U169 *rec*A1 *end*A1 *hsd*R17(r_k^-, m_k^+) *pho*A*sup*E44 *thi*-1 *gyr*A96 *rel*A1 λ^-) (Thermo Fisher Scientific).
- Sterile LB medium (for 1 L, add 10 g bactotryptone, 10 g NaCl and 5 g yeast extract. For LB agar, add 15 g agar-agar per 1 L of medium).
- Antibiotic stock.
- Ethanol.

2.3 Procedure

1. Design primers including restriction sites suitable to get three ligation-compatible pieces of DNA. For instance, C1 fragment including NotI and SalI, C2 fragment including BamHI and NotI and, C3 fragment including SalI and BamHI. If C3 fragment is PCR produced, these sequences can be incorporated in the primers. Note that the selected restriction sites cannot be present anywhere else along each specific DNA fragment (Fig. 1).
2. Dissolve primer stocks to 100 μM in nuclease free water. Store at −20 °C.
3. Set up PCR reactions for C1 and C2 fragments on ice with a high-fidelity DNA polymerase enzyme such as Phusion high-fidelity DNA polymerase, following the manufacturer's instructions (Table 1).
4. Transfer the tubes to the PCR thermocycler and start the PCR program (Table 2). Set the lid at 100 °C to prevent condensation.
5. Analyze the PCR products by 1% agarose gel (30 mL) in 1 × TBE buffer. Load each well with a mix of 5 μL of PCR sample and 1 μL

Table 1 Materials to produce DNA fragments by PCR.

Components	Volume	Final concentration
Nuclease-Free Water	X μL	
5 × Phusion Buffer (HF)	20 μL	1 ×
10 mM dNTP mixture[a]	2 μL	0.2 mM[a]
100 μM FW primer	0.5 μL	0.5 μM
100 μM RV primer	0.5 μL	0.5 μM
DNA template	X μL	40 ng
Phusion DNA Polymerase	1 μL	0.02 U/μL
Final volume	*100 μL*	

[a]cc of each dATP, dCTP, dGTP and dTTP.

Table 2 Thermocycling conditions to produce DNA fragments by PCR.

Step	Time	Temperature
1	2 min	98 °C
2	10 s	98 °C
3	20 s	65 °C
4	X[a]	72 °C
		Repeat steps 2–4 for 35 cycles
5	5 min	72 °C
6	hold	4 °C

[a]Adjust: 1 min per 2.5 kbp extension.

of 6 × loading dye. Run electrophoresis for 50 min at 90 V. Place the gel in a plastic container and stain with 1 × SYBR® Safe DNA Gel Stain in 1 × TBE solution for 30 min. Cover the container with aluminum foil to protect from light and gently agitate on a Mini-Rocker Shaker. Examine the gel under UV light in a UV transilluminator. If the expected band is observed, continue to the next step. If PCR did not work, then repeat **step 3** but decreasing the annealing temperature to 60 °C (step 3 in Table 2) or adding to the PCR reaction a final 2% DMSO and 1 mM $MgCl_2$, supplied in the kit.

6. Purify the PCR products with a commercial PCR purification kit according to manufacturer's instructions and elute in 30 μL of supplied Elution Buffer (EB).
7. Digest PCR fragments with the proper restriction enzymes (10–20 units of each enzyme) in buffer provided by the manufacturer at recommended temperature for 1 h.
8. Purify the digested-PCR products (C1 and C2 fragments). If no additional bands were observed in **step 5**, purify with a PCR purification kit the whole PCR product and elute in 30 μL of EB. However, it is highly recommended to purify the fragment of interest with a commercial gel extraction kit after running the whole sample into a 1% agarose gel. If the fragment has to be purified from a gel, the DNA must not be exposed to UV light nor intercalant dyes. To avoid that, load a small amount of the sample with 6 × loading dye in one lane next to the 1 kb DNA ladder and the rest of the sample in another lane. After running the gel, cut the gel to separate both lanes and stain only the section that contains the low amount of the sample and the DNA ladder. Then reconstruct the gel in the blue light LED transilluminator and cut the fragment of interest from the unstained part of the gel. Elute the sample in 30 μL of EB.
9. Determine DNA concentration with a NanoDrop spectrophotometer. Store at 4 °C.
10. If C3 fragment is obtained by PCR, follow **steps 3–9**. However, if C3 fragment is obtained by digestion of a plasmid of interest, follow **steps 11–13**.
11. Cut the plasmid of interest (1–2 μg) with the appropriate restriction enzymes (e.g., SalI and BamHI) using 5 units of each enzyme per μg of DNA in 50 μL of buffer provided by the manufacturer at the recommended temperature for 1 h. Add 1 unit per 1 pmol of DNA ends of rSAP into the reaction to dephosphorylate DNA ends to prevent re-ligation of the plasmid in subsequent steps. This will generate two linear segments (one of them is the C3 fragment) with the proper overhangs for ligation with C1 and C2 fragments. The phosphatase can be added directly into the digestion reaction, during, or after DNA digestion.
12. Purify the C3 fragment. If the other linear segment produced in the digestion of the plasmid is smaller than 100 bp, purify the fragment of interest with a PCR purification kit. If it is longer than 100 bp, it is necessary to purify the fragment of interest with a gel extraction

kit after running the whole sample into a 1% agarose gel as described in **step 8**. Elute the sample in 30 μL of EB.

13. Determine C3 fragment concentration with a NanoDrop spectrophotometer. Store at 4 °C.
14. It is recommended to visualize the purified C1, C2 and C3 fragments in a 1% agarose gel. Load around 40 ng of each fragment and proceed as described in **step 5**.
15. Mix the three fragments in a 1:1:1 M ratio using a final concentration of each fragment of between 0.3 and 2 nM, and then ligate with 400 units of T4 DNA ligase in 20 μL of buffer provided by the manufacturer. Transfer the tube to the PCR thermocycler and incubate at 16 °C for 15 h. Store at 4 °C. Although it is recommended to perform overnight ligation, after 2–3 h it is possible to make a first attempt of transformation with the ligation mix.
16. Set a water bath to the recommended temperature for heat shock transformation. Pre-warm 1 mL LB media to 37 °C in the thermo block, then thaw an aliquot of chemically competent DH5α cells on ice and transform them with the ligation product following manufacturer's instructions. Plate the whole culture onto an LB agar plate containing the appropriate antibiotic. Grow overnight at 37 °C (see **Note 1**).
17. To check if some of the colonies are positive, perform colony PCR of 10–20 colonies. It is highly recommended to perform colony PCR with 2 oligonucleotides that amplify a fragment comprising C1 and C2 pieces. Note that the fact that colonies have grown in the LB agar plate containing antibiotic confirm C3 fragment is already present.
18. Take two PCR tubes per tested colony and put them in two parallel rows. The first row must contain each colony in LB (see **step 19**) and the second row should contain the corresponding sample of master mix for the colony PCR reaction (see **step 22**).
19. Transfer 50 μL LB into one row of PCR tubes. Do not add any antibiotic.
20. Pick a single colony into each tube and mix by pipetting. Let sit for 10 min at room temperature.
21. Prepare a master mix of 100 μL for every set of 10 colonies (10 μL/colony) to perform PCR with, for example, GoTaq® G2 Flexi DNA polymerase (Table 3). Mix well and gently.
22. Add 9.5 μL of master mix into each empty PCR tube of the second row. An extra tube is needed for the control. It is highly recommended to look for a plasmid that works as a positive or a negative control of the

Table 3 Materials for colony PCR.

Components	Volume	Final concentration
Nuclease-Free Water	68.5 μL	
5 × Taq Buffer (without $MgCl_2$)	20 μL	1 ×
25 mM $MgCl_2$	6 μL	1.5 mM
10 mM dNTP mixture[a]	2 μL	0.2 mM[a]
100 μM FW primer	0.5 μL	0.5 μM
100 μM RV primer	0.5 μL	0.5 μM
Polymerase Go Taq flexi	0.5 μL	0.025 U/μL
Final volume	*100 μL*	

[a]cc of each dATP, dCTP, dGTP and dTTP.

Table 4 Thermocycling conditions for colony PCR.

Step	Time	Temperature
1	2 min	94 °C
2	30 s	94 °C
3	30 s	55 °C
4	X[a]	72 °C
		Repeat steps 2–4 for 30 cycles
5	5 min	72 °C
6	hold	4 °C

[a]Adjust: 1 min 30 s per 1 kbp extension.

PCR to confirm that the PCR is working properly but also to identify that a colony has incorporated the desired plasmid.

23. Transfer 1 μL of template (colony in LB from **step 20**) and mix by pipetting. Picked colonies can be stored for around 4 days in LB at 4 °C.
24. Transfer the tubes to the PCR thermocycler and start the PCR program (Table 4). Set the lid at 100 °C to prevent condensation.
25. Analyze the PCR products in a 1% agarose gel. Mix 10 μL of PCR sample with 2 μL of 6 × loading dye and load the gel. Run the gel, stain it and image as described in **step 5**. If the analyzed colonies are negative and there are still some colonies in the plate, repeat from **step 19** selecting a new collection of colonies. If any of the analyzed colonies seems to be positive, continue with the next step.

26. To screen for positive transformants, pick 10 μL of a possible positive colony (from tubes of **step 20**) and grow them separately in 5 mL LB supplemented with the appropriate antibiotic. Usually pick 2 or 3 colonies. Grow overnight at 37 °C at 170 rpm in a shaking incubator.
27. Isolate large plasmids from each culture using a commercial miniprep plasmid purification kit, according to the manufacturer's instructions. Determine the DNA concentration with a NanoDrop spectrophotometer.
28. Perform screening by digestion with selected restriction endonucleases. For each screening, digest 100 ng of plasmid combining two restriction enzymes (1–2 units of each enzyme) in 10 μL of appropriate buffer at recommended temperature for 30–60 min.
29. Analyze the restriction digestions in a 1% agarose gel as described in **step 5** and look for appropriate sizes dependent on the restriction enzyme chosen
30. Perform DNA sequence analysis to any preliminary positive plasmid in a commercially available service. For sequencing, select oligonucleotides that allow you to sequence fragments of DNA that comprises junctions between fragments C1-C2, C2-C3 and C3-C1, and any region of interest that need to be present in the large final plasmid.
31. Once the large plasmid is confirmed, it is ready to be used to prepare long DNA substrates for single molecule experiments (see **Note 2**).

2.4 Notes

1. For bacterial work, sterile conditions should be maintained. Unless you have access to a laminar flow cabinet, you can work on an open bench near the flame generated by a Bunsen burner or the small Butane/propane gas bottle. The updraft from the heat should create a relatively sterile field with which to work. Clean your work space with 70% ethanol, wear gloves and "flame" the openings of bottles to remove any contaminant.
2. The final large plasmid can be used to clone a new large plasmid by introducing a new DNA fragment of interest as shown in Sections 5 and 6.

3. Fabrication of long double-stranded DNA constructs from large plasmids

This section describes the general procedure to incorporate modified handles to the previously linearized large plasmid described in Section 2. This procedure results in a long dsDNA construct (>17 kbp) suitable for single-molecule experiments that require immobilization of the molecule from their ends. These can be OT or MT experiments, but also

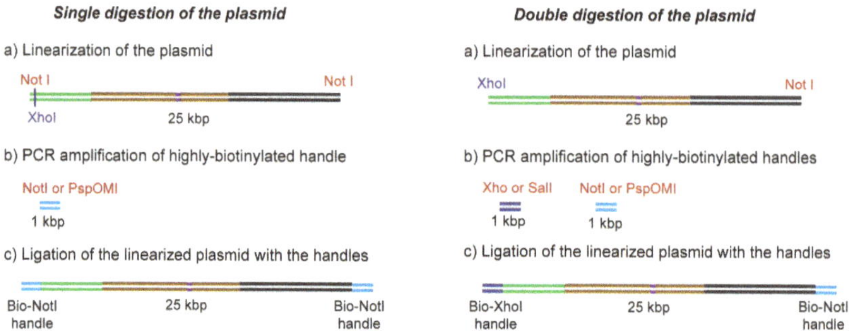

Fig. 2 Fabrication of long DNA constructs. The large plasmid shown in Fig. 1 can be linearized by single or double digestion (left and right side, respectively) for later ligation with highly-biotinylated handles. Sticky ends obtained by restriction digestion with NotI or PspOMI are compatible, and therefore a biotinylated handle obtained by digestion with either NotI or PspOMI can be ligated to the NotI-sticky end of the central part. The same is true for sticky ends obtained after digestion with either XhoI or SalI. This can help to avoid generating tandem (double-length) tethers as mentioned in Section 3.4.

flow-stretched experiments that require immobilization of the DNA on a surface for further TIRF imaging. The procedure we describe below is based on the cleavage of the plasmid by one or two restriction enzymes and, without further purification, the ligation to functionalized fragments of ~1 kbp (Fig. 2). These handles are typically produced by PCR and can incorporate either biotins or digoxigenins. This methodology is extremely efficient and versatile: depending on the plasmid sequence design, one can choose different restriction enzymes to locate the region of interest at a particular position with respect to the immobilized ends.

An alternative to ligation of PCR-labeled handles is the filling of 5′-overhangs with modified deoxynucleotides using the Klenow Fragment of DNA polymerase, which lacks $3' \rightarrow 5'$ exonuclease activity (Derbyshire et al., 1988) as performed by (Candelli et al., 2013, 2014; Sánchez et al., 2021; Spakman et al., 2020; van den Broek, Noom, & Wuite, 2005). This is a simple procedure often employed with Lambda DNA because its *cos* end sites are 12-nt long and therefore allow labeling with several biotins or digoxigenins (Balaguer et al., 2021; Brouwer et al., 2017; Forget, Dombrowski, Amitani, & Kowalczykowski, 2013; Gross, Farge, Peterman, & Wuite, 2010; Smith et al., 1996; Wasserman et al., 2019). However, we have found this procedure to be limited, compared to the use of long handles, in several aspects. On one hand, only 5′-overhangs (3′-recessed ends) can be filled in by Klenow Fragment *exo*−. On the other hand, the same kind of modified nucleotide, in terms of biotins or digoxigenins, is incorporated in both sides of the DNA construct, so differential end-labeling, if needed, will require extra steps. Additionally, DNA ends are labeled on different strands, and this makes it impossible to prepare in a single step pure ssDNA or hybrid ssDNA-dsDNA constructs (see next sections). For the same reason, molecules cannot be torsionally constrained, if required. Finally, the amount of labeled nucleotides per DNA end is very low compared to a PCR-produced handle as usually a restriction digestion overhang is 4-nt long. This reduces the force that can be tolerated in a pulling experiment and makes in-flux fishing more difficult in dual-trap OT experiments.

Other approaches to label DNA ends are based on the annealing and ligation of biotinylated oligonucleotides to the overhang generated after restriction digestion. Single-stranded regions are then filled with biotinylated deoxynucleotides by Klenow Fragment *exo*-, rendering a DNA construct with biotin labels at both strands on either end (Van Mameren et al., 2009). Slight modifications of this strategy allows one to obtain a final dsDNA construct labeled in both ends and on the same strand with the aim of generating ssDNA tethers (Candelli et al., 2013, 2014). Drawbacks of these methods are the price of biotinylated nucleotides and potential difficulties in annealing and ligating a ssDNA-oligonucleotide to an only 4-nt overhang. To overcome this issue, Lambda DNA is usually employed because it has longer 12-nt *cos* ends (Camunas-Soler et al., 2013; Candelli et al., 2013, 2014; Madariaga-Marcos et al., 2018; Smith et al., 1996; Van Oijen et al., 2003). Similarly, expensive internally-labeled oligonucleotides can be replaced by tailed oligonucleotides produced by terminal transferase (Camunas-Soler et al., 2013; Madariaga-Marcos et al., 2018). The problem of direct ligation of ssDNA over dsDNA can be also overcome by the use of a biotinylated end-cap hairpin oligonucleotide that self-anneals to generate a cohesive dsDNA end (Paik & Perkins, 2011). The use of biotinylated end-cap hairpin oligonucleotides with a 12-nt overhang compatible to *cos* end sites of Lambda DNA has also been reported (King et al., 2013; King, Burla, Peterman, & Wuite, 2019; King, Peterman, & Wuite, 2016). These methods have the inconvenience of using rather expensive biotinylated oligonucleotides together with having a reduced amount of labeled nucleotides per DNA end.

As mentioned above, we opted for production of modified handles by PCR. This is an easy and inexpensive method with additional advantages. It incorporates a high frequency of labeled nucleotides per DNA end, and this is important for in-flow capture of DNA molecules in OT experiments. The ligation process is very efficient and allows one to easily modify the position of a region of interest by just selecting another pair of restriction sites. It allows for the generation of torsionally constrained or non-torsionally constrained tethers, simply by using either phosphorylated or dephosphorylated handles, respectively. Importantly, it facilitates simple preparation of pure ssDNA or hybrid ssDNA-dsDNA constructs as described in following sections.

3.1 Equipment

- Thin-walled 0.2-mL PCR tubes and 1.5-mL tubes (DNAse free and/or sterile).
- PCR thermocycler machine (Mastercycler nexus gradient, Eppendorf).
- Centrifuge (5424R, Eppendorf).
- Agarose gel electrophoresis chamber (Mini-Sub® Cell GT Cell, Bio-Rad) and Electrophoresis power supply.
- Mini-rocker shaker (Mini-Rocker Shaker MR-1, BOECO).
- UV transilluminator (Gel Doc™ 2000, Bio-Rad).
- Nano Drop spectrophotometer (NanoDrop 1000, Thermo Scientific).
- Vortex.

3.2 Buffers and reagents

- PCR primers (Sigma-Aldrich).
- DNA plasmid as template for preparation of handles.
- Large DNA plasmid prepared as described in Section 2
- Biotin-16-dUTP (Roche).
- Restriction enzymes (NEB).
- Shrimp Alkaline phosphatase (rSAP, NEB).
- T4 DNA ligase (NEB).
- Nuclease-free water (Sigma).
- 50 mM EDTA pH 8.0.
- GoTaq® PCR core system II kit (Promega).
- UltraPure agarose (Invitrogen).
- 10 × TBE buffer (Promega).
- 6 × Blue/Orange loading dye (Promega).
- 1 kb DNA ladder (NEB).
- 10,000 × SYBR® safe DNA gel stain (Invitrogen).
- QIAquick PCR purification kit (Qiagen).
- 4.38 μm SPHERO streptavidin-coated polystyrene beads (Spherotech).
- TE buffer (10 mM Tris-HCl pH 8.0, 1 mM EDTA).
- TEN buffer (10 mM Tris-HCl pH 8.0, 1 mM EDTA, 50 mM NaCl).
- SYTOX™ green nucleic acid stain (Thermo Fisher Scientific).

3.3 Procedure

3.3.1 Preparation of the central part

The plasmid design and the position of its unique restriction sites will determine the choice of appropriate restriction enzymes to locate any region of interest at a particular position with respect to the immobilized ends. Additionally, depending on the final purpose of these DNA constructs, the large plasmid can be linearized by digestion with a single restriction enzyme or with two different ones, which will render two non-complementary overhangs (Fig. 2). In the case of digestion with two enzymes, it is recommended that the restriction sites are close to one another (within 200 bp), because this will avoid an extra purification step. For the same reason, it is also recommended to use restriction enzymes that can be heat inactivated. It is important to keep purification steps to a minimum to avoid undesired damage to the DNA and dilute samples. Finally, it is important to keep in mind that long DNAs can easily break due to shear forces created by pipetting. Use wide-bore pipette tips and mix solutions containing large DNAs gently, avoiding high-speed vortexing and repetitive pipetting.

1. Cut the plasmid of interest (1 μg) with the appropriate restriction enzyme/s (5 units of each enzyme) in 6.25 μL buffer provided by the manufacturer for 1 h 30 min at the recommended temperature. This will create the necessary overhangs for ligation with highly-biotinylated handles. In this particular example, we linearized the plasmid shown in Fig. 1 with NotI (Fig. 2, left).
2. Heat-inactivate the restriction enzyme/s by incubating at the temperature and time recommended by the manufacturer.
3. Store at 4 °C. Do not purify the central fragment at this stage. The final concentration should be around 160 ng/μL.

Additional steps are required for fabrication of the central part of the flap DNA construct (Section 6).

3.3.2 Preparation of the highly-biotinylated handles

The handles are DNA fragments of ~1 kbp produced by PCR and labeled with biotins or digoxigenins. The procedure below uses the pSP73-JY0 (Fili et al., 2010) or pBlueScript SK+ (Stratagene) plasmids as a template, but any other plasmid or linear DNA could be used.

1. Design primers including in one of them the desired restriction site that generates a sticky end for ligation to the central fragment of the DNA construct. In this particular example, we fabricated a NotI handle or a compatible PspOMI handle (Fig. 2, left).

2. Dissolve primer stocks to 100 μM in nuclease free water. Store at −20 °C.
3. Set up PCR reactions to prepare highly-labeled handles, biotins in the present example, with GoTaq® G2 Flexi DNA polymerase following the manufacturer's instructions (Table 5). If the same handle is ligated to both ends of the central part, just perform a single PCR reaction.
4. Transfer the tubes to the PCR thermocycler and start the PCR program (Table 6). Set the lid at 100 °C to prevent condensation.
5. It is recommended to analyze the PCR products by loading 5 μL of PCR sample into a 1% agarose gel as described in Section 2.3, **step 5**. If the expected band is observed, continue with the next step.
6. Purify the PCR products with a commercial PCR purification kit, and then elute products in 30 μL of supplied EB.
7. Perform digestion of the PCR fragments with 20–30 units of the proper restriction enzyme for 1 h 30 min in its recommended buffer and at indicated temperature.
8. Perform a final purification of the digested-PCR products with a commercial PCR Purification kit and elute in 30 μL of supplied EB.
9. Determine DNA concentration with a NanoDrop spectrophotometer. Store at 4 °C.

Table 5 Materials to produce highly-biotinylated handles by PCR.

Components	Volume	Final concentration
Nuclease-Free Water	X μL	
5 × Taq Buffer (without $MgCl_2$)	20 μL	1 ×
25 mM $MgCl_2$	6 μL	1.5 mM
10 mM dATP, dCTP, dGTP mixture[a]	2 μL	0.2 mM[a]
10 mM dTTP	1.4 μL	0.14 mM
1 mM Biotin-16-dUTP	6.6 μL	0.066 mM
DNA template	X μL	40 ng
100 μM FW primer	0.5 μL	0.5 μM
100 μM RV primer	0.5 μL	0.5 μM
Polymerase Go Taq flexi	0.5 μL	0.025 U/μL
Final volume	*100 μL*	

[a]cc of each dATP, dCTP and dGTP.

Table 6 Thermocycling conditions to produce highly-biotinylated handles by PCR.

Step	Time	Temperature
1	30 s	94 °C
2	30 s	94 °C
3	30 s	55 °C
4	X[a]	72 °C
		Repeat steps 2–4 for 35 cycles
5	5 min	72 °C
6	hold	4 °C

[a]Adjust: 1 min 30 s per 1 kbp extension.

An additional step of dephosphorylation could be required depending on the specific final DNA constructs as shown in Sections 4 and 5.

3.3.3 Ligation of central part and handles

1. Mix 1 μg of the central part of the DNA construct (**Step 3**, Section 3.3.1) with 10 ×-15 × excess of each highly-biotinylated handle/s (**Step 9**, Section 3.3.2), and ligate with 200 units of T4 DNA ligase in 12.5 μL of buffer provided by the manufacturer. Transfer the tube to the PCR thermocycler and incubate at 16 °C for 15 h followed by heat inactivation at 65 °C for 20 min (see **Note 1 and Note 2**).
2. Add EDTA pH 8.0 to 1 mM final concentration to preserve. Store the final long DNA construct at 4 °C, without further purification. The final concentration should be around 80 ng/μL.

Additional steps could be required depending on the specific DNA constructs as shown in next sections.

3.3.4 Test of production of long dsDNA constructs in C-Trap

Standard quality control assays of the fabricated long dsDNA construct involve force-extension curves to characterize the contour length of the tethers. Although we have performed these studies using the C-Trap instrument from LUMICKS (Candelli, Wuite, & Peterman, 2011), the procedure should be easy adaptable to any other home-made or commercial optical tweezers. Our equipment incorporates a confocal microscope that allows

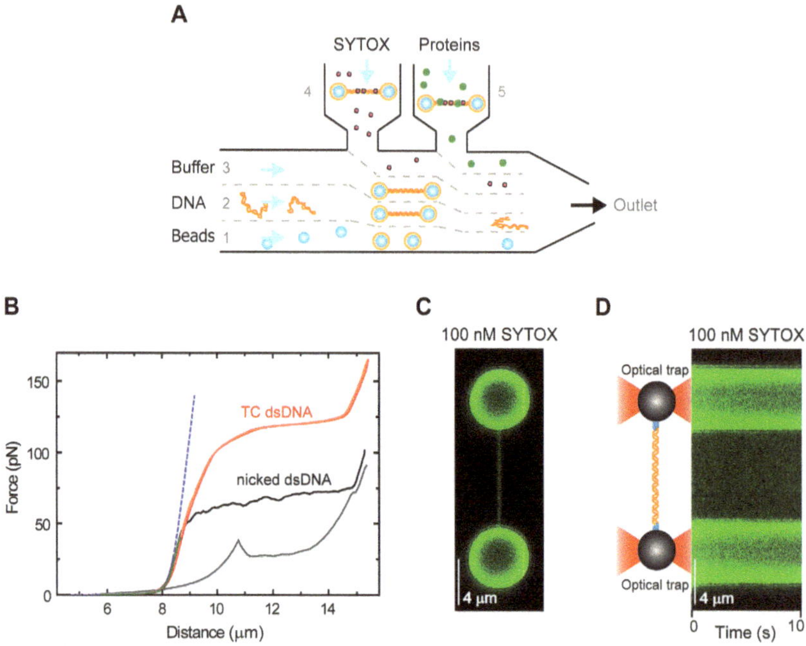

Fig. 3 Optical tweezers and confocal experiments to test the production of dsDNA constructs. (A) Individual DNA tethers are assembled in channels 1–3 separated by laminar flow containing streptavidin-coated beads, the biotinylated DNA construct and buffer, respectively. Channel 3 is filled with low-salt TE buffer if hybrid ssDNA-dsDNA or pure ssDNA constructs are prepared or with TEN buffer if the DNA will not be subjected to force-induced melting, as for dsDNA or flap DNA constructs. Force-extension curves are performed in this channel to discard double tethers and confirm the single-tether has the expected contour length. In this channel, pure ssDNA or hybrid ssDNA-dsDNA constructs are also formed. Once the quality of the DNA construct has been checked, the traps are moved to channels 4 and 5 for labeling with SYTOX or for protein loading and imaging. (B) Force-extension curve of a torsionally-constrained, and nicked 25 kbp dsDNA (TC dsDNA, stretching (red) and relaxation (light red) curves; nicked, stretching (black and green) and relaxation (gray) curves). Blue dashed line is a fit to the eWLC model. (C) 2D confocal scan of the tethered dsDNA construct stained with 100 nM SYTOX. The experiment was performed at 40 pN. (D) Representative kymograph of the dsDNA construct stained with 100 nM SYTOX. The experiment was performed at 30 pN.

visualization of duplex DNA by SYTOX intercalation or labeling of proteins with organic dyes or quantum dots. Experiments were done using a microfluidic cell with five laminar flow separated channels (Fig. 3A). This design without physical barriers allows the direct transfer of the trapped DNA between different flow channels and the *in situ* assembly and characterization of the DNA constructs.

3.3.4.1 Mechanical characterization of dsDNA molecules

1. Prepare flow cell for experiments (cleaning and passivation) as recommended by manufacturer
2. Fill in syringes of flow cell channels as follows:

   Channel 1: 10 μL polystyrene beads (previously resuspended by gently vortexing) in 990 μL TEN buffer.

   Channel 2: 1–2 μL of stock DNA diluted in 500 μL TEN buffer.

   Channel 3: 1 mL of TEN buffer for regular dsDNA.

   Channel 4: 1 mL of TEN buffer.

   Channel 5: 1 mL of TEN buffer.
3. Open channels 1–3 and start flowing their content by applying a pressure of 0.2–0.4 bar.
4. Move optical traps to a mid-position into channel 1. Trap one bead in each trap. Adjust the laser power to achieve a trap stiffness of ~0.4 pN/nm.
5. Move the traps to channel 2 and capture a single DNA tether between the beads. To do that, steer the bead in the mobile trap back and forth with respect to the other bead while keeping the flow open. If the DNA concentration is adequate, the biotinylated construct will ensemble between the streptavidin-covered beads after a few attempts, and a response in the force sensor will be noticed when moving one bead with respect to the other. The presence of moderate salt in the buffer helps to capture DNA using this procedure.
6. Move the trapped DNA to channel 3. Perform a force extension curve at 1 μm/s to discard double tethers and detect if the molecule is torsionally-constrained (TC) or nicked (Fig. 3B). As expected for the protocol described above, most of the molecules were TC (Fig. 3B, TC dsDNA, stretching (red) and relaxation (light red) curves), but occasionally, we also observed nicked molecules (Fig. 3B, nicked dsDNA, stretching (black and green) and relaxation (gray) curves). To assure a homogeneous population of nicked DNA tethers some strategies can be considered (see **Note 3**).
7. Check the contour length of the tether by fitting to a polymer extension model. In this case, a fit to the extensible Worm-Like-Chain (eWLC) model (Odijk, 1995) (Fig. 3B, blue dashed line) provided a contour length of 24,940 bp with a persistence length, P_{dsDNA}, of 19 nm (using a crystallographic length for dsDNA, $L_{0,\ dsDNA}$, of 0.34 nm bp^{-1}, and a stretch modulus, S_{dsDNA}, of 1500 pN).

3.3.4.2 Direct visualization of dsDNA molecules

1. Once confirmed the integrity of a dsDNA tether, fill in channel 4 with 200 μL of 100 nM SYTOX Green dissolved in TEN buffer.

2. Move the molecule to channel 4 and hold the molecule at 20–40 pN.
3. Take a scan with the confocal microscope using excitation light at 488 nm. The whole molecule should be visible as SYTOX Green is a dsDNA labeling dye (Fig. 3C).
4. Alternatively, take a kymograph to capture florescence between the beads versus time (Fig. 3D). Standard parameters are a pixel size of 50 nm and a line scan speed of 50–100 ms. It is recommended to work with the minimum laser power (~2%) to avoid the stained DNA from breaking during the kymograph acquisition. The emission is detected employing a green filter 585/75 nm.

It is important to note that the use of intercalant dyes like SYTOX are useful to visualize dsDNA and check proper production of molecules as described above, but they can interfere with the binding and activity of proteins, and therefore should be avoided in those experiments.

3.4 Notes

1. Tandem products (double length) are often produced in the final ligation step (Section 3.3.3) because restriction sites are often palindromic. To avoid the generation of tandem molecules, one can increase the excess of handles or use compatible sticky ends obtained by restriction with two different enzymes. Sites generated by compatible restriction enzymes lose the original restriction sequence once they are ligated. In this case, add the enzyme employed to linearize the plasmid to the ligation mix. Note however, that there are occasions where it is desirable to promote the generation of tandem molecules to obtain longer constructs. In this case, decrease the excess of handles (**Step 1**, Section 3.3.3).
2. Concentration of cleaved handles (**Step 1**, Section 3.3.3.) might be lower than expected, if the restriction site contains a deoxythymidine. In this case, a Bio-dUTP might be incorporated instead of a dTTP during the PCR fabrication, and the modified restriction site will not be cut by the restriction enzyme. If the restriction site contains a Thymine, a higher excess of that handle will be required.
3. Homogeneous population of nicked DNA tethers instead of torsionally constrained molecules can be obtained by linearizing the large plasmid with two different restriction enzymes as shown in Fig. 2, right, e.g., XhoI and NotI, and ligation with two different labeled handles, being one of them dephosphorylated.

4. Case 1: A hybrid ssDNA-dsDNA construct for single-molecule experiments

This section describes a specific procedure to produce ssDNA-dsDNA hybrids. These constructs are useful to assess the substrate specificity (ssDNA or dsDNA) of a DNA-binding protein, to mimic a single ssDNA-dsDNA junction, such as those found in the nucleus at replication forks and R-loops, or to study the activity of DNA polymerases at the junction (Hoekstra et al., 2017). Moreover, ssDNA-dsDNA junctions are loading sites for some translocases that can initiate translocation from such a position (Tomko et al., 2010). Some proteins like CST bind preferentially to these junctions, an activity that can explain the incremental nature of telomeric C-strand synthesis following telomerase action (Bhattacharjee, Wang, Diao, & Price, 2017).

The first attempts to produce ssDNA-dsDNA hybrids were based on the use of exonucleases (Lee, Balci, Jia, Lohman, & Ha, 2013; Wuite et al., 2000). However, these approaches are inconvenient for several reasons. They do not allow full control of the position or the orientation of the junction because they can degrade DNA from an internal nick and it is difficult to control the length of the final ssDNA fragment. This results in a non-homogeneous population of molecules that may not be suitable for single-molecule studies. Additionally, if digestion of the dsDNA is produced *in situ*, then inactivation of the enzyme and/or an extra cleaning step of the flow cell is required.

Alternative methods based on the separation of one of the duplex strands by applying a stretching force to its complementary partner strand are currently preferred. In these methods one needs to produce nicks in the strand to be removed. Most approaches use the asymmetric BbvCI restriction site (Heiter, Lunnen, & Wilson, 2005) that allows nicking of either the bottom or top strand using Nb.BbvCI or Nt.BbvCI. This nicking combined with end-labeling using biotinylated/non-biotinylated oligonucleotides and sometimes Klenow *exo-* enzyme, can produce DNA molecules labeled at both ends but on only one of the strands (Candelli et al., 2013). This method obviously requires the engineering of a plasmid to contain those asymmetric restriction sites and relies on the use of oligonucleotides despite the difficulty of ligating a ssDNA oligonucleotide to a dsDNA end that is usually 4-nt long. A more recent approach to introduce nicks and gaps in a long DNA substrate uses site-specific Cas9 nicking (Anand et al., 2022; Belan, Barroso, et al., 2021;

Belan, Moore, et al., 2021). The length and position of the gaps is determined by the choice of the guide RNA sequences. If one is looking for gaps of a specific-length, this is a very attractive strategy, although, caution must be taken in designing guide RNAs to avoid undesired off-target effects.

In the approach described below, we ligate a dephosphorylated biotinylated handle and use nicking enzymes to create a single nick along the DNA construct. This creates a region flanked by two nicks (depicted with red and green dots in Fig. 4A, respectively). The ssDNA strand between the nicks is then removed *in situ* by tension as described before (Candelli et al., 2013). This method is simple because it only requires the engineering of one BbvCI restriction site in one of the three fragments employed to generate the large plasmid (Fig. 4B) and the second nick is easily introduced by the use of a dephosphorylated handle. Additionally, our method does not require the use of biotinylated/non-biotinylated oligonucleotides. This gives complete control over the position and orientation of the ss-dsDNA junction. Moreover, the position of the junction with respect to one DNA end can be easily modified by linearizing the plasmid with different enzymes and ligating with different handles (Fig. 4C).

4.1 Design of the DNA plasmid

This method requires a large plasmid that has one BbvCI restriction site (Fig. 4B). An example of oligonucleotides employed for PCR-amplification of C1 and C2 fragments from Lambda DNA is shown in Table 7. In this particular example, the BbvCI site is contained in the C1 fragment. Production of the C3 fragment by plasmid digestion is described in Section 2. The general procedure to produce this plasmid as well as the equipment, buffers, strain and reagents are described in Section 2.

4.2 Assembly of the ssDNA-dsDNA construct

The equipment, buffers, reagents, and procedure for the assembly of this DNA construct, are described in Section 3. However, the following specific points need to be considered.

4.2.1 Preparation of the central part

The central part must be linearized with two different restriction enzymes to generate non-complementary overhangs. The ss-dsDNA junction can be positioned with respect to the beads by selecting particular pairs of restriction enzymes (Fig. 4C). This is convenient because it allows one to have different

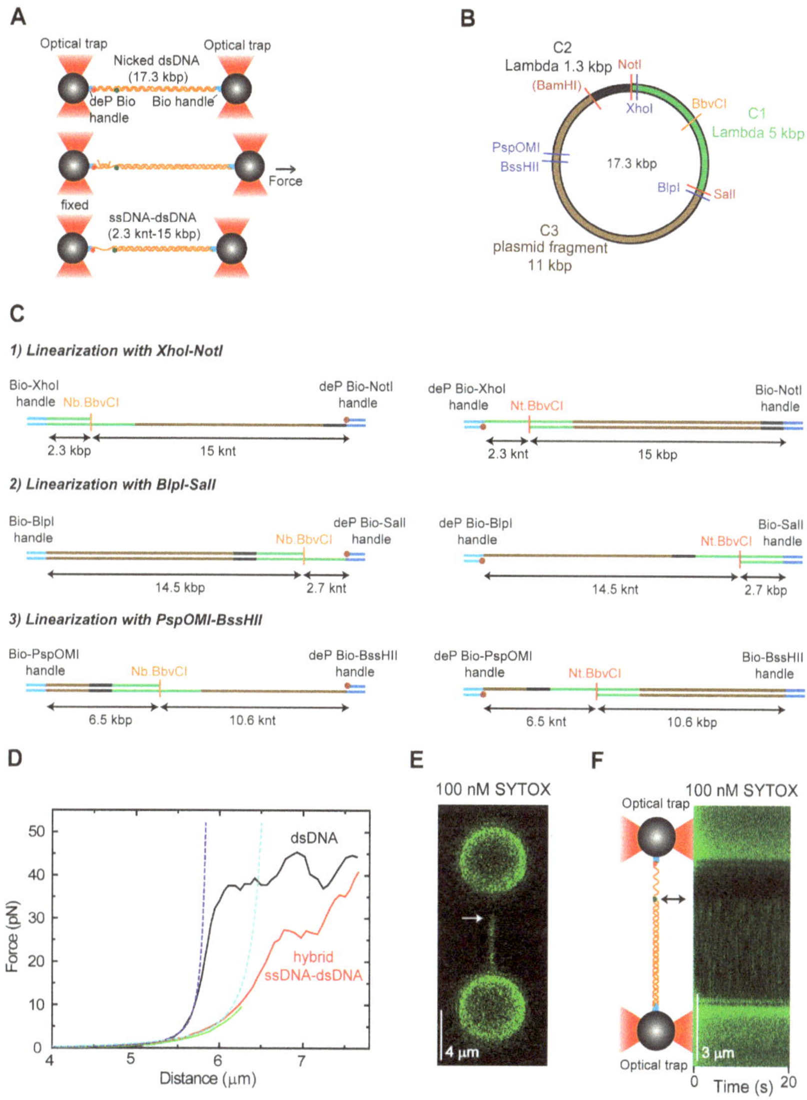

Fig. 4 Fabrication of a long hybrid ssDNA-dsDNA construct. (A) Scheme illustrating in situ formation of a ss-dsDNA hybrid tether corresponding to DNA construct 1 (panel C, right side), using dual-trap optical tweezers. A functionalized dsDNA molecule with two nicks, one due to Nt.BbvCI nicking and the other to ligation to a dephosphorylated handle (depicted with green and red dots, respectively), is captured with the optical traps. Then, the fragment of ssDNA between the two nicks is removed by force-induced melting of the duplex. (B) Schematic representation of the DNA plasmid employed to produce the ss-dsDNA hybrid constructs. The indicated restriction sites are unique ones

(Continued)

Table 7 List of primers used in the design of the ssDNA-dsDNA plasmid.

Name	PCR fragment	Lambda overlapping region	Sequence
FWC1	C1	33,464–33,489	GCGTAAGTGCGGCCGC GGTGTGCTCCTTATTTAT ACATAACG
RVC1	C1	38,453–38,474	TAAGTGTCGAC *GAATTC*ATCCGA*GAATTC* CGATAG*GAATTC*CACGTT *GAATTC*TACGAC *GAATTC*GCGACTTATCA ACGCCCACAGC
FWC2	C2	13,254–13,274	TAAGTGGATCC*GAATTC* ATCCGA*GAATTC*CGATAG *GAATTC*CACGTT*GAATT* *C*TACGAC*GAATTC*GTGA GGAGGCCACGGTACTGG
RVC2	C2	14,489–14,509	GCGTAAGTGCGGCCGC CCGGAACAGTGTGTAA CAGTC

Restriction sites for ligation are underlined: NotI and SalI in C1 fragment, BamHI and NotI in C2 fragment. Shorter primers between 35 and 42 nt can be employed. These very long primers that contain several EcoRI restriction sites (italic) were employed in (Balaguer et al., 2021).

Fig. 4—Cont'd among others not included. The unique BbvCI restriction site is necessary to produce the ss-dsDNA junction. (C) Different ssDNA-dsDNA constructs, among others, can be prepared by digestion of the plasmid shown in (B) with different pairs of two close restriction enzymes and later ligation with two different highly biotinylated handles, with one of them being dephosphorylated (marked with a red dot). The nicking with Nb.BbvCI (left side) or Nt.BbvCI (right side) in combination with force-induced melting, would allow the creation of two different oriented ss-dsDNA junctions at well-defined positions for the same dsDNA fragment. Not only the junction but also the sequence of interest if present in C3 fragment, are differently positioned with respect to the beads. (D) Force-extension curves of DNA construct 1 (panel C, right side) that allows the transition from pure dsDNA (black) to a ssDNA-dsDNA hybrid construct (red and green). Dark blue and light blue dashed lines are representations of the Petrosyan equation for dsDNA and hybrid ss-dsDNA, respectively (see main text). (E) 2D confocal scan of a tethered ss-dsDNA construct stained by 100 nM SYTOX. White arrow denotes the ss-dsDNA junction. The experiment was performed at 40 pN. (F) Representative kymograph of the hybrid construct stained with 100 nM SYTOX. The black arrow denotes the ss-dsDNA junction. The experiment was performed at 44 pN.

proportions of ssDNA and dsDNA without the need to fabricate a new plasmid (Fig. 4B). The linearized central part is next ligated to two different highly biotinylated handles, with one of them being dephosphorylated. In this particular example, we chose to linearize the plasmid of Fig. 4B with XhoI and NotI (Fig. 4C, option 1, right).

4.2.2 Preparation of the highly-biotinylated handles

1. Select the handle to be dephosphorylated (in this example, the XhoI handle). This handle should include the restriction site compatible with one end of the linearized central part. Note that the 5′-sticky end of this handle and the nick after the digestion with the nicking BbvCI restriction enzyme must be on the same strand. Follow next step of dephosphorylation after **step 9** of Section 3.3.2.
2. Dephosphorylation reaction. Incubate the purified handle with 1 unit per 1 pmol of DNA ends of rSAP in buffer provided by the manufacturer for 2 h at 37 °C. Inactivate the enzyme by heating at 65 °C for 10 min. This will remove the 5′-phosphate from the 5′-end of that specific handle, but it still can be ligated to the central part. No further purification is required.

4.2.3 Ligation

Follow and adapt the procedure described in Section 3.3.3.

1. Mix half of the central part fragment (0.5 μg) with 10 ×-15 × excess of each highly-biotinylated handle (one of them dephosphorylated), and ligate with 100 units of T4 DNA ligase in 6.5 μL of buffer provided by the manufacturer. No further purification is required. The other half of the central part can be kept at 4 °C for an additional preparation. It is recommended not to prepare a large volume of this construct because random nicking or damage of the central part DNA will lead to unwanted products or the loss of the molecule in the pulling process to generate the gap.

4.2.4 Nicking with Nb.BbvCI or Nt.BbvCI restriction enzyme

1. Digest the ligated DNA molecule from Section 4.2.3 with the nicking enzyme Nb.BbvCI or Nt.BbvCI (Nt.BbvCI in this particular example). Incubate 6.5 μL of ligated product with 6 units of nicking enzyme in 13 μL of buffer provided by manufacturer for 2 h at 37 °C. Inactivate the enzyme by heating at 80 °C for 20 min. No further purification is required.

2. Add EDTA pH8.0 to 1 mM final concentration to preserve. Store the final DNA construct at 4 °C. The final concentration should be around 38 ng/μL.

The final DNA construct contains 2 nicks that define a region that can be further removed *in situ* by force as described below.

4.3 Test of production of ssDNA-dsDNA constructs in the C-Trap instrument

To illustrate the procedure, we fabricated a ssDNA-dsDNA tether that contains a ssDNA region of 2.3 knt and a dsDNA region of 15 kbp (Fig. 4C, option 1, right). The molecule was produced *in situ* by applying a force of ~40–50 pN in a low salt buffer as detailed in Section 3.3.4 (Fig. 4A and D). This assay also determines the contour length of the tethers. Additional tests using confocal imaging with fluorescent intercalant SYTOX Green were conducted.

4.3.1 Mechanical generation of ssDNA-dsDNA hybrids

1. Prepare the C-trap fluid cell as described in Section 3.3.4, with:
   Channel 2: 2–4 μL of stock DNA diluted in 300 μL TEN buffer.
   Channel 3: 1 mL of low-salt TE buffer for generation of ss-dsDNA hybrid.
2. Capture a duplex DNA molecule with a dual optical tweezers following the procedure described in Section 3.3.4.
3. Check the contour length of the duplex DNA by performing a force-extension curve up to ~40 pN. In this case, it is convenient to use the Petrosyan model that provides an approximation for the extension-force dependence for the WLC model that can be easily adapted to both dsDNA and ssDNA (Petrosyan, 2017). The extension given by the Petrosyan equation correctly described the data using a contour length of 17,430 bp with a crystallographic rise per base pair, $L_{0,\ dsDNA}$, of 0.34 nm bp^{-1}, and a persistence length, P_{dsDNA}, of 50 nm (Fig. 4D, dark blue dashed line).
4. Remove the nicked region by performing a force-extension curve up to ~40–50 pN. Detachment of the nicked region is produced when subsequent pulling cycles show stable and longer extensions than the expected for dsDNA at forces above 6 pN (Dessinges, Lionnet, Xi, Bensimon, & Croquette, 2004). The extension given by a hybrid ssDNA-dsDNA model (Petrosyan, 2017) using 15,130 bp of dsDNA ($L_{0,\ dsDNA}$ = 0.34 nm bp^{-1} and P_{dsDNA} = 50 nm) and 2300 nt of ssDNA ($L_{0,\ ssDNA}$ = 0.69 nm nt^{-1}, and P_{ssDNA} = 2 nm) correctly described the data and is included in

Fig. 4D (light blue dashed line). These data confirmed the correct generation of the designed hybrid ss-dsDNA molecule.

4.3.2 Direct visualization of ssDNA-dsDNA hybrids

1. Once confirmed the generation of the ss-dsDNA hybrid by force, fill in channel 4 with 200 μL of 100 nM SYTOX Green dissolved in TEN buffer.
2. Move the molecule to channel 4 and hold it at 40 pN.
3. Take a scan with the confocal microscope using excitation light at 488 nm. SYTOX is a dsDNA labeling dye and only the duplex DNA region should be visible (Fig. 4E). The junction between ssDNA and dsDNA is marked with an arrow.
4. Alternatively, take a kymograph to capture florescence between the beads versus time as described in Section 3.3.4. In our case, we observe a clear non-labeled region corresponding to the 2.3-knt ssDNA fragment (Fig. 4F).

5. Case 2: A pure single-stranded DNA construct for single molecule experiments

This section describes the specific procedure to clone a large plasmid to produce a long pure single-stranded DNA construct for optical tweezers experiments. The procedure is very similar to that described in Section 4, but with positioning of the nicking site close to the intact handle (the one with a 5′-phosphate). This already defines a complete region of the DNA construct flanked by two nicks (depicted with red and green dots in Fig. 5A) located in the same strand that can be removed *in situ* by force in a low-salt buffer (Candelli et al., 2013). Note that additional BbvCI sites, although not essential, would help the process, but they would need to have the same orientation to generate nicks on the same strand. A key advantage of this method is that it employs a customized large plasmid, and therefore particular sequences, e.g., a region that could fold into G-quadruplex structures or contain telomeric sequences, can easily be incorporated. These constructs are useful to study the mechanical properties of ssDNA or the activity of translocases (see Section 7) (Hormeno et al., 2022; Lee et al., 2013).

5.1 Design of the DNA plasmid

This method requires a large plasmid with at least one BbvCI restriction site (underlined in Fig. 5B) located next to the restriction sites employed to linearize the plasmid. We used the plasmid of Fig. 4B as starting material and

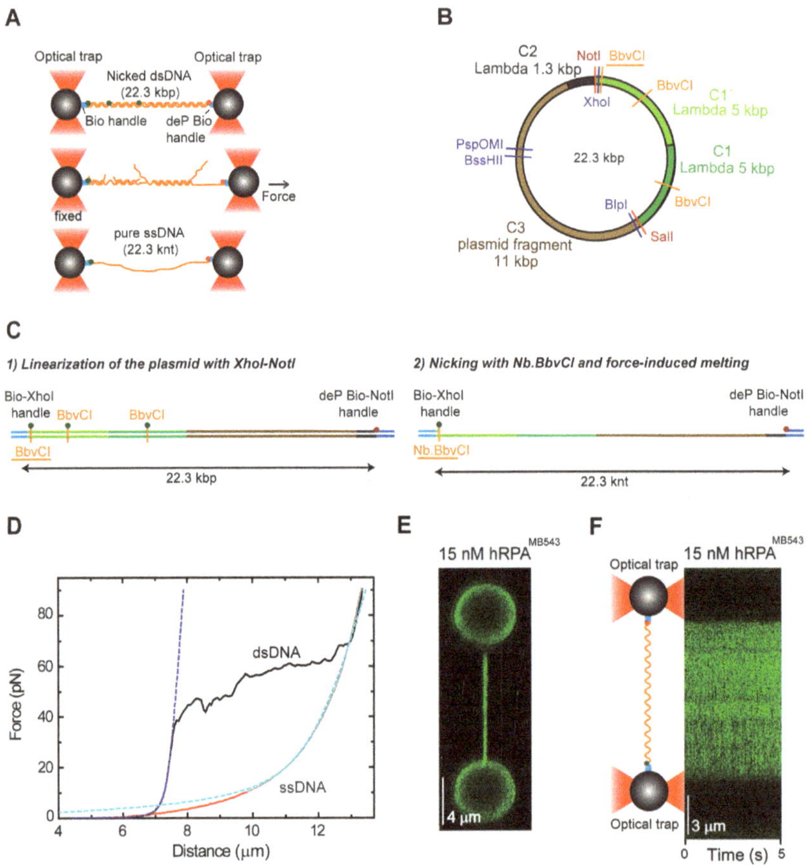

Fig. 5 Fabrication of a long pure ssDNA construct. (A) Scheme illustrating in situ formation of a pure ssDNA tether corresponding to the DNA construct shown in panel (C), using dual-trap optical tweezers. A functionalized dsDNA molecule with four nicks, three due to Nb.BbvCI nicking (green dots) and the fourth due to ligation to a dephosphorylated handle (red dot), is captured with the optical traps. The essential BbvCI nick is the one next to the phosphorylated handle. The complete upper strand is removed by force-induced melting of the duplex. (B) Schematic representation of the large DNA plasmid employed to produce a pure ssDNA construct. In this particular example, a new C1´ fragment (light green fragment) has been included in the plasmid shown in Fig. 4B, in between NotI and XhoI. The essential BbvCI site (underlined) is absolutely required at that position. Additional copies of BbvCI can be present, but they need to be in the same orientation as the essential site to obtain nicks in the same strand. (C) Scheme of the DNA construct obtained after digestion of plasmid shown in (B) with NotI and XhoI and ligation with handles. Once the final nicked construct is melted by force as described in (A), the upper strand is completely released. Dephosphorylated handle is depicted with a red dot, and Nb.BbvCI nicks with green dots. The essential BbvCI restriction site is underlined. (D) Force-extension curves indicate the transition from dsDNA (black) to ssDNA (gray and red). Fits to the eWLC and eFJC models are included (see main text). (E) 2D confocal scan of a tethered ssDNA covered by 15 nM hRPA^{MB543}. The experiment was performed at 17 pN. (F) Kymograph of the same experiment shown in (E).

Table 8 List of primers used in the design of the pure ssDNA plasmid.

Name	PCR fragment	Lambda overlapping region	Sequence
FWC1b	C1'	33530-33552	TAAAATGCGGCCGCAATTAACTCGAGGCTGAGG CCGCACTCAGGCGGCCTTGATAG
RVC1b	C1'	38454-38474	GCGTAAGTGTCGACGCGACTTATCAACGCCCAC AG

Restriction sites for ligation are underlined: NotI and SalI.
This specific position of restriction sites NotI (underlined), XhoI (red) and BbvCI (purple) is required to produce the pure ssDNA construct as detailed in main text.

produced a new C1' fragment by PCR with the primers shown in Table 8. The C1' fragment was then digested with NotI and SalI and introduced into the previous plasmid between NotI and XhoI. This new plasmid is 5-kbp larger and contains three BbvCI sites with the same orientation (Fig. 5B). An alternative, if the plasmid of Fig. 4B has not been previously cloned, is to follow the method described in Section 4.1, but replacing the C1 fragment with the C1´ fragment. The general procedure to produce this plasmid as well as the equipment, buffers, strain and reagents are described in Section 2.

5.2 Assembly of the pure ssDNA construct

The equipment, buffers, reagents, and procedure for the assembly of this DNA construct, are described in Section 3. However, the following specific points need to be considered.

5.2.1 Preparation of the central part

The central part is obtained by linearization of the plasmid shown in Fig. 5B with NotI and XhoI (Fig. 5C). The linearized central part is next ligated to two different highly biotinylated handles, being one of them dephosphorylated (the NotI handle). This creates a nick in the upper strand (red dot). This construct contains an essential BbvCI restriction site next to the XhoI cohesive end, where the phosphorylated handle is ligated, and two additional BbvCI restriction sites with the same orientation, leading altogether to three nicks on the same strand after cleavage with the Nb.BbvCI enzyme (green dots).

5.2.2 Preparation of the highly-biotinylated handles

Follow the procedure described in Section 4.2.2.

5.2.3 Ligation

Follow the procedure described in Section 4.2.3.

5.2.4 Nicking with Nb.BbvCI or Nt.BbvCI restriction enzyme

Follow the procedure for nicking as described in Section 4.2.4. In this particular example, we used Nb.BbvCI (Fig. 5C) leading to a final DNA construct with 4 nicks, which will facilitate the process of removal of the upper strand by pulling in the optical tweezers setup.

5.3 Test of production of pure ssDNA construct in the C-Trap instrument

To illustrate the procedure, we fabricated a DNA construct of 22.3 kbp (Fig. 5C) and nicked the substrate with Nb.BbvCI, as described above. Detachment of the nicked strand was achieved with a dual optical tweezers *in situ* by applying a force of ~80–90 pN in a low salt buffer, as detailed in Section 3.3.4 (Fig. 5A and D). This assay also determines the contour length of the tethers. Additional tests using confocal imaging with a labeled single-stranded binding protein (MB543 labeled hRPA, $hRPA^{MB543}$, with emission at 570 nm (Kuppa, Pokhrel, Corless, Origanti, & Antony, 2021)) were conducted.

5.3.1 Mechanical generation of pure ssDNA

1. Prepare the C-trap fluid cell as described in Section 3.3.4, with:
   Channel 2: 2 μL of stock DNA diluted in 300 μL TEN buffer.
   Channel 3: 1 mL of low-salt TE buffer for generation of pure ssDNA.
   Channel 4: 1 mL of reaction buffer (20 mM Tris pH 7.5, 30 mM NaCl, 4 mM $MgCl_2$, 5 mM DTT).
2. Capture a duplex DNA molecule with a dual optical tweezers following the procedure described in Section 3.3.4.
3. Check the contour length of the duplex DNA by performing a force-extension curve up to 40 pN (black line). In this case, a fit to the eWLC model (Odijk, 1995) (Fig. 5D, dark blue dashed line) provided a contour length of 22,164 bp with a persistence length, P_{dsDNA}, of 17 nm (using $L_{0,\ dsDNA} = 0.34\ nm\ bp^{-1}$, and $S_{dsDNA} = 1500$ pN).
4. Completely remove the nicked upper strand by performing a force-extension curve beyond the overstretching transition up to ~80–90 pN. Detachment of the nicked strand is produced when subsequent pulling cycles show stable and longer extensions than those expected for dsDNA at forces above 6 pN (Fig. 5D, gray and red lines). A fit

to the extensible Freely Jointed Chain (eFJC) model (Smith et al., 1996) for single-stranded DNA is included in Fig. 5D (light blue dashed line) and provided a contour length of 12 μm for the ssDNA tether ($0.54\,\text{nm}\,\text{nt}^{-1}$) and a Kuhn length of 1.8 nm (using $S = 600$ pN).

5.3.2 Direct visualization of pure ssDNA using fluorescently labeled RPA

1. Once confirmed the production of the ssDNA tether, introduce 200 μL of 15 nM $\text{hRPA}^{\text{MB543}}$ in reaction buffer in channel 4.
2. Move the molecule to channel 4 and hold it at 20 pN.
3. Take a scan with the confocal microscope using excitation light at 532 nm. RPA is a ssDNA binding protein and will visually provide the coverage of the ssDNA tether (Fig. 5E).
4. Alternatively, take a kymograph to capture florescence between the beads versus time as described in Section 3.3.4. In our case, we observe a clear full-length labeled ssDNA tether (Fig. 5F).

6. Case 3: A flap dsDNA construct for single molecule experiments

This section describes the specific procedure to clone a large plasmid to produce a flap dsDNA construct. This DNA construct contains a single-stranded DNA tail, the flap, flanked by duplex DNA. The flap can be 5′- or 3′-terminated and can provide a loading site for SF1–6 helicases (Gorbalenya & Koonin, 1993; Singleton, Dillingham, & Wigley, 2007). Similar substrates to the one described here have been employed to monitor the unwinding activity of different helicases (Hormeno et al., 2022; Levikova et al., 2013; Wilkinson et al., 2020).

The method described below is based on the generation of a single-stranded gap region by nicking enzymes acting on closely spaced restriction sites (Luzzietti et al., 2011; Luzzietti, Knappe, Richter, & Seidel, 2012; Wang & Hays, 2001). The gap can be filled in with an oligonucleotide to create a flap (our case below) or with other more complex tertiary structures such as hairpins, bulges, T-shape junctions, etc. The gap can be also used to incorporate oligomers carrying a desired modification such as damaged bases, a fluorophore, or biotin-labeled deoxythymidines to facilitate the attachment of streptavidin-conjugated quantum dots for example (Brutzer et al., 2010; Dekker et al., 2002; Hormeno et al., 2022; Levikova et al., 2013; Luzzietti et al., 2011; Wilkinson et al., 2020). This strategy needs precise positioning of the nicking restriction sites and therefore necessarily involves the production of plasmids. In the procedure described below, a large plasmid containing several closely-spaced BbvCI restriction sites was cloned based on the design of Luzzietti et al. (2011, 2012). After cleavage, a single-stranded gap of 63 nt was generated and partially filled with a long oligonucleotide. To generate a 5′-terminated flap we designed an oligo with 37 deoxythymidines (dT) at the 5′-end and a homologous region of 26 nt. This leaves a ssDNA gap of 37 nt. To produce a 3′-terminated flap the homologous region should be at the 5′-end and include a non-homologous region at the 3′-end. Note that in this case, the 5′-end should be phosphorylated to be ligated to the main body of the construct. This approach based on customized plasmids allows the combination of a flap structure with the presence of sequences of interest located along the dsDNA. The procedure below describes the production of a 5′-terminated flap substrate.

6.1 Design of the DNA plasmid

This method requires a large plasmid with several closely-spaced BbvCI restriction sites (Fig. 6A). First, fabricate a large plasmid similar to the one shown in Fig. 4B with appropriate restriction sites. Then, introduce a poly-BbvCI region into this plasmid. To do this, we used the plasmid reported in Wilkinson et al. (2020) and extracted a region with 5 BbvCI sites by XhoI-SalI cleavage. This fragment of 1428 bp was introduced once in the XhoI site of the large plasmid shown in Fig. 4B, resulting in new, large modified plasmid of 18,772 bp (Fig. 6A). Any plasmid with a BbvCI region would serve this purpose, but if this is not available, one can employ complementary oligonucleotides to generate a region with several BbvCI restriction site repetitions and introduce that region into the large plasmid. In any case, the orientation of the insert should be determined because this will ultimately have consequences for the orientation of the flap (see Section 2.3). Additional copies of BbvCI restriction sites along the plasmid in either orientation should not be a problem, as isolated nicks should be repaired in the final ligation step (see below). The general procedure to produce this plasmid as well as the equipment, buffers, strain and reagents are described in Section 2.

6.2 Assembly of the flap dsDNA construct

The equipment, buffers, reagents, and procedure for the assembly of this DNA construct, are described in Section 3. However, the following specific points need to be considered.

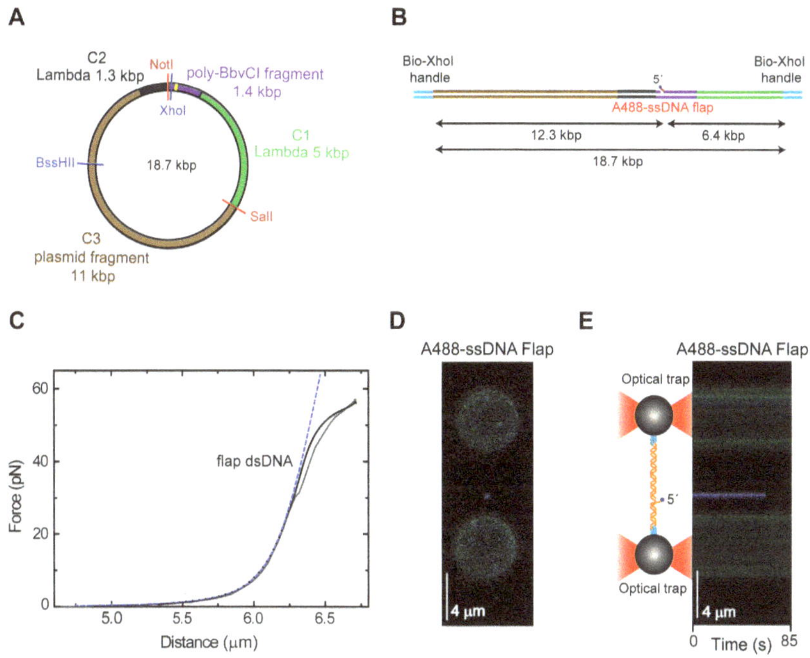

Fig. 6 Fabrication of a long DNA construct containing a ssDNA flap. (A) Schematic representation of the large plasmid employed to produce a flap DNA construct. A new fragment (purple) that contains five closely-spaced BbvCI restriction sites (yellow) has been introduced into the plasmid shown in Fig. 4B, in between XhoI restriction site. (B) Scheme of the DNA construct obtained after digestion of the plasmid with SalI and ligation with compatible XhoI handles. The construct includes a 5′-ssDNA flap sequence of 37 poly(dT)-nts labeled with Alexa 488 fluorophore on the 5′end. (C) Force-extension curves of the flap dsDNA construct (stretching (black) and relaxation (gray) curves). The blue dashed line is a fit to the eWLC (see main text). (D) 2D confocal scan of a tethered fluorophore-labeled flap dsDNA construct. The dsDNA molecule is free of nucleic acid stain, and the blue dot is consistent with the position of the Alexa488 fluorophore in the flap. The experiment was performed at 15 pN. (E) Kymograph of the same experiment shown in (D).

6.2.1 Preparation of the central part

The central part can be linearized with a single restriction enzyme. The choice of this enzyme will define the position of the poly-BbvCI region and ultimately the position of the flap with respect to the beads (see **Note 1**). This flexibility facilitates the production of different constructs with different proportions of dsDNA available for unwinding from the flap without the need to fabricate a new plasmid (Fig. 6B). In our particular example, we chose to linearize the plasmid with SalI.

6.2.1.1 Generation of the ssDNA gap by nicking with Nb.BbvCI or Nt.BbvCI restriction enzyme

1. Select the nicking enzyme. In our example, we chose Nt.BbvCI. Digest 0.5 µg of the linearized plasmid with 6 units of Nt.BbvCI in 12.5 µL of buffer provided by the manufacturer at 37 °C for 2 h. This will create 5 nicks in the poly BbvCI region on the same strand. If desired, linearization and nicking of the DNA plasmid can be combined in a single step.
2. Heat inactivate the enzyme for 20 min at 80 °C. This should also release the nicked fragments from the body of the molecule.
3. Store the nicked central part at 4 °C, without further purification.

6.2.1.2 Annealing of the oligonucleotide that constitutes the flap to the gap

4. Design an oligonucleotide with a poly(dT) of desired length, the flap, and a homologous region to the gap. In our example, we designed a 63-nt oligonucleotide, that after the annealing of 26 nt into the Nt.BbvCI-gap, leaves a flap of 37-poly (dT) nucleotides labeled with an Alexa 488 fluorophore (A488) in the 5′-end (Table 9 and **Note 2**) and a gap of 37 nt. To avoid this gap, another complementary oligonucleotide should be designed (blocking oligonucleotide). This oligonucleotide must be phosphorylated in its 5′-end. In this example, we designed an oligonucleotide of 32 nt that would leave a small gap of 5 nt after annealing (Table 9 and **Note 2**).
5. Dissolve oligonucleotide stocks to 100 µM and prepare a 12.5 µM stock aliquots in nuclease free water. Store at −20 °C. If the oligonucleotide is fluorescently-labeled, minimize the light exposure.
6. Mix half of the sample produced at step 3 (0.25 µg) with a 200 × excess of poly(dT)-flap and 200 × excess of blocking oligonucleotide in 10 µL of annealing buffer (50 mM Tris pH 8.0, 1 mM EDTA, 100 mM NaCl).

Table 9 Oligonucleotides required to produce the flap dsDNA construct.

Name	Sequence
A488–37 Poly(dT) 5'-ssDNA flap	A488- TTTTTTTTTTTTTTTTTTTTTTTTTTTTT TTTTTTTTCAGCTAGCCTCAGCCTACAATCACC
Blocking oligonucleotide	[Pho]TCAGCTCATGTCATCCTCAGCACACTT GACCC

In this particular example, after Nt.BbvCI nicking the gap has this 5′-to-3′ sequence: GGTGATTGTA GGCTGAGGCTAGCTGAGCTGAGGGTCAAGTGTGCTGAGGATGACATGAGCTGA

Heat for 5 min at 80 °C, and slowly cool down to 30 °C at a − 0.5 °C min^{-1} rate (Hormeno et al., 2022; Wilkinson et al., 2020).

7. Store the long flap DNA central part at 4 °C, without further purification.

6.2.2 Preparation of the highly-biotinylated handles

Follow the procedure described in Section 3.3.2.

6.2.3 Ligation

Follow and adapt the procedure described in Section 3.3.3.

1. Mix the long flap DNA central part (0.25 μg) with 15 × excess of highly-biotinylated handle (XhoI in this example), and ligate with 50 units of T4 DNA ligase in 12.5 μL of buffer provided by the manufacturer in the presence of SalI restriction enzyme to prevent tandem products (see Section 3.4). No further purification is required.
2. Add EDTA pH 8.0 to 1 mM final concentration to preserve. Store the final flap DNA construct at 4 °C. The final concentration should be around 20 ng/μL.

The final dsDNA construct contains a 5'-ssDNA flap at a defined position that could act as a loading site for a protein of interest (see section 7).

6.3 Test of production of a flap dsDNA construct in the C-Trap instrument

To illustrate the procedure, we fabricated a dsDNA construct of 18.7 kbp containing a 5′-terminated 37-nt poly(dT) ssDNA flap with an Alexa 488 fluorophore in the 5′-end positioned at 12.3 kbp from one end (Fig. 6B). Initially, a force extension curve was performed to determine the contour length of the tethers.

6.3.1 Mechanical characterization and visualization of a flap dsDNA construct

This procedure is identical to the one described in Section 3.3.4, with slight differences. The force-extension curve should lead to molecules around 18,700 bp. In our example (Fig. 6C), a fit to the eWLC model provided a contour length of 18,559 bp with a persistence length, P_{dsDNA}, of 20 nm (using $L_{0,dsDNA} = 0.34\,nm\,bp^{-1}$, and a stretch modulus, S_{dsDNA}, of 1500 pN). Similarly, a confocal scan and kymograph following the procedure of Section 3.3.4 but in the absence of SYTOX Green, can be obtained to identify the presence of the flap in the DNA construct (Fig. 6D and E). A blue dot consistent with the position of the fluorophore-labeled flap was identified.

6.4 Notes

1. In the process of linearization of the plasmid, it is recommended to choose a restriction enzyme that generates compatible sticky ends with another restriction enzyme. This should avoid the formation of tandem molecules in the ligation step if the original restriction enzyme is present in the ligation reaction (see Section 3.4). In this manuscript we have used the pairs XhoI/SalI and NotI/PspOMI.
2. Nicking with Nb.BbvCI will require the design of different poly(dT) tailed and blocking oligonucleotides to anneal into that gap.

7. Long DNA constructs for the study of human HELB in C-trap experiments

The specific long DNA constructs described in previous sections were used as substrates to study the activity of human DNA helicase B (HELB). HELB has a central domain sharing homology with the Superfamily 1 (SF1) helicase RecD. RPA protein and DNA binding activities of HELB have been linked to interactions with this central domain (Hazeslip, Zafar, Chauhan, & Byrd, 2020). We and others have shown that HELB possesses single-stranded DNA dependent ATPase activity and 5′-to-3′ helicase activity (Hormeno et al., 2022; Saikrishnan, Griffiths, Cook, Court, & Wigley, 2008; Taneja et al., 2002).

We performed translocation and unwinding experiments using the three kinds of substrates fabricated in this manuscript. The activity of the protein was measured as changes in extension of the DNA tethers or by direct imaging with the confocal microscope coupled to our dual optical tweezers. To visualize HELB protein in confocal fluorescent experiments, we labeled HELB with quantum dots (QDs). Direct conjugation of proteins with biochemically modified QDs has become very useful for single-molecule studies (Nelson, Ali, & Warshaw, 2011). The only requirement is that the protein has a tag (i.e., biotin, Strep-tag, His-tag), to serve as the attachment point for the QDs. The use of QDs has multiple advantages, such as a higher photo stability compared to organic dyes, a higher brightness, and a wide variety of commercially available QDs with different excitation/emission properties or surface coatings. The procedures detailed below describe the process of conjugation of HELB with QDs, and the experiments with the confocal microscope.

7.1 Procedure

7.1.1 Conjugation of biotinylated-HELB with streptavidin-coated QDs

1. Incubate a 1:5 M ratio solution of the biotinylated protein and streptavidin-coated QDs (Qdot™ 525 Streptavidin, Invitrogen) for 30 min on ice. Keep the volume small, 5–10 μL. Gently mix two or three times during the incubation.
2. Add 10 μL 1 mM biotin (Sigma) to neutralize the streptavidin molecules of the QDs not bound to the protein. Incubate 10 min on ice.
3. Dilute the protein-QD mixture in the reaction buffer (20 mM Tris pH 7.5, 30 mM NaCl, 4 mM $MgCl_2$, 5 mM DTT) supplied with 2 mM ATP, to a final volume of 200–300 μL. Keep this volume low to avoid wasting proteins. Typically, the final concentration is 5–20 nM protein-QD.

7.1.2 Mechanical characterization of the specific DNA tether

This procedure is identical to that described for the different types of molecules fabricated: hybrid ss-dsDNA, pure ssDNA, or flap dsDNA (Sections 4.3, 5.3, and 6.3). Fill in channel 4 and 5 with reaction buffer.

7.1.3 Direct visualization of HELB-QD protein on the specific DNA tether

1. Once confirmed the capture in the C-trap of the appropriate DNA tether, fill in channel 4 with 200 μL of fluorescent 5–20 nM HELB-QD in reaction buffer supplied with 2 mM ATP.
2. Move the molecule to channel 4 and hold it at 8–25 pN.
3. Take a kymograph to capture florescence between the beads versus time as described in Section 3.3.4, using excitation light at 488 nm. Labeled HELB particles should be visible mainly in the blue channel employing the blue filter 512/25 nm. Force and position data are recorded simultaneously. Typical line velocity was 50 ms $line^{-1}$.

7.2 Experimental results

7.2.1 Hybrid ss-dsDNA construct

A hybrid ss-dsDNA construct is particularly useful to study translocation on ssDNA and potential unwinding activities beyond the ss-dsDNA junction by DNA helicases and translocases. As an example, we built a hybrid substrate of 2.3 knt ssDNA and 15 kbp dsDNA (Section 4.2) and held the construct at 25 pN force, to avoid the formation of secondary structures on the single-stranded region. At this force, about 1/3 of the distance between the beads corresponds to single-stranded DNA and 2/3 to dsDNA. Once the substrate was successfully produced and its integrity checked (Section 4.3), we moved it to the protein channel and started recording a kymograph following the procedure described in Section 3.3.4. These experiments showed that HELB binds to the ssDNA region and confirmed that it translocates unidirectionally from 5′-to-3′ (towards the dsDNA junction) as reported in (Hormeno et al., 2022) (Fig. 7A). HELB was also able to overcome the junction and unwind the duplex DNA, although occasionally we also observed stalling events at the junction. The unwinding activity was detected by an increase in the extension of the tether at 25 pN, because the extension of single-stranded DNA is larger than dsDNA at that force (Dessinges et al., 2004). The binding and translocation of other HELB proteins on the exposed single-stranded region was also observed.

7.2.2 Pure ssDNA construct

A pure ssDNA construct was produced to study the translocation activity of HELB following the procedure described in Section 5.2. Then, we moved the tether to a channel containing 5 nM HELB-QD and 2 mM ATP and kymographs were recorded at 50 ms line^{-1}. As expected, HELB particles always moved in the same direction on each particular ssDNA molecule (Fig. 7B). This assay cannot define protein translocation polarity—a hybrid substrate is best for this—but it allowed us to determine a translocation rate of $72 \pm 40\,\text{nt}\,\text{s}^{-1}$ within a force range of 13–30 pN (Hormeno et al., 2022).

7.2.3 Flap dsDNA construct

An Alexa 488 labeled 37 nt-flap dsDNA construct was produced to study unwinding activity of HELB following the procedure described in Section 6.2. HELB showed a complex behavior on this kind of substrate. We observed that the protein could unwind the duplex starting from the 5′-tail but it was also able to translocate without preventing re-annealing (Fig. 7C), in a process that resembles a translocation property of the AddAB helicase (Yeeles, van Aelst, Dillingham, & Moreno-Herrero, 2011). Slight modifications of the procedure to fabricate the flap substrate (Section 6.2) should allow us to incorporate other potential loading structures, like overlapping 5′ and 3′-flaps, or a T-junction. A detailed study will be subject of future work.

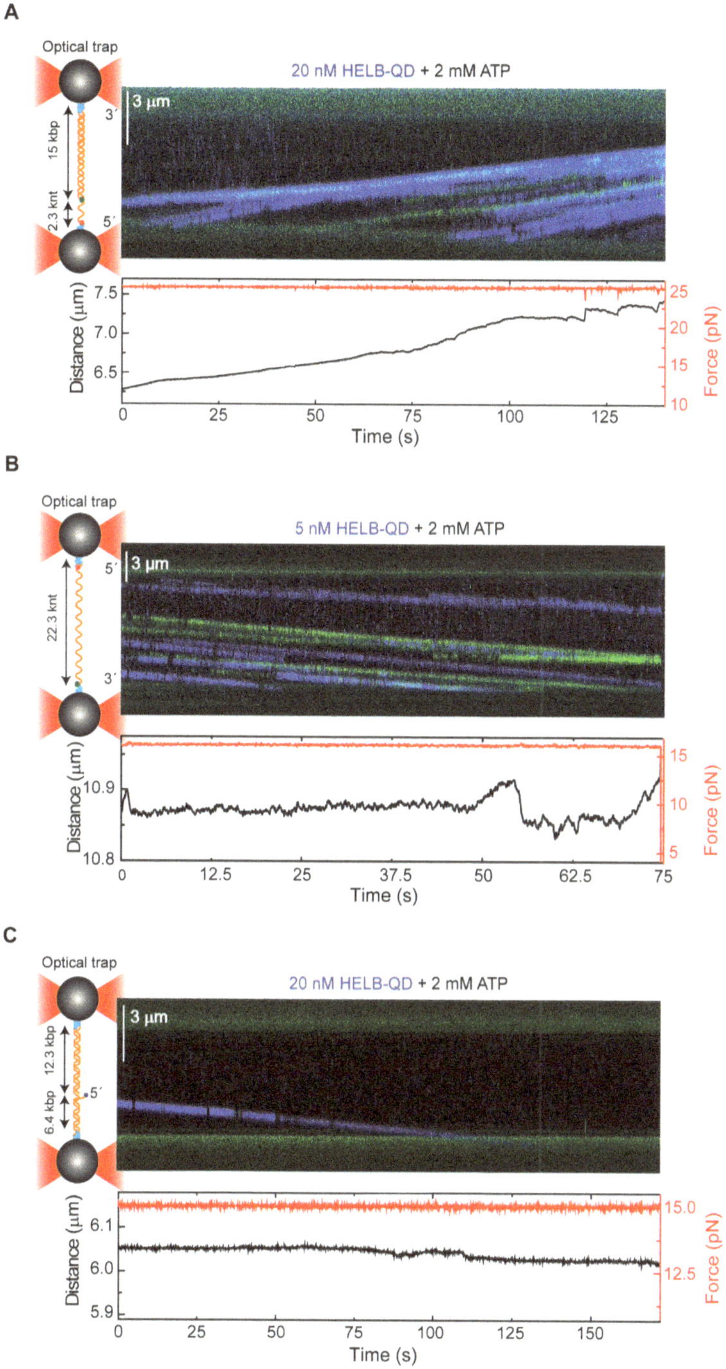

Fig. 7 See figure legend on next page.

8. Summary and conclusions

The DNA construct itself is a central component of any single-molecule experiment aimed at understanding the mechanism of action of DNA-interacting proteins. Moreover, the detection of the activity of multiple motor proteins is based on the mechanical properties of DNA either in its duplex or single-stranded form. The length and sequence of these substrates are two key elements that must be controlled during their design and fabrication. The length is a particularly important issue in assays that involve fluorescence because of the inherent spatial resolution limitations imposed by diffraction.

This chapter describes a general molecular design for customized large DNA plasmids of up to 25 kbp. These plasmids are used to prepare long DNA constructs by simple linearization with selected restriction enzyme/s and ligation to labeled handles. Along similar lines, large plasmids have recently been exploited to produce equivalent customized DNA constructs (Mueller et al., 2020; Sánchez et al., 2021; Spakman et al., 2020; Van Loenhout et al., 2012). However, the reported methods did not address in detail the procedures required to generate long ssDNA or ssDNA-dsDNA hybrid molecules with customized sequences, because most of them were based on Lambda DNA (Belan, Barroso, et al., 2021; Belan, Moore, et al., 2021; Brouwer et al., 2017; Candelli et al., 2013, 2014; King et al., 2013; Smith et al., 1996). The method presented here is very flexible and

Fig. 7 Examples of the use of the DNA constructs to study HELB helicase in C-trap experiments. Biotinylated-human HELB is firstly conjugated with streptavidin-coated QDs and then its activity is studied using the fabricated DNA constructs. The 5′-to-3′ configuration is supported by bulk experiments shown (Hormeno et al., 2022). In each case, the lower panel shows the corresponding distance between the beads and force. (A) Representative kymograph of HELB-QDs movement (blue) on hybrid ss-dsDNA construct shown in Fig. 4C (DNA construct 1, right side) in the presence of 2 mM ATP at 25 pN. HELB requires a ssDNA region to bind the DNA and then translocates in the 5′-to-3′ direction towards the ss-dsDNA junction. Once it reaches the junction, HELB is also able to overcome the junction and unwind the duplex DNA, although sometimes we also observed stalling events at the junction. (B) Representative kymograph of HELB-QDs translocation (blue) on the pure ssDNA construct shown in Fig. 5C in the presence of 2 mM ATP under 16 pN of tension. All HELB molecules move in the same direction. (C) Representative kymograph of HELB-QDs movement on the 37 nt- and Alexa 488 labeled-flap dsDNA construct shown in Fig. 6B. We observed that the protein could unwind the duplex starting from the 5′-tail but cannot prevent re-annealing of the strands, as the movement is not accompanied by a concomitant increase in tether length.

easy to implement in any molecular biology laboratory with a minimum infrastructure. For instance, a new hybrid ss-dsDNA construct with a different proportion of single and double-stranded DNA could be easily generated by choosing other restriction sites to linearize the central part of the molecule. Similarly, following the same strategy, a particular sequence of interest could be positioned at a new location. In addition to pure ssDNA and hybrid ss-dsDNA molecules, we also present a procedure to fabricate long dsDNA molecules containing a flap ssDNA, which are of particular interest to study the activity of helicases. Overall, our method fulfilled the two key requirements of long length and sequence specificity for optical tweezers experiments. Indeed, one of the examples included multiple copies of EcoRI restriction sites (see Table 7), which can be used as helicase roadblocks using EcoRI^{E111G}, a catalytically inactive mutant with exceptionally tight binding (Balaguer et al., 2021; Brüning, Howard, Myka, Dillingham, & McGlynn, 2018; King, Benkovic, & Modrich, 1989). The three substrates presented here were quality-controlled in force extension and fluorescence imaging experiments and validated for use in single molecule experiments using the human HELB protein, demonstrating their enormous potential to study DNA helicases and translocases.

Acknowledgments

We thank Edwin Antony for kindly providing labeled human RPA protein to visualize ssDNA. F.M.-H. acknowledges support from the European Research Council (ERC) under the European Union Horizon 2020 Research and Innovation Program (grant agreement 681299). Work in the Moreno-Herrero laboratory was also supported by Spanish Ministry of Economy and Competitiveness grants PID2020-112998GB-100 (AEI/10.13039/501100011033, to F. M.-H), Comunidad de Madrid grants Tec4-Bio – S2018/NMT-4443 and NanoBioCancer – Y2018/BIO-4747 (to F.M.-H.). Work in the Dillingham laboratory was supported by a Wellcome Trust Investigator Grant (100401/Z/12/Z to M.S.D.) and an Elizabeth Blackwell Early Career Fellowship from the University of Bristol (to O.J.W.).

References

Anand, R., Buechelmaier, E., Belan, O., Newton, M., Vancevska, A., Kaczmarczyk, A., et al. (2022). HELQ is a dual-function DSB repair enzyme modulated by RPA and RAD51. *Nature*, *601*(7892), 268–273.

Balaguer, F.d. A., Aicart-Ramos, C., Fisher, G. L. M., de Bragança, S., Martin-Cuevas, E. M., Pastrana, C. L., et al. (2021). Ctp promotes efficient parb-dependent dna condensation by facilitating onedimensional diffusion from pars. *eLife*, *10*, e67554.

Basu, A., Bobrovnikov, D. G., & Ha, T. (2021). DNA mechanics and its biological impact. *Journal of Molecular Biology*, *433*(6), 166861.

Belan, O., Barroso, C., Kaczmarczyk, A., Anand, R., Federico, S., O'Reilly, N., et al. (2021). Single-molecule analysis reveals cooperative stimulation of Rad51 filament nucleation and growth by mediator proteins. *Molecular Cell, 81*(5), 1058–1073.e7.

Belan, O., Moore, G., Kaczmarczyk, A., Newton, M. D., Anand, R., Boulton, S. J., et al. (2021). Generation of versatile ss-dsDNA hybrid substrates for single-molecule analysis. *STAR Protocols, 2*(2), 100588.

Bell, J. C., & Kowalczykowski, S. C. (2016). Mechanics and single-molecule interrogation of DNA recombination. *Annual Review of Biochemistry, 85*(1), 193–226.

Bell, J. C., Plank, J. L., Dombrowski, C. C., & Kowalczykowski, S. C. (2012). Direct imaging of RecA nucleation and growth on single molecules of SSB-coated ssDNA. *Nature, 491*(7423), 274–278.

Bhattacharjee, A., Wang, Y., Diao, J., & Price, C. M. (2017). Dynamic DNA binding, junction recognition and G4 melting activity underlie the telomeric and genome-wide roles of human CST. *Nucleic Acids Research, 45*(21), 12311–12324.

Bianco, P. R., & Kowalczykowski, S. C. (1997). The recombination hotspot Chi is recognized by the translocating RecBCD enzyme as the single strand of DNA containing the sequence 5′-GCTGGTGG-3′. *Proceedings of the National Academy of Sciences of the United States of America, 94*(13), 6706–6711.

Block, S. M., Goldstein, L. S. B., & Schnapp, B. J. (1990). Bead movement by single kinesin molecules studied with optical tweezers. *Nature, 348*(6299), 348–352.

Bocanegra, R., Plaza, G. A. I., Pulido, C. R., & Ibarra, B. (2021). DNA replication machinery: Insights from in vitro single-molecule approaches. *Computational and Structural Biotechnology Journal, 19*, 2057–2069.

Brahmachari, S., & Marko, J. F. (2018). DNA mechanics and topology. *Advances in Experimental Medicine and Biology, 1092*, 11–39.

Brouwer, I., Zhang, H., Candelli, A., Normanno, D., Peterman, E. J. G., Wuite, G. J. L., et al. (2017). Human RAD52 captures and holds DNA strands, increases DNA flexibility, and prevents melting of duplex DNA: Implications for DNA recombination. *Cell Reports, 18*(12), 2845–2853.

Brown, M. W., De La Torre, A., & Finkelstein, I. J. (2017). Inserting Extrahelical structures into long DNA substrates for single-molecule studies of DNA mismatch repair. *Methods in Enzymology, 582*, 221–238.

Brüning, J. G., Howard, J. A. L., Myka, K. K., Dillingham, M. S., & McGlynn, P. (2018). The 2B subdomain of rep helicase links translocation along DNA with protein displacement. *Nucleic Acids Research, 46*(17), 8917–8925.

Brutzer, H., Luzzietti, N., Klaue, D., & Seidel, R. (2010). Energetics at the DNA supercoiling transition. *Biophysical Journal, 98*(7), 1267–1276.

Bustamante, C., Alexander, L., Maciuba, K., & Kaiser, C. M. (2020). Single-molecule studies of protein folding with optical tweezers. *Annual Review of Biochemistry, 89*(1), 443–470.

Bustamante, C., Bryant, Z., & Smith, S. B. (2003). Ten years of tension: Single-molecule DNA mechanics. *Nature, 421*(6921), 423–427.

Bustamante, C., Chemla, Y. R., Liu, S., & Wang, M. D. (2021). Optical tweezers in single-molecule biophysics. *Nature Reviews Methods Primers, 1*(1), 25.

Camunas-Soler, J., Frutos, S., Bizarro, C. V., De Lorenzo, S., Fuentes-Perez, M. E., Ramsch, R., et al. (2013). Electrostatic binding and hydrophobic collapse of peptide-nucleic acid aggregates quantified using force spectroscopy. *ACS Nano*, 7(6), 5102–5113.

Camunas-Soler, J., Ribezzi-Crivellari, M., & Ritort, F. (2016). Elastic properties of nucleic acids by single-molecule force spectroscopy. *Annual Review of Biophysics, 45*(1), 65–84.

Candelli, A., Hoekstra, T. P., Farge, G., Gross, P., Peterman, E. J. G., & Wuite, G. J. L. (2013). A toolbox for generating single-stranded DNA in optical tweezers experiments. *Biopolymers, 99*(9), 611–620.

Candelli, A., Holthausen, J. T., Depken, M., Brouwer, I., Franker, M. A. M., Marchetti, M., et al. (2014). Visualization and quantification of nascent RAD51 filament formation at single-monomer resolution. *Proceedings of the National Academy of Sciences of the United States of America*, *111*(42), 15090–15095.

Candelli, A., Wuite, G. J. L., & Peterman, E. J. G. (2011). Combining optical trapping, fluorescence microscopy and micro-fluidics for single molecule studies of DNA-protein interactions. *Physical Chemistry Chemical Physics*, *13*(16), 7263–7272.

Dekker, N. H., Rybenkov, V. V., Duguet, M., Crisona, N. J., Cozzarelli, N. R., Bensimon, D., et al. (2002). The mechanism of type IA topoisomerases. *Proceedings of the National Academy of Sciences of the United States of America*, *99*(19), 12126–12131.

Derbyshire, V., Freemont, P. S., Sanderson, M. R., Beese, L., Friedman, J. M., Joyce, C. M., et al. (1988). Genetic and crystallographic studies of the 3′,5′- exonucleolytic site of DNA polymerase I. *Science*, *240*(4849), 199–201.

Dessinges, M. N., Lionnet, T., Xi, X. G., Bensimon, D., & Croquette, V. (2004). Single-molecule assay reveals strand switching and enhanced processivity of UvrD. *Proceedings of the National Academy of Sciences of the United States of America*, *101*(17), 6439–6444.

Doyle, P. S., Ladoux, B., & Viovy, J. L. (2000). Dynamics of a tethered polymer in shear flow. *Physical Review Letters*, *84*(20), 4769–4772.

Dufrêne, Y. F., Ando, T., Garcia, R., Alsteens, D., Martinez-Martin, D., Engel, A., et al. (2017). Imaging modes of atomic force microscopy for application in molecular and cell biology. *Nature Nanotechnology*, *12*(4), 295–307.

Dulin, D., Lipfert, J., Moolman, M. C., & Dekker, N. H. (2013). Studying genomic processes at the single-molecule level: Introducing the tools and applications. *Nature Reviews Genetics*, *14*(1), 9–22.

Eckert, K. A., & Kunkel, T. A. (1991). DNA polymerase fidelity and the polymerase chain reaction. *Genome Research*, *1*(1), 17–24.

Fazal, F. M., Meng, C. A., Murakami, K., Kornberg, R. D., & Block, S. M. (2015). Real-time observation of the initiation of RNA polymerase II transcription. *Nature*, *525*(7568), 274–277.

Fazio, T., Visnapuu, M. L., Wind, S., & Greene, E. C. (2008). DNA curtains and nanoscale curtain rods: High-throughput tools for single molecule imaging. *Langmuir*, *24*(18), 10524–10531.

Fili, N., Mashanov, G. I., Toseland, C. P., Batters, C., Wallace, M. I., Yeeles, J. T. P., et al. (2010). Visualizing helicases unwinding DNA at the single molecule level. *Nucleic Acids Research*, *38*(13), 4448–4457.

Forget, A. L., Dombrowski, C. C., Amitani, I., & Kowalczykowski, S. C. (2013). Exploring protein-DNA interactions in 3D using in situ construction, manipulation and visualization of individual DNA dumbbells with optical traps, microfluidics and fluorescence microscopy. *Nature Protocols*, *8*(3), 525–538.

Gilhooly, N. S., Gwynn, E. J., & Dillingham, M. S. (2013). Superfamily 1 helicases. *Frontiers in Bioscience - Scholar*, *5*(*S*(1)), 206–216.

Gorbalenya, A. E., & Koonin, E. V. (1993). Helicases: Amino acid sequence comparisons and structure-function relationships. *Current Opinion in Structural Biology*, *3*(3), 419–429.

Gross, P., Farge, G., Peterman, E. J. G., & Wuite, G. J. L. (2010). Combining optical tweezers, single-molecule fluorescence microscopy, and microfluidics for studies of DNA-protein interactions. *Methods in Enzymology*, *475*, 427–453.

Gutierrez-Escribano, P., Newton, M. D., Llauró, A., Huber, J., Tanasie, L., Davy, J., et al. (2019). A conserved ATP- and Scc2/4-dependent activity for cohesin in tethering DNA molecules. *Science Advances*, *5*(11).

Ha, T., Kaiser, C., Myong, S., Wu, B., & Xiao, J. (2022). Next generation single-molecule techniques: Imaging, labeling, and manipulation in vitro and in cellulo. *Molecular Cell*, *82*(2), 304–314.

Hashemi Shabestari, M., Meijering, A. E. C., Roos, W. H., Wuite, G. J. L., & Peterman, E. J. G. (2017). Recent advances in biological single-molecule applications of optical tweezers and fluorescence microscopy. *Methods in Enzymology*, *582*, 85–119.

Hazeslip, L., Zafar, M. K., Chauhan, M. Z., & Byrd, A. K. (2020). Genome maintenance by DNA helicase B. *Genes*, *11*(5), 578.

Heiter, D. F., Lunnen, K. D., & Wilson, G. G. (2005). Site-specific DNA-nicking mutants of the heterodimeric restriction endonuclease R.BbvCI. *Journal of Molecular Biology*, *348*(3), 631–640.

Hoekstra, T. P., Depken, M., Lin, S. N., Cabanas-Danés, J., Gross, P., Dame, R. T., et al. (2017). Switching between Exonucleolysis and replication by T7 DNA polymerase ensures high Fidelity. *Biophysical Journal*, *112*(4), 575–583.

Hormeno, S., Wilkinson, O. J., Aicart-Ramos, C., Kuppa, S., Antony, E., Dillingham, M. S., et al. (2022). Human HELB is a processive motor protein that catalyzes RPA clearance from single-stranded DNA. *Proceedings of the National Academy of Sciences of the United States of America*, *119*(15), e2112376119.

Kellermayer, M. S. Z., Smith, S. B., Granzier, H. L., & Bustamante, C. (1997). Folding-unfolding transitions in single titin molecules characterized with laser tweezers. *Science*, *276*(5315), 1112–1116.

Kim, Y., De La Torre, A., Leal, A. A., & Finkelstein, I. J. (2017). Efficient modification of λ-DNA substrates for single-molecule studies. *Scientific Reports*, 7(1), 2071.

King, K., Benkovic, S. J., & Modrich, P. (1989). Glu-111 is required for activation of the DNA cleavage center of EcoRI endonuclease. *Journal of Biological Chemistry*, *264*(20), 11807–11815. Elsevier.

King, G. A., Burla, F., Peterman, E. J. G., & Wuite, G. J. L. (2019). Supercoiling DNA optically. *Proceedings of the National Academy of Sciences of the United States of America*, *116*(52), 26534–26539.

King, G. A., Gross, P., Bockelmann, U., Modesti, M., Wuite, G. J. L., & Peterman, E. J. G. (2013). Revealing the competition between peeled ssDNA, melting bubbles, and S-DNA during DNA overstretching using fluorescence microscopy. *Proceedings of the National Academy of Sciences of the United States of America*, *110*(10), 3859–3864.

King, G. A., Peterman, E. J. G., & Wuite, G. J. L. (2016). Unravelling the structural plasticity of stretched DNA under torsional constraint. *Nature Communications*, 7(1), 11810.

Kong, M., & Greene, E. C. (2021). Mechanistic insights from single-molecule studies of repair of double Strand breaks. *Frontiers in Cell and Developmental Biology*, *9*.

Kuppa, S., Pokhrel, N., Corless, E., Origanti, S., & Antony, E. (2021). Generation of fluorescent versions of Saccharomyces cerevisiae RPA to study the conformational dynamics of its ssDNA-binding domains. *Methods in Molecular Biology*, *2281*, 151–168.

Lee, K. S., Balci, H., Jia, H., Lohman, T. M., & Ha, T. (2013). Direct imaging of single UvrD helicase dynamics on long single-stranded DNA. *Nature Communications*, *4*(1), 1878.

Levikova, M., Klaue, D., Seidel, R., & Cejka, P. (2013). Nuclease activity of Saccharomyces cerevisiae Dna2 inhibits its potent DNA helicase activity. *Proceedings of the National Academy of Sciences of the United States of America*, *110*(22), E1992–E2001.

Li, S., Wasserman, M. R., Yurieva, O., Bai, L., O'Donnell, M. E., & Liu, S. (2021). ORC binds and remodels nucleosomes to specify MCM loading onto DNA. *bioRxiv*. 2021.08.17.456647.

Luzzietti, N., Brutzer, H., Klaue, D., Schwarz, F. W., Staroske, W., Clausing, S., et al. (2011). Efficient preparation of internally modified single-molecule constructs using nicking enzymes. *Nucleic Acids Research*, *39*(3), e15.

Luzzietti, N., Knappe, S., Richter, I., & Seidel, R. (2012). Nicking enzyme-based internal labeling of DNA at multiple loci. *Nature Protocols*, 7(4), 643–653.

Madariaga-Marcos, J., Hormeño, S., Pastrana, C. L., Fisher, G. L. M., Dillingham, M. S., & Moreno-Herrero, F. (2018). Force determination in lateral magnetic tweezers combined with TIRF microscopy. *Nanoscale*, *10*(9), 4579–4590.

Maier, B., Bensimon, D., & Croquette, V. (2000). Replication by a single DNA polymerase of a stretched single-stranded DNA. *Proceedings of the National Academy of Sciences of the United States of America, 97*(22), 12002–12007.

Marin-Gonzalez, A., Vilhena, J. G., Perez, R., & Moreno-Herrero, F. (2021). A molecular view of DNA flexibility. *Quarterly Reviews of Biophysics, 54*, e8.

Miller, H., Zhou, Z., Shepherd, J., Wollman, A. J. M., & Leake, M. C. (2018). Single-molecule techniques in biophysics: A review of the progress in methods and applications. *Reports on Progress in Physics, 81*(2), 024601.

Mueller, S. H., Spenkelink, L. M., van Oijen, A. M., & Lewis, J. S. (2020). Design of customizable long linear DNA substrates with controlled end modifications for single-molecule studies. *Analytical Biochemistry, 592*, 113541.

Nelson, S. R., Ali, M. Y., & Warshaw, D. M. (2011). Quantum dot labeling strategies to characterize single-molecular motors. *Methods in Molecular Biology, 778*, 111–121.

Odijk, T. (1995). Stiff chains and filaments under tension. *Macromolecules, 28*(20), 7016–7018.

Ordu, O., Lusser, A., & Dekker, N. H. (2016). Recent insights from in vitro single-molecule studies into nucleosome structure and dynamics. *Biophysical Reviews, 8*(S1), 33–49.

Paik, D. H., & Perkins, T. T. (2011). Overstretching DNA at 65 pN does not require peeling from free ends or nicks. *Journal of the American Chemical Society, 133*(10), 3219–3221.

Perkins, T. T., Smith, D. E., & Chu, S. (1997). Single polymer dynamics in an elongational flow. *Science, 276*(5321), 2016–2021.

Perkins, T. T., Smith, D. E., Larson, R. G., & Chu, S. (1995). Stretching of a single tethered polymer in a uniform flow. *Science, 268*(5207), 83–87.

Petrosyan, R. (2017). Improved approximations for some polymer extension models. *Rheologica Acta, 56*(1), 21–26.

Saikrishnan, K., Griffiths, S. P., Cook, N., Court, R., & Wigley, D. B. (2008). DNA binding to RecD: Role of the 1B domain in SF1B helicase activity. *EMBO Journal, 27*(16), 2222–2229.

Sánchez, H., McCluskey, K., van Laar, T., van Veen, E., Asscher, F. M., Solano, B., et al. (2021). DNA replication origins retain mobile licensing proteins. *Nature Communications, 12*(1).

Singleton, M. R., Dillingham, M. S., & Wigley, D. B. (2007). Structure and mechanism of helicases and nucleic acid translocases. *Annual Review of Biochemistry, 76*(1), 23–50.

Sitters, G., Kamsma, D., Thalhammer, G., Ritsch-Marte, M., Peterman, E. J. G., & Wuite, G. J. L. (2015). Acoustic force spectroscopy. *Nature Methods, 12*(1), 47–50.

Smith, S. B., Cui, Y., & Bustamante, C. (1996). Overstretching B-DNA: The elastic response of individual double-stranded and single-stranded DNA molecules. *Science, 271*(5250), 795–799.

Smith, S. B., Finzi, L., & Bustamante, C. (1992). Direct mechanical measurements of the elasticity of single DNA molecules by using magnetic beads. *Science, 258*(5085), 1122–1126.

Smith, D. E., Tans, S. J., Smith, S. B., Grimes, S., Anderson, D. L., & Bustamante, C. (2001). The bacteriophage φ29 portal motor can package DNA against a large internal force. *Nature, 413*(6857), 748–752.

Spakman, D., King, G. A., Peterman, E. J. G., & Wuite, G. J. L. (2020). Constructing arrays of nucleosome positioning sequences using Gibson assembly for single-molecule studies. *Scientific Reports, 10*(1), 9903.

Spinks, R. R., Spenkelink, L. M., Dixon, N. E., & van Oijen, A. M. (2021). Single-molecule insights into the dynamics of replicative helicases. *Frontiers in Molecular Biosciences, 8*.

Strick, T. R., Allemand, J. F., Bensimon, D., Bensimon, A., & Croquette, V. (1996). The elasticity of a single supercoiled DNA molecule. *Science, 271*(5257), 1835–1837.

Strick, T. R., Allemand, J. F., Bensimon, D., & Croquette, V. (1998). Behavior of supercoiled DNA. *Biophysical Journal, 74*(4), 2016–2028.

Strick, T. R., Kawaguchi, T., & Hirano, T. (2004). Real-time detection of single-molecule DNA compaction by condensin I. *Current Biology, 14*(10), 874–880.
Svoboda, K., Schmidt, C. F., Schnapp, B. J., & Block, S. M. (1993). Direct observation of kinesin stepping by optical trapping interferometry. *Nature, 365*(6448), 721–727.
Taneja, P., Gu, J., Peng, R., Carrick, R., Uchiumi, F., Ott, R. D., et al. (2002). A dominant-negative mutant of human DNA helicase B blocks the onset of chromosomal DNA replication. *Journal of Biological Chemistry, 277*(43), 40853–40861.
Thorpe, H. M., & Smith, M. C. M. (1998). In vitro site-specific integration of bacteriophage DNA catalyzed by a recombinase of the-resolvase/invertase family. *Proceedings of the National Academy of Sciences of the United States of America, 95*(10), 5505–5510.
Tomko, E. J., Jia, H., Park, J., Maluf, N. K., Ha, T., & Lohman, T. M. (2010). 5'-Single-stranded/duplex DNA junctions are loading sites for E. coli UvrD translocase. *EMBO Journal, 29*(22), 3826–3839.
van den Broek, B., Noom, M. C., & Wuite, G. J. L. (2005). DNA-tension dependence of restriction enzyme activity reveals mechanochemical properties of the reaction pathway. *Nucleic Acids Research, 33*(8), 2676–2684.
Van Loenhout, M. T. J., De Grunt, M. V., & Dekker, C. (2012). Dynamics of DNA supercoils. *Science, 338*(6103), 94–97.
Van Mameren, J., Gross, P., Farge, G., Hooijman, P., Modesti, M., Falkenberg, M., et al. (2009). Unraveling the structure of DNA during overstretching by using multicolor, single-molecule fluorescence imaging. *Proceedings of the National Academy of Sciences of the United States of America, 106*(43), 18231–18236.
Van Oijen, A. M., Blainey, P. C., Crampton, D. J., Richardson, C. C., Ellenberger, T., & Xie, X. S. (2003). Single-molecule kinetics of λ exonuclease reveal base dependence and dynamic disorder. *Science, 301*(5637), 1235–1238.
Wang, H., & Hays, J. B. (2001). Simple and rapid preparation of gapped plasmid DNA for incorporation of oligomers containing specific DNA lesions. *Applied Biochemistry and Biotechnology - Part B Molecular Biotechnology, 19*(2), 133–140.
Wang, M. D., Yin, H., Landick, R., Gelles, J., & Block, S. M. (1997). Stretching DNA with optical tweezers. *Biophysical Journal, 72*(3), 1335–1346.
Wasserman, M. R., Schauer, G. D., O'Donnell, M. E., & Liu, S. (2019). Replication fork activation is enabled by a single-stranded DNA gate in CMG helicase. *Cell, 178*(3), 600–611.e16.
Wen, J. D., Lancaster, L., Hodges, C., Zeri, A. C., Yoshimura, S. H., Noller, H. F., et al. (2008). Following translation by single ribosomes one codon at a time. *Nature, 452*(7187), 598–603.
Wilkinson, O. J., Carrasco, C., Aicart-Ramos, C., Moreno-Herrero, F., & Dillingham, M. S. (2020). Bulk and single-molecule analysis of a bacterial DNA2-like helicase-nuclease reveals a single-stranded DNA looping motor. *Nucleic Acids Research, 48*(14), 7991–8005.
Wilson, M. D., & Costa, A. (2017). Cryo-electron microscopy of chromatin biology. *Acta Crystallographica Section D: Structural Biology, 73*(6), 541–548.
Wuite, G. J. L., Smith, S. B., Young, M., Keller, D., & Bustamante, C. (2000). Single-molecule studies of the effect of template tension on T7 DNA polymerase activity. *Nature, 404*(6773), 103–106.
Yeeles, J. T. P., van Aelst, K., Dillingham, M. S., & Moreno-Herrero, F. (2011). Recombination hotspots and single-stranded DNA binding proteins couple DNA translocation to DNA unwinding by the AddAB helicase-nuclease. *Molecular Cell, 42*(6), 806–816.
Yodh, J. G., Schlierf, M., & Ha, T. (2010). Insight into helicase mechanism and function revealed through single-molecule approaches. *Quarterly Reviews of Biophysics, 43*(2), 185–217.
Zhao, Y., Jiang, Y., & Qi, Z. (2017). Visualizing biological reaction intermediates with DNA curtains. *Journal of Physics D: Applied Physics, 50*(15), 153001.

CHAPTER THIRTEEN

Novel approaches to study helicases using magnetic tweezers

Nadia Ruiz-Gutierrez[a,†], Martin Rieu[a,b,†], Jimmy Ouellet[c], Jean-François Allemand[a,b], Vincent Croquette[a,b,d,*], and Hervé Le Hir[a,*]

[a]Institut de Biologie de l'Ecole Normale Supérieure (IBENS), Ecole normale supérieure, CNRS, INSERM, PSL Research University, Paris, France
[b]Laboratoire de Physique de L'Ecole Normale Supérieure de Paris, CNRS, ENS, Université PSL, Sorbonne Université, Université de Paris, Paris, France
[c]Depixus SAS, 3-5 impasse Reille, Paris, France
[d]ESPCI Paris, Université PSL, Paris, France
*Corresponding authors: e-mail address: vincent.croquette@espci.psl.eu; lehir@bio.ens.psl.eu

Contents

Abstract

Helicases form a universal family of molecular motors that bind and translocate onto nucleic acids. They are involved in essentially every aspect of nucleic acid metabolism: from DNA replication to RNA decay, and thus ensure a large spectrum of functions in the cell, making their study essential. The development of micromanipulation techniques

† These authors participated equally in the writing of this paper.

Methods in Enzymology, Volume 673
ISSN 0076-6879
https://doi.org/10.1016/bs.mie.2022.03.035

such as magnetic tweezers for the mechanistic study of these enzymes has provided new insights into their behavior and their regulation that were previously unrevealed by bulk assays. These experiments allowed very precise measures of their translocation speed, processivity and polarity. Here, we detail our newest technological advances in magnetic tweezers protocols for high-quality measurements and we describe the new procedures we developed to get a more profound understanding of helicase dynamics, such as their translocation in a force independent manner, their nucleic acid binding kinetics and their interaction with roadblocks.

1. Introduction

Helicases are ubiquitous nanometric molecular motors involved in every step of nucleic acid (NA) metabolism and are present in nearly every living branch. They are NA-dependent NTPases that convert chemical energy from nucleoside triphosphates (NTP, mainly ATP) binding and hydrolysis to mechanical energy. This induces conformational changes that lead not only to translocation on single-stranded NA (ssNA) or unwinding double-stranded NA (dsNA) as previously thought, but also to remodeling NA-bound protein structures, or simply binding to NA and serving as protein recruitment scaffolds or NA clamps (Gao & Yang, 2020; Linder & Jankowsky, 2011). Initially, the role of helicases had largely been restricted to duplex unwinding and translocation during DNA replication, repair and recombination (Abdelhaleem, 2010; Brosh & Matson, 2020; Mendoza, Bourdoncle, Boulé, Brosh, & Mergny, 2016), yet, helicases are also required for RNA metabolism such as splicing, maturation, export, translation and decay (Cordin & Beggs, 2013; Khemici & Linder, 2018; Linder & Fuller-Pace, 2015).

It is estimated that around 1% of eukaryotic genes encode a panoply of RNA and DNA helicases, that are tightly regulated by intra- and intermolecular interactions (Lohman, Tomko, & Wu, 2008; Patel & Donmez, 2006; Umate, Tuteja, & Tuteja, 2011; Wu, 2012). Host and viral-encoded helicases play important roles in virus life cycle and infection, and given their vital roles in all cellular aspects, these enzymes are also linked to numerous genetic diseases and cancers. Thus, helicases constitute excellent targets for drug therapies and diagnostic tools (Abdelkrim, Banroques, & Kyle Tanner, 2021; Brosh & Matson, 2020; Datta & Brosh, 2018; Dhar, Datta, & Brosh, 2020; Steimer & Klostermeier, 2012).

The substantial coverage of helicase functions illustrates the importance of studying their mechanisms and regulation. Helicases are classified into six superfamilies (SF1-SF6) differing from their primary sequence, NA preference and polarity, cellular function and structure (Gorbalenya & Koonin, 1993; Singleton, Dillingham, & Wigley, 2007). Superfamilies 1 and 2 share the same characteristic sequence motifs in their helicase core and generally act as monomers or dimers, unlike SF3-SF6 helicases that form hexameric toroidal ring-like structures. SF1 and SF2 include helicases with 5'-3' or 3'-5' polarities, and some subfamilies recognize either exclusively DNA, RNA or both NA.

Despite their structural homologies, each helicase dynamic is unique. Indeed, when studying helicase dynamics many questions arise regarding their behavior: does it act locally or does it translocate onto NA and if so, at which speed, processivity (number of consecutive base pairs translocated) and polarity? What is their affinity and how strong is their grip on double-stranded (ds), single-stranded (ss) or duplex NAs? How are their properties coupled to their function? How are they regulated and how do they interact with roadblocks such as NA-binding proteins or NA tertiary structures?

In order to answer these questions, a combination of bulk assays with single-molecule micromanipulation techniques such as optical tweezers and magnetic tweezers have been employed (Bockelmann, 2004; Greenleaf, Woodside, & Block, 2007; Neuman, Lionnet, & Allemand, 2007; Neuman & Nagy, 2008). Contrary to bulk assays, single-molecule approaches consist in observing in real time the unwinding of individual dsNA by individual helicases. In magnetic tweezers, a NA is tethered on both sides and stretched by applying an external force through a paramagnetic bead subjected to the gradient field of magnets (Fig. 1A). The NA's extension is measured by optically tracking the distance between the surface and the bead. When a helicase translocates along the NA, a portion of dsNA is unwound (Fig. 1B). The real-time position of the helicase along the NA can then be deduced, and thus its velocity, its processivity, and its directionality can be estimated.

Useful practical handbooks for these methods (Bustamante, Chemla, Liu, & Wang, 2021; Vilfan, Lipfert, Koster, Lemay, & Dekker, 2009) as well as several reviews of their application to helicase study have been published (Bianco, 2021; Hodeib et al., 2016; Manosas et al., 2010; Sun et al., 2010; Yodh, Schlierf, & Ha, 2010). Here, we describe new protocols that we designed to investigate further properties of helicases such as their translocation independently of the applied force, their binding grip, and the impact of roadblocks.

The article is organized as follows: we will first provide some tips for successful synthesis of DNA/RNA hairpins and focus on general progress

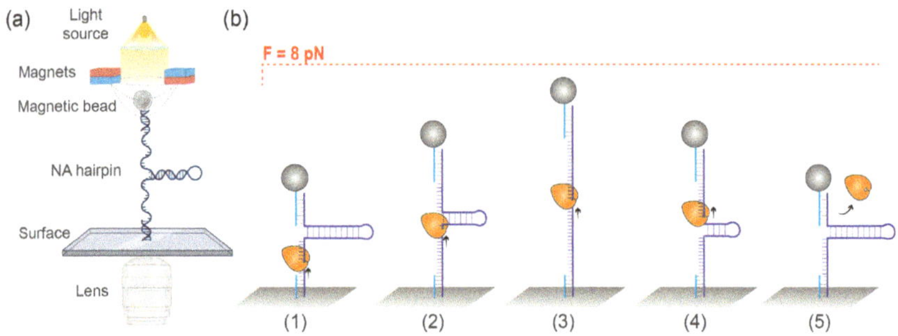

Fig. 1 Standard setup for the single-molecule characterization of helicases with magnetic tweezers. (A) A nucleic acid hairpin is tethered between a coverslip and a superparamagnetic bead. Magnets create a magnetic field gradient applying a force on the molecule through the bead. (B) Progression of a helicase (in orange) through the tethered hairpin at constant force (F = 8 pN) and bead position variation. *Panel (A): Image created with BiorRender.com.*

regarding surface chemistry, NA tethering techniques and *in vitro* purification of recombinant helicases before we detail magnetic tweezers force protocols and the associated data analysis procedures that underlie our most recent advances in helicase properties characterization.

2. Advances in synthesis, attachment and calibration of nucleic acid hairpins for magnetic tweezer experiments

As mentioned before, bulk assays and single-molecule micromanipulations have been largely used for the study of helicases. The latter allow unique measurements on individual molecules, accounting for population diversity in protein batch purifications, contrary to bulk experiments which result in average measurements. Here, we focus on magnetic tweezers experiments for helicase characterization, where a NA hairpin is attached on one side to a glass surface (coverslip) and on the other side to a paramagnetic bead subjected to a force imposed by magnets. The bead's position (Δz, in nanometers) reflects the molecule's extension under the constraint (Fig. 1A). Thus, magnetic tweezers are force-clamps while optical tweezers are position clamps. Many DNA/RNA synthesis and attachment strategies have been developed to study single NA mechanics under force (Janissen, Oberbarnscheidt, & Oesterhelt, 2009; Papini, Seifert, & Dulin, 2019; Riener et al., 2003; Walsh, Wang, & Weimer, 2001; Wildling et al., 2011; Yang et al., 2019) and are now routinely used in the single-molecule community.

All of the helicase experiments that we describe here are based on NA hairpins, self-complementary sequences that unfold at forces between 13 and 20 pN (depending on the NA nature and sequence) and that close between 8 and 14 pN. While short hairpins can now be directly ordered from various providers (up to ~40 base pairs), longer hairpins need to be constructed *in vitro* by attaching tethers and a loop to a dsNA segment. Short synthetic hairpins should be preferred for low-noise experiments while long constructed hairpins can be used when a precision of 2 nm is sufficient. Indeed they allow a better statistical yield, since the larger distance of the bead from the surface reduces the risk of unspecific attachment and allows better identification of well-folded hairpins. As we explain in (Valle-Orero et al., 2022), short hairpins should also be used when the biological question requires that the analysis of a low-processivity enzyme is performed during the closing of the hairpin. In this case the hairpin should be smaller than the enzyme's processivity.

The NA is generally attached to the bead through streptavidin-biotin interactions, due to their biocompatibility, fast association and extremely high dissociation time at zero force ($\tau_0 = 10^5$ to 10^6 s at neutral pH and room temperature (Chilkoti & Stayton, 1995; Wilchek & Bayer, 1990)). This can be estimated using a simple Arrhenius model that gives τ_F the dissociation time under a force F as function of its value τ_0 at 0 pN force, and the distance Δx to the transition state: $\tau_F = \tau_0\, e^{-F\Delta x/kBT}$, $\Delta x \sim 0.2$–1 nm, kBT being the thermal energy (Merkel, Nassoy, Leung, Ritchie, & Evans, 1997; Pincet & Husson, 2005; Williams et al., 2000). This dissociation time can even be made virtually infinite for this range of force compared to experimental timescales by inserting two or three biotins in series (Guo, Ray, Kirkpatrick, Lad, & Akhremitchev, 2008).

For the surface-binding end, there are two choices of NA tethering: a fast and relatively weak digoxygenin/anti-digoxygenin (DIG/AntiDIG) binding (Fig. 2A and B) (fluorescein/anti-fluorescein have also been used, (Bryant et al., 2003)) or a slow (~1 h) but strong covalent binding, the most biocompatible and specific being dibenzylcyclooctyne (DBCO)-azide Click chemistry (Eeftens, van der Torre, Burnham, & Dekker, 2015) (Fig. 2C and D). DIG/AntiDIG binding can be reinforced by functionalizing the NA with several DIG groups but this is not compatible with high-resolution experiments such as helicase stepping. Indeed, the transient binding/unbinding of single groups at the timescale of a few tens of seconds create spurious steps in the nanometer range that pollute the discrete signal of helicases. For faster attachment using Click chemistry covalent bonds, the surface

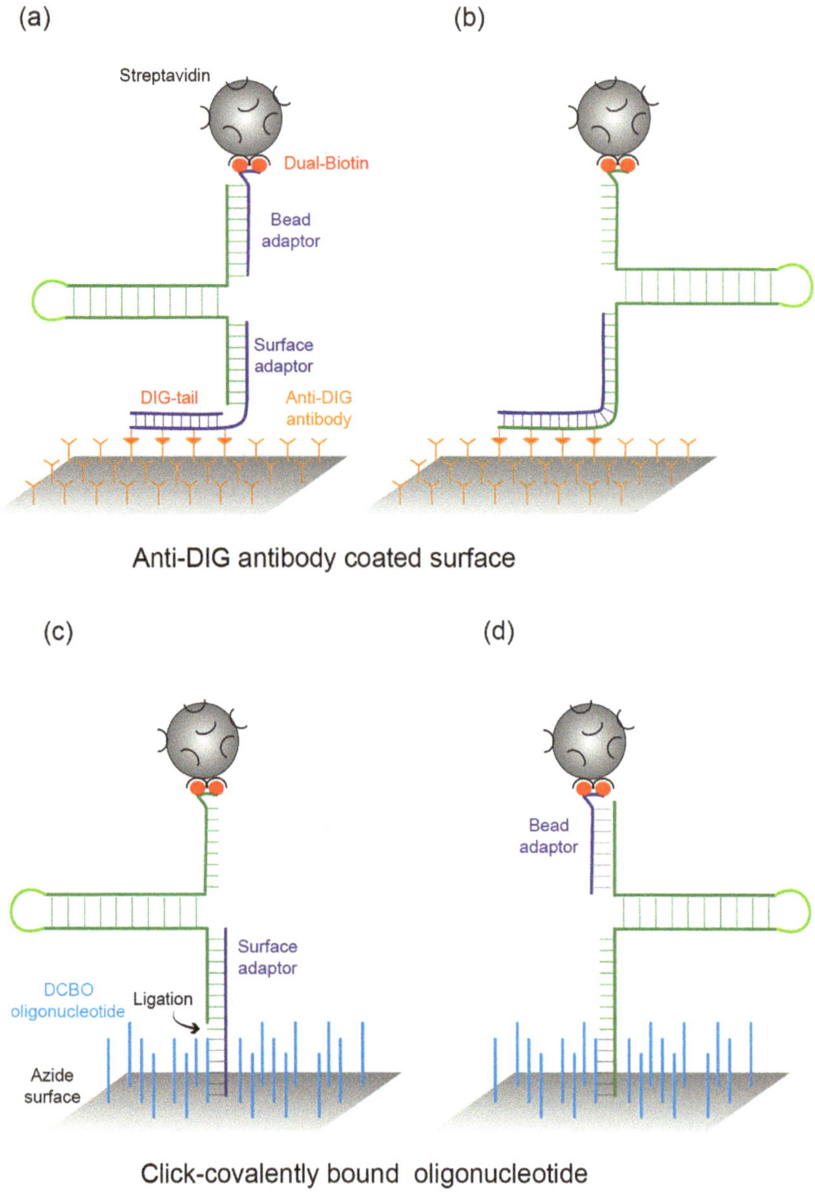

Fig. 2 Different configurations for tethering a NA hairpin between a surface and a magnetic bead in magnetic tweezers. (A, B) DIG-AntiDIG hairpin tethering. The surface is first coated with Anti-DIG antibodies and the adaptor (A) or hairpin (B) are labeled with multiple DIG groups. The hairpin may also be indirectly (A) or directly (B) tethered to the bead through dual-biotinylation. (C, D) Click chemistry hairpin tethering. The azide surface is first coated with a DBCO-labeled ssDNA that is complementary to the adaptor (C) or the hairpin (D). The hairpin may also be directly (C) or indirectly (D) tethered to the bead through dual-biotinylation. In configuration (C), a ligation of the hairpin and the surface oligonucleotide can be performed *in situ* to covalently attach the hairpin to the surface.

can be coated with a DBCO-azide bound oligonucleotide to which the NA of interest can be attached with a complementary sequence of more than 30 base pairs (bp). This allows performing the attachment step only once per microfluidic chamber and then rapidly hybridizing hairpins (Fig. 2D) (or adaptors, Fig. 2C) and removing them by applying denaturing conditions. This solution is tractable for forces smaller than the shearing force of NA duplexes (~45–60 pN (Hatch, Danilowicz, Coljee, & Prentiss, 2008; Li et al., 2021)). For higher forces, the hairpin can also be ligated *in situ* if it is attached through a NA adaptor to the surface oligonucleotide (Fig. 2C). The different attachment options are summarized in Fig. 2.

If the hairpin is indirectly attached by hybridization, the adaptors should be long enough (>50 bp) so as they close behind the helicase passage, avoiding the molecule's detachment by the helicase. When high spatiotemporal resolution is required, such as for helicase stepping experiments or binding assays based on fluctuating hairpins, the tethers should be integrally made of dsNA so their higher stiffness increases the spatio-temporal resolution of the experiment.

Table 1 summarizes the optimal choices of hairpin lengths and attachment modes while the following section describes the protocols to follow to synthesize them.

2.1 Construction of hairpins

All NA sequences used in this section are referenced in Supporting Information Data 1 in the online version at https://doi.org/10.1016/bs.mie.2022.03.035 and their names are indicated in italic.

2.1.1 DNA hairpins

DNA hairpins are built from dsDNA obtained either by direct synthesis of gene fragments by a suitable provider, PCR or plasmid amplification. For applications where it is important to ensure the absence of epigenetic modifications, the first two methods shall be preferred. The dsDNA is then digested by Esp3I (BsmBI), a type IIS restriction enzyme that cleaves outside its recognition site leaving a 5' overhang, generating two different overhangs with a single cutting enzyme. The digested dsDNA is then ligated to the loop structure and the 5' biotinylated Y-shape adaptor which bears the complementary sequences needed for tethering. The obtained DNA hairpin is then ready to be coupled to the streptavidin-coated beads and hybridized to the Click-bound DNA coated on the surface (Fig. 3A).

Table 1 Preferred hairpin lengths and attachment methods as a function of the main objective behind the single-molecule helicase experiment.

Specific aim of the helicase experiment	Preferred hairpin lengths	Preferred attachment mode
Good statistics of velocity and processivity	200 bp ~ several kbp	Hybridized to ssDNA coated on the surface through azide-DBCO covalent binding
High spatio-temporal resolution	10–100 bp	Fully hybridized to ssDNA coated on the surface (no ssDNA left after hybridization). Avoid DIG-antiDIG since it can produce spurious steps in the data
Good control of helicase concentration	Not relevant	DIG-antiDIG attachment of the hairpin to the surface or direct covalent bond of the hairpin to the surface (Eeftens et al., 2015) (to avoid DNA-coating effects on local concentration of enzyme)
Screening of a large number of different sequences (sequence effect on helicase translocation)	10–50 bp: library of synthetic hairpins can be directly bought from providers, preventing time-consuming lab constructions	Hybridized to ssDNA coated on the surface through azide-DBCO covalent binding. (no need of hairpin functionalization and reusable surface for each sequence)

Materials

- Plasmid of interest or synthesized gene fragment (*pPS006*).
- Two primers designed for PCR amplification (*PrimerHP1*, *PrimerHP2*) containing the Esp3I recognition site and leaving 5' overhangs compatible with the Y-shape (*OliBiotin* + *OliCompAdaptor*) and the loop (*OliLoop*).
- Q5 PCR kit.
- Wizard ® SV Gel and PCR Clean-Up System (Promega) or any other DNA purification kit
- 5' Dual-biotin oligonucleotide (*OliBiotin*, chemically synthesized).
- 5' Phosphorylated oligonucleotide complementary to the adaptor to the surface (*OliCompAdaptor*)
- Hairpin loop (*OliLoop*, chemically synthesized).
- T4 DNA ligase kit.

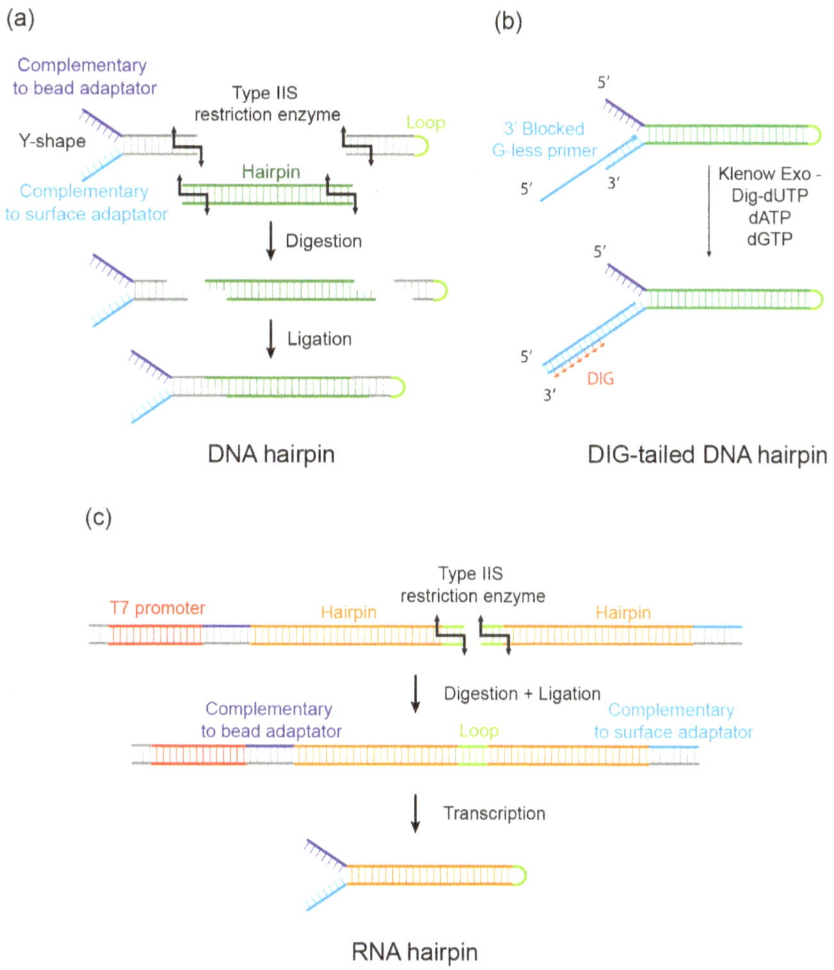

Fig. 3 Representation of the biochemical steps for hairpin construction. (A) Construction of a DNA hairpin. (B) DIG-tailing of a DNA hairpin for DIG/AntiDIG surfaces. (C) Construction of an RNA hairpin.

Protocol

1. Perform PCR protocol according to the primers' characteristics using Q5 polymerase and gel purify the PCR product.
2. Digest the purified fragment with Esp3I one hour at 37 °C.
3. Purify the digestion by column (no gel purification needed).
4. Ligate 100 nM of the product with 1 μM of Y-shape (*OliBiotin* + *OliCompAdaptor*) and 2 μM of *OliLoop* (final concentrations) with T4 DNA Ligase in its buffer in a final volume of 20 μL for one hour at 25 °C.

5. Run the ligation product on a 1.5% agarose gel containing EtBr at 100 V in TAE 1X
6. Cut out the bands corresponding to the product using a blue light table.
7. Extract the DNA from the agarose band either by using an extraction kit or by electroelution to maximize elution efficiency (Davis, Dibner, & Battey, 1986) and resuspend in water.
8. Measure concentration by spectrophotometry.

Tips

- Primers must be designed to have a melting temperature close to 60 °C and to avoid primer dimers.
- For time saving, digestion and ligation can be coupled in the digestion mix using NEB CutSmart Buffer: after digestion at 37 °C, add 1 mM ATP and 0.5 µL T4 DNA Ligase and incubate 2 h at room temperature.
- Other types IIS restriction enzymes such as BsaI can be used.
- Avoid any heating steps during the purification since the melted hairpins may hybridize head-to-tail as dimers, especially when they are highly concentrated.

2.1.2 DIG-tailing DNA hairpins

The following protocol allows the preparation of hairpins compatible with DIG/AntiDIG attachment in microfluidic chambers (Fig. 2C). Briefly, an oligonucleotide complementary to the ss 3' end of the hairpin is hybridized to the hairpin. Its long tail lacking guanosines is polymerized using a Klenow Exo-, a polymerase without strand-displacement activity, in presence of DIG-coupled uridines and results in a poly-DIG-tailed hairpin (Fig. 3B).

Materials

- DNA hairpin synthesized as described in Section 2.1.1.
- Filling DIG oligonucleotide (*Fill-in-Dig-NoG*, chemically synthesized) designed to be complementary to the 3' single-stranded end of the DNA hairpin and a long 5'-end flap containing regularly spaced adenines and no guanosines. The 3' of the oligonucleotide may be blocked with a 3' phosphate in order to avoid 3' polymerization.
- Klenow Exo- and NEBuffer2 (NEB).
- dATP, dGTP and dUTP-Digoxigenin (ThermoFisher).

Protocol

1. Mix 10–100 ng of hairpin with 5 nM of Filling DIG oligonucleotide, 25 nM dATP, dGTP and dUTP-Digoxygenin, 1X NEBuffer 2, with 5U Klenow Exo- in a final volume of 20 µL.

Note: No dCTP nor dUTP are added to the mix.

2. Incubate 1 h at 37 °C.
3. Run the product in a 1.5% agarose gel containing EtBr, cut the band and purify the hairpin by gel extraction column kit.

2.1.3 DNA fluctuating probe

Short DNA hairpins (*HPKinLock*) used for kinetic locking assays (Section 3.4.2) were ordered from Eurogentec in order to have a 10 bp hairpin with a 4 bp loop, making a total of 24 bp. For more information refer to (Rieu, Valle-Orero, Ducos, Allemand, & Croquette, 2021).

2.1.4 RNA hairpin

Here, we describe a fast and reliable construction of an RNA hairpin inspired from (Desai et al., 2019). It consists of the *in vitro* transcription of a dsDNA palindromic sequence. Since a palindromic sequence cannot be easily chemically synthesized, it is obtained by the ligation of two gene blocks, both containing the hairpin sequence. One of the gene blocks contains the T7 promoter for RNA synthesis and the gene block junction corresponds to the tetra-thymidine loop (Fig. 3C). This one-step protocol reduces risks of RNAse contamination.

Materials

- Gene block 1 (*GbforRNA1*, chemically synthesized) designed in order to have a T7 promoter in 5' of the 180 bp hairpin sequence followed by a type IIS restriction enzyme recognition site (BsaI) that leaves a tetra-thymidine overhang.
- Gene block 2 (*GbforRNA2*, chemically synthesized) designed to have the BsaI recognition site in 5' of the 180 bp hairpin sequence.
- BsaI-HF enzyme and CutSmart buffer.
- T7 transcription kit.
- Monarch RNA purification kit (NEB, T2040L).
- Murine RNAse inhibitor (NEB, M0314L).

Protocol

1. Mix 2 pmol of each gene block with 0.5 U of the restriction enzyme, 200 U T4 DNA ligase in 1 mM ATP, 1x CutSmart Buffer at a final volume of 12 μL.
2. Incubate 2 h at 37 °C.
3. Transcribe 2 μL of ligated palindromic DNA according to the T7 Transcription kit manufacturer's protocol in a final volume of 20 μL.

4. Purify the product using by column.
5. Elute with 20 μL RNAse free water.
6. Measure concentration by spectrophotometry.
7. Add 1 μL of the murine RNAse inhibitor before storing at −80 °C.

Tips

- Avoiding RNAse contamination is key to perform long single-molecule experiments with the RNA hairpin. We recommend using gloves and RNAse free water.
- The palindromic DNA and the RNA hairpin can be loaded in a 1.5% Agarose gel with EtBr in order to check their size and integrity.
- Enzymes are not heat-inactivated to avoid RNA/DNA duplexes during renaturation.

2.2 Surface preparation

Here, we describe the two main ways to tether hairpins (or adaptors) to the surface: through Click chemistry or DIG/AntiDIG interaction. Click chemistry corresponds to the attachment of DNA oligonucleotides coupled to a DCBO group on an azide-functionalized surface by producing a stable triazole link (Eeftens et al., 2015). The hairpins (or adaptors) can be then indirectly tethered to the DNA-coated surface by hybridization through complementary sequences and washed away with NaOH, rendering reusable and long-lasting surfaces. The two main disadvantages of Click attachment are that hairpins may dehybridize after helicase passage and, when working with ssDNA-binding helicases, the latter are titrated by the surface DNA resulting in protein concentration underestimation. On the other hand, hairpin attachment to the surface through DIG/Anti-DIG antibody is weaker but can be improved by using poly DIG-tailed hairpins that are directly tethered to anti-DIG antibody coated surfaces (Manosas et al., 2010).

For both preparations a 2.5 mL syringe, a 80 x 120 mm Mylar sheet with two 2 mm holes (50 μm thick, 48-2F-OC, CSHyde), a 35 x 15 mm double-sided channel-containing tape 50 μm thick (ARcare 92712, Adhesive Research), two 10 x 10 mm double-sided tapes with 2 mm holes (ARcare 92712, Adhesive Research), an injection sink and an output connector (homemade in Altuglass) are needed (Fig. 4).

2.2.1 Preparation of ssDNA-coated surfaces through Click chemistry

Materials

- Azide functionalized coverslip (PolyAn 2D Azide).
- 3'-DBCO oligonucleotide 100 nM (*OliDBCO*, chemically synthesized, resuspended in 500 mM NaCl, 10 mM PBS).

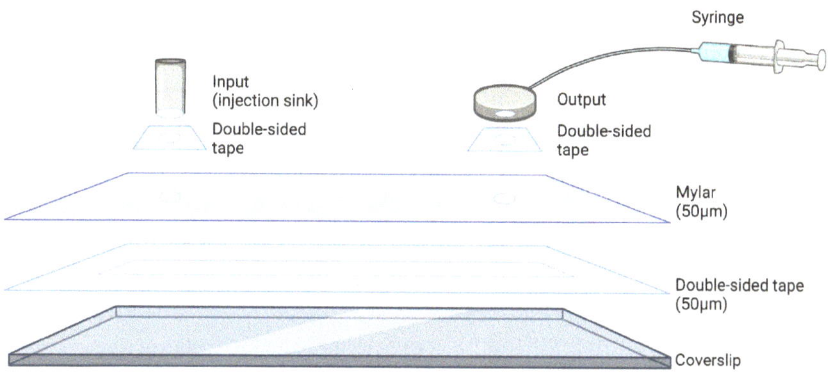

Fig. 4 Assembly of a microfluidic chamber for magnetic tweezers experiments. Double-sided tapes allow the assembly of the microfluidic chamber between a coverslip and a Mylar sheet, bearing a channel through which a constant flow passes. The buffers and enzymes are injected through the input and a syringe connects the output to an outside pump. *Image created with BiorRender.com.*

– Passivation buffer 1X (PB) (140 mM NaCl, 3 mM KCl, 10 mM Na_2HPO_4, 1.76 mM KH_2PO_4, 2% BSA, 2% Pluronic F-127, 5 mM EDTA, 10 mM NaN_3 pH 7.4).

Protocol

1. Thoroughly wash the injection sink, the output connector, the Mylar sheet and the double-sided tapes with water then with Ethanol 70%. Rinse the output tube with water to ensure it is not blocked.
2. Stick one face of the double-sided tape directly onto the coverslip without removing the other side's protection (Fig. 4).
3. Put a 20–40 µL drop of 3'-DBCO oligonucleotide in the middle of the channel. Incubate at room temperature for 2 h until dry.
4. Rinse thoroughly the channel with PB then with water.
5. Once everything is dry, stick the Mylar sheet on the other side of the double-sided tape, making sure the holes are aligned with the channel (Fig. 4).
6. Stick the two double-sided tapes with 2 mm holes aligned with the holes in the Mylar sheet then stick the input sink and the output connector (Fig. 4).

   Microfluidic chambers can be used immediately or up to a month if stored at 4 °C and protected from light.

Tips

– To avoid RNase contamination, do not add BSA in the PB as we noted it often introduces RNases, and beware of dust. Buffers may be filtered, and gloves must be used at all times.

- To wash away hybridized molecules from the surface it is possible to rinse with 15 mM NaOH.
- DNA-binding protein concentrations may be underestimated due to binding to the surface DNA.
- Working with helicases can cause dehybridization of the attached molecule. This can be avoided by adding longer hybridization sequences (>50 bp) that re-hybridize and encircle the helicase after its passage.
- Packs of PolyAn 2D Azide coverslips should be used within a week. If opened, the pack should be stocked at 4 °C, protected from light.
- Never use a sodium azide (NaN_3) containing solution to dilute the DBCO-containing oligonucleotide or it will reduce the yield of the Click-chemistry reaction between the azide surface and the DBCO-labeled DNA.

2.2.2 Preparation of AntiDIG-coated surfaces

Materials

- Non-functionalized coverslip (Schott, NEXTERION® Coverslip custom, #1.5H).
- Sigmacote (Sigma-Aldrich).
- Monoclonal mouse Anti-Digoxin IgG (Jackson Immuno Research, 200-002-156).
- Passivation buffer 1X (PB) (140 mM NaCl, 3 mM KCl, 10 mM Na_2HPO_4, 1.76 mM KH_2PO_4, 2% BSA, 2% Pluronic F-127, 5 mM EDTA, 10 mM NaN_3 pH 7.4).

Protocol

1. Thoroughly wash the injection sink, the output connector, the Mylar sheet and the double-sided tapes with water then with Ethanol 70%. Rinse the output tube with water to ensure it is not blocked.
2. Put a drop of Sigmacote on the coverslip and spread it with a pipette tip. Wait until dry in a covered place to avoid dust from falling on the coverslip.
3. Stick one side of the double-sided tape directly onto the coverslip. Stick the Mylar sheet on the other side. The holes in the Mylar sheet should be aligned in the middle of the tape's channel (Fig. 4).
4. Stick the two double-sided tapes with 2 mm holes aligned with the holes in the Mylar sheet then stick the input sink and the output connector. Check if there are any leaks by flowing water through the channel with the syringe (Fig. 4).
5. Once the microfluidic chamber is ready, inject 40 μL of the antiDIG in the input and gently flow it through the channel with the syringe in

order to cover the whole channel with the antibody. Incubate 3h at room temperature and cover the input sink to avoid evaporation.

6. Rinse the channel with 200 μL PB and let it incubate overnight at 4 °C to saturate unspecific sites.

   Note: Microfluidic chambers can be used for up to a couple of weeks if kept with PB at 4 °C.

Tips

- When sticking the tape, apply slight pressure to ensure proper adhesion of the tape and avoid leaks, but beware of breaking the coverslip.
- Bubbles are an indicator of a leak. If there is a leak, locate it and plug it with glue.
- To avoid evaporation, always cover the input sink with a 1.5 mL Eppendorf tube lid.
- To covalently attach the anti-DIG antibodies to the surface, an epoxy-coated coverslip (Schott, NEXTERION® Coverslip E @1.5H) may be used.

2.3 Bead preparation

Tethering the hairpin to the streptavidin-coated bead can be achieved either indirectly through a dual-biotin labeled oligonucleotide (chemically synthesized) complementary to the hairpin's end, or directly by biotinylating the hairpin itself. The use of dual-biotin increases the interaction affinity in order to stabilize the hairpin's tethering.

2.3.1 Pre-hybridization of hairpin to adaptors

Materials

- Non-biotinylated hairpin (100 nM) produced as described in Section 2.1.
- Biotinylated bead adaptor (*OliBiotin*, 100 nM, chemically synthesized)
- NaCl 1 M or KCl 1 M.
- Passivation buffer 1X (PB) (140 mM NaCl, 3 mM KCl, 10 mM Na_2HPO_4, 1.76 mM KH_2PO_4, 2% BSA, 2% Pluronic F-127, 5 mM EDTA, 10 mM NaN_3 pH 7.4).

Protocol

1. Mix 1 μL of the hairpin with 1 μL of the adaptor and 1 μL of 1 M salt (NaCl or KCl).
2. Incubate 10 min at room temperature for hybridization.
3. Add 97 μL of PB to dilute the concentration to 1 nM.
4. The pre-hybridized hairpin can be immediately hybridized to the beads or stored at −20 °C.

Tips

- Biotinylated adaptors must be added at the same concentration as the hairpin since they cannot be washed after bead hybridization.

2.3.2 Hairpin/bead hybridization

Materials

- Dynabeads MyOne T1 (Thermofisher).
- Hairpin (pre-hybridized to adaptor or not) (10 pM).
- Passivation buffer 1X (PB) (140 mM NaCl, 3 mM KCl, 10 mM Na_2HPO_4, 1.76 mM KH_2PO_4, 2% BSA, 2% Pluronic F-127, 5 mM EDTA, 10 mM NaN_3 pH 7.4).
- If needed (Fig. 2C) adaptor between the hairpin and the surface (*OliAdaptor*, 100 nM)

Protocol

1. Wash 5 μL of beads three times with 200 μL of PB.
2. Resuspend the beads in 19 μL of PB.
3. Add 2 μL of the diluted pre-hybridized hairpin (10 pM), and if needed, 1 μL of *OliAdaptor*.
4. Incubate at least 10 min at room temperature in rotation.
5. Wash three times with 200 μL of PB in order to remove unbound NA.
6. Resuspend in 20 μL of PB (final concentration of DNA, 1 pM).
7. Stock beads on a rotator (10 rpm) at room temperature.

Note: Beads can be used for up to several months if sedimentation does not occur. For RNA beads this estimation may be shorter.

Tips

- The concentration of pre-hybridized hairpin and incubation time may need to be adjusted. Too high attachment to the surface without the possibility to open the hairpin indicates that the beads have more than one NA tethered. On the contrary, low attachment indicates insufficient hairpin tethering to the beads. We sometimes increase the final concentration of hairpins from 1 pM up to 25 pM (stock concentration up to 250 pM).
- Washing away the adaptor oligonucleotide is crucial or free adaptors will saturate the surface. To ensure thorough washing, make sure that less than 5 μL of liquid are left in the tube between each washing step.

2.3.3 Bead injection onto the surface

Materials

- Beads tethered to the hairpin as prepared in Section 2.3.
- Click or DIG surface prepared as described in Section 2.2.

- Passivation buffer 1X (PB) (140 mM NaCl, 3 mM KCl, 10 mM Na_2HPO_4, 1.76 mM KH_2PO_4, 2% BSA, 2% Pluronic F-127, 5 mM EDTA, 10 mM NaN_3 pH 7.4).

Protocol

1. Lift the magnets to remove any magnetic force.
2. Put 200 μL of PB into the input inlet.
3. Apply a flow of 10 μL per minute with the microscope pump.
4. Inject 1 μL of the beads directly onto the cell.
5. When the beads are seen in the field of view, quickly lower the magnets to 10 pN and stop the flow.
6. When the flow is fully stopped, lift the magnets and let the beads sediment for 10 min.
7. Wash the surface with PB at 10 μL/s until all the beads seen in the screen are attached to the surface.

Tips

- Before injection, verify that the syringe of the microscope pump is empty. Avoid emptying it during the experiment which induces flows that detach most of the molecules.
- To avoid bubbles, when injecting the beads, do not press until the second stop of the pipette and regularly check that the input sink is not dry when the flow is on.
- The incubation time of sedimented beads may have to be adjusted if too few or too many beads are attached.
- Once the beads are attached, lowering the magnets to a higher force (20 pN) for a few seconds may help the release of untethered beads.

2.4 Bead selection and calibration

Once the beads are prepared and injected onto the surface, their correct folding must be checked and their position calibrated in order to begin the experimental process.

2.4.1 Selection of properly folded hairpins

As described in (Manosas et al., 2010), beads tethered by a single NA hairpin must be distinguished from ones either non-specifically bound to the surface, not properly folded in hairpins or bound by two or more molecules. The latter need twice as much force to be opened than a single bound hairpin. Typically, DNA hairpins rapidly open around 15 pN and RNA hairpins at 18 pN, so the double-bound beads will not be opened even at the microscope's maximal force. Improperly folded hairpins will show progressive opening and closing, contrary to properly folded ones.

Protocol

1. Select a field of view that includes between 30 and 60 beads that are not aggregated.
2. While alternating the force between 18 and 7 pN for example, visualize the beads that show a z-position change between high and low forces.
3. If the z-position variation of the bead corresponds to the size of the hairpin (around 1 nm for 1 bp) and the opening is abrupt, the bead is bound by a properly folded hairpin and should be selected.
4. If the bead doesn't show movement with force variation it should not be selected.

2.4.2 Bead-to-bead force correction using hairpin opening forces

The force applied on the NA hairpins is crucial for the reproducibility of helicase single-molecule experiments, especially if the helicase's behavior is strongly force-dependent, as seen in the case of the Nsp13 Sars-CoV-2 helicase (Mickolajczyk et al., 2021). Force variation can be calibrated as a function of the distance between the magnets and the microfluidic chamber using the fluctuation of a bead tethered to a relatively long NA (Allemand, 1997; Gosse & Croquette, 2002; Ostrofet, Papini, & Dulin, 2018; Yu et al., 2014). For closed hairpins, where the bead is closer to the surface, these calibration protocols are not accurate due to the change of viscous drag near the surface (Brenner, 1961) and the random attachment points of the molecule with regard to their magnetic axis (Klaue & Seidel, 2009).

Thus, magnets need to be calibrated with long dsDNA molecules which then serve for subsequent experiments with shorter hairpin molecules. However, an uncertainty remains, due to the variation of magnetization from bead to bead, which depends on the heterogeneity of the magnetic bead preparation. This variation of magnetization induces a multiplicative factor between the true force applied on the bead and the one that is estimated through the calibration which is calculated as follows.

Protocol

1. The calibration with long molecules is performed on a large sample of different molecules and beads in order to get an accurate average of the force applied on a typical bead of the preparation.
2. Each subsequent experiment with hairpins should start with a few (~20) cycles of opening and closing of the hairpins at a slow magnet speed (typically 0.2 mm/s) (Fig. 5A).

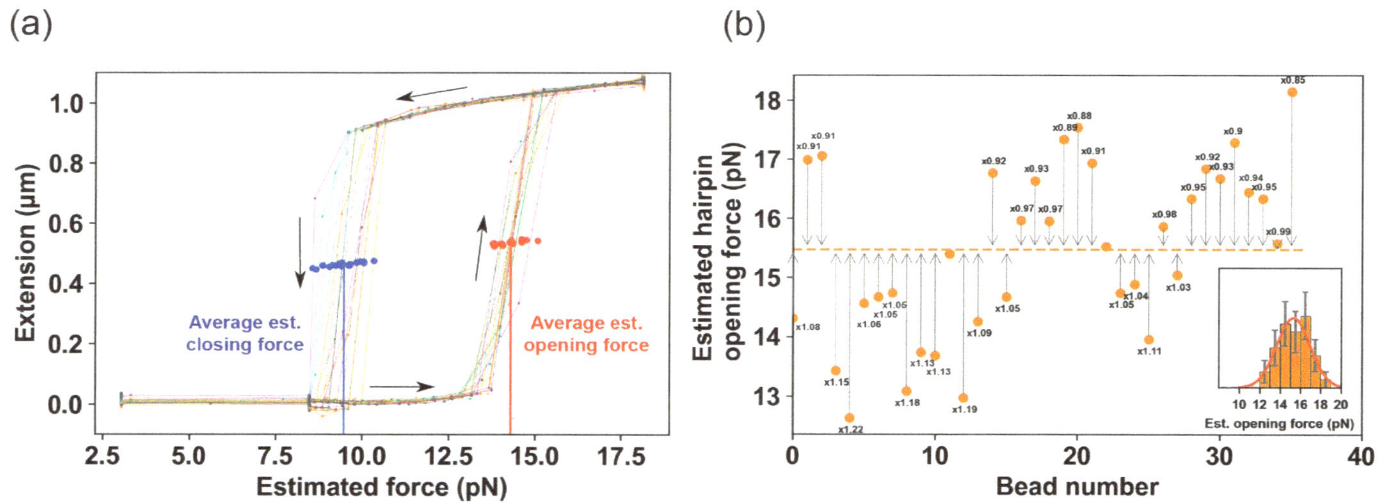

Fig. 5 Force adjustment protocol to correct for the bead-to-bead magnetic heterogeneity. (A) Force extension curve over several force cycles of a 1 kb hairpin tethered to a Dynabead MyOne microbead. The red (blue) points represent the points with the highest time derivative of the extension change during the opening (closing). The average of the abscissa of these points over several cycles gives the estimated (i.e. before the bead-to-bead correction) opening (closing) force for that bead. (B) The estimated opening forces are compared for all beads tracked during a particular experiment. They follow a Gaussian distribution (inset), here of average 15 pN and standard deviation 1.6 pN. The average of the force applied over all beads gives the true opening force F_{true} (dashed orange line) and allows deducing the corrective multiplicative factor for each bead. Buffer: 70 mM KCl, 5 mM $MgCl_2$, 20 mM Tris–HCl pH 7.6.

3. Draw the extension as a function of force during these opening cycles for each bead (Fig. 5B), which follows hysteresis cycles. Then, measure and average the magnet position for which the opening and the closing of the hairpin happen. This value is then translated into estimated opening and closing forces using the force calibration of step 1.
4. Draw the distribution of opening forces for the whole assembly of beads (Fig. 5B inset). The average of the distribution gives the true opening force of the hairpin in the buffer used and at this magnet speed.
5. The multiplicative correction factor is assessed independently for each bead and corresponds to the ratio between the estimated opening force of each individual hairpin and the average opening force of the whole sample. For Dynabeads MyOne, it is typically 10% but sometimes reaches 20% (Fig. 5B). Taking this factor into account is important to avoid force errors.

2.4.3 Nanometers to base pairs conversion

The elasticity of ssDNA is involved in many aspects of the energetic interactions with enzymes. For single-molecule helicase experiments, it also allows the conversion from nanometers (nm) to base pairs (bp). Contrary to dsDNA where this elasticity curve is in complete agreement with the Worm Like Chain model (Bouchiat et al., 1999) and where only two parameters, which have been accurately measured, fully characterize it. ssDNA elasticity is not universal, it depends on sequence and salt (McIntosh, Duggan, Gouil, & Saleh, 2014) and there are fewer experimental characterizations with somewhat diverging parameters (Jacobson, McIntosh, Stevens, Rubinstein, & Saleh, 2017; Viader-Godoy, Pulido, Ibarra, Manosas, & Ritort, 2021). The measurement of the extension of single molecules is made harder by the propensity of ssDNA to unspecifically attach to the bead or the surface. As a consequence, when the extension of a ssDNA is measured, it is possible that a shorter molecule is being analyzed and thus, less bases than expected are being stretched.

To overcome this difficulty, we use a thorough protocol for the measurement of ssDNA force-extension curves (Fig. 6A). This consists of measuring the extension of a hairpin as a function of the force while preventing its closure with a small oligonucleotide complementary to the loop (Fig. 6B). This allows removing the contribution of the handles to the total extension and thus getting rid of the surface effects (Fig. 6C). The protocol shown in Fig. 6 needs to be performed before any single-molecule helicase experiment, in the enzyme's activity buffer, to ensure a thorough estimation of the nm-to-bp conversion coefficient.

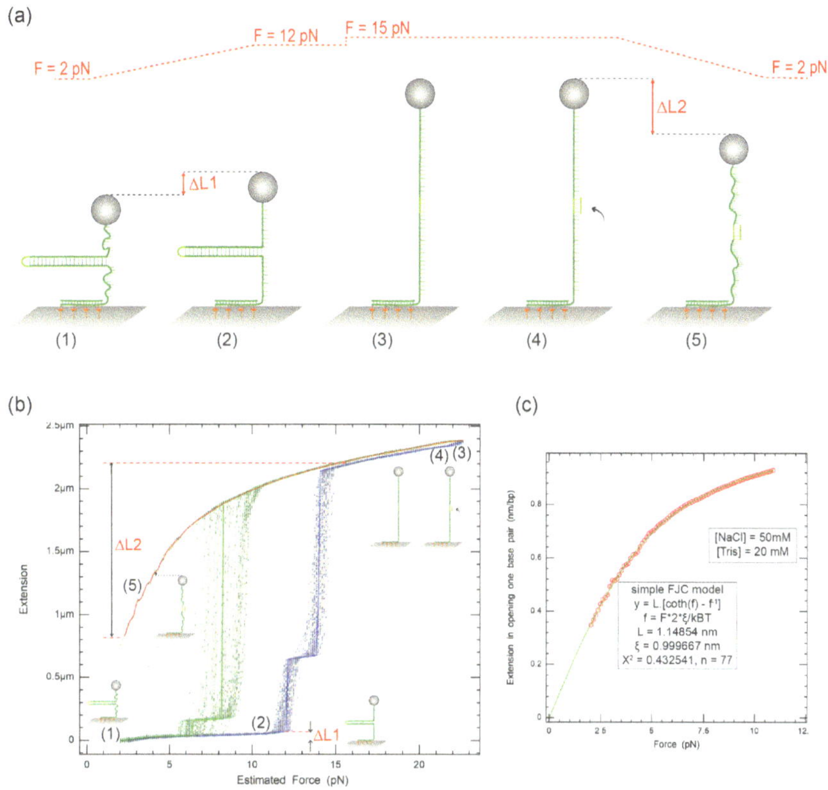

Fig. 6 Protocol for nanometer to base pair conversion. (A) Sketch of the force protocol and the DNA behavior. (B) Superposition of curves representing the extension of a 2129 bp hairpin vs force in a series of cycles with force sweeping from 2 pN to 22 pN. Light gray curves with dots showcase the raw data with typically 100 cycles. The molecule is placed in a 50 mM NaCl buffer in the presence of 100 nM of a 12 bp oligonucleotide complementary to the apex region of the hairpin. The blue curve from (1) to (2) is the average extension of the closed hairpin which reflects the elasticity of the handles used to attach the hairpin to the bead and the surface. The blue curve from (2) to (3) represents the averaged opening behavior of the hairpin. The green curve from (4) to (8) pN corresponds to the elastic behavior of ssDNA, while the red curve represents the same behavior until 2 pN. The red curve corresponds to the situation where the oligonucleotide is hybridized and prevents the hairpin from refolding before the force is low enough to encircle the oligonucleotide or to eject it. (C) The final elasticity curve extracted from the curve in (A). The extension shown in the blue averaged curve from (1) to (2) was subtracted from the red averaged curve and normalized by the number of base pairs in the hairpin (2129 bp). This curve thus corresponds to the extension of two ssDNA nucleotides that were fitted with the Langevin model used for Freely Jointed Chain (Smith, Finzi, & Bustamante, 1992).

2.5 Fast *in situ* three-way junction assembly by hairpin invasion

Assembling three-way junctions directly in the microfluidic chamber has been proven useful to study helicases or helicase-like complexes that are able to regress stalled replication forks into Holliday junctions (Amit, Gileadi, & Stavans, 2004; Bétous et al., 2013; Bianco & Lu, 2021; Dawid, Croquette, Grigoriev, & Heslot, 2004; Manosas et al., 2013; Tan et al., 2017). However, the study of the processivity of such proteins is limited by the size of the synthetic oligonucleotides used to form the junctions. In order to overcome this limitation, we developed a protocol for the fast and simple assembly of three-way junctions of arbitrary size. The protocol is described in Fig. 7A. It consists in using a modified primer (*PrimerSpacer*) to amplify by PCR a segment of dsDNA with a flap. The flap and ds segments are complementary to the hairpin that is tethered between the surface and the bead. Force cycles are applied to the hairpin with the presence in solution of the flapped dsDNA. When the force is lowered below 10 pN, the DNA hairpin can close and the flap of the PCR product can hybridize to the remaining ss region on the handle of the hairpin. Then, the PCR product can invade the tethered hairpin. Indeed, while the hairpin opens and the PCR product hybridizes, the number of hybridized bases stays constant but the bead moves in the direction of the force, which is energetically favored. Thus, the invasion spontaneously occurs, as shown in Fig. 7A, resulting in the formation of the three-way junction. Fig. 7B shows a typical trace of hairpin invasion and the three way junction resolution by RecG as an example.

2.6 Recombinant protein production by double affinity purification

Recombinant proteins used in single-molecule experiments usually go through column chromatography steps (gel filtration, size exclusion, ion exchange…), in order to ensure RNAse free, pure and properly folded active proteins (Bianco, 2021). Yet, due to low yields, this strategy is hard to implement for unstable and lowly expressed proteins bearing unstructured domains. Even if low amounts of protein are needed (10–100 nM) to obtain single events in magnetic tweezers, the double affinity purification strategy is a good compromise between yield and purity. Here, we describe an optimized protocol for this strategy, based on our previous work (Fiorini, Bonneau, & Le Hir, 2012).

Briefly, coding sequences of interest were cloned into a pET28a vector allowing fusing a Calmodulin Binding Peptide (CBP) sequence in 5’ and a

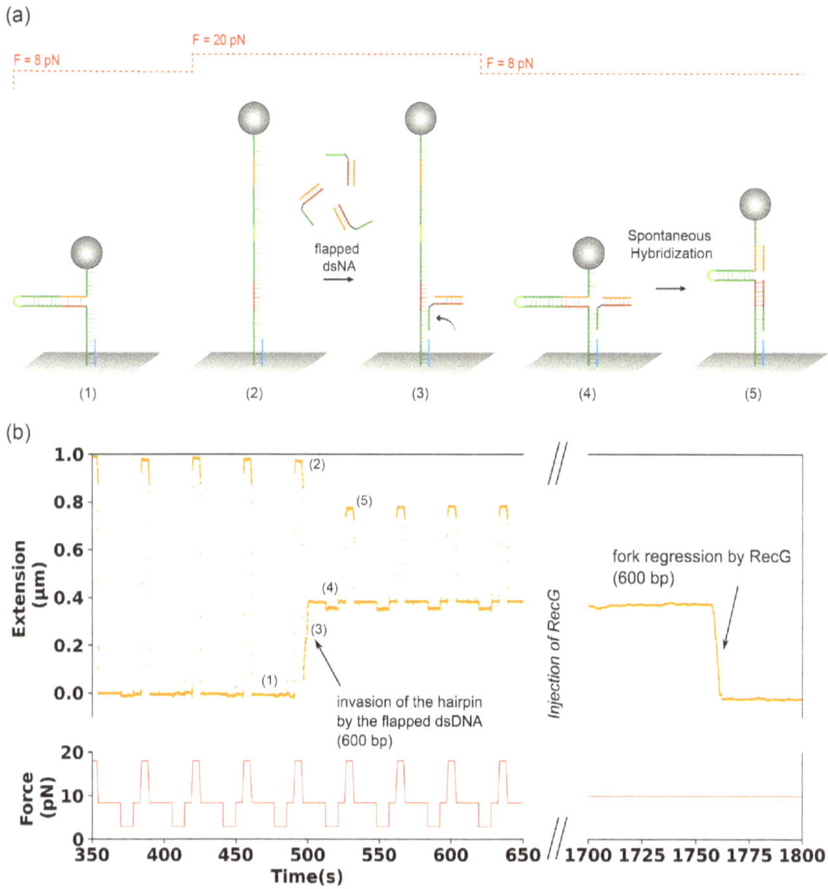

Fig. 7 Protocol for a fast *in situ* assembly of arbitrary long three-way junctions onto a tethered hairpin. (A) Sketch of the experiment. (B) Corresponding experimental traces of RecG. Force cycles (1)–(2) are applied to repeatedly open and close a hairpin in presence of 18 nM of a flapped 600 bp dsDNA complementary to the hairpin, until the dsDNA binds to the hairpin (3) and spontaneously invades the hairpin, at time 510 s, to form a three-way junction (4)–(5). The substrate is then ready to study fork regression by an enzyme such as RecG (1750 s) (Manosas et al., 2013). 5 nM RecG, 25 mM TrisOAc (pH 7.50), 150 mM KOAc, 5 mM MgOAc, 1 mM dithiothreitol and 1 mM ATP.

hexahistidine sequence in 3' of the expressed recombinant proteins. The plasmids were then transformed into BL21(DE3)-CodonPlus *Escherichia coli* competent bacteria by heat shock and selected on kanamycin containing Luria Broth (LB) plates. Bacteria were then pre-cultivated in LB liquid media containing kanamycin for 4 h at 37 °C and then used for inoculation of 1 L of LB Broth Auto Inducible (LBAI) media containing metals.

Bacteria were then left to grow overnight at 18 °C. After bacterial lysis, the double purification was performed first on Nickel-NTA resin then on Calmodulin-coupled resin. Proteins were dialyzed against a storage buffer and frozen at −80 °C. One advantage of this technique is that, by purifying by both extremities, only full-length proteins are selected and thus, no size exclusion chromatography is needed. This protocol can be adapted for the use of other tags such as GST or Twin-Strep instead of CBP, the fusion of highly structured tags may help the solubility of unstructured proteins.

2.6.1 Protein overexpression

1. Transform 100 μL of BL21(DE3)-CodonPlus competent *Escherichia coli* bacteria (Agilent) with 500 ng of plasmid by 2 min at 42 °C heat shock. Add 1 mL liquid LB media and incubate 1 h at 37 °C. Pellet the cells by 4 min centrifugation at 3500 × *g*. Resuspend the pellet in 50 μL LB media, plate in kanamycin containing LB plates and leave to grow overnight at 37 °C.

   Tip: Other competent bacteria designed for protein overexpression such as BL21(DE3)-Rosetta strain may be tested depending on the expressed proteins.
2. Resuspend 2–3 colonies in 25 mL LB liquid media containing kanamycin in a 250 mL Erlenmeyer flask and incubate at 37 °C, 200 rpm for 4 h.
3. Inoculate the culture in 1 L of LBAI containing metals (Formedium) added with kanamycin and chloramphenicol in a 5 L Erlenmeyer flask and incubate at 18 °C, 200 rpm overnight.

   Tip: Other culture media may be used depending on the expressed proteins. For non-inducible media, induction is performed by addition of 0.2–0.5 mM of IPTG when optic density reaches the exponential phase. Although we highly recommend the use of auto-inducible media (Luria Broth or Terrific Broth), with or without metals, protein yields may vary upon culture conditions.
4. Pellet the culture by 10 min, 4000 × *g* centrifugation at 4 °C.

   Tip: Pellets can be stored at −20 °C. However, we recommend using fresh cultures for purification.

2.6.2 First affinity purification

All following steps must be performed at 4 °C.

1. Resuspend the pellet in 25 mL of Lysis buffer (1.5X PBS, 1 mM $MgAc_2$, 0.1% NP-40, 20 mM imidazole, 10% glycerol, 2 mM DTT).

2. Add 1X of protease inhibitors (Apoprotinine, Leupeptine, PMSF, Pepstatin).
3. Sonicate in a metal beaker at 30% amplitude and 1 s pulse for 4 min.
   Tip: So temperature does not exceed 15 °C, the beaker can be placed on a mix of dry ice and ice.
4. Pellet the bacterial debris for 30 min at 18,000 rpm.
5. Retrieve the supernatant and mix in a Falcon tube with 1 mL of Nickel-NTA Agarose slurry (Qiagen) equilibrated with Lysis Buffer. Incubate in a rotator for 1 h at 4 °C.
6. Centrifuge at 4,000 rpm for 5 min at 4 °C to pellet the beads.
7. Remove the supernatant (Flow Through) and wash with 25 mL of Lysis Buffer.
8. Transfer into a Poly-Prep chromatography column (Bio-Rad) pre-equilibrated with 1 mL of Lysis Buffer.
9. Wash the column with 5 mL of Lysis Buffer then with 5 mL of Wash Buffer (1.5X PBS, 250 mM NaCl, 1 mM $MgAc_2$, 0.1% NP-40, 50 mM imidazole, 10% glycerol, 2 mM DTT) then again with 5 mL of Lysis Buffer.
10. Close the column and incubate the resin with 1 mL of Elution Buffer (1.5X PBS, 1 mM $MgAc_2$, 0.1% NP-40, 250 mM imidazole, 10% glycerol, 2 mM DTT) for 10 min at 4 °C.
11. Elute successively with 1 mL of Elution Buffer and measure protein concentration by Bradford Assay.
12. Pool the fractions with highest concentration and dialyze overnight against 1 L of Calmodulin Binding Buffer (CBB) (10 mM Tris–HCl pH 7.5, 250 mM NaCl, 1 mM $MgAc_2$, 4 mM $CaCl_2$, 0.05% NP-40, 10% glycerol, 2 mM DTT) at 4 °C.

2.6.3 Second affinity purification

1. Retrieve the dialyzed fractions and mix with 500 μL of Calmodulin Sepharose 4B resin (50% Slurry, GE Healthcare) washed with CBB. Incubate in a rotor for 2 h at 4 °C.
2. Transfer into a Poly-Prep chromatography column (Bio-Rad) pre-equilibrated with 1 mL of CBB.
3. Wash with 5 mL of CBB.
4. Close the column and incubate with 500 μL of Calmodulin Elution Buffer (10 mM Tris–HCl pH 7.5, 250 mM NaCl, 1 mM $MgAc_2$, 0.05% NP-40, 20 mM EGTA, 10% glycerol, 2 mM DTT) for 10 min at 4 °C.

5. Elute successively with 500 μL of Calmodulin Elution Buffer and measure protein concentration by Bradford Assay.
6. Pool the highest concentrated fractions and dialyze overnight against 1 L of Storage Buffer (1.5X PBS, 1 mM $MgAc_2$, 10% glycerol, 2 mM DTT).
7. Retrieve the dialyzed proteins and aliquot them in small volumes (10–100 μL, depending on the final protein concentration). Store at −80 °C.

Tips

- 10 μL of Supernatant, Pellet, Flow Through Histidine, Wash Histidine, Elution Histidine, Dialysis Histidine, Flow Through Calmodulin, Elution Calmodulin and Dialysis Calmodulin may be kept for loading a 10% SDS-PAGE gel, revealed by Coomassie Blue coloring in order to visualize the different purification steps.
- Reducing agents must be added to the buffers extemporaneously.
- Buffers may have to be adapted for each purified protein depending on their stability. Some proteins may need more or less salt, no glycerol, or different pH.
- If the protein precipitates during dialysis, centrifuge and measure the concentration again.
- For higher reproducibility during single molecule experiments, aliquots should be only thawed once before each experiment. Leftovers should be discarded after thawing.

3. Advanced protocols for helicase characterization using magnetic tweezers

In previous reviews (Hodeib et al., 2016; Manosas et al., 2010), we described classical magnetic tweezers protocols for helicase studies and explained how the processivity, velocity and directionality can be extracted by dynamically measuring the extension of a hairpin unwound by the helicase. Here, we will focus on recent experimental advances we developed for measuring other helicase properties.

First, we will describe two assays for measuring translocation of a helicase in a force-independent manner: the peeling assay and the bubble assay. Then, we will focus on the binding capacities of helicases by describing a qualitative binding assay using force cycles and a more quantitative assay using a fluctuating short hairpin. The latter allows measuring binding kinetics independently of their unwinding activity and in a fluorescence-free manner. Lastly, we will focus on the impact of roadblocks, either NA-binding proteins or NA tertiary structures, on the translocation of helicases by embedding the roadblock inside a hairpin and monitoring the time they take to either resolve or bypass the obstacle.

3.1 Mitigating the impact of force on helicase translocation

In classical magnetic tweezer assays, a constant force is applied. Typically, in order to observe a good quality signal, the applied force lies between 6 pN and 12/18 pN (DNA/RNA). At higher forces, the hairpin spontaneously opens, preventing the observation of helicase unwinding activity and at lower forces, the signal-to-noise ratio is too low. This assay draws criticism concerning passive (force-dependent activity) and active (force-independent activity) helicases. Indeed, if the helicase is passive (such as polymerase-coupled helicases) it is possible that at zero force, no activity would be measured. Active helicases, on the other hand, which would display activity even at zero force, may also be influenced by the applied force. Hence, the observed activity in these assays may differ from the helicase's behavior as measured in bulk assays. In order to go beyond this assumption, unwinding activity can be probed in other configurations. Here, we propose two assays to mitigate the impact of force on helicase assays in magnetic tweezers: the peeling assay and the bubble assay.

3.2 Peeling assay

The peeling assay consists of hybridizing a lateral oligonucleotide on the side of the main hairpin, on which the helicase translocates (Fig. 8A). In this configuration, an increasing force hinders the unwinding, instead of favoring it since the bead moves opposite to the force when the enzyme unwinds the hairpin. The unwinding helicase activity is thus in an energetic context closer to zero-force bulk experiments. It is noteworthy that this configuration induces a specific biochemical context in which the helicase is confronted with two forks, which might affect its translocation. Fig. 8B (inset, 1235 s) shows an example of peeling assay of the catalytic core of *E.coli* RecQ, RecQ-ΔC. Note that after the oligonucleotide is peeled, the helicase is able to translocate directly onto the hairpin as in classical constant-force experiments (1425–1432 s).

Protocol

1. Using a microfluidic chamber (prepared as described in Section 2.2), run a force cycle for successive hairpin openings and closings (for example 2 s at 18 pN, 4 s at 8 pN) and start the recording of the beads' positions.

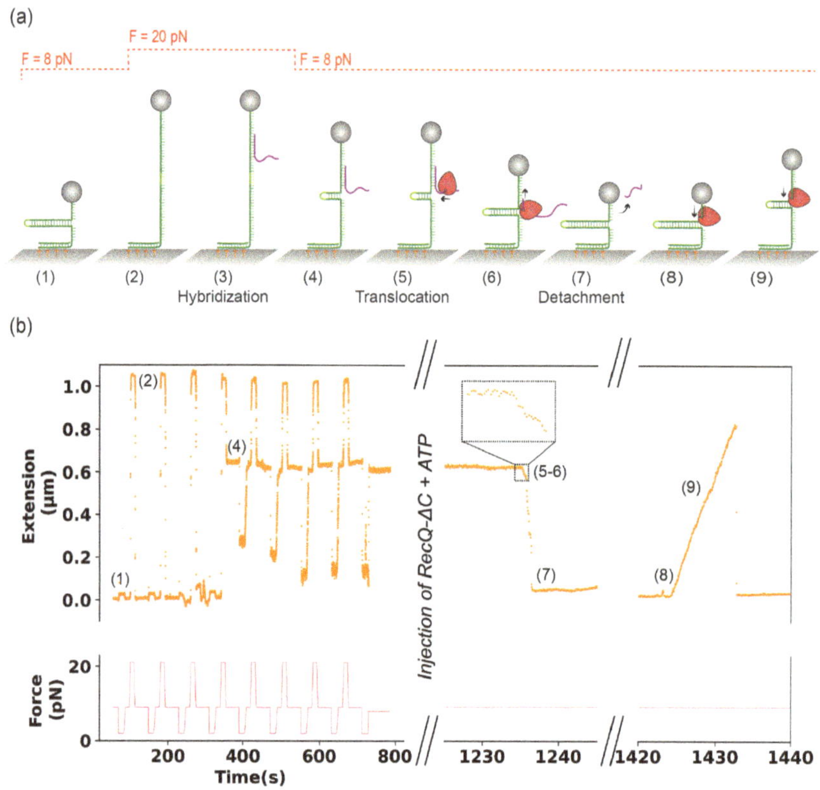

Fig. 8 Peeling assay: an alternative experiment to mitigate the impact of the force on helicase translocation. (A) Sketch of the experiment. (B) Corresponding experimental traces of ReqQ-ΔC. Force cycles (1)–(2) are applied to repeatedly open and close a hairpin in presence of 100 nM of a 3′ flapped ssDNA complementary to the hairpin, until the ssDNA binds to the hairpin (3). The hairpin is then closed (4) and RecQ-ΔC injected at a concentration of 100 pM (5). The helicase binds to the flap and unwinds the oligonucleotide, resulting in the progressive closing of the hairpin (6 and inset of the trace), until it completely removes the oligonucleotide at 1235 s with an unwinding velocity of 97 ± 10 bp/s (7). At 1425 s, another enzyme binds and unwinds the hairpin in the standard configuration with an unwinding velocity of 95 ± 0.3 bp/s (8–9). Tris–HCl (pH 7.60), 50 mM NaCl, 3 mM $MgCl_2$, 1 mM DTT and 1 mM ATP.

2. Remove the buffer from the microfluidic chamber inlet with a pipette. Rinse the inlet three times with 200 µL of reaction buffer (no ATP, no enzyme).
3. Flow 160 µL of reaction buffer at a constant speed of 10 µL/min to rinse the microfluidic channel.
4. Flow 100 µL of reaction buffer + 100 nM of complementary oligonucleotide.

5. Wait for the oligonucleotide to be hybridized to all hairpins. This results in a partial blocking of the hairpins at 7 pN in opening/closing cycles.
6. Repeat steps 2–3 to remove the excess oligonucleotides.
7. Stop the force cycles and measure the beads' positions at constant medium force (F = 8 pN).
8. Add the buffer mixed with the enzyme and 1 mM ATP to the inlet of the chamber. Flow at a constant speed of 5 μL/min.
9. Record data during the desired period of time.

Tips

- Start the thermalization of your setup early enough before the recording, at the desired temperature to avoid slow measurement drifts due to temperature changes.
- In addition to adding reducing agents such as TCEP or DTT to the solution, you may degas the buffer to minimize protein oxidation: leave the tube containing the buffer without ATP nor enzyme in a vacuum desiccator at room temperature for 10 min.
- Between each rinsing step, carefully remove most of the buffer from the inlet. Only a few μL should remain in the inlet so that each rinsing step allows the dilution of the previous buffer by a factor ~100.
- When preparing the enzyme mix, gently stir with the pipette tip. Avoid the insertion of air and bubble formation when mixing in order to avoid the buffer's fast re-oxygenation as well as the concentration of proteins at air-water interfaces.

Data analysis

1. Find the unwinding events in the data: they consist of a progressive decrease of the bead extension followed by an abrupt rezipping of the hairpin once the full oligonucleotide has been removed.
2. Measure the speed of the enzyme in nm/s for each event by dividing the total loss of extension during the progressive phase by the duration of this phase.
3. Convert the speed from nm/s to bp/s. The conversion factor is $l_{ssNA}(F) + l_{dsNA}(F)$, since one unwound base results in removing one ds base and one ss base from the extended molecule.

3.3 Bubble assay

This assay is performed using a long hairpin (>1 kb) that is able to rehybridize after the helicase's passage and thus encircle the helicase in a "bubble." By periodically changing the force as shown in Fig. 9A, the hairpin is open at high force (>15 pN), then at medium force (~7–8 pN), it starts

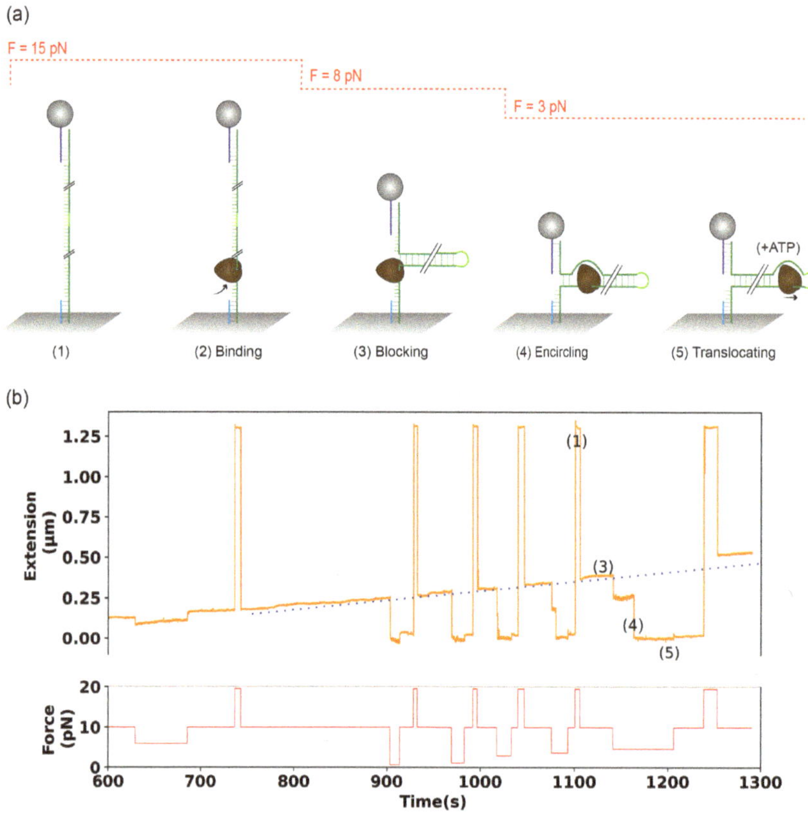

Fig. 9 Bubble assay: an alternative protocol to measure helicase progression inside a dsDNA at zero force. (A) Sketch of the experiment. (B) Corresponding experimental traces of human UPF1 (hUPF1) helicase core. Force cycles are applied to repeatedly open (1), close (3) and encircle the helicase (4). The translocation velocity of the helicase stays approximately constant over the whole experiment (blue dashed line, linear fit of steps (3) where the position of the bead depends on the position of the helicase). Thus, hUPF1 keeps translocating roughly at the same speed when it is encircled. Note that the time it takes for hUPF1 to be encircled increases with the encircling force. Buffer: 20 mM Tris–HCl pH 7.5, 75 mM KOAc, 3 mM $MgCl_2$, 1% BSA, 1 mM DTT, 1 mM ATP, 30 nM yUPF1-HD.

to close. When the enzyme is bound, the hairpin will only partially close until the enzyme's position. By imposing a low force (3–4 pN), the hairpin completely closes and encircles the helicase (if it's not ejected, see Section 3.4.1). At this step, no force is applied inside the hairpin, meaning that the helicase's behavior while it is encircled is measured at zero force. By comparing the bead's position over successive cycles, it is possible to extrapolate the helicase's translocation while encircled. Fig. 9B, illustrates an example of peeling assay on the human helicase core of Upf1.

Protocol

1. Follow steps 1–3 from protocol in Section 3.2.
2. Stop the force cycles and choose the force cycles for the experiment. For example 2 s at 15 pN, 5 s at 8 pN, 5 s at 3 pN.
3. Add the buffer mixed with the enzyme and 1 mM ATP to the inlet of the chamber. Flow at a constant speed of 5 μL/min.
4. When the inlet is almost empty, repeat step 3.
5. Record data during the desired period of time.

Data analysis

1. Measure the binding rate k_{on} of the enzyme to the substrate by drawing the distribution of times elapsed between the injection of the enzyme and the first blockage or the first unwinding event. These times will typically follow exponential distributions, of parameter $\tau = 1/k_{on}$.
2. If the binding rate is large compared to the time spent at high force, where the presence of the enzyme cannot be probed, decrease the enzyme concentration to ensure that a signal observed in two successive cycles is the same enzyme.
3. Compute the probability of encircling the enzyme during re-hybridization by dividing the number of events where the enzyme is present both before and after the re-hybridization step by the total number of events where the enzyme is present before the re-hybridization step.

3.4 Estimating NA binding dynamics of helicases

3.4.1 Qualitative single-molecule binding assay (SMBA)

One important aspect of helicase dynamics is their binding to NA and their residence time. Standard single-molecule unwinding assays allow measuring the dissociation time of the helicase from its NA substrate in the presence of ATP, when the enzyme translocates. However, in order to decouple this dissociation from the conformational changes induced by ATP hydrolysis, it is sometimes useful to be able to measure and compare the grip of the helicase on its NA substrate in absence of ATP.

In Kanaan et al. (2018), we described an ATP-free qualitative single-molecule binding assay (SMBA, Fig. 10B and C), to estimate helicase binding to NA and their residence time. In this assay, the protein is injected without ATP allowing binding but not translocation. Force cycles are then applied to test the persistence of the helicase despite the mechanical constraints of a closing hairpin. First, a high force opening the hairpin (>15 pN for DNA and

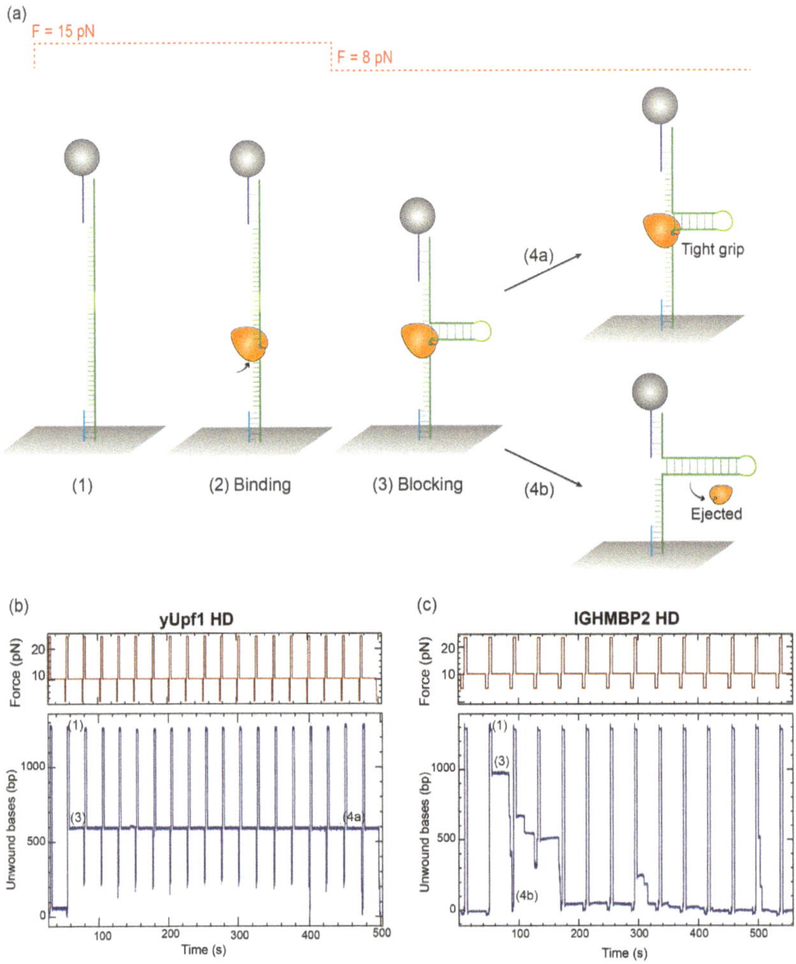

Fig. 10 Sketch of the single molecule binding assay (SMBA) protocol allowing qualitative estimation of a helicase's grip onto nucleic acids. (A) Sketch of the experiment. (B, C) Corresponding experimental traces. The numbers corresponding to the hairpin state in panel (A) were added to the traces. Yeast UPF1 catalytic helicase core (yUpf1 HD) and IGHMBP2 helicase domain were subjected to force cycles without the addition of ATP. yUPF1 HD blocks the hairpins closure with its tight grip, unlike IGHMBP2 HD that slides and is ejected by the hairpins closure. *Panels (A) and (B): Reproduced from Kanaan, J., Raj, S., Decourty, L., Saveanu, C., Croquette, V., & Hir, H. L. (2018). UPF1-like helicase grip on nucleic acids dictates processivity.* Nature Communications, 9*(1), 3752. doi:10.1038/s41467-018-06313-y, in agreement with the Creative Commons Attribution 4.0 International License (http://creativecommons.org/licenses/by/4.0/)*

>18 pN for RNA) is applied for a couple of seconds, the helicase is thus able to bind randomly anywhere in the single-stranded regions. Then, a medium force (~7–8 pN) induces the hairpin's closure. The closing fork pressing against the bound helicase can have two outcomes: if the helicase's grip is strong enough, the hairpin is blocked at the helicase's position, else, the helicase is ejected and the hairpin completely closes (Fig. 10A). The ejection time can be almost immediate or last a few tenths of seconds. Thus, this step is held for 20–30 s or more. Fig. 10B and C shows a qualitative comparison of the grip of two different catalytic cores from Superfamily 1 helicases: yeast UPF1 and yeast IGHMBP2 (Kanaan et al., 2018).

Protocol

1. Follow steps 1–3 from protocol in Section 3.2.
2. Stop the force cycles and choose the force cycles for the experiment: 1–3 s at 15–18 pN, 20–30 s at 7–8 pN.
3. Add the buffer mixed with the enzyme without ATP to the inlet of the chamber. Flow at a constant speed of 5 μL/min.
4. When the inlet is almost empty, repeat step 3.
5. Record data during the desired period of time.

Tips

- In order to avoid multiple helicases binding to the same substrate, the lowest concentration possible is injected and can vary from 1 to 20 nM, depending on protein activity.

Data analysis

1. The binding time corresponds to the inverse of the ejection rate. Sum the blocking time of the closing hairpin at intermediate force and divide it by the binding events during the recording.
2. Relative errors are calculated by computing the square root of the number of unbinding events.

3.4.2 Quantitative binding kinetics using kinetic locking assay

In some cases, the association of a helicase with its substrate does not cause changes of NA extension that are sufficiently large to be measured with single-molecule micromanipulation techniques. Measurement of helicases' binding kinetics is thus often performed with bulk stopped-flow assays (Kocsis, Sarlós, Harami, Martina, & Kovács, 2014). However, such bulk assays involve complex fitting procedures, and may require to label the proteins and/or to introduce chemical competitors to evaluate dissociation kinetics.

We recently showed (Rieu, Valle-Orero, et al., 2021) that it is possible to use a fluctuating hairpin to perform single-molecule measurements of helicases' binding kinetics at a few tens of milliseconds timescales. Indeed, small hairpins (10–15 bases) fluctuate rapidly when they are pulled at the force that equilibrates the free energy level of the open and closed states ($F_{1/2}$) (Woodside et al., 2006). When the helicase binds to the hairpin, it keeps the hairpin open and prevents its fluctuations (Fig. 11A). The association and dissociation rates of the enzymes from the substrates can be extracted from the duration and the frequencies of these long pauses. Fig. 11B and C shows the binding kinetics of the catalytic core of RecQ helicase in this configuration (Rieu, Valle-Orero, et al., 2021).

Protocol

1. Using a microfluidic chamber (prepared as described in Section 2, run a force ramp (5 s per force level, steps of 0.5 pN).
2. Find beads which fluctuate rapidly between the open and closed states. Change the force to $F_{1/2}$, the force where the duration of the closed and open states are equal.
3. Increase the camera frequency up to a few kHz.
4. Rinse the microfluidic chamber with the reaction buffer as explained in previous protocols.
5. Add the buffer mixed with the enzyme to the inlet of the chamber. Flow at a constant speed of 5 μL/min.
6. Record data during the desired period of time.
7. Repeat step 5 with increasing helicase concentrations.

Tips

1. Since the fluctuations only happen in a limited range of force (~0.5–1 pN around $F_{1/2}$) and beads have some variability in their magnetization (~10–20%), it is hard to find many beads that fluctuate at the same magnet position, especially at high acquisition frequencies where the field of view is limited by the camera's transfer frequency. However, since the fluctuations are fast, one or two beads are usually sufficient to sample several hundreds of events.
2. It is important to unambiguously distinguish between the two states of the hairpins (change of 10 nm at the millisecond timescale). A magnetic tweezer setup with high-spatio-temporal resolution is crucial (Dulin et al., 2015; Huhle et al., 2015; Kim & Saleh, 2009; Rieu et al., 2021). A signal of poor quality may be improved by removing the spurious correlations between the axial and transverse movements of the beads, as described in the supplementary materials of Rieu, Vieille, et al. (2021).

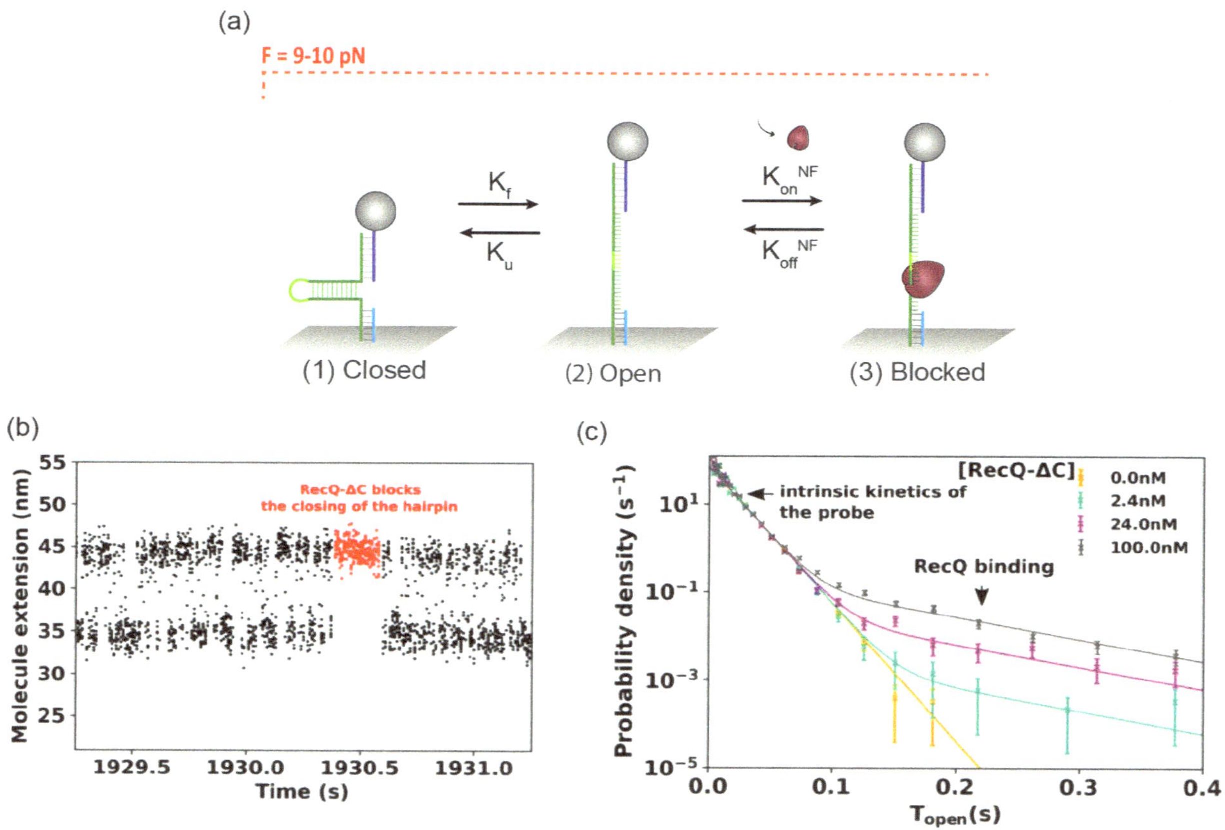

Fig. 11 See figure legend on next page.

3. The temperature of the sample must be precisely controlled (~0.01 °C) since the kinetics of the hairpin fluctuations are sensitive to it.
4. Maintaining a constant concentration of helicase is critical to assess the association time. For this reason, keeping a constant flow during the experiments is important. If the enzyme has a strong affinity to DNA, we recommend not using DNA-coated azide surfaces but other attachment chemistry, such as DIG, to avoid biasing the local concentration of helicase close to the surface.

Data analysis

Analysis of the freely fluctuating hairpin (no helicase):

1. Compute the standard deviation σ of the measured position in both closed states and open states. At 1 kHz of acquisition frequency and a 10 bp hairpin with 120 bp dsDNA handles, they typically lie between 1 and 2 nm.
2. Automatically detect the transitions between the closed and open states. A criterion of 3–4 σ (~6–8 nm) can be used by detecting a jump if the difference of position measured between two successive frames is larger than 3 σ. To avoid biases caused by the finite velocity of the bead during the transition, sum up all differences of position among successive frames as long as the movement is monotonous (series of frames with the position of the bead either increasing or decreasing).
3. Compute the times spent in the closed states (difference between two successive closing and opening events) and the times spent in the open states (difference between two successive closing and opening events).

Fig. 11 Single-molecule measurement of the binding kinetics of a helicase without ATP through kinetic locking. (A) Sketch of the experiment. A rapidly fluctuating 10 bp hairpin allows measuring short binding times of RecQ-ΔC through the alteration of the folding/unfolding rates induced by the association of the helicase. (B) Corresponding force-extension curve of RecQ-ΔC. When the helicase binds to the hairpin, it blocks it in its open state for a longer time (red). (C) Distributions of open times as a function of the concentration of helicases. Fitting to double-exponential distributions allows retrieving the association and dissociation kinetic rates. *Panels (B) and (C): Reproduced and unmodified from Rieu, M., Valle-Orero, J., Ducos, B., Allemand, J.-F., & Croquette, V. (2021). Single-molecule kinetic locking allows fluorescence-free quantification of protein/nucleic-acid binding.* Communications Biology, *4(1), 1083. doi:10.1038/s42003-021-02606-z, in agreement with the Creative Commons Attribution 4.0 International License (http:/creativecommons.org/licenses/by/4.0/).*

4. Draw the histograms of multiple (>100) closed and open times (the bin size should be a multiple of the acquisition frequency to avoid spurious irregularities). Fit the histograms by an exponential distribution and infer the typical times spent in the close and open states.
5. Average the closing and opening times by chunks of N points (N > 100) and draw the averages as a function of the acquisition time. Check that the averages are constant within the statistical error (square root of N). If an important drift is observed over time, this might be due to a flow that is not constant or a change of temperature. Wait until this parameter is stabilized and restart the measurement.

Derivation of the binding kinetics

1. When the helicase is present in solution and can bind to the substrate, the association and dissociation processes of the helicase of rates k_{on} and k_{off} adds up to the internal fluctuating rates of the hairpins. This results in the distribution of the open times of the hairpin changing from single exponential to double exponential. By fitting the time spent in the open times at the different concentrations of helicases as described in (Rieu, Valle-Orero, et al., 2021), one can deduce k_{on} and k_{off}.
2. Noteworthy, if k_{off} is much smaller (~10 fold) than the spontaneous closing rate of the hairpin, k_{off} will be very close to the inverse of the long-time of the double-exponential distribution (either the hairpin closes spontaneously—short time—or it stays open due to the presence of the helicase—a long time). In this case, the distributions are particularly easy to interpret.

3.4.3 Helicase interaction with roadblocks

We have previously described how some helicases are capable of remodeling ribonucleoprotein complexes (Fiorini, Bagchi, Le Hir, & Croquette, 2015) by covering a ssDNA hairpin with NA-binding proteins and analyzing the progression of the helicase in this context. In Fiorini et al. (2015), we demonstrated that the human UPF1 helicase core is able to translocate onto a Gp32-covered DNA at a normal rate, by remodeling NA-protein interactions. Recently, we developed an encirclement-based protocol to embed roadblocks, such as tertiary NA structures, in the NA hairpin (Tran et al., 2021) shown in Fig. 12A. The dynamics of a helicase interaction with such roadblocks as well as their probability of removing or bypassing them can be studied by injecting the enzyme and analyzing its stalling at the roadblock's position. The time spent at this position can be calculated in order to analyze

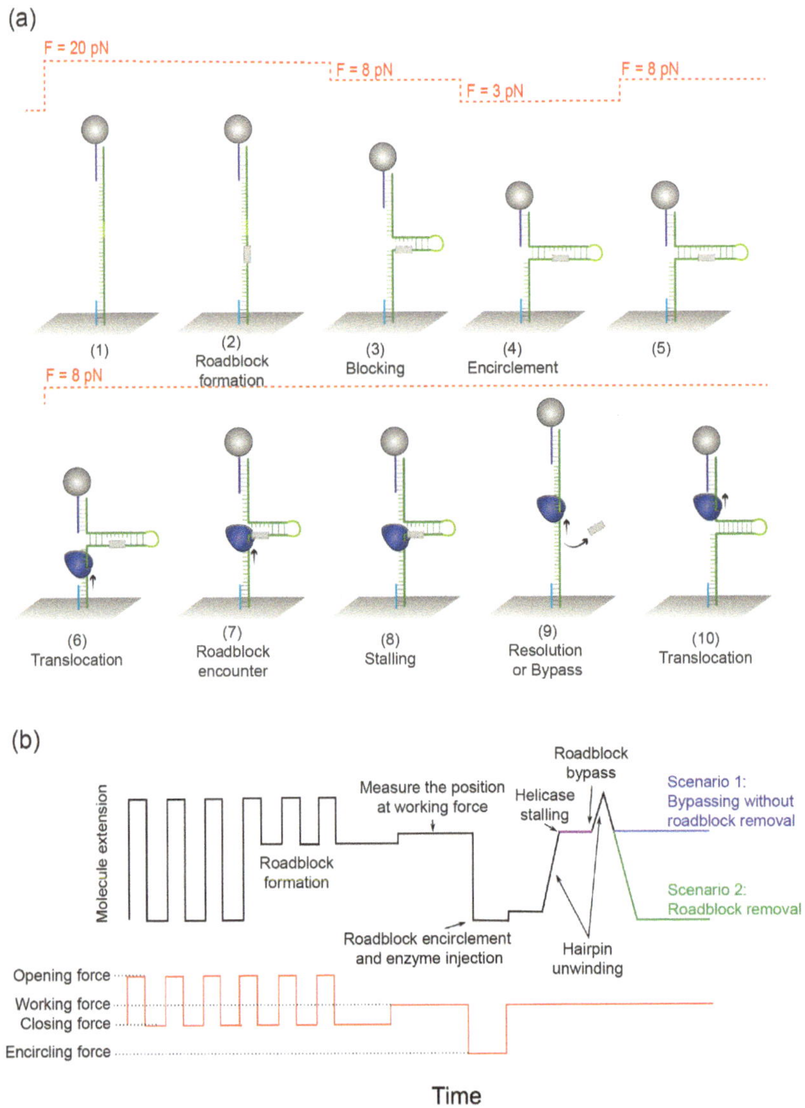

Fig. 12 Protocol for the single-molecule observation of a helicase's interaction with a roadblock. (A) Sketch of the experiment. (B) Typical extension traces and interpretation. (1) A hairpin is opened by applying a high force on the magnetic bead (20 pN). (2) The roadblock is formed. It can consist in the deposition of a NA-binding protein present in solution or in the formation of a NA-tertiary structure. (3) At a force of ~8 pN, the hairpin can only partially close due to the presence of the roadblock. This allows probing the presence of the roadblock. (4) The force is then lowered to a force (generally 2–5 pN) allowing encircling the roadblock inside a double-stranded hairpin. (5) The force is then

how the helicase interacts with the roadblock. This protocol has been recently used to study the interaction of the *Saccharomyces cerevisiae* helicase Pif1 with G-quadruplexes (Valle-Orero et al., 2022). Fig. 12B shows a schematic trace of the roadblock embedding and a helicase's translocation in this configuration.

Protocols

1. Follow steps 1–3 from protocol in Section 3.1.
2. Add the roadblock in solution and observe its binding/formation at the 7–8 pN phase (it will cause a transient blockage of the closing of the hairpin). Check that the roadblock is not expelled during the low force phase and is stable when encircled in the hairpin.
3. Rinse the chamber with the reaction buffer.
4. Stop the force cycles.
5. Apply a high force to open the hairpin, then go to 7 pN to close the hairpin until the roadblock position and finally to your working force at which the helicase activity is analyzed.
6. Apply a low force (2–4 pN) to encircle the roadblock.
7. From the low force, go to the working force, where the hairpin should be closed.
8. Add the buffer mixed with the enzyme to the inlet of the chamber. Flow at a constant speed of 5 μL/min.
9. Record data during the desired period of time.

Data analysis

1. Detect unwinding events corresponding to molecules whose extensions increase.
2. Check whether the helicase stalls at the roadblock position by comparing the time spent by the helicase at the roadblock position at the working force (checked at step 6 of the protocol) with the average time spent at any other position during unwinding.

---

increased to the working force of the helicase. The helicase is injected in solution. While the helicase unwinds the hairpin, the extension of the molecule increases. If the roadblock prevents the translocation of the helicase, a pause is observed in this increase of extension. If the helicase stalls and passes through the roadblock, one records the bypass time (in purple), which probability distribution is the main quantitative parameter provided by the assay. After the passage of the helicase, either the roadblock is still present (blue) or it has been removed (green). If it is still here, provided that the working force is too high to allow the encirclement of the roadblock, the hairpin will only partially close. Measuring several events thus allows accessing the probability of resolution.

3. If the helicase stalls and then resumes translocation, draw the distribution of the stalling duration over multiple events. It will typically follow an exponential distribution, whose characteristic time will be the "bypass time" i.e. the time needed for the helicase to go through the roadblock.
4. Check whether the helicase removes the roadblock. For this purpose, choosing a working force that prevents the full closing of the hairpin in the presence of the roadblock is recommended. In the absence of a roadblock, the hairpin will close completely after the passage of the helicase. If the roadblock remained and was just bypassed by the enzyme, the hairpin will only partially close after the passage of the helicase.

4. Discussion

The various elements that we have described demonstrate the wealth of possibilities that have emerged to study helicases in a single molecule configuration. This panel of approaches is in part the result of the richness of NA structures (ssDNA, dsDNA, three-way or four-way junction). The only limitation seems to be our imagination. As a result, depending upon the question investigated, it is possible to choose among various possibilities to either achieve the strongest signal of the simplest experimental design. What we have described for magnetic tweezers can also be used in most situations with optical tweezers.

We have not discussed nanopore-related technologies applied to study helicases, although they offer very interesting features: thanks to their very high spatial resolution, single-molecule measurements of helicase stepping are extremely promising (Caldwell & Spies, 2017; Craig et al., 2019; Laszlo, Derrrington, & Gundlach, 2017). Their main advantage is the ability to study protein translocation on hairpin-free ssNA while measuring rates of translocation at a nucleotide resolution. Micromanipulation techniques require a hairpin that converts translocation into a change of extension and recent studies have achieved the accuracy of a single base (Cheng, Arunajadai, Moffitt, Tinoco, & Bustamante, 2011; Qi, Pugh, Spies, & Chemla, 2013; Rieu, Vieille, et al., 2021), but they are not yet as precise as nanopores.

Altogether, the single molecule approaches described here nicely complement bulk biological tools to study helicases' activities which are not simple to characterize since their unwinding effect can easily be masked by the re-hybridization of NA if nothing is done to prevent it.

References

Abdelhaleem, M. (2010). Helicases: An overview. *Methods in Molecular Biology (Clifton, N.J.), 587*, 1–12. https://doi.org/10.1007/978-1-60327-355-8_1.

Abdelkrim, Y. Z., Banroques, J., & Kyle Tanner, N. (2021). Known inhibitors of RNA helicases and their therapeutic potential. *Methods in Molecular Biology (Clifton, N.J.), 2209*, 35–52. https://doi.org/10.1007/978-1-0716-0935-4_3.

Allemand, J.-F. (1997). *Micromanipulation d'une molécule individuelle d'ADN*. Université Pierre et Marie Curie Paris VI.

Amit, R., Gileadi, O., & Stavans, J. (2004). Direct observation of RuvAB-catalyzed branch migration of single Holliday junctions. *Proceedings of the National Academy of Sciences of the United States of America, 101*(32), 11605–11610. https://doi.org/10.1073/pnas.0404332101.

Bétous, R., Couch, F. B., Mason, A. C., Eichman, B. F., Manosas, M., & Cortez, D. (2013). Substrate-selective repair and restart of replication forks by DNA translocases. *Cell Reports, 3*(6), 1958–1969.

Bianco, P. R. (2021). Insight into the biochemical mechanism of DNA helicases provided by bulk-phase and single-molecule assays. *Methods*. https://doi.org/10.1016/j.ymeth.2021.12.002.

Bianco, P. R., & Lu, Y. (2021). Single-molecule insight into stalled replication fork rescue in Escherichia coli. *Nucleic Acids Research, 49*(8), 4220–4238. https://doi.org/10.1093/nar/gkab142.

Bockelmann, U. (2004). Single-molecule manipulation of nucleic acids. *Current Opinion in Structural Biology, 14*(3), 368–373.

Bouchiat, C., Wang, M. D., Allemand, J.-F., Strick, T., Block, S. M., & Croquette, V. (1999). Estimating the persistence length of a worm-like chain molecule from force-extension measurements. *Biophysical Journal, 76*(1), 409–413. https://doi.org/10.1016/S0006-3495(99)77207-3.

Brenner, H. (1961). The slow motion of a sphere through a viscous fluid towards a plane surface. *Chemical Engineering Science, 16*(3), 242–251. https://doi.org/10.1016/0009-2509(61)80035-3.

Brosh, R. M., & Matson, S. W. (2020). History of DNA helicases. *Genes, 11*(3), 255. https://doi.org/10.3390/genes11030255.

Bryant, Z., Stone, M. D., Gore, J., Smith, S. B., Cozzarelli, N. R., Bustamante, C., et al. (2003). Structural transitions and elasticity from torque measurements on DNA. *Nature, 424*(6946), 338–341.

Bustamante, C. J., Chemla, Y. R., Liu, S., & Wang, M. D. (2021). Optical tweezers in single-molecule biophysics. *Nature Reviews Methods Primers, 1*(1), 1–29. https://doi.org/10.1038/s43586-021-00021-6.

Caldwell, C. C., & Spies, M. (2017). Helicase SPRNTing through the nanopore. *Proceedings of the National Academy of Sciences of the United States of America, 114*(45), 11809–11811. https://doi.org/10.1073/pnas.1716866114.

Cheng, W., Arunajadai, S. G., Moffitt, J. R., Tinoco, I., & Bustamante, C. (2011). Single-base pair unwinding and asynchronous RNA release by the hepatitis C virus NS3 helicase. *Science, 333*(6050), 1746–1749. https://doi.org/10.1126/science.1206023.

Chilkoti, A., & Stayton, P. S. (1995). Molecular origins of the slow streptavidin-biotin dissociation kinetics. *Journal of the American Chemical Society, 117*(43), 10622–10628. https://doi.org/10.1021/ja00148a003.

Cordin, O., & Beggs, J. D. (2013). RNA helicases in splicing. *RNA Biology, 10*(1), 83–95. https://doi.org/10.4161/rna.22547.

Craig, J. M., Laszlo, A. H., Nova, I. C., Brinkerhoff, H., Noakes, M. T., Baker, K. S., et al. (2019). Determining the effects of DNA sequence on Hel308 helicase translocation along single-stranded DNA using nanopore tweezers. *Nucleic Acids Research, 47*(5), 2506–2513.

Datta, A., & Brosh, R. M. (2018). New insights into DNA helicases as druggable targets for cancer therapy. *Frontiers in Molecular Biosciences*, *5*, 59. https://doi.org/10.3389/fmolb.2018.00059.

Davis, L. G., Dibner, M. D., & Battey, J. F. (1986). Section 9-3—Electroelution. In L. G. Davis, M. D. Dibner, & J. F. Battey (Eds.), *Basic methods in molecular biology* (pp. 112–114). Elsevier. https://doi.org/10.1016/B978-0-444-01082-7.50035-7.

Dawid, A., Croquette, V., Grigoriev, M., & Heslot, F. (2004). Single-molecule study of RuvAB-mediated Holliday-junction migration. *Proceedings of the National Academy of Sciences of the United States of America*, *101*(32), 11611–11616.

Desai, V. P., Frank, F., Lee, A., Righini, M., Lancaster, L., Noller, H. F., et al. (2019). Co-temporal force and fluorescence measurements reveal a ribosomal gear shift mechanism of translation regulation by structured mRNAs. *Molecular Cell*, *75*(5), 1007–1019, e5. https://doi.org/10.1016/j.molcel.2019.07.024.

Dhar, S., Datta, A., & Brosh, R. M. (2020). DNA helicases and their roles in cancer. *DNA Repair*, *96*, 102994. https://doi.org/10.1016/j.dnarep.2020.102994.

Dulin, D., Cui, T. J., Cnossen, J., Docter, M. W., Lipfert, J., & Dekker, N. H. (2015). High spatiotemporal-resolution magnetic tweezers: Calibration and applications for DNA dynamics. *Biophysical Journal*, *109*(10), 2113–2125.

Eeftens, J. M., van der Torre, J., Burnham, D. R., & Dekker, C. (2015). Copper-free click chemistry for attachment of biomolecules in magnetic tweezers. *BMC Biophysics*, *8*(1), 9. https://doi.org/10.1186/s13628-015-0023-9.

Fiorini, F., Bagchi, D., Le Hir, H., & Croquette, V. (2015). Human Upf1 is a highly processive RNA helicase and translocase with RNP remodelling activities. *Nature Communications*, *6*, 7581. https://doi.org/10.1038/ncomms8581.

Fiorini, F., Bonneau, F., & Le Hir, H. (2012). Chapter twelve—Biochemical characterization of the RNA helicase UPF1 involved in nonsense-mediated mRNA decay. In E. Jankowsky (Ed.), *511. Methods in enzymology* (pp. 255–274). Academic Press. https://doi.org/10.1016/B978-0-12-396546-2.00012-7.

Gao, Y., & Yang, W. (2020). Different mechanisms for translocation by monomeric and hexameric helicases. *Current Opinion in Structural Biology*, *61*, 25–32. https://doi.org/10.1016/j.sbi.2019.10.003.

Gorbalenya, A., & Koonin, E. V. (1993). Helicases: Amino acid sequence comparisons and structure-function relationships. *Current Opinion in Structural Biology*. https://doi.org/10.1016/S0959-440X(05)80116-2.

Gosse, C., & Croquette, V. (2002). Magnetic tweezers: Micromanipulation and force measurement at the molecular level. *Biophysical Journal*, *82*(6), 3314–3329. https://doi.org/10.1016/S0006-3495(02)75672-5.

Greenleaf, W. J., Woodside, M. T., & Block, S. M. (2007). High-resolution, single-molecule measurements of biomolecular motion. *Annual Review of Biophysics and Biomolecular Structure*, *36*, 171–190.

Guo, S., Ray, C., Kirkpatrick, A., Lad, N., & Akhremitchev, B. B. (2008). Effects of multiple-bond ruptures on kinetic parameters extracted from force spectroscopy measurements: Revisiting biotin-streptavidin interactions. *Biophysical Journal*, *95*(8), 3964–3976. https://doi.org/10.1529/biophysj.108.133900.

Hatch, K., Danilowicz, C., Coljee, V., & Prentiss, M. (2008). Demonstration that the shear force required to separate short double-stranded DNA does not increase significantly with sequence length for sequences longer than 25 base pairs. *Physical Review E*, *78*(1), 011920. https://doi.org/10.1103/PhysRevE.78.011920.

Hodeib, S., Raj, S., Manosas, M., Zhang, W., Bagchi, D., Ducos, B., et al. (2016). Single molecule studies of helicases with magnetic tweezers. *Methods*, *105*, 3–15.

Huhle, A., Klaue, D., Brutzer, H., Daldrop, P., Joo, S., Otto, O., et al. (2015). Camera-based three-dimensional real-time particle tracking at kHz rates and Ångström accuracy. *Nature Communications*, *6*(1). https://doi.org/10.1038/ncomms6885.

Jacobson, D. R., McIntosh, D. B., Stevens, M. J., Rubinstein, M., & Saleh, O. A. (2017). Single-stranded nucleic acid elasticity arises from internal electrostatic tension. *Proceedings of the National Academy of Sciences of the United States of America*, *114*(20), 5095–5100. https://doi.org/10.1073/pnas.1701132114.

Janissen, R., Oberbarnscheidt, L., & Oesterhelt, F. (2009). Optimized straight forward procedure for covalent surface immobilization of different biomolecules for single molecule applications. *Colloids and Surfaces B: Biointerfaces*, *71*(2), 200–207. https://doi.org/10.1016/j.colsurfb.2009.02.011.

Kanaan, J., Raj, S., Decourty, L., Saveanu, C., Croquette, V., & Hir, H. L. (2018). UPF1-like helicase grip on nucleic acids dictates processivity. *Nature Communications*, *9*(1), 3752. https://doi.org/10.1038/s41467-018-06313-y.

Khemici, V., & Linder, P. (2018). RNA helicases in RNA decay. *Biochemical Society Transactions*, *46*(1), 163–172. https://doi.org/10.1042/BST20170052.

Kim, K., & Saleh, O. A. (2009). A high-resolution magnetic tweezer for single-molecule measurements. *Nucleic Acids Research*, *37*(20), e136. https://doi.org/10.1093/nar/gkp725.

Klaue, D., & Seidel, R. (2009). Torsional stiffness of single superparamagnetic microspheres in an external magnetic field. *Physical Review Letters*, *102*(2), 028302.

Kocsis, Z. S., Sarlós, K., Harami, G. M., Martina, M., & Kovács, M. (2014). A nucleotide-dependent and HRDC domain-dependent structural transition in DNA-bound RecQ helicase. *The Journal of Biological Chemistry*, *289*(9), 5938–5949. https://doi.org/10.1074/jbc.M113.530741.

Laszlo, A. H., Derrington, I. M., & Gundlach, J. H. (2017). Subangstrom measurements of enzyme function using a biological nanopore, SPRNT. *Methods in Enzymology*, *582*, 387–414. Elsevier.

Li, H., Zhang, C., Hu, Y., Liu, P., Sun, F., Chen, W., et al. (2021). A reversible shearing DNA probe for visualizing mechanically strong receptors in living cells. *Nature Cell Biology*, *23*(6), 642–651. https://doi.org/10.1038/s41556-021-00691-0.

Linder, P., & Fuller-Pace, F. (2015). Happy birthday: 25 years of DEAD-box proteins. *Methods in Molecular Biology (Clifton, N.J.)*, *1259*, 17–33. https://doi.org/10.1007/978-1-4939-2214-7_2.

Linder, P., & Jankowsky, E. (2011). From unwinding to clamping—The DEAD box RNA helicase family. *Nature Reviews Molecular Cell Biology*, *12*(8), 505–516. https://doi.org/10.1038/nrm3154.

Lohman, T. M., Tomko, E. J., & Wu, C. G. (2008). Non-hexameric DNA helicases and translocases: Mechanisms and regulation. *Nature Reviews. Molecular Cell Biology*, *9*(5), 391–401. https://doi.org/10.1038/nrm2394.

Manosas, M., Meglio, A., Spiering, M. M., Ding, F., Benkovic, S. J., Barre, F.-X., et al. (2010). Magnetic tweezers for the study of DNA tracking motors. *Methods in Enzymology*, *475*, 297–320. https://doi.org/10.1016/S0076-6879(10)75013-8.

Manosas, M., Perumal, S. K., Bianco, P. R., Ritort, F., Benkovic, S. J., & Croquette, V. (2013). RecG and UvsW catalyse robust DNA rewinding critical for stalled DNA replication fork rescue. *Nature Communications*, *4*(1), 2368. https://doi.org/10.1038/ncomms3368.

McIntosh, D. B., Duggan, G., Gouil, Q., & Saleh, O. A. (2014). Sequence-dependent elasticity and electrostatics of single-stranded DNA: Signatures of base-stacking. *Biophysical Journal*, *106*(3), 659–666. https://doi.org/10.1016/j.bpj.2013.12.018.

Mendoza, O., Bourdoncle, A., Boulé, J.-B., Brosh, R. M., & Mergny, J.-L. (2016). G-quadruplexes and helicases. *Nucleic Acids Research*, *44*(5), 1989–2006. https://doi.org/10.1093/nar/gkw079.

Merkel, R., Nassoy, P., Leung, A., Ritchie, K., & Evans, E. (1997). Energy landscapes of receptor-ligand bonds explored with dynamic force spectroscopy. *Nature*, *397*, 50–53.

Mickolajczyk, K. J., Shelton, P. M. M., Grasso, M., Cao, X., Warrington, S. E., Aher, A., et al. (2021). Force-dependent stimulation of RNA unwinding by SARS-CoV-2 nsp13 helicase. *Biophysical Journal*, *120*(6), 1020–1030. https://doi.org/10.1016/j.bpj.2020.11.2276.

Neuman, K. C., Lionnet, T., & Allemand, J.-F. (2007). Single-molecule micromanipulation techniques. *Annual Review of Materials Research*, *37*, 33–67.

Neuman, K. C., & Nagy, A. (2008). Single-molecule force spectroscopy: Optical tweezers, magnetic tweezers and atomic force microscopy. *Nature Methods*, *5*(6), 491–505.

Ostrofet, E., Papini, F. S., & Dulin, D. (2018). Correction-free force calibration for magnetic tweezers experiments. *Scientific Reports*, *8*(1), 15920. https://doi.org/10.1038/s41598-018-34360-4.

Papini, F. S., Seifert, M., & Dulin, D. (2019). High-yield fabrication of DNA and RNA constructs for single molecule force and torque spectroscopy experiments. *Nucleic Acids Research*, *47*(22), e144. https://doi.org/10.1093/nar/gkz851.

Patel, S. S., & Donmez, I. (2006). Mechanisms of helicases. *Journal of Biological Chemistry*, *281*(27), 18265–18268. https://doi.org/10.1074/jbc.R600008200.

Pincet, F., & Husson, J. (2005). The solution to the streptavidin-biotin paradox: The influence of history on the strength of single molecular bonds. *Biophysical Journal*, *89*(6), 4374–4381. https://doi.org/10.1529/biophysj.105.067769.

Qi, Z., Pugh, R. A., Spies, M., & Chemla, Y. R. (2013). Sequence-dependent base pair stepping dynamics in XPD helicase unwinding. *ELife*. https://doi.org/10.7554/eLife.00334.

Riener, C. K., Kienberger, F., Hahn, C. D., Buchinger, G. M., Egwim, I. O. C., Haselgrübler, T., et al. (2003). Heterobifunctional crosslinkers for tethering single ligand molecules to scanning probes. *Analytica Chimica Acta*, *497*(1), 101–114. https://doi.org/10.1016/j.aca.2003.08.041.

Rieu, M., Valle-Orero, J., Ducos, B., Allemand, J.-F., & Croquette, V. (2021). Single-molecule kinetic locking allows fluorescence-free quantification of protein/nucleic-acid binding. *Communications Biology*, *4*(1), 1083. https://doi.org/10.1038/s42003-021-02606-z.

Rieu, M., Vieille, T., Radou, G., Jeanneret, R., Ruiz-Gutierrez, N., Ducos, B., et al. (2021). Parallel, linear, and subnanometric 3D tracking of microparticles with Stereo Darkfield Interferometry. *Science Advances*, 7(6), eabe3902.

Singleton, M. R., Dillingham, M. S., & Wigley, D. B. (2007). Structure and mechanism of helicases and nucleic acid translocases. *Annual Review of Biochemistry*, *76*(1), 23–50. https://doi.org/10.1146/annurev.biochem.76.052305.115300.

Smith, S. B., Finzi, L., & Bustamante, C. (1992). Direct mechanical measurements of the elasticity of single DNA molecules by using magnetic beads. *Science*, *258*(5085), 1122–1126. https://doi.org/10.1126/science.1439819.

Steimer, L., & Klostermeier, D. (2012). RNA helicases in infection and disease. *RNA Biology*, *9*(6), 751–771. https://doi.org/10.4161/rna.20090.

Sun, Y., Sato, O., Ruhnow, F., Arsenault, M. E., Ikebe, M., & Goldman, Y. E. (2010). Single-molecule stepping and structural dynamics of myosin X. *Nature Structural & Molecular Biology*, *17*(4), 485–491. https://doi.org/10.1038/nsmb.1785.

Tan, H. Y., Wilczek, L. A., Pottinger, S., Manosas, M., Yu, C., Nguyenduc, T., et al. (2017). The intrinsically disordered linker of E. coli SSB is critical for the release from single-stranded DNA. *Protein Science*, *26*(4), 700–717. https://doi.org/10.1002/pro.3115.

Tran, P. L. T., Rieu, M., Hodeib, S., Joubert, A., Ouellet, J., Alberti, P., et al. (2021). Folding and persistence times of intramolecular G-quadruplexes transiently embedded in a DNA duplex. *Nucleic Acids Research*, *49*(9), 5189–5201. https://doi.org/10.1093/nar/gkab306.

Umate, P., Tuteja, N., & Tuteja, R. (2011). Genome-wide comprehensive analysis of human helicases. *Communicative & Integrative Biology*, *4*(1), 118–137. https://doi.org/10.4161/cib.4.1.13844.

Valle-Orero, J., Rieu, M., Tran, P. L. T., Joubert, A., Allemand, J.-F., Croquette, V., et al. (2022). Strand-switching mechanism of Pif1 helicase induced by its collision with a G-quadruplex embedded in dsDNA (p. 2022.01.11.475363). *bioRxiv*. https://doi.org/10.1101/2022.01.11.475363.

Viader-Godoy, X., Pulido, C. R., Ibarra, B., Manosas, M., & Ritort, F. (2021). Cooperativity-dependent folding of single-stranded DNA. *Physical Review X*, *11*(3), 031037. https://doi.org/10.1103/PhysRevX.11.031037.

Vilfan, I. D., Lipfert, J., Koster, D. A., Lemay, S. G., & Dekker, N. H. (2009). Magnetic tweezers for single-molecule experiments. In P. Hinterdorfer, & A. Oijen (Eds.), *Handbook of single-molecule biophysics* (pp. 371–395). US: Springer. https://doi.org/10.1007/978-0-387-76497-9_13.

Walsh, M. K., Wang, X., & Weimer, B. C. (2001). Optimizing the immobilization of single-stranded DNA onto glass beads. *Journal of Biochemical and Biophysical Methods*, *47*(3), 221–231. https://doi.org/10.1016/S0165-022X(00)00146-9.

Wilchek, M., & Bayer, E. A. (1990). Introduction to avidin-biotin technology. *Methods in Enzymology*, *184*, 5–13. https://doi.org/10.1016/0076-6879(90)84256-g.

Wildling, L., Unterauer, B., Zhu, R., Rupprecht, A., Haselgrübler, T., Rankl, C., et al. (2011). Linking of sensor molecules with amino groups to amino-functionalized AFM tips. *Bioconjugate Chemistry*, *22*(6), 1239–1248. https://doi.org/10.1021/bc200099t.

Williams, P. M., Moore, A., Stevens, M. M., Allen, S., Davies, M. C., Roberts, C. J., et al. (2000). On the dynamic behaviour of the forced dissociation of ligand–receptor pairs. *Journal of the Chemical Society, Perkin Transactions*, *2*(1), 5–8. https://doi.org/10.1039/A907750B.

Woodside, M. T., Anthony, P. C., Behnke-Parks, W. M., Larizadeh, K., Herschlag, D., & Block, S. M. (2006). Direct measurement of the full, sequence-dependent folding landscape of a nucleic acid. *Science*, *314*(5801), 1001–1004.

Wu, Y. (2012). Unwinding and rewinding: Double faces of helicase? *Journal of Nucleic Acids*, *2012*, 140601. https://doi.org/10.1155/2012/140601.

Yang, Y.-J., Song, L., Zhao, X.-C., Zhang, C., Wu, W.-Q., You, H.-J., et al. (2019). A universal assay for making DNA, RNA, and RNA-DNA hybrid configurations for single-molecule manipulation in two or three steps without ligation. *ACS Synthetic Biology*, *8*(7), 1663–1672. https://doi.org/10.1021/acssynbio.9b00241.

Yodh, J. G., Schlierf, M., & Ha, T. (2010). Insight into helicase mechanism and function revealed through single-molecule approaches. *Quarterly Reviews of Biophysics*, *43*(2), 185–217. https://doi.org/10.1017/S0033583510000107.

Yu, Z., Dulin, D., Cnossen, J., Köber, M., van Oene, M. M., Ordu, O., et al. (2014). A force calibration standard for magnetic tweezers. *The Review of Scientific Instruments*, *85*(12), 123114. https://doi.org/10.1063/1.4904148.

CHAPTER FOURTEEN

Reconstitution and biochemical characterization of the RNA-guided helicase-nuclease protein Cas3 from type I-A CRISPR–Cas system

Chunyi Hu and Ailong Ke*
Department of Molecular Biology and Genetics, Cornell University, Ithaca, NY, United States
*Corresponding author: e-mail address: ailong.ke@cornell.edu

Contents

Abstract

Type I is the most prevalent CRISPR system found in nature. It can be further defined into six subtypes, from I-A to I-G. Among them, the Type I-A CRISPR-Cas systems are almost exclusively found in hyperthermophilic archaeal organisms. The system achieves RNA-guided DNA degradation through the concerted action of a CRISPR

Methods in Enzymology, Volume 673
ISSN 0076-6879
https://doi.org/10.1016/bs.mie.2022.03.059

RNA containing complex Cascade and a helicase-nuclease fusion enzyme Cas3. Here, we summarize assays to characterize the biochemical behavior of Cas3. A steep temperature-dependency was found for the helicase component of Cas3HEL, but not the nuclease component HD. This finding enabled us to establish the correct experimental condition to carry out I-A CRISPR-Cas based genome editing in human cells with extremely high efficiency.

1. Introduction

CRISPR-Cas is a RNA-guided adaptive immunity system in prokaryotes against foreign genetic elements (Barrangou et al., 2007; Bolotin, Quinquis, Sorokin, & Ehrlich, 2005; Makarova, Grishin, Shabalina, Wolf, & Koonin, 2006; Marraffini & Sontheimer, 2008; Mojica, García-Martínez, & Soria, 2005; Pourcel, Salvignol, & Vergnaud, 2005). The CRISPR operon is transcribed and processed into mature guide RNA (or named crRNAs), which assemble with one or more of the CRISPR-associated (Cas) proteins into RNA-guided effector complexes. Among them, the DNA-targeting CRISPR-Cas systems define a target by promoting R-loop formation at a complementary sequence flanked by a protospacer adjacent motif (PAM). The effector then either activates the associated nuclease (s) or recruits one *in trans* to generate double-strand breaks (DSBs).

CRISPR-Cas systems can be grouped into Class 1 and Class 2, based on whether a multi- or single-subunit effector complex, respectively, is utilized during CRISPR interference (Makarova et al., 2020). Each Class is further classified into three types and multiple subtypes therein. Class 1 Type I is the most prevalent CRISPR system. It can be further divided into six subtypes, from I-A to I-G (Makarova et al., 2020). Among them, the most extensively studied and mechanistically understood system is the Type I-E system, based on work with *Escherichia coli* (Brouns et al., 2008; Hochstrasser et al., 2014; Jackson et al., 2014; Mulepati & Bailey, 2013; Mulepati, Heroux, & Bailey, 2014; Rutkauskas et al., 2015; Sashital, Wiedenheft, & Doudna, 2012; Sinkunas et al., 2013; Westra et al., 2012; Wiedenheft et al., 2011; Zhao et al., 2014) and *Thermobifida fusca* Type I-E systems (Dillard et al., 2018; Hayes et al., 2016; Huo et al., 2014; Xiao et al., 2017; Xiao, Luo, Dolan, Liao, & Ke, 2018). Results suggest that the Type I-E systems shred the DNA target into pieces through a two-step mechanistic process. First, a multi-subunit ribonucleoprotein (RNP) complex called Cascade (CRISPR associated complex for antiviral defense) relies on the associated crRNA to define a matching DNA target flanked by the Protospacer Adjacent Motif (PAM) (Brouns et al., 2008; Westra et al., 2012; Wiedenheft et al., 2011). This results in the formation of an R-loop, which then triggers a large conformational change in Cascade (Hochstrasser et al., 2014; Huo et al., 2014; Wiedenheft et al., 2011; Xiao et al., 2017). The helicase-nuclease fusion enzyme Cas3 is then recruited *in trans* to the R-loop-forming Cascade (Xiao et al., 2017, 2018). Upon binding, Cas3 nicks the non-target strand (NTS) DNA in the R-loop region and processively degrades the PAM-proximal side of the dsDNA in an ATP-dependent fashion (Dillard et al., 2018; Hochstrasser et al., 2014; Kunne et al., 2016; Mulepati & Bailey, 2013; Redding et al., 2015; Sinkunas et al., 2013). This stepwise activation mechanism minimizes the possibility of mis-targeting a near-cognate DNA target.

Type I-A CRISPR-Cas system differs from other systems in several aspects. It is exclusively found in archaeal organisms growing in extreme high temperature environment (higher than 85 °C). The helicase and nuclease portions of I-A Cas3 (Cas3HEL and Cas3HD) are encoded by separate open reading frames. In a recent study focusing on a *Pyrococcus furiosus* (*Pfu*) I-A CRISPR-Cas system (Elmore, Deighan, Westpheling, Terns, & Terns, 2015; Garrett et al., 2020; Majumdar & Terns, 2019), we found that the *Pfu* Cas3HEL and Cas3HD assemble into a stable complex and function as a unit. Interestingly, the helicase and nuclease activities of Cas3 show drastically different temperature-dependent behaviors. We further revealed that Cas3 associates with the I-A *Pfu* Cascade in a constitutive rather than R-loop dependent fashion, and that Cas3's nuclease activity is allosterically rather than through trans-recruitment (manuscript under review). These findings suggest that the Type I-A CRISPR-Cas utilizes a very different set of mechanisms from the previously characterized Type I systems to achieve RNA-guided DNA degradation (Fig. 1).

In this chapter, we compile the protocols for the purification of the *Pfu* Cas3HEL and CasHD, and the detailed protocol to assay for the RNA-guided DNA degrading activity from the *Pfu* Cascade-Cas3 complex. Finally, we provide the assay protocol for *Pfu* Cascade-Cas3 mediated genome editing in human cells. The parameters in the genome editing experiments were derived from the early biochemical assays, which is a nice illustration of the power of biochemistry.

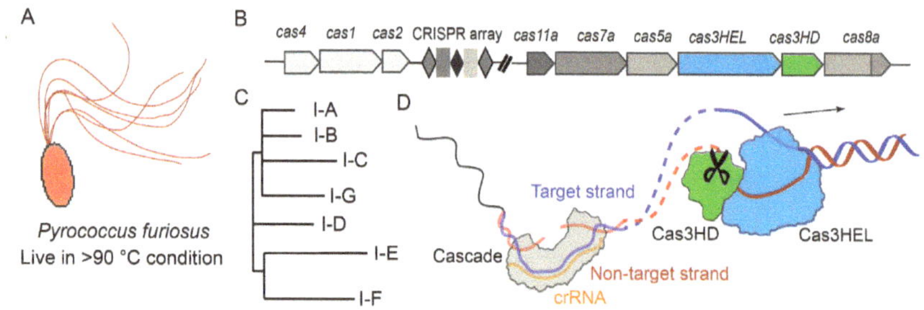

Fig. 1 Schematic illustration of the Type I-A CRISPR-*Cas* system from *Pyrococcus furiosus*. (A) A cartoon model depicts the hyperthermophile *Pyrococcus furiosus*. (B) Gene cluster of the Type I-A CRISPR-Cas system from *P. furiosus*. The helicase *cas3HEL* and nuclease *cas3HD* genes are highlighted by cyan and green. (C) Phylogeny tree of different Type I CRISPR-Cas systems based on homology of Cas1 family classification. (D) A mechanistic model illustrating how Cas3 is sent off by Cascade from an RNA-defined target to shred DNA into pieces, causing long-range deletion.

2. Materials and equipment

2.1 Key resources table

Reagent or resource	Source	Identifier
Electroporation buffers	Thermo Fisher	MPK10096
HisPur™ Ni-NTA Resin	Thermo Fisher	88223
Imidazole	Sigma	I2399
β-mercaptoethanol (BME)	Sigma	M6250-100ML
Agarose	Sigma	A4718
HEPES	Sigma	
Sodium hydroxide (NaOH)	Sigma	
PMSF inhibitor powder	Roche	
NaCl	VWR	
Trizma base	VWR	
$CoCl_2$•7H2O	Sigma	
5x SDS-loading dye	NEB	B7703S
2x DNA loading dye	NEB	

—cont'd

Reagent or resource	Source	Identifier
TBE buffer (10X)	Thermo Scientific™	B52
DMEM culture media	Sigma	
Fetal Bovine Serum	Gibco	A4766801
PBS	VWR	
Trypsin/EDTA Solution	Gibco™	R001100
LB media powder	Teknova	L9145
IPTG	Goldbio	I2481C500
$MgCl_2 \cdot 6H_2O$	Sigma	
ATP powder	Sigma	
Kanamycin powder	Thermo Fisher	11815032
3C protease	Homemade (Ullah et al., 2016)	5 mg/ml
SDS-PAGE commercial gel	Genscript	M00654
UltraPure™ Phenol	Invitrogen™	15593031
GelRed® DNA staining 10000x	Biotium	41002

2.2 Buffers and chemicals

1. Lysis buffer: 50 mM HEPES (pH 7.5), 500 mM NaCl, 10% Glycerol, 2 mM β-mercaptoethanol (BME), 1 mM PMSF (protease inhibitor, added immediately before use).
2. Binding and wash buffer: 50 mM HEPES (pH 7.5), 500 mM NaCl, 10% Glycerol, 10 mM Imidazole, 2 mM BME.
3. Elute buffer: 50 mM HEPES (pH 7.5), 500 mM NaCl, 10% Glycerol, 200 mM Imidazole.
4. FPLC buffer: 25 mM HEPES (pH 7.5), 300 mM NaCl.
5. Cleavage buffer: 50 mM HEPES (pH 7.5), 150 mM NaCl, 5% Glycerol, 10 mM $MgCl_2$, 200 μM $CoCl_2$.
6. Degrading buffer: 50 mM HEPES (pH 7.5), 150 mM NaCl, 5% Glycerol, 10 mM $MgCl_2$, 200 μM $CoCl_2$, 2 mM ATP.
7. 100x Kanamycin solution: 50 mg/mL
8. IPTG solution: 1 M in autoclaved water.
9. 100x PMSF protease inhibitor solution: 100 mM in 100% isopropanol.

2.3 Strain, plasmid and guide

Plasmids and cell lines	Source
Type I-A Cascade plasmids	Published in Hu et al. (2022)
pRSFDuet-Cas3HEL	Published in Hu et al. (2022)
pRSFDuet-Cas3HD	Published in Hu et al. (2022)
pRSFDuet-Target DNA	Published in Hu et al. (2022)
HAP1_GFP cell line	Published in Dolan et al. (2019)
BL21 competent cell	NEB
pRSFDuet-GFP guide	This study
GFP guide sequence	TGAGCAAAGACCCCAACGAGAA GCGCGATCACATGGT

2.4 Equipment

1. Ice machine and box
2. 42 °C water bath incubator
3. Bacteria culture shaker
4. Bacteria culture dish and flask
5. 37 °C incubator
6. Autoclave
7. Gravity column for purification
8. Sonicator device to disrupt bacteria cell (Vivtek, SKU: USP250-HT3-115V)
9. GE AKTA™ FPLC systems and Superdex 200 column
10. Syringe, 0.22 μm and 0.45 μm filter unit
11. Protein concentrator (30 kD cut off)
12. −80 °C freezer
13. Agarose gel and SDS PAGE electrophoresis apparatus
14. Pipettes and tips
15. PCR machine and PCR tubes.
16. Centrifuge and tubes (1.5 mL, 50 mL)
17. Baker (50 mL)
18. ChemiDoc gel scanning device (Bio-Rad)

19. Neon™ Transfection System (Invitrogen: MPK5000)
20. Vertical superclean bench (KL-CR-CWB-0035)
21. Cell culture plate (10 cm dish)
22. CO_2 Cell incubator (Thermo Scientific: IMC18 50125590)
23. The Incucyte® Live-Cell Analysis System (Incucyte® S3)
24. Rotate Incubator for purification.
25. Heat block at 90 °C.
26. SDS-PAGE staining and destaining machine (eStain L1 Protein Staining Device)
27. Vortex
28. Cell counter or auto-counter machine
29. Vacuum pump for cell manipulation

3. Step-by-step method details

3.1 Plasmid transformation

Timing: 1–1.5 h
Preparation: 5 min
Protocol: 1–1.5 h

Preparation

1. Prepare some ice in box from ice machine.
2. Prepare the plasmids of pRSFDuet-Cas3HEL and pRSFDuet-Cas3HEL.
3. Prepare two agarose-LB plate with Kanamycin at 50 μg/mL.
4. Prepare a bacteria culture shaker at 37 °C.
5. Prepare a bacteria culture incubator at 37 °C.
6. Prepare a water bath at 42 °C.

Protocol

Recommend doing this step at 3–6 pm

1. Take two tubes BL21 (DE3) competent cell and label them as HEL and HD respectively. Then put them in ice box for 3 min to thaw them.
2. Transfer 100 ng amount into the competent for each plasmid respectively.
3. Incubate them on ice for 25 min.
4. Put the competent cell tubes in 42 °C water bath for 60 s.
5. Put them in ice again immediately for 3 min.
6. Add 600 μL LB media and put in 37 °C shaker to recover bacteria for 30 min.

7. Take 150 μL bacteria solution from the tube and spread it on the agarose-LB plate.
8. Transfer the two plates into 37 °C incubator and culture it for overnight (14–20 h is good).

3.2 Cas3HEL purification

Timing: 4–5 days
Preparation: 1 days
Protocol: 3–4 days

Preparation

1. Prepare a 250 mL flask with 50 mL LB and autoclave.
2. Prepare 2 L LB media for large culture in a big flask and autoclave.
3. Prepare a bacteria culture shaker at 37 °C.
4. Prepare the buffers and reagents as step 2.
5. Prepare 2 mL Ni-NTA resin in a gravity column and equilibrium it by lysis buffer.
6. Equilibrate the Superdex 200 10/300 GL column with the FPLC buffer on AKTA.

Protocol

1. *Day 1 afternoon*: pick a single colony from the agarose plate and inoculate a 50 mL LB media with Kanamycin at 50 μg/mL. (*Recommend doing this step at 4–6 pm.)*
2. Grow the culture overnight in a shaker set to 37 °C and 100 rpm.
3. *Day 2 morning*: transfer 25 mL of the overnight culture into 4 L flasks containing 2 L sterilized LB media with 50 μg/mL Kanamycin added.
4. Place the large flasks into the holders of the shaker and grow the culture at 37 °C and 180 rpm until the $O.D._{600nm}$ reaches ~0.6. This will usually take 3–6 h.
5. Adjust the setting of the shaker to 16 °C and 120 rpm, continue to shake the culture for an extra 2 h. This step cools down the cell culture media, reduces the protein production speed, and supposedly results in better protein folding.
6. Add IPTG solution into the media to 0.5 mM final concentration.
7. Reset the shaker to 180 rpm and grow the culture overnight at 16 °C.
8. *Day 3 morning*: harvest the bacteria using a Beckmann J6 centrifuge. 1 L centrifuge bottles are used. Each spin is set to 4000 rpm for 15 min.
9. Discard the clear supernatant.
10. Add 35 mL lysis buffer to resuspend the cell pellet from the bottom of the centrifuge bottle. Transfer the cell solution to the next bottle and repeat the process.

11. Transfer the resuspended bacteria solution into a 50 mL beaker and add 0.4 mL PMSF protease inhibitor solution.
12. Place the beaker in an ice-filled bucket. Place the sonicator probe in the middle of the beaker, 1 cm above the bottom. Start the sonicator to disrupt the cells. We use 10 min for the whole program at 50% power, repeating a 2-s-on, 6-s-off cycle.
13. Transfer the sonicated solution into 50 mL centrifuge tubes. Balance them in pairs.
14. Place the balanced tubes into a high-speed centrifuge. Spin for 50 min at 12000 rpm.
15. Carefully transfer the supernatant into a clean 50 mL conical tube, add 2 mL of pre-equilibrated Ni-NTA resin, fill the tube with the binding buffer to almost full capacity to avoid air bubbles in the next incubation step.
16. Place the conical tube onto a nutator, rock gently for 1 hour in cold room.
17. Transfer the slurry in from the conical tube to a gravity column, let it drain.
18. When the liquid level is about to enter the resin, add 20 mL of wash buffer to wash the resin. Do so without disturbing the resin. Repeat this step for three times.
19. Add 4x5 mL of elute buffer and collect the elution in a clean 50 mL tube.
20. Add 50 μL of 5 mg/mL 3C protease into the protein solution and incubate for 3 h in the cold room (or overnight in cold room).
21. Transfer the eluted sample into a protein concentrator and spin it down at 4000 rpm for 10 min at a 4 °C centrifuge. Agitate the solution to avoid a concentration gradient. Repeat the process until the sample is concentrated to 1 mL.
22. Remove any protein precipitation by centrifuging the microfuge tube at 16,000 rcf for 10 min in cold room.
23. Inject the ~1 mL supernatant to a Superdex 200 10/300 GL column pre-equilibrated in the FPLC buffer on an AKTA Pure chromatography system.
24. Run the gel-filtration chromatography column at 0.4 mL/min and collect the eluted fractions as 0.5 mL per fraction.
25. Take 10 μL of each peak fractions (for example, A7-A12 fractions in the profile in Fig. 2), mix with 5 μL of SDS loading dye, heat at 95 °C for 5 min, cool at room temperature for 10 min, Spin the sample for 1 min at 10000 rpm. And load 5 μL of sample on the 12% SDS-PAGE gel and run the gel at 150 V for 60 min.

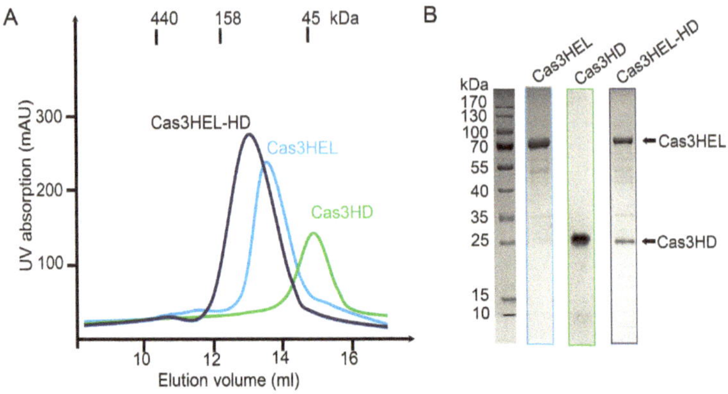

Fig. 2 Purification of Cas3HEL and HD. (A) Representative size-exclusion chromatography profile of Cas3HEL, Cas3HD and Cas3 complex. The elution volumes for HEL, HD and HEL-HD complex are 13.8 mL, 15.1 mL and 13.2 mL, respectively. This result indicates Cas3HEL and HD assembles into a stable complex. (B) Representative SDS-PAGE for these proteins.

26. Take off the SDS-PAGE gel and stain it on the eStain L1 Protein Staining Device in 10 min.
27. Based on the SDS-PAGE result, pool the peak fractions containing the pure proteins, following the previously described procedure to concentrate the protein to a final concentration of 5–10 mg/mL. Measure the concentration of HEL by this formula: C (mg/mL) = $A_{280} * 10^6 * 75/94{,}325\,M^{-1}\,cm^{-1}$. Flash-freeze the concentrated proteins in 50 μL aliquots for long-term storage in the −80 °C freezer. Typically, the yield of Cas3HEL from a 2 L culture volume is 2–5 mg. (Fig. 2)

3.3 Cas3HD purification

The protein expression and purification protocol for Cas3HD is identical to that for Cas3HEL. The estimated yield is 5–10 mg (10 mg/mL, 500–1000 μL) yield for HD from 2 L LB media. Measure the concentration of HEL by this formula: C (mg/mL) = $A_{280} * 10^6 * 27/14{,}565\,M^{-1}\,cm^{-1}$.

3.4 Complex formation of Cas3 (HEL-HD)

Timing: 3-4 h
Preparation: 1–2 h
Protocol: 2 h

Preparation

1. Prepare 1 mg Cas3HEL and 1 mg Cas3HD proteins. For example, 200 μL of 5 mg/mL HEL and 200 μL of 5 mg/mL HD.

2. Equilibrate the Superdex 200 10/30 column in FPLC buffer on AKTA chromatography system.

Protocol

1. Mix 1 mg HEL (10 mg/mL, 100 μL) with 1 mg HD (10 mg/mL, 100 μL) in a 1.5 mL microfuge tube (HD is two times excess than HEL)
2. Incubate on ice for 15 min.
3. Spin down any protein precipitates at 16,000 rcf for 10 min in the cold room.
4. Inject the cleared protein sample into Superdex 200 10/30 column on AKTA.
5. Run the gel-filtration chromatography and collect all elution fractions using the AKTA machine.
6. One should expect two peaks in the elution profile. The assembled Cas3 (HEL-HD) resides in the early peak; the second peak contains the excess subunit. Run a SDS-PAGE to decide which fractions to pool. For example, take 10 μL sample from each peak fractions and add 5 μL SDS loading dye. Heat 95 °C for 5 min, then spin it down for 1 min at 16,000 rcf, load 5 μL sample on the SDS-PAGE gel and run the gel at 150 V for 60 min.
7. Concentrate the protein to 250 μL and measure the concentration. Calculate the molar concentration from this formula: [Cas3] = A280 number x 10^6/108890 (μM). (Fig. 2)

3.5 Assay for RNA-guided DNA degradation activity in type I-A cascade-Cas3

Timing: 1.5–2.5 h
Preparation: 30 min
Protocol: 1–2 h

Preparation

1. Prepare type I-A Cascade complex (target on eGFP gene, the sequence refer to table 2.3) at 1 μM in the cleavage buffer. Here, we describe a 14-reaction assay, with 2 μL of Cascade sample per reaction. Prepare slight excess of samples, 30 μL.
2. Prepare Cas3HEL-HD complex protein at 1 μM by cleavage buffer. Prepare 30 μL.
3. Prepare couples of 0.2 mL PCR tubes.
4. Use a PCR machine with gradient function to different reaction temperatures. For example, the six rows can be set to 25, 30, 37, 42, 50 and 60 °C.
5. Prepare one 1% agarose gel in 0.5x TBE buffer with 20 lanes.

6. Prepare enough target plasmid (step 2.3). Each reaction requires 150ng, so the required amount should be 15 reactions x 150ng/reactions = 2250ng. We usually dilute the plasmid to 50ng/μL concentration. So, the target plasmid should be 45μL at 50ng/μL.
7. Prepare 14 PCR tubes and separate them into two sets. One set is for the cleavage assays without ATP and one with ATP. Cleavage assay tubes can be labeled as Ctrl, 25, 30, 37, 42, 50, and 60min. Degrading assay tubes can be labeled as Ctrl-ATP, 25-ATP, 30-ATP, 37-ATP, 42-ATP, 50-ATP and 60-ATP.

Protocol

The reaction volume is 20μL. And this table is the recipe for the reaction:

	Ctrl	**25**	**30**	**37**	**42**	**50**	**60**
pTarget (μL)	3	3	3	3	3	3	3
Cascade (μL)		2	2	2	2	2	2
Cas3HEL-HD (μL)		2	2	2	2	2	2
Cleavage buffer (μL)	17	13	13	13	13	13	13
Total (μL)	20						
	Ctrl-ATP	**25-ATP**	**30-ATP**	**37-ATP**	**42-ATP**	**50-ATP**	**60-ATP**
pTarget (μL)	3	3	3	3	3	3	3
Cascade (μL)		2	2	2	2	2	2
Cas3HEL-HD (μL)		2	2	2	2	2	2
Degrading buffer (μL)	17	13	13	13	13	13	13
Total (μL)	20						

1. Set up the reactions by mixing components according to the tables above.
2. Incubate the tubes at the pre-established temperatures in PCR machine (Ctrl tubes can be put at the bench).
3. Incubate the tubes for 30min.
4. Take all tubes out of PCR machine, transfer each sample to a new 1.5mL centrifuge tube (label the tubes clearly), add 30μL UltraPure™ Phenol, Vortex tubes for 2min.
5. Centrifuge the tubes for 15min at 15,000 rcf.

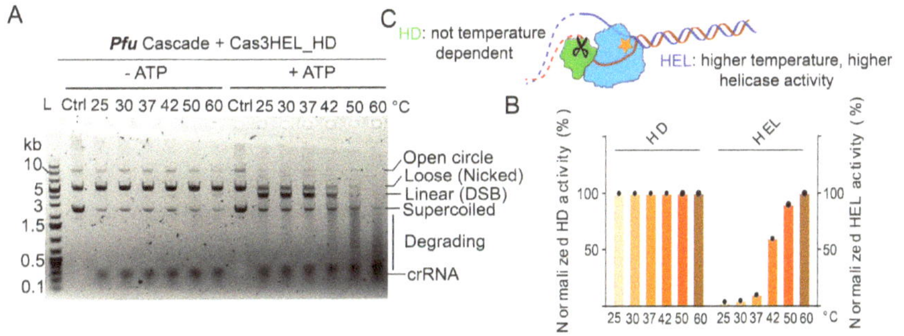

Fig. 3 RNA-guided processive DNA degradation activity inside the I-A Cascade-Cas3. (A) Plasmid degrading assay at different temperatures. DNA are targeted by the RNA guide inside the *Pfu*Cascade-Cas3 complex. Assays done without ATP present reveal the HD nuclease activity of Cas3, whereas those with ATP present reveal the helicase-facilitated processive DNA degradation by Cas3 HEL-HD complex. The downward smears reflect the DNA substrate is shredded into pieces. (B) Normalized DNA cleavage activity in Cas3 at different temperature. While the HD nuclease activity remains constant over the temperature range, the HEL-facilitated degradation is steeply temperature-dependent. (C) A schematic model summarizing the experimental findings.

6. Take 15 μL of sample from the aqueous phase (top layer) in each tube and transfer it into a new labeled 1.5 mL centrifuge tube.
7. Add 5 μL of DNA loading dye into each tube.
8. Load the sample onto a 1% agarose gel and run the gel in 0.5x TBE at 90 V for 30 min at room temperature.
9. Use the GelRed® Nucleic Acid Gel Stain at 1x diluted concentration (stock is 10,000 fold) to stain the gel for 5 min at room temperature.
10. Scan the gel using a ChemiDOC device. Record the data through the GelRed channel (Fig. 3).

4. Applications

4.1 Gene knock-out by temperature-dependent type I-A Cascade-Cas3HEL-HD system

Timing: 7–9 days hours

Preparation: 1 day

Protocol: 6–8 days

Preparation

Day 1:

1. Prepare one 10-cm tissue culture petri dish of HAP1_GFP cell at 50–60% confluency. This cell was originated from HAP1 cell which is a near-haploid cell line derived from the patient with

chronic myeloid leukemia (CML) expressing GFP under control of a CMV promoter. The cells showed have been thawed from liquid nitrogen storage and recovered for one to two passages.

Day 1:

2. Prepare type I-A Cascade (GFP guided) and Cas3 proteins at 10 µM (purified in Section 3). Proteins for genome editing experiments should have been sterilized by passing through the 0.22 µm micro-filter syringe into sterilized tubes before long-term cold storage.

28. Turn on UV-light to sterilize the clean bench.
29. Prewarm the DMEM media, PBS, and trypsin solutions.
30. Prewarm the Incucyte® Live-Cell Analysis System.
31. Prewarm the Neon Transfection System.
32. Prewarm a 24 wells cell culture plate. Aliquot Each 1 mL of DMEM media to each well.
33. Store plenty sterilized 1.5 mL centrifuge tubes inside the hood.

Protocol

1. Take the cell plate and use a sterile long-neck glass pipet connected to a vacuum pump to aspirate most of the media.
2. Add 2 mL 1X PBS, gently wash the cell and aspirate. Repeat the wash step 3 times.
3. Add 1.5 mL of trypsin solution, incubate the plate in 37 °C cell incubator for 10 min.
4. Add 2 mL DMEM (with 10% FBS) and resuspend the cell gently.
5. Transfer cell solution into a new 15 mL centrifuge tube.
6. Spin it down at 1000 *g* for 5 min.
7. Aspirate the media. Add 1 mL of 1X PBS to resuspend the cells.
8. Calculate the cell density. One dish typically provides 1×10^8 cells for editing experiments.
9. Dilute the cell density to 5×10^3 cells/µL using Neon electroporation buffer R.
10. Mix 2 µL Cascade, 2 µL Cas3 and 10 µL cell solution (from step 9).
11. Aspirate the mixture using the Neon Pipette into the gold tip. Follow the instructions to place the pipette into the Neon transfection device, with the tip submerged in buffer R.
12. Perform electroporation using the following setting: 1450 V, 13 ms, and 4 pulses.
13. Remove the pipette and dispense the cells into one well of the 24-well plate.
14. Repeat step 11–13 two more times to establish biological triplicates.
15. Place the plate into 37 °C incubator for temporary storage.

16. Repeat steps 11–15 for controls and other editing experiments. Afterwards, incubate the plate at 42 °C for 8 h, then transfer the plate to a 37 °C incubator.
17. Continue to culture the cells for 5 days before evaluating the genome editing efficiency at the GFP locus (Fig. 4). It takes ~4 days to deplete the GFP protein after gene knockout.

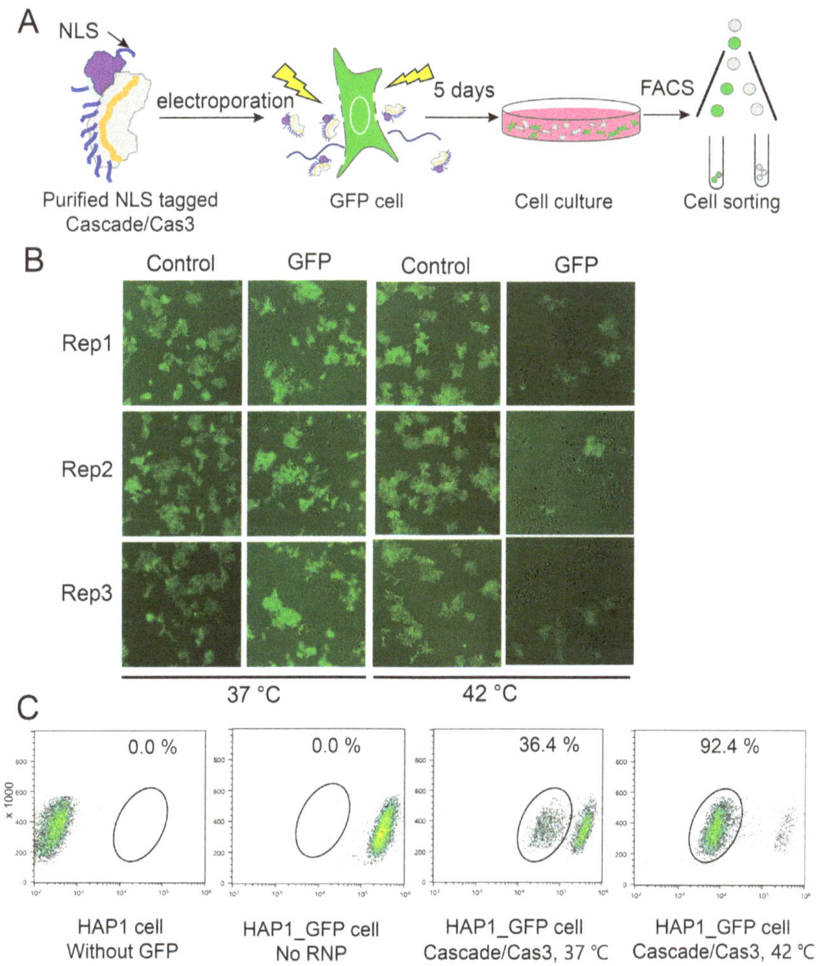

Fig. 4 Temperature-dependent genome editing in human cells by I-A Cascade-Cas3. (A) Workflow of the type I-A Cascade-Cas3 genome editing experiment. (B) Representative images for the genome editing experiments. Cascade was programmed with a GFP-targeting guide for the GFP knockout experiments, or a non-targeting guide for the control experiments. (C) FACS sorting charts showing the GFP editing efficiency by I-A Cascade-Cas3 at 37 and 42 °C.

18. GFP targeting efficiency is quantified using the fluorescence-activated cell sorting (FACS) instrument per the instructions. Editing efficiency = dark cell counts/total cell counts (dark + GFP^+).

5. Conclusions

Type I-A CRISPR-*Cas* systems are almost exclusively found in hyperthermophiles. We show that the biochemical properties of the *Pyrococcus furiosus* I-A system clearly reflect adaptations to the thermophilic environment, particularly in the helicase module of the Cas3 protein. The *Pfu*Cas3 is assembled from the HD nuclease subunit and the helicase subunit. The helicase activity enables Cas3 to translocate along the guide RNA-specified DNA substrate, shredding the DNA substrate into pieces. While the nuclease activity remained constant over a wide temperature range, we found that the helicase-facilitated processive DNA degradation activity in *Pfu*Cas3 was significantly more active at 42 °C or higher. This knowledge motivated us to introduce a transient 42 °C incubation step (8 h) when carrying out the I-A CRISPR-Cas mediated genome editing experiments in human cells, which improved the editing efficiency from 36% to 92%. Overall, this demonstrates the power of enzymology studies for applications in cell biology.

References

Barrangou, R., Fremaux, C., Deveau, H., Richards, M., Boyaval, P., Moineau, S., et al. (2007). CRISPR provides acquired resistance against viruses in prokaryotes. *Science*, *315*, 1709–1712.

Bolotin, A., Quinquis, B., Sorokin, A., & Ehrlich, S. D. (2005). Clustered regularly interspaced short palindrome repeats (CRISPRs) have spacers of extrachromosomal origin. *Microbiology*, *151*, 2551–2561.

Brouns, S. J., Jore, M. M., Lundgren, M., Westra, E. R., Slijkhuis, R. J., Snijders, A. P., et al. (2008). Small CRISPR RNAs guide antiviral defense in prokaryotes. *Science*, *321*, 960–964.

Dillard, K. E., Brown, M. W., Johnson, N. V., Xiao, Y., Dolan, A., Hernandez, E., et al. (2018). Assembly and Translocation of a CRISPR-Cas Primed Acquisition Complex. *Cell*.

Dolan, A. E., Hou, Z., Xiao, Y., Gramelspacher, M. J., Heo, J., Howden, S. E., et al. (2019). Introducing a spectrum of long-range genomic deletions in human embryonic stem cells using Type I CRISPR-Cas. *Molecular Cell*.

Elmore, J., Deighan, T., Westpheling, J., Terns, R. M., & Terns, M. P. (2015). DNA targeting by the type I-G and type I-A CRISPR-Cas systems of Pyrococcus furiosus. *Nucleic Acids Research*, *43*, 10353–10363.

Garrett, S., Shiimori, M., Watts, E. A., Clark, L., Graveley, B. R., & Terns, M. P. (2020). Primed CRISPR DNA uptake in Pyrococcus furiosus. *Nucleic Acids Research*, *48*, 6120–6135.

Hayes, R. P., Xiao, Y., Ding, F., van Erp, P. B., Rajashankar, K., Bailey, S., et al. (2016). Structural basis for promiscuous PAM recognition in type I-E cascade from E. coli. *Nature*, *530*, 499–503.

Hochstrasser, M. L., Taylor, D. W., Bhat, P., Guegler, C. K., Sternberg, S. H., Nogales, E., et al. (2014). CasA mediates Cas3-catalyzed target degradation during CRISPR RNA-guided interference. *Proceedings of the National Academy of Sciences of the United States of America*, *111*, 6618–6623.

Huo, Y., Nam, K. H., Ding, F., Lee, H., Wu, L., Xiao, Y., et al. (2014). Structures of CRISPR Cas3 offer mechanistic insights into Cascade-activated DNA unwinding and degradation. *Nature Structural & Molecular Biology*, *21*, 771–777.

Jackson, R. N., Golden, S. M., van Erp, P. B., Carter, J., Westra, E. R., Brouns, S. J., et al. (2014). Crystal structure of the CRISPR RNA-guided surveillance complex from Escherichia coli. *Science*, *345*, 1473–1479.

Kunne, T., Kieper, S. N., Bannenberg, J. W., Vogel, A. I., Miellet, W. R., Klein, M., et al. (2016). Cas3-derived target DNA degradation fragments fuel primed CRISPR adaptation. *Molecular Cell*, *63*, 852–864.

Majumdar, S., & Terns, M. P. (2019). CRISPR RNA-guided DNA cleavage by reconstituted Type I-A immune effector complexes. *Extremophiles*, *23*, 19–33.

Makarova, K. S., Grishin, N. V., Shabalina, S. A., Wolf, Y. I., & Koonin, E. V. (2006). A putative RNA-interference-based immune system in prokaryotes: computational analysis of the predicted enzymatic machinery, functional analogies with eukaryotic RNAi, and hypothetical mechanisms of action. *Biology Direct*, *1*, 7.

Makarova, K. S., Wolf, Y. I., Iranzo, J., Shmakov, S. A., Alkhnbashi, O. S., Brouns, S. J. J., et al. (2020). Evolutionary classification of CRISPR-Cas systems: A burst of class 2 and derived variants. *Nature Reviews. Microbiology*, *18*, 67–83.

Marraffini, L. A., & Sontheimer, E. J. (2008). CRISPR interference limits horizontal gene transfer in staphylococci by targeting DNA. *Science*, *322*, 1843–1845.

Mojica, F. J., García-Martínez, J., & Soria, E. (2005). Intervening sequences of regularly spaced prokaryotic repeats derive from foreign genetic elements. *Journal of Molecular Evolution*, *60*, 174–182.

Mulepati, S., & Bailey, S. (2013). In vitro reconstitution of an Escherichia coli RNA-guided immune system reveals unidirectional, ATP-dependent degradation of DNA target. *Journal of Biological Chemistry*, *288*, 22184–22192.

Mulepati, S., Heroux, A., & Bailey, S. (2014). Crystal structure of a CRISPR RNA-guided surveillance complex bound to a ssDNA target. *Science*, *345*, 1479–1484.

Pourcel, C., Salvignol, G., & Vergnaud, G. (2005). CRISPR elements in Yersinia pestis acquire new repeats by preferential uptake of bacteriophage DNA, and provide additional tools for evolutionary studies. *Microbiology*, *151*, 653–663.

Redding, S., Sternberg, S. H., Marshall, M., Gibb, B., Bhat, P., Guegler, C. K., et al. (2015). Surveillance and processing of foreign DNA by the Escherichia coli CRISPR-Cas system. *Cell*, *163*, 854–865.

Rutkauskas, M., Sinkunas, T., Songailiene, I., Tikhomirova, M. S., Siksnys, V., & Seidel, R. (2015). Directional R-loop formation by the CRISPR-Cas surveillance complex cascade provides efficient off-target site rejection. *Cell Reports*, *10*(9), 1534–1543.

Sashital, D. G., Wiedenheft, B., & Doudna, J. A. (2012). Mechanism of foreign DNA selection in a bacterial adaptive immune system. *Molecular Cell*, *46*, 606–615.

Sinkunas, T., Gasiunas, G., Waghmare, S. P., Dickman, M. J., Barrangou, R., Horvath, P., et al. (2013). In vitro reconstitution of Cascade-mediated CRISPR immunity in Streptococcus thermophilus. *EMBO Journal*, *32*, 385–394.

Ullah, R., Shah, M. A., Tufail, S., Ismat, F., Imran, M., Iqbal, M., et al. (2016). Activity of the human rhinovirus 3C protease studied in various buffers, additives and detergents solutions for recombinant protein production. *PLoS One*, *11*(4), e0153436.

Westra, E. R., van Erp, P. B., Kunne, T., Wong, S. P., Staals, R. H., Seegers, C. L., et al. (2012). CRISPR immunity relies on the consecutive binding and degradation of negatively supercoiled invader DNA by Cascade and Cas3. *Molecular Cell*, *46*, 595–605.

Wiedenheft, B., Lander, G. C., Zhou, K., Jore, M. M., Brouns, S. J. J., van der Oost, J., et al. (2011). Structures of the RNA-guided surveillance complex from a bacterial immune system. *Nature*, *477*, 486–489.

Xiao, Y., Luo, M., Dolan, A. E., Liao, M., & Ke, A. (2018). Structure basis for RNA-guided DNA degradation by Cascade and Cas3. *Science*, 361.

Xiao, Y., Luo, M., Hayes, R. P., Kim, J., Ng, S., Ding, F., et al. (2017). Structure basis for directional R-loop formation and substrate handover mechanisms in Type I CRISPR-Cas system. *Cell*, *170*(48-60), e11.

Zhao, H., Sheng, G., Wang, J., Wang, M., Bunkoczi, G., Gong, W., et al. (2014). Crystal structure of the RNA-guided immune surveillance Cascade complex in Escherichia coli. *Nature*, *515*, 147–150.

Further reading

Caliando, B. J., & Voigt, C. A. (2015). Targeted DNA degradation using a CRISPR device stably carried in the host genome. *Nature Communications*, *6*, 6989.

Cameron, P., Coons, M. M., Klompe, S. E., Lied, A. M., Smith, S. C., Vidal, B., et al. (2019). Harnessing type I CRISPR-Cas systems for genome engineering in human cells. *Nature Biotechnology*, *37*, 1471–1477.

Carte, J., Wang, R., Li, H., Terns, R. M., & Terns, M. P. (2008). Cas6 is an endoribonuclease that generates guide RNAs for invader defense in prokaryotes. *Genes Development*, *22*, 3489–3496.

Chen, Y., Liu, J., Zhi, S., Zheng, Q., Ma, W., Huang, J., et al. (2020). Repurposing type I-F CRISPR-Cas system as a transcriptional activation tool in human cells. *Nature Communications*, *11*, 3136.

Chen, J. S., Ma, E., Harrington, L. B., Da Costa, M., Tian, X., Palefsky, J. M., et al. (2018). CRISPR-Cas12a target binding unleashes indiscriminate single-stranded DNase activity. *Science*, *360*, 436–439.

Csorgo, B., Leon, L. M., Chau-Ly, I. J., Vasquez-Rifo, A., Berry, J. D., Mahendra, C., et al. (2020). A compact Cascade-Cas3 system for targeted genome engineering. *Nature Methods*, *17*, 1183–1190.

Elmore, J. R., Sheppard, N. F., Ramia, N., Deighan, T., Li, H., Terns, R. M., et al. (2016). Bipartite recognition of target RNAs activates DNA cleavage by the Type III-B CRISPR-Cas system. *Genes Development*, *30*, 447–459.

Foster, K., Gruschow, S., Bailey, S., White, M. F., & Terns, M. P. (2020). Regulation of the RNA and DNA nuclease activities required for Pyrococcus furiosus Type III-B CRISPR-Cas immunity. *Nucleic Acids Research*, *48*, 4418–4434.

Gong, B., Shin, M., Sun, J., Jung, C. H., Bolt, E. L., van der Oost, J., et al. (2014). Molecular insights into DNA interference by CRISPR-associated nuclease-helicase Cas3. *Proceedings of the National Academy of Sciences of the United States of America*, *111*, 16359–16364.

Gootenberg, J. S., Abudayyeh, O. O., Kellner, M. J., Joung, J., Collins, J. J., & Zhang, F. (2018). Multiplexed and portable nucleic acid detection platform with Cas13, Cas12a, and Csm6. *Science*, *360*, 439–444.

Govindarajan, S., Borges, A., & Bondy-Denomy, J. (2020). Distinct subcellular localization of a Type I CRISPR complex and the Cas3 nuclease in bacteria. *bioRxiv*. 2020. 2009.2029.318501.

He, L., Matosevic, Z. J., Mitic, D., Markulin, D., Killelea, T., Matkovic, M., et al. (2021). A tryptophan 'Gate' in the CRISPR-Cas3 nuclease controls ssDNA entry into the nuclease site, that when removed results in nuclease hyperactivity. *International Journal of Molecular Sciences*, *22*.

Hochstrasser, M. L., Taylor, D. W., Kornfeld, J. E., Nogales, E., & Doudna, J. A. (2016). DNA targeting by a minimal CRISPR RNA-guided cascade. *Molecular Cell*, *63*, 840–851.

Jia, N., Mo, C. Y., Wang, C., Eng, E. T., Marraffini, L. A., & Patel, D. J. (2019). Type III-A CRISPR-Cas Csm complexes: assembly, periodic RNA cleavage, DNase activity regulation, and autoimmunity. *Molecular Cell*, *73*(264-277), e265.

Li, Y., Pan, S., Zhang, Y., Ren, M., Feng, M., Peng, N., et al. (2016). Harnessing Type I and Type III CRISPR-Cas systems for genome editing. *Nucleic Acids Research*, *44*, e34.

Lin, J., Fuglsang, A., Kjeldsen, A. L., Sun, K., Bhoobalan-Chitty, Y., & Peng, X. (2020). DNA targeting by subtype I-D CRISPR-Cas shows type I and type III features. *Nucleic Acids Research*, *48*, 10470–10478.

Luo, M. L., Mullis, A. S., Leenay, R. T., & Beisel, C. L. (2015). Repurposing endogenous type I CRISPR-Cas systems for programmable gene repression. *Nucleic Acids Research*, *43*, 674–681.

McBride, T. M., Schwartz, E. A., Kumar, A., Taylor, D. W., Fineran, P. C., & Fagerlund, R. D. (2020). Diverse CRISPR-Cas complexes require independent translation of small and large subunits from a single gene. *Molecular Cell*, *80*(971-979), e977.

Morisaka, H., Yoshimi, K., Okuzaki, Y., Gee, P., Kunihiro, Y., Sonpho, E., et al. (2019). CRISPR-Cas3 induces broad and unidirectional genome editing in human cells. *Nature Communications*, *10*, 5302.

O'Brien, R. E., Santos, I. C., Wrapp, D., Bravo, J. P. K., Schwartz, E. A., Brodbelt, J. S., et al. (2020). Structural basis for assembly of non-canonical small subunits into type I-C cascade. *Nature Communications*, *11*, 5931.

Osakabe, K., Wada, N., Miyaji, T., Murakami, E., Marui, K., Ueta, R., et al. (2020). Genome editing in plants using CRISPR type I-D nuclease. *Communications Biology*, *3*, 648.

Osakabe, K., Wada, N., Murakami, E., Miyashita, N., & Osakabe, Y. (2021). Genome editing in mammalian cells using the CRISPR type I-D nuclease. *Nucleic Acids Research*, *49*, 6347–6363.

Pickar-Oliver, A., Black, J. B., Lewis, M. M., Mutchnick, K. J., Klann, T. S., Gilcrest, K. A., et al. (2019). Targeted transcriptional modulation with type I CRISPR-Cas systems in human cells. *Nature Biotechnology*, *37*, 1493–1501.

Punjani, A., Rubinstein, J. L., Fleet, D. J., & Brubaker, M. A. (2017). cryoSPARC: algorithms for rapid unsupervised cryo-EM structure determination. *Nature Methods*, *14*, 290–296.

Rath, D., Amlinger, L., Hoekzema, M., Devulapally, P. R., & Lundgren, M. (2015). Efficient programmable gene silencing by cascade. *Nucleic Acids Research*, *43*, 237–246.

Rollins, M. F., Chowdhury, S., Carter, J., Golden, S. M., Miettinen, H. M., Santiago-Frangos, A., et al. (2019). Structure reveals a mechanism of CRISPR-RNA-guided nuclease recruitment and Anti-CRISPR viral mimicry. *Molecular Cell*, *74*(132-142), e135.

Sofos, N., Feng, M., Stella, S., Pape, T., Fuglsang, A., Lin, J., et al. (2020). Structures of the Cmr-beta complex reveal the regulation of the immunity mechanism of type III-B CRISPR-Cas. *Molecular Cell*, *79*(741-757), e747.

Spilman, M., Cocozaki, A., Hale, C., Shao, Y., Ramia, N., Terns, R., et al. (2013). Structure of an RNA silencing complex of the CRISPR-Cas immune system. *Molecular Cell*, *52*, 146–152.

Szczelkun, M. D., Tikhomirova, M. S., Sinkunas, T., Gasiunas, G., Karvelis, T., Pschera, P., et al. (2014). Direct observation of R-loop formation by single RNA-guided Cas9 and Cascade effector complexes. *Proceedings of the National Academy of Sciences of the United States of America, 111*, 9798–9803.

Taylor, D. W., Zhu, Y., Staals, R. H., Kornfeld, J. E., Shinkai, A., van der Oost, J., et al. (2015). Structural biology. Structures of the CRISPR-Cmr complex reveal mode of RNA target positioning. *Science, 348*, 581–585.

Wang, L., Mo, C. Y., Wasserman, M. R., Rostol, J. T., Marraffini, L. A., & Liu, S. (2019). Dynamics of Cas10 govern discrimination between self and non-self in Type III CRISPR-Cas immunity. *Molecular Cell, 73*(278–290), e274.

Yoshimi, K., Takeshita, K., Kodera, N., Shibumura, S., Yamauchi, Y., Omatsu, M., et al. (2021). Dynamic mechanisms of CRISPR interference by <em>Escherichia coli</em> CRISPR-Cas3. *bioRxiv*. 2021.2007.2018.452824.

You, L., Ma, J., Wang, J., Artamonova, D., Wang, M., Liu, L., et al. (2019). Structure studies of the CRISPR-Csm complex reveal mechanism of co-transcriptional interference. *Cell, 176*(239-253), e216.

Young, J. K., Gasior, S. L., Jones, S., Wang, L., Navarro, P., Vickroy, B., et al. (2019). The repurposing of type I-E CRISPR-Cascade for gene activation in plants. *Communications Biology, 2*, 383.

Zhang, J., Rouillon, C., Kerou, M., Reeks, J., Brugger, K., Graham, S., et al. (2012). Structure and mechanism of the CMR complex for CRISPR-mediated antiviral immunity. *Molecular Cell, 45*, 303–313.

CHAPTER FIFTEEN

Purification and characterization of Mtr4 and TRAMP from *S. cerevisiae*

Matthew K. Yim, Joshua M. Denson, Mark D. Gold, and Sean J. Johnson*
Department of Chemistry & Biochemistry, Utah State University, Logan, Utah, United States
*Corresponding author: e-mail address: sean.johnson@usu.edu

Contents

Methods in Enzymology, Volume 673
ISSN 0076-6879
https://doi.org/10.1016/bs.mie.2022.03.042

Abstract

The Ski2-like RNA helicase, Mtr4, plays a central role in nuclear RNA surveillance pathways by delivering targeted substrates to the RNA exosome for processing or degradation. RNA target selection is accomplished by a variety of Mtr4-mediated protein complexes. In *S. cerevisiae*, the Trf4/5-Air1/2-Mtr4 polyadenylation (TRAMP) complex prepares substrates for exosomal decay through the combined action of polyadenylation and helicase activities. Biophysical and structural studies of Mtr4 and TRAMP require highly purified protein components. Here, we describe robust protocols for obtaining large quantities of pure, active Mtr4 and Trf4-Air2 from *S. cerevisiae*. The proteins are recombinantly expressed in *E. coli* and purified using affinity, ion exchange, hydrophobic exchange and size exclusion chromatography. Care is taken to remove nuclease contamination during the prep. Assembly of TRAMP is achieved by combining individually purified Mtr4 and Trf4-Air2. We further describe a strand displacement assay to characterize Mtr4 helicase unwinding activity.

1. Introduction

Mtr4 belongs to the Ski2-like family of RNA helicases. This family functions in diverse biological roles, including RNA processing and turnover, pre-mRNA splicing, translation, and DNA recombination and repair (Jankowsky & Fairman-Williams, 2010). Mtr4 plays a central role in nuclear RNA surveillance pathways by providing a bridge between RNA targeting and RNA decay (Johnson & Jackson, 2013; Olsen & Johnson, 2021; Weick & Lima, 2021). Mtr4 identifies RNA substrates directly or in coordination with adaptor proteins and complexes (Schmid & Jensen, 2019). Targeted substrates are threaded through the helicase and delivered to the RNA exosome, which processes or degrades the RNA (Schuller, Sebastian, Lisa, Ed, & Conti, 2018; Weick et al., 2018).

Mtr4 is conserved throughout eukaryotic systems and is essential in yeast (de la Cruz, Kressler, Tollervey, & Linder, 1998). Mutation of the conserved DExH motif in the RecA1 domain abolishes ATPase and unwinding activity in vitro (Wang, Jia, Jankowsky, & Anderson, 2008) and is lethal (Taylor et al., 2014). Mtr4 translocates across RNA substrates in a 3′-5′ direction (Bernstein, Patterson, Wilson, & Toth, 2008; Wang et al., 2008), but is unable to translocate across DNA (Jia, Wang, Anderson, & Jankowsky, 2012; Weick et al., 2018). A short 3′ single stranded overhang is required for productive binding and translocation (Jia et al., 2011, 2012). In *S. cerevisiae*, Mtr4 (*Sc*Mtr4) has a modest preference for a 3′ polyadenylated overhang (Bernstein et al., 2008; Jia et al., 2012). Mtr4 moves more slowly over longer single stranded regions. A single molecule study showed that Mtr4 only translocates along an artificial substrate when a duplex region is located immediately upstream of the *Sc*Mtr4 binding site (Patrick, Srinivasan, Jankowsky, & Comstock, 2017).

Mtr4 engages with a variety of adaptor proteins and complexes to identify RNA substrates for degradation (Ogami, Chen, & Manley, 2018; Olsen & Johnson, 2021; Schmid & Jensen, 2019). These interactions can affect Mtr4 helicase activity. For example, the *S. cerevisiae* Trf4/5-Air1/2-Mtr4 polyadenylation (TRAMP) complex is composed of an RNA polymerase (Trf4 or Trf5), a zinc knuckle containing protein (Air1 or Air2), and Mtr4 (LaCava et al., 2005; Vanacova et al., 2005; Wyers et al., 2005) (Fig. 1). The complex targets RNA substrates for degradation by adding a short poly(A) tail to the 3′ end, which then passes through Mtr4 en route to the exosome (Fig. 1C). Different combinations of Trf4/5 and Air1/2 target different substrates and appear to direct localization within the nucleus (Delan-Forino, Spanos, Rappsilber, & Tollervey, 2020; Houseley, Lacava, & Tollervey, 2006). Notably, Trf4-Air2-Mtr4 (for simplicity, hereafter referred to as *Sc*TRAMP) formation enhances Mtr4 unwinding on a polyadenylated substrate by four- to sixfold (Jia et al., 2012).

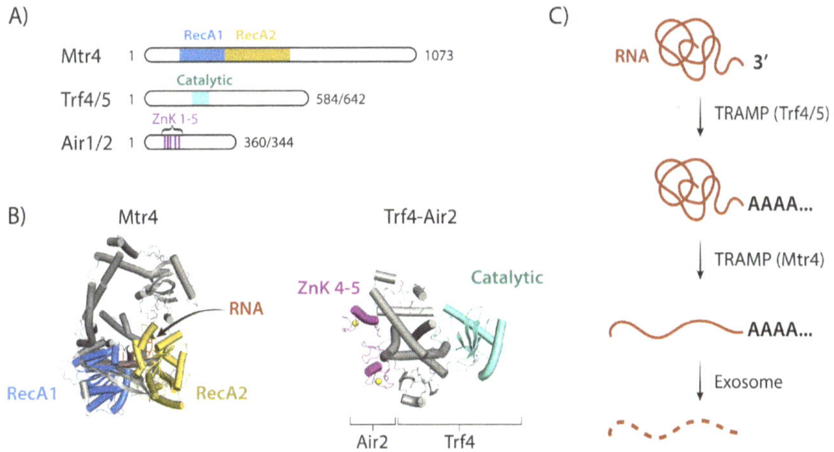

Fig. 1 The *S. cerevisiae* TRAMP complex. (A) The complex is composed of an RNA helicase (Mtr4), poly(A) polymerase (Trf4 or Trf5), and a zinc knuckle containing protein (Air1 or Air2). The position of the Mtr4 RecA domains, Trf4/5 catalytic domain, and Air1/2 zinc knuckles is highlighted. Numbers indicate amino acid length. (B) Structures of *Sc*Mtr4 with a 5 nt RNA (PDB ID: 2XGJ) and *Sc*Trf4-Air2 (PDB ID: 3NYB), colored as in panel A. (C) TRAMP targets RNA substrates for exosomal decay by adding a short poly(A) tail to the RNA 3′ end and unwinding substrates in a 3′-5′ direction.

Here, we describe protocols for recombinant expression and purification of Mtr4, Trf4-Air2, and TRAMP from *S. cerevisiae*. These protocols work well for wild-type and mutant forms of Mtr4, Trf4, and a C-terminal truncation of Air2 (see explanation of Air2 truncation in *Expression and Purification of ScTrf4-Air2*). Recombinant expression and purification have also been reported for Mtr4 and Mtr4 homologs from *S. pombe* (Weick et al., 2018), *N. crassa* (Conrad et al., 2016; Morales, Olsen, Bulcher, & Johnson, 2018), and *H. sapiens* (Puno & Lima, 2018). An important aspect of the *Sc*Mtr4 purification presented here is elimination of contaminating nucleases. Nuclease contamination is often associated with compromised helicase activity even in the presence of RNase inhibitors. It also can prevent the productive assembly of the *Sc*TRAMP complex. The protocols presented here consistently yield high levels of RNase-free, active protein.

We also describe a method for measuring *Sc*Mtr4 unwinding using a minimal RNA substrate. An important consideration in setting up the assay is the presence of competing polyadenylation activity in the TRAMP complex. We therefore measure *Sc*Mtr4 unwinding activity within *Sc*TRAMP using a catalytically inactive mutant of *Sc*Trf4 (*Sc*Trf4$^{D236A/D238A}$). Methods for reconstituting the RNA exosome, including exosome-Mtr4 complexes, have been described in detail previously (Januszyk, Weick, & Lima, 2020).

2. Expression and purification of *Sc*Mtr4

2.1 Before you begin

*Sc*Mtr4 is typically expressed from an ampicillin resistant pET151-D-topo *E. coli* expression vector (Invitrogen) with an N-terminal, TEV-protease cleavable, hexahistidine-tag. *Sc*Mtr4 can tolerate other N-terminal tags such as MBP (Bernstein et al., 2008), TAP (LaCava et al., 2005), or HA tags (Lim et al., 2017), but C-terminal tags should be avoided. In our experience, C-terminal tags cause significant degradation during purification and deleterious effects have been similarly noted in vivo (Schuch et al., 2014).

*Sc*Mtr4 expresses optimally in an *E. coli* BL21-CodonPlus (DE3) cell line. The expression of Mtr4 is robust using an autoinduction method over a 48–56 h time course (Studier, 2005). However, overgrowth of cells (beyond 56 h) can result in increased RNase contamination and protein degradation.

Cells are lysed by sonication and clarified by centrifugation. *Sc*Mtr4 is purified by Ni-NTA immobilized metal affinity, heparin affinity, ion-exchange, and size exclusion chromatography (SEC). Nuclease contamination is monitored throughout the prep using a commercial kit (RNaseAlert).

The high salt wash over Ni-NTA and the stepped elution off heparin described below is particularly important for removing nuclease contamination. Protein expression and purity are evaluated by SDS-PAGE. We typically grow 2 L of cell culture, which yields 18–25 g of cell pellet and ~7 mg of purified *Sc*Mtr4 (3.5 mg of purified *Sc*Mtr4/L of cell culture). The prep can be scaled down as needed.

2.2 Prepare the following reagents

1. Sterile Luria-Bertani (LB) medium (For 1 L: 10 g tryptone, 10 g NaCl, 5 g yeast extract. Bring up to volume in water)
2. Sterile ZY medium (For 1 L: 10 g tryptone, 5 g yeast extract. Bring up to volume in water)
3. Sterile 50 × 5052 (2.5% (*w*/w) glucose, 10% (w/w) α-lactose, 25% glycerol)
4. Sterile 20 × NPS (500 mM ammonium sulfate, 1 M potassium dihydrogen phosphate, 1 M sodium phosphate dibasic anhydrous)
5. Sterile 1 M $MgSO_4$
6. Lysis buffer (50 mM HEPES pH 7.5, 500 mM NaCl, 5% glycerol, 7 mM imidazole, 2 mM BME)
7. Wash 1 (50 mM HEPES pH 7.5, 2.5 M NaCl, 5% glycerol, 15 mM imidazole, 2 mM BME)
8. Wash 2 (50 mM HEPES pH 7.5, 1 M NaCl, 5% glycerol, 20 mM imidazole, 2 mM BME)
9. Wash 3 (50 mM HEPES pH 7.5, 160 mM NaCl, 5% glycerol, 30 mM imidazole, 2 mM BME)
10. Nickel elution buffer (50 mM HEPES pH 7.5, 160 mM NaCl, 5% glycerol, 500 mM imidazole (don't re-pH after addition of imidazole, see note below), 2 mM BME)
11. Buffer A (50 mM HEPES pH 7.5, 160 mM NaCl, 5% glycerol, 2 mM BME)
12. Buffer B (50 mM HEPES pH 7.5, 1 M NaCl, 5% glycerol, 2 mM BME)
13. Buffer C (50 mM HEPES pH 7.5, 50 mM NaCl, 5% glycerol, 2 mM BME)
14. Storage buffer (50 mM HEPES pH 7.5, 160 mM NaCl, 55% glycerol)

Note: Filter nickel elution buffer and buffers A-C before running over the chromatography columns. The high concentration of imidazole in the nickel elution buffer increases the pH, but in our in our experience this improves Mtr4 elution off the nickel column.

Note: For a single *Sc*Mtr4 purification prep, you will need approximately 200 mL lysis buffer, 2 L buffer A, 500 mL buffer B, 1 L buffer C, and 100 mL each of wash 1–3.

2.3 Key resources table

Reagent or resource	Source	Identifier
Bacterial and Virus Strains		
BL21-CodonPlus (DE3)-RIPL or Rosetta 2 (DE3) cells		
Chemicals, Peptides, and Recombinant Proteins		
4-(2-hydroxyethyl)piperazine-1-ethanesulfonic acid (HEPES)	Gold Biotechnology	7365-45-9
Glycerol	PHARMCO	56-81-5
Sodium Chloride (NaCl)	Thermo Fisher Scientific	7647-14-5
Isopropyl ß-D-1-thiogalactopyranoside (IPTG)	Gold Biotechnology	367-93-1
Lysozyme	Gold Biotechnology	12650-88-3
Aprotinin	Gold Biotechnology	9087-70-1
Pepstatin	Gold Biotechnology	p-020-25
Leupeptin Hemisulfate	Gold Biotechnology	103476-89-7
Phenylmethylsulfonyl fluoride (PMSF)	Thermo Fisher Scientific	329-98-6
Imidazole	Alfa Aesar	288-32-4
Ampicillin	Gold Biotechnology	69-52-3
Chloramphenicol	Gold Biotechnology	56-75-7
2-Mercaptoethanol (BME)	Sigma-Aldrich	60-24-2
0.45 μM syringe filters	Millipore Sigma	SLHVR33RS
Magnesium sulfate ($MgSO_4$)	Millipore Sigma	10034-99-8
Ammonium sulfate	Thermo Fisher Scientific	7783-20-2
Potassium dihydrogen phosphate	Thermo Fisher Scientific	7778-77-0

—cont'd

Reagent or resource	Source	Identifier
Sodium phosphate dibasic anhydrous	VWR Life Science	7558-79-4
α-Lactose	Millipore Sigma	5989-81-1
Tryptone	Thermo Fisher Scientific	91079-40-2
Yeast extract	Thermo Fisher Scientific	8013-01-2
Glucose	Thermo Fisher Scientific	50-99-7
Zinc sulfate ($ZnSO_4$)	Thermo Fisher Scientific	7446-20-0
Critical Commercial Assays		
RNaseAlert	Integrated DNA Technologies	11-02-01-02

2.4 Materials and equipment

- Temperature controlled shaker
- Refrigerated centrifuge
- Bottle-top filter apparatus and 0.45 μm PVDF membrane (Millipore Sigma)
- Sonicator (Branson sonifier 450)
- Electrophoresis system (Power Pac HC and Mini-Protean Tetra system, Bio-Rad)
- Gel shaker
- Ni-NTA agarose resin (Gold Biotechnology)
- Flex chromatography column (Kimble)
- Chromatography pump system
- 5 mL HiTrap Heparin column (Cytiva)
- 5 mL HiTrap Phenyl column (Cytiva)
- 15 mL HiPrep 26/10 Desalting column (Cytiva)
- 5 mL HiTrap DEAE column (Cytiva)
- HiLoad 26/60 Superdex 200 column (Cytiva)
- Centrifugal concentrator (Vivaspin or Vivaspin Turbo*)
- 12–14 kDa MWCO dialysis tubing
- NanoDrop spectrophotometer (ND-1000, Thermo Scientific)
- Spectrophotometer to measure OD_{600}
- Imager for SDS-PAGE gel and RNaseAlert (Bio-Rad ChemiDoc MP, or equivalent)

2.5 Step-by-step method details

2.5.1 *ScMtr4* Cell Growth

Timing: 2–3 days

Day 1:

1. From a single colony of freshly transformed cells or glycerol stock, inoculate 5 mL of LB media containing appropriate antibiotics.
2. Incubate overnight at 200 RPM, 37 °C.

Day 2:

3. Prepare 500 mL autoinduction growth media in 2.8 L baffled Fernbach Flask by mixing ~464 mL ZY media, 0.5 mL 1 M $MgSO_4$, 10 mL 50 × 5052, 25 mL 20 × NPS, and appropriate antibiotics.
4. Spin down overnight culture. Discard media and resuspend in fresh LB media. Inoculate the 500 mL autoinduction growth media with 1 mL of the resuspended overnight culture.
5. Incubate for 6–8 h at 200 RPM, 37 °C.
6. Move to room temperature and continue shaking at 100 RPM for 42–48 h. Room temperature growth results in optimal expression of soluble Mtr4, while poor expression is observed if cells are kept at 37 °C.
7. Harvest cells by centrifugation at 6000–9000 RPM, 4 °C for 25 min. Discard supernatant and store pellet at −80 °C.

2.6 Purification of Recombinantly Expressed *ScMtr4*

Timing: 2 days

Critical: *Sc*Mtr4 tends to degrade significantly if chromatography steps are not completed within 24 h. The entire purification process from cell lysis to final concentration should take no longer than 36 h. Protein samples should be kept cool on ice or in a cold room throughout the entire purification process.

Day 1:

1. Thaw cell pellet and transfer to metal beaker on ice.
2. Resuspend in lysis buffer (2 × cell mass volume). For example, resuspend 25 g of cells in 50 mL of lysis buffer.
3. Add protease inhibitors and lysozyme: 1 μg/mL leupeptin, 1.4 μg/mL pepstatin A, 1 μg/mL aprotinin, 175 μg/mL PMSF, and 4 mg/mL lysozyme. Incubate on ice for 10–30 min.
4. Sonicate cells with 60% amplitude and 80% duty cycle for 30 pulses/cycle using a 1/2″ (12.7 mm) probe. Keep sample on ice and allow to cool between cycles. Cell lysis is deemed complete when the lysate has a consistency similar to water when dispensing droplets from a transfer pipette (~10 cycles).

5. Clarify the lysate by centrifugation at 18,000 RPM, 4 °C for 25 min.
6. Equilibrate 2–5 mL Ni-NTA resin with lysis buffer. Mix clarified lysate with equilibrated resin. Gently shake or stir at 4 °C for 1 h.

Critical: Incubation with the Ni-NTA resin for longer than 1 h can result in excessive binding of contaminating proteins to the resin.

7. Apply lysate/resin to a Flex column. Elute lysate and wash with ~50 mL of lysis buffer followed by 100 mL each of washes 1–3.
8. Elute protein from Ni-NTA resin with 100 mL of nickel elution buffer.
9. Run elution samples on SDS-PAGE to verify presence of *Sc*Mtr4.
10. Run elution fractions containing *Sc*Mtr4 over a 5 mL HiTrap Heparin column equilibrated with buffer A:
    a. Inject sample at 1.5 mL/min. Subsequent washes and elutions run at 2.5 mL/min.
    b. Wash with 5 column volumes (CV) of buffer A.
    c. Elute using a 10 CV linear gradient from buffer A to buffer B. Hold gradient when the A_{280} peak begins to rapidly increase (usually between 25% and 27% buffer B). This fraction will contain *Sc*Mtr4 (Fig. 2, peak 1).

Critical: RNase and protein contamination are likely to occur if the hold starts later than the recommended range.

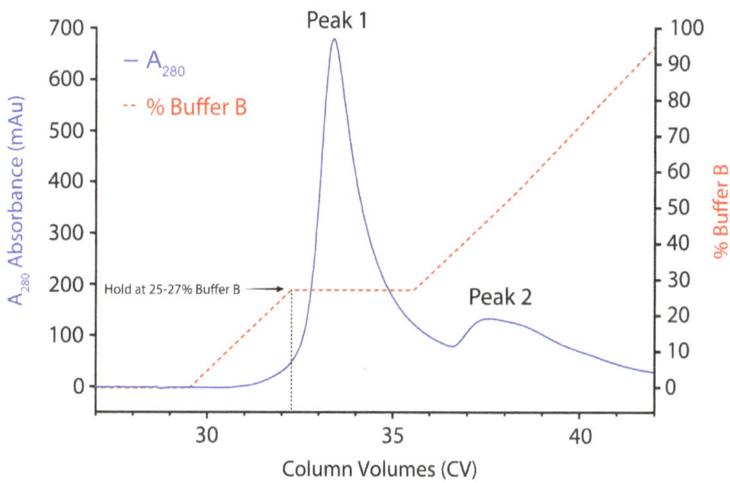

Fig. 2 *Sc*Mtr4 purification by heparin chromatography. A typical heparin purification step is shown in the chromatogram. The A_{280} trace (blue line) shows two *Sc*Mtr4-containing peaks that elute during the salt gradient (red line). The gradient is held when peak 1 begins to elute (between 25% and 27% buffer B, black line). This hold effectively separates relatively RNase free *Sc*Mtr4 (peak 1) from heavily RNase contaminated *Sc*Mtr4 (peak 2).

d. Continue hold until peak returns to baseline, resume gradient to 100% buffer B.

Note: A second A_{280} peak will come off upon continuation of the gradient. This peak contains *Sc*Mtr4, but it is highly contaminated with nucleases and should be discarded (Fig. 2, peak 2).

11. Run peak 1 (step 10c) fractions on SDS-PAGE and test for RNase contamination using RNaseAlert. RNaseAlert substrate is prepared per manufacturer's instructions in clear 0.65 mL microcentrifuge tubes and mixed with Mtr4 fractions. A negative control (nuclease-free water) should be made fresh and may be used for up to 1 day. A positive control (RNase I) can be used for up to a month if it is stored in foil. Image all RNaseAlert tubes using a UV transilluminator (Epi-blue setting on Bio-Rad ChemiDoc XP, 460–490 nm excitation). To reduce background signal from tubes, use a 530/28 emission/bandpass filter (Bio-Rad ChemiDoc XP setting). Fluorescence indicates nuclease-contamination. Some RNase contamination will likely be observed in all fractions. However, fractions with higher molecular weight protein contaminants (which are unlikely to be separated over size exclusion) or heavy RNase contamination should be discarded. We typically discard fractions with fluorescence intensity comparable to the positive control, as judged by eye.

12. Pool the selected fractions and buffer exchange the sample into buffer C using a HiPrep 26/10 desalting column. For larger preps, we use two 15 mL desalting columns connected in tandem.

13. Load buffer exchanged *Sc*Mtr4 over a 5 mL HiTrap DEAE column, equilibrated with buffer C:

a. Inject sample at 1.5 mL/min. Subsequent washes and elutions run at 2.5 mL/min.

b. Wash with 5 CV of buffer C.

c. Elute using a 10 CV linear gradient from buffer C to buffer B. We typically hold the gradient when *Sc*Mtr4 begins to elute at 12–16% buffer B until the A_{280} returns to baseline.

14. Run peak fractions on SDS-PAGE to verify *Sc*Mtr4 presence and test for RNase contamination using RNaseAlert. Discard highly contaminated fractions (see step 11).

15. Pool selected *Sc*Mtr4 fractions and run sample onto HiLoad 26/60 Superdex 200 size exclusion chromatography column (sample volume capacity $\leq$13 mL), pre-equilibrated with buffer A. Elute sample with buffer A at 1.25 mL/min.

Day 2:

16. Run peak fractions on SDS-PAGE and test for RNase contamination using RNaseAlert.
17. Pool clean fractions and concentrate using 50,000 MWCO centrifugal concentrator at 1600 RPM (410 × *g*), 4 °C, 12 min intervals, mixing by pipette in between intervals.
18. Dilute concentrated protein with *Sc*Mtr4 storage buffer in a 1:1 ratio. Addition of the storage buffer increases the glycerol content of the protein sample to 30% (cryoprotectant).
19. Flash freeze 10–20 μL aliquots of *Sc*Mtr4 in liquid nitrogen and store at −80 °C.

3. Expression and purification of *Sc*Trf4-Air2

3.1 Before you begin

*Sc*Trf4 and *Sc*Air2 are co-expressed on an ampicillin resistant pETDuet-1 vector as originally designed by the Jankowsky lab (Jia et al., 2011). *Sc*Air2 is placed in multiple cloning site 1 with an N-terminal hexahistidine tag. *Sc*Trf4 is in cloning site 2 with a C-terminal FLAG-tag. *Sc*Air2 (344 aa) is typically expressed with a large C-terminal truncation because the C-terminus is largely unstructured and expression of the full-length protein is highly susceptible to degradation. Our construct removes 121 aa from the C-terminus, while retaining all 5 zinc knuckles, to give *Sc*Air2$^{1-223}$. *Sc*Trf4 and *Sc*Air2 generally do not express well in isolation, but when co-expressed they form a tight complex that is resistant to salt concentrations as high as 1 M NaCl (Vanacova et al., 2005).

*Sc*Trf4-Air2 expresses optimally in an *E. coli*, BL21-CodonPlus (DE3) cell line. Zinc is added to the growth to facilitate assembly of the *Sc*Air2 zinc knuckles. Protein expression is induced by IPTG induction. The optical density of the culture before induction cannot exceed 0.8–1.0 OD_{600} and the induction phase cannot exceed 3 h as late induction or excessive induction time will result in degradation products that are difficult to remove during purification.

Cells are lysed by sonication and clarified by centrifugation. *Sc*Trf4-Air2 is purified by Ni-NTA immobilized metal affinity, heparin affinity, reverse-phase, and size exclusion chromatography. Previously published protocols utilized an anti-FLAG purification step (Jia et al., 2011; Taylor et al., 2014). However, the protocol described below eliminates that step, making it more cost effective. Protein expression and purity are evaluated by SDS-PAGE. We typically grow 8 L of cell culture, which yields ~20 g of cell pellet and ~2 mg of purified *Sc*Trf4-Air2$^{1-223}$ (0.26 mg of purified *Sc*Trf4-Air2$^{1-223}$/L of cell culture). The prep can be scaled down as needed.

3.2 Prepare the following reagents

1. Sterile Luria-Bertani (LB) medium (For 1 L: 10 g tryptone, 10 g NaCl, 5 g yeast extract. Bring up to volume in water)
2. Sterile 1 M $ZnSO_4$
3. Lysis buffer (50 mM HEPES pH 7.5, 500 mM NaCl, 10% glycerol, 10 mM imidazole, 2 mM BME)
4. Nickel elution buffer (50 mM HEPES pH 7.5, 150 mM NaCl, 10% glycerol, 500 mM imidazole, 2 mM BME)
5. Buffer A (50 mM HEPES pH 7.5, 150 mM NaCl, 10% glycerol, 2 mM BME)
6. Buffer B (50 mM HEPES pH 7.5, 1 M NaCl, 10% glycerol, 2 mM BME)
7. Buffer C (50 mM HEPES pH 7.5, 1 M ammonium sulfate, 10% glycerol, 2 mM BME)
8. Buffer D (50 mM HEPES pH 7.5, 100 mM ammonium sulfate, 10% glycerol, 2 mM BME)

Note: Filter nickel elution buffer and buffers A-D before running over chromatography columns.

3.3 Step-by-step method details

3.3.1 *ScTrf4-Air2 Cell Growth*

Timing: 2 days

1. From a single colony of freshly transformed cells or glycerol stock, inoculate 25 mL of LB media containing appropriate antibiotics.
2. Incubate overnight at 200 RPM, 37 °C.
3. Prepare 800 mL of growth media in 2.8 L baffled Fernbach flask by mixing 775 mL LB medium, 200 μL 1 M $ZnSO_4$ (250 μM final concentration), and appropriate antibiotics.
4. Inoculate the growth media with 25 mL of overnight culture (from step 2).
5. Incubate at 37 °C, 200 RPM, until OD_{600} reaches 0.8–1.0.
6. Add IPTG to a final concentration of 0.5 mM.
7. Continue shaking at room temperature for 3 h.
8. Harvest cells by centrifugation at 6000–9000 RPM, 4 °C for 25 min. Discard supernatant and store pellet at −80 °C.

3.4 Purification of *Sc*Trf4-Air2

Timing: 3 days

Critical: Protein samples should be kept cool on ice or in the cold room throughout the entire purification process.

Day 1:

1. Thaw cell pellet and transfer to metal beaker on ice.
2. Resuspend in lysis buffer (2 × cell mass volume). For example, resuspend 20 g of cells in 40 mL of lysis buffer.
3. Add protease inhibitors and lysozyme: 1 μg/mL leupeptin, 1.4 μg/mL pepstatin A, 1 μg/mL aprotinin, and 0.5 mg/mL lysozyme. Incubate on ice for 10–30 min.
4. Sonicate cells with 60% amplitude and 60% duty cycle for 15 pulses/cycle using a 1/2" (12.7 mm) probe. Keep sample on ice and allow to cool between cycles. Cell lysis is deemed complete when the lysate has a consistency similar to water when dispensing droplets from a transfer pipette (~10 cycles).
5. Clarify the lysate by centrifugation at 18,000 RPM, 4 °C for 25 min.
6. Equilibrate 2–5 mL Ni-NTA resin with lysis buffer. Mix clarified lysate with equilibrated resin. Gently shake or stir at 4 °C for 1 h.

Critical: Incubation with the Ni-NTA resin for longer than 1 h can result in excessive binding of contaminating proteins to the resin.

7. Apply lysate/resin to a Flex column. Elute lysate and wash with 400 mL lysis buffer, followed by 200 mL buffer B (high salt) and then 100 mL buffer A (low salt).
8. Elute protein from Ni-NTA resin with 60 mL of nickel elution buffer.
9. Run elution samples on SDS-PAGE to verify presence of *Sc*Trf4-Air2.

Note: *Sc*Air2 typically runs high on an SDS-PAGE gel. For example, *Sc*Air2$^{1-223}$ (27 kDa) runs at an apparent molecular weight of 35 kDa.

10. Run elution fractions containing *Sc*Trf4-Air2 over a 5 mL HiTrap Heparin column equilibrated with buffer A:
    a. Inject sample at 1.5 mL/min. Subsequent washes and elutions run at 2.5 mL/min.
    b. Wash with 5 CV of buffer A.
    c. Elute using a 10 CV linear gradient from buffer A to buffer B.
11. Run peak fractions on SDS-PAGE. Pool fractions containing both *Sc*Trf4 and *Sc*Air2.
12. Pool the selected fractions and dialyze the sample into buffer C using 12–14 kDa MWCO dialysis tubing at 4 °C, overnight.

Day 2:

13. Filter the dialyzed sample with a 0.45-μm syringe filter to remove any precipitated protein. Run filtered sample over a 5 mL HiTrap Phenyl FF reverse phase column equilibrated with buffer C:

   a. Inject sample at 1.5 mL/min. Subsequent washes and elutions run at 2.5 mL/min.

   b. Wash with 5 CV of buffer C.

   c. Wash with 3 CV of 30% buffer D. Continue wash until A_{280} peak returns to baseline (this is a contaminating peak).

   d. Start a 10 CV linear gradient from 30% buffer D to 100% buffer D. A peak containing *Sc*Trf4-Air2 will begin soon after gradient starts. Allow the gradient to continue to 60% buffer D and hold the gradient until the *Sc*Trf4-Air2 peak has completely eluted.

   Note: This hold effectively separates full-length *Sc*Trf4 from a slightly smaller contaminating protein band.

   e. Resume gradient to 100% buffer D.

14. Run peak fractions on SDS-PAGE. Pool clean fractions and concentrate to ≤10 mL using a 30,000 MWCO centrifugal concentrator at 3000 RPM (1650 × *g*), 4 °C for 10 min intervals, mixing in between intervals. (The concentrating step improves resolution over size exclusion column.)

15. Run protein sample over a HiLoad 26/60 Superdex 200 column equilibrated with buffer A. Elute sample with buffer A at 1.25 mL/min.

Day 3:

16. Run peak fractions on SDS-PAGE and test for RNase contamination using RNaseAlert.

17. Pool clean fractions and concentrate using a 30,000 MWCO centrifugal concentrator at 3000 RPM, 4 °C for 10 min intervals, mixing in between intervals.

18. Flash freeze 10–20 μL aliquots of *Sc*Trf4-Air2 in liquid nitrogen and store at −80 °C. (Since buffer A contains 10% glycerol, no additional glycerol is added for the freezing step.)

4. Reconstitution of the TRAMP complex

4.1 Before you begin

The *Sc*TRAMP complex is reconstituted by combining frozen stocks of purified *Sc*Mtr4 and *Sc*Trf4-Air2. Alternatively, *Sc*Mtr4 and *Sc*Trf4-Air2 can be combined at the final stages of purification prior to size exclusion chromatography, which may be preferred when performing structural studies or other experiments where freezing of samples is undesirable. The protocols described below consistently produce *Sc*TRAMP complexes with a 1:1:1 M ratio of *Sc*Trf4, *Sc*Air2, and *Sc*Mtr4. Note that *Sc*Mtr4 interaction with *Sc*Trf4-Air2 is sensitive to salt concentration and generally dissociates above 200 mM NaCl (LaCava et al., 2005; Losh et al., 2015; Taylor et al., 2014). In practice, we keep the salt concentration at or below 160 mM NaCl.

4.2 Materials and equipment

- Superdex 200 10/300 GL column (Cytiva)
- HiLoad 26/60 Superdex 200 column (Cytiva)
- Refrigerated centrifuge
- Centrifugal concentrator

4.3 Step-by-step method details

4.3.1 Method 1: Reconstitution of ScTRAMP Complex from Purified Frozen Components

Timing: 1 Day

*Sc*TRAMP is reconstituted the day of use by running samples over a size exclusion column.

1. Thaw concentrated protein stocks of *Sc*Trf4-Air2 and *Sc*Mtr4 on ice and mix in 2:1 M ratio of *Sc*Trf4-Air2:*Sc*Mtr4.
2. Purify *Sc*Trf4-Air2 + *Sc*Mtr4 sample over a size exclusion column with Mtr4 buffer A (from *Sc*Mtr4 purification protocol). We typically run samples over a 24 mL Superdex 200 10/300 GL column at 0.5 mL/min.

Note: To avoid the difficulty of loading small samples volumes on a column, the complex can be diluted in Mtr4 buffer A. For example, when loading onto a 24 mL size exclusion column, we bring ~100 μL of *Sc*Trf4-Air2 + *Sc*Mtr4 sample up to 200–500 μL in Mtr4 buffer A.

Optional: RNA substrates can be added to the protein sample prior to size exclusion. Alternatively, RNA can be added after final purification and concentration of *Sc*TRAMP.

3. Concentrate sample as needed using a 30,000 MWCO centrifugal concentrator at 1600 RPM (410 × *g*), 4 °C for 10 min intervals, mixing in between intervals.

4.3.2 Method 2: Reconstitution of ScTRAMP during ScTrf4-Air2 and ScMtr4 Purification

Timing: 3 Days

This method is used when needing to avoid freezing samples (e.g., for structural studies).

1. Start purification of *Sc*Trf4-Air2 on day 1 as outlined in Purification of *Sc*Trf4-Air2.
2. On day 2, continue *Sc*Trf4-Air2 purification and start *Sc*Mtr4 purification as outlined in Purification of *Sc*Mtr4.
3. Following the phenyl column for *Sc*Trf4-Air2 (Purification of *Sc*Trf4-Air2, step 13) and the DEAE column for *Sc*Mtr4 (Purification of *Sc*Mtr4, step 13), concentrate the individual samples to ~15 µM and then combine so that *Sc*Trf4-Air2 is in 2 × molar concentration compared to *Sc*Mtr4 (e.g., 5 mL of 15 µM *Sc*Trf4-Air2 and 2.5 mL of 15 µM *Sc*Mtr4).
4. Run *Sc*Trf4-Air2 + *Sc*Mtr4 over a HiLoad 26/60 Superdex 200 column (sample volume ≤13 mL) equilibrated with Mtr4 buffer A (from Purification of *Sc*Mtr4). Elute sample with buffer A at 1.25 mL/min. *Sc*TRAMP will elute as the first peak with excess *Sc*Trf4-Air2 in the second peak.

Optional: RNA substrates can be added to the protein sample prior to size exclusion. Alternatively, RNA can be added after final purification and concentration of *Sc*TRAMP.

5. Concentrate sample as needed using a 30,000 MWCO centrifugal concentrator at 1600 RPM (410 × *g*), 4 °C for 10 min intervals, mixing in between intervals.

4.4 Quantification and statistical analysis

4.4.1 Determining the concentration of ScTrf4-Air2 and ScTRAMP complexes

Since *Sc*Trf4-Air2 and *Sc*TRAMP form 1:1 (or 1:1:1) stoichiometric complexes, the extinction coefficient for *Sc*Trf4-Air2 or *Sc*TRAMP can be simply calculated by combining the sequences of the individual components into a single file and submitting them to the ExPASy ProtParam tool (https://web.expasy.org/protparam/) (Gasteiger et al., 2005). This can be used to calculate the concentration of purified complexes by A_{280} in the absence of nucleic acid. Since RNA and/or DNA will interfere with the A_{280} signal, concentrations of complexes bound to nucleic acid are determined by gel densitometry, using known concentrations of *Sc*Mtr4 or *Sc*Trf4 as a control. This concentration can be used to calculate an extinction coefficient at A_{230}, which is not influenced by nucleic acid.

4.5 Advantages

Earlier *Sc*TRAMP purification protocols relied on combining *Sc*Trf4-Air2 and *Sc*Mtr4 cell lysates before purification steps (Jia et al., 2011; Taylor et al., 2014). Since *Sc*Mtr4 dissociates from the complex at low salt (>200 mM NaCl), it was difficult to purify the complex and frequently resulted in significant nuclease contamination. Additionally, each *Sc*TRAMP prep required significant optimization to ensure success. These challenges are overcome by the protocols presented here, which enable purification of individual components (*Sc*Trf4-Air2 and *Sc*Mtr4) that can be easily assembled into *Sc*TRAMP.

5. Helicase unwinding assay

5.1 Before you begin

Helicase unwinding activity of *Sc*Mtr4 or *Sc*TRAMP is determined by monitoring displacement of a fluorescently labeled oligonucleotide from a complementary RNA strand containing a 3′ single-stranded overhang (Fig. 3). This protocol is similar to those used to assay a variety of DEAD box, DExH box, and Ski2-like helicases (Jia et al., 2012; Weick et al., 2018; Yang & Jankowsky, 2005, 2006). Unwinding rate constants are calculated using the method described by Jankowsky (Yang & Jankowsky, 2005). A 3′ overhang of at least 4–6 nucleotides on the loading strand is required for *Sc*Mtr4 binding and translocation (Jia et al., 2012). *Sc*Mtr4 only translocates across an RNA loading strand (Jia et al., 2012; Weick et al., 2018), but can displace either RNA or DNA. A DNA trap that is complementary to the displacement strand is included to prevent reannealing to the loading strand. Our standard substrate contains a 16 base pair (bp) duplex region and a 6-nucleotide overhang (Fig. 3A). While the sequence of the duplex region may vary, we have recently shown that longer and more highly structured duplex regions can lead to increased Mtr4-substrate interactions and can also affect unwinding activity (Zhang, Olsen, Ball, Johnson, & D'Arcy, 2022). As a precaution, assays are performed in the presence of an RNase inhibitor (RiboLock).

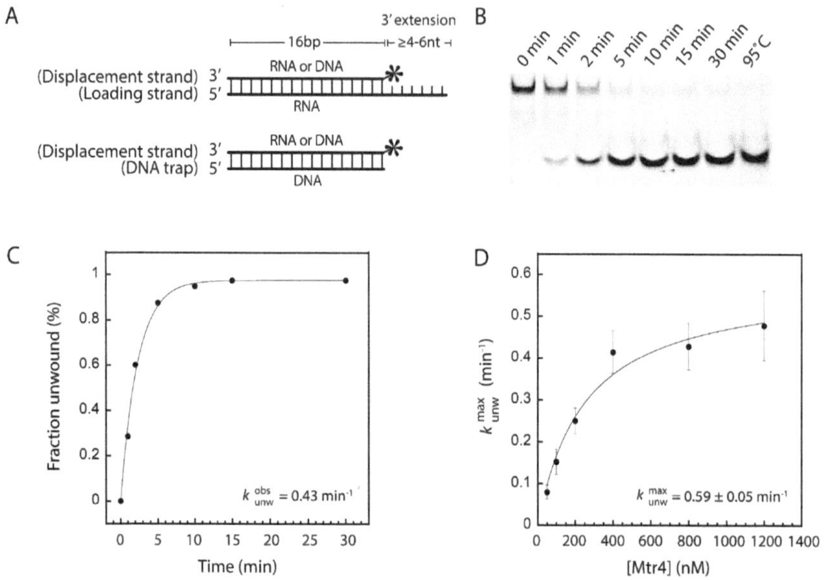

Fig. 3 *Sc*Mtr4 helicase unwinding assay. (A) Standard substrate design for unwinding assay includes a 16 bp duplex with a short 3′ extension on the loading strand and a fluorescent (FAM-labeled, *) probe on the displacement strand. The optimal length of the 3′ extension on the loading strand is between 4 and 6 nucleotides. The loading strand on which *Sc*Mtr4 translocates must be entirely RNA. The displacement strand can be either RNA or DNA. Visualization of substrate unwinding is monitored by tracking the fluorescent probe. When *Sc*Mtr4 moves across the loading strand, it displaces the displacement strand. A DNA trap base pairs with the displacement strand, preventing re-annealing to the loading strand. (B) Example gel of *Sc*Mtr4 unwinding on a 16 bp RNA:RNA substrate with a 6 nucleotide overhang (at 1200 nM *Sc*Mtr4). Quenched samples at each time point are loaded into separate lanes and run on a 15% native acrylamide TBE gel to separate substrate (loading strand + displacement strand, top band) from reaction product (displacement strand + DNA trap, bottom band). The last lane on the right is a fully "unwound" control (95 °C). (C) Quantification of gel from panel B, displaying fraction unwound substrate as a function of time at 1200 nM *Sc*Mtr4. The curve fit gives a $k_{unw}^{obs} = 0.43\ \text{min}^{-1}$. (D) Plot of k_{unw}^{obs} values obtained at varying protein concentrations. The curve fit gives a $k_{unw}^{max} = 0.59 \pm 0.05\ \text{min}^{-1}$ for *Sc*Mtr4 on the 16 bp substrate shown in panel A.

Time: 1 day

1. RNA substrate preparation:
   a. Resuspend lyophilized single stranded loading strand and displacement strand oligonucleotides in nuclease-free water to 200 μM. To avoid quenching, fluorescently labeled substrates are placed in dark microcentrifuge tubes and shielded from light during preparation.
   b. Prepare 20× annealing buffer (0.6 M HEPES pH 7.5, 2 M NaCl in nuclease-free water, filter using 0.45 μm syringe filter).

c. Make a 100 μL solution containing 50 μM of each oligonucleotide and 1 × annealing buffer in nuclease free water (25 μL 200 μM loading strand, 25 μL 200 μM displacement strand, 5 μL 20 × annealing buffer, 45 μL nuclease free water).
d. Heat tube with oligonucleotide mixture to 95 °C for 5 min.
e. Turn off heat block and allow to cool to ambient temperature.
f. Briefly centrifuge to collect condensation and run sample over 24 mL Superdex 200 10/300 GL column, equilibrated in 1 × helicase buffer. Elute sample with 1 × helicase buffer at 0.5 mL/min.
g. Concentrate and prepare working stocks of 1.2 μM duplex with nuclease-free water in separate tubes.
h. Flash freeze and store oligonucleotides at −80 °C.

2. Prepare 15% native acrylamide 1 × TBE gels. Ensure that the gel solution is degassed and gel plates and combs are free from RNase contamination.
3. Prepare 5 × helicase buffer (200 mM MOPS pH 6.5, 500 mM NaCl, 2.5 mM $MgCl_2$, 25% glycerol, 0.05% nonidet P-40, 10 mM dithiothreitol) using nuclease free water. Filter buffer using 0.45 μm syringe filter and store at −20 °C in 1 mL aliquots.
4. Prepare 10 mM ATP/$MgCl_2$ stock solution:
   a. Make 100 mM ATP in nuclease free water. Adjust to pH 7.0 with NaOH.
   b. Make 100 mM $MgCl_2$ in nuclease free water.
   c. Combine 0.5 mL of 100 mM ATP with 0.5 mL of 100 mM $MgCl_2$. Bring up to 5 mL in nuclease free water.
   d. Freeze in 0.5 mL aliquots.
5. Prepare 2 × quench buffer (1% SDS, 10 mM EDTA, 20% glycerol).

5.2 Key resources table

Reagent or resource	Source	Identifier
Chemicals, Peptides, and Recombinant Proteins		
3-(N-morpholino)propanesulfonic acid (MOPS)	Gold Biotechnology	1132-61-2
RiboLock RNase inhibitor 40 U/μL	Thermo Scientific	EO0381
40% acrylamide and bis-acrylamide	Bio-Rad	1610148
Adenosine triphosphate (ATP)	Sigma-Aldrich	51963-61-2

Continued

—cont'd

Reagent or resource	Source	Identifier
Boric acid	Fisher scientific	10043-35-3
Dithiothreitol (DTT)	Gold Biotechnology	27565-41-9
Ethylenediaminetetraacetic acid (EDTA)	Fisher Chemical	6381-92-6
Glycerol	PHARMCO	56-81-5
Magnesium chloride ($MgCl_2$)	MP biomedicals	155334
Nonidet P-40	Sigma-Aldrich	9036-19-5
Nuclease Decontamination Solution	Integrated DNA Technologies	11-05-01-01
Sodium chloride (NaCl)	Fisher Chemical	7647-14-5
Sodium dodecyl sulfate (SDS)	Fisher bioreagents	151-21-3
SYBR Gold nucleic acid gel stain	Invitrogen	S11494
Tris base	Sigma-Aldrich	77-86-1
Tetramethylethylenediamine (TEMED)	Acros Organics	110-18-9
Ammonium persulfate (APS)	Affymetrix	7727-54-0
Oligonucleotides		
DNA trap	Integrated DNA Technologies	
RNA substrates	Integrated DNA Technologies	
Software and Algorithms		
Fluorescent imaging software		
Data analysis software (e.g. KaleidaGraph or GraphPad Prism)		

5.3 Materials and equipment

- Sterile 10 × TBE buffer (For 1 L: 108 g Tris base, 55 g boric acid, 40 mL 0.5 M EDTA pH 8.0. Bring up to volume in nuclease-free water)
- 20 × annealing buffer, filtered using 0.45 μm syringe filter
- 5 × helicase buffer, filtered using 0.45 μm syringe filter

- 2 × quench buffer
- Gel caster
- 1.5 mm spacer plates and short plates (Bio-Rad)
- 1.5 and 0.65 mL microcentrifuge tubes
- Imaging system to image fluorescent probes (Bio-Rad ChemiDoc XP, or equivalent)
- Dry bath incubator (Thermo Scientific)
- Electrophoresis system
- NanoDrop spectrophotometer (ND-1000, Thermo Scientific)
- Temperature-controlled heating block: Mini cooler (Temperature incubator MS-0203, Major Science)

5.4 Step-by-step method details

Note: Care should be taken to prevent over-exposure of fluorescent substrate stocks from light. Cover reaction and quench tubes with foil to protect from exposure to light. During reactions, other precautions may be taken like turning off lights while setting up the reaction or taking time points. While running gel, turn off lights in room or cover with a cardboard box.

1. Pre-run 15% native acrylamide TBE gels in 1 × TBE buffer at 15–17 V/cm in cold room (4 °C) for at least 80 min prior to running samples.
2. Thaw 5 × helicase buffer, substrates, and protein (*Sc*Mtr4 or *Sc*TRAMP, prepared as described above) on ice. Thaw 10 mM ATP/$MgCl_2$ stock and preheat to 30 °C. Keep RiboLock at −20 °C until needed. (Initial stock concentration of each reagent is outlined in Table 1.)
3. Once the protein is thawed, measure the absorbance by NanoDrop at 280 nm to accurately determine the concentration using Beer's law.
4. In a 1.5 mL microcentrifuge tube, combine the following reagents:
   1.19 × helicase buffer (23.80 μL of 5 × helicase buffer)
   1.19 U/μL RiboLock (2.98 μL of RiboLock stock)
   *Sc*Mtr4 or *Sc*TRAMP (concentration and volume will vary depending on desired experimental conditions)
   Nuclease-free water (added to 97.82 μL)
5. Incubate at 30 °C for 5 min.
6. Add RNA substrate and DNA trap:
   11.88 nM RNA substrate (0.99 μL of 1200 nM working stock)
   119.00 nM DNA trap (1.19 μL of 10 μM working stock)

Note: The total volume at this point should be 100 μL. This is the pre-reaction sample and each reagent concentration is outlined in Table 1.

Table 1 Helicase unwinding assay concentrations.

Reagent	Initial stock concentration	Pre-reaction sample concentration	Final reaction concentration (after addition of $ATP/MgCl_2$)
Helicase buffer	5×	1.2×	1×
RiboLock	40 U/μL	1.2 U/μL	1 U/μL
RNA duplex	1200 nM	12 nM	10 nM
DNA trap	10 μM	119 nM	100 nM
$ATP/MgCl_2$	10 mM	–	1.6 mM

7. Mix and incubate at 30 °C for 5–10 min.
8. Prepare quench tubes to collect time points. For each non-zero time point, aliquot 10 μL of 2× quench buffer into a 0.65 mL microcentrifuge tube. Typical time points include 0, 1, 2, 5, 10, 15, and 30 min. Faster reactions may require shorter time points (<1 min). Slower reactions may require longer time points.
9. In a separate 0.65 mL microcentrifuge tube, remove 16 μL of the pre-reaction sample and mix with 19 μL (in 1:1 ratio) of 2× quench buffer and 3 μL of nuclease-free water. Remove 19 μL and place in a second 0.65 mL microcentrifuge tube. One of the tubes will be used as the $t = 0$ min time point and the other will be the 95 °C control. The dilution step brings the solution to the same final concentrations as the non-zero reaction time points.
10. Incubate the 95 °C control sample at 95 °C for 4–6 min.
11. Add 16 μL of 10 mM $ATP/MgCl_2$ to the remaining pre-reaction sample to initiate the unwinding reaction. The final reaction concentration of each reagent (after the addition of $ATP/MgCl_2$) is outlined in Table 1.

Critical: Mix the $ATP/MgCl_2$ into the reaction well by pipetting up and down to initiate the reaction properly. Keep all reaction tubes in 30 °C heat block for the duration of the assay.

12. Quench each time point by removing 10 μL of the reaction and placing it in the pre-prepared quench tubes. Mix well.

Pause Point: Completed reactions can be covered with foil and stored at room temperature while other reactions are being performed.

13. Once all time points are collected, load 15 μL of each sample into a 15% native acrylamide TBE gel. Run the gel at 15–17 V/cm for 80–120 min at 4 °C. Run the gel until good separation between duplex and displaced bands is achieved.

Note: No dye is added to the gel samples to avoid interference with the fluorescent signal. Since the samples are clear, it may be helpful to use a Sample Loading Guide (Bio-Rad) to load the samples into the gel.

14. Image gel using a fluorescent imager and the fluorescein mode setting (Epi-blue light source, 460–490 nm excitation, and 530/28 filter using Bio-Rad ChemiDoc XP imager). Minimize background noise and do not overexpose image.

Note: The gel can be imaged while still in the gel plates to determine if good separation between the duplex and displaced strands has been achieved. If needed, the gel can then be placed back into the running module to resume electrophoresis.

5.5 Expected outcomes

The final gel should show good separation between the duplex substrate and trapped displacement strand as shown in Fig. 3B.

5.6 Quantification and statistical analysis

1. Measure the intensity of each fluorescent band on the gel using imaging software (Image Lab (Bio-Rad) or similar). Ensure that uniform boxes (same dimensions) are used to quantify each band. The box size is selected to capture the entire signal of the largest band while minimizing background signal. Boxes should not overlap with adjacent fluorescent bands.

   a. Determine the total signal for each time point by adding the signal of the substrate (top) strand and the displaced (bottom) strand.

   b. Calculate the fraction unwound by dividing the signal of the displaced strand by the total signal of each time point.

   c. To account for partially unwound product at time zero (due to incomplete annealing), the fraction unwound for each time point is adjusted using Eq. 1 (Fiorini, Bonneau, & Le Hir, 2012):

$$F_t = (P_t - F_0(P_t + R_t))/((1 - F_0)(P_t + R_t)) \tag{1}$$

where F_t is the fraction unwound at a given time point, P_t is the intensity of the reaction product (displacement strand + DNA trap, bottom strand, Fig. 3A and B) at a given time point, R_t is the intensity of the reaction substrate (loading strand + displacement strand, top band, Fig. 3A and B) at a given time point, $(P_t + R_t)$ is the total signal, and F_0 is the fraction unwound (unwound/total) at time point zero.

Note: The F_t values are simply adjusted fraction unwound values accounting for any initial product "unwound" due to incomplete annealing of top and bottom strands or other factors.

2. Plot the F_t values vs time and fit the data to the integrated rate law:

$$\text{Frac ss} = A^*\left(1 - e^{-k_{unw}^{obs} * t}\right) \tag{2}$$

where Frac ss is the fraction unwound (single-stranded product), A is the amplitude, k_{unw}^{obs} is the observed rate constant, and t is time.

3. The observed rate constant k_{unw}^{obs} is expressed in units of reciprocal time (min^{-1}) (Fig. 3C).
4. To allow comparison of Mtr4 activity on different substrates, a rate constant k_{unw}^{max} can be determined by measuring the observed rate constants at varying protein concentrations and plotting the calculated k_{unw}^{obs} against protein concentration (Yang & Jankowsky, 2005) (Fig. 3D).

Note: Typically, observed rate constants (k_{unw}^{obs}) are determined at 50, 100, 200, 400, 800, and 1200 nM *Sc*Mtr4. Enzyme saturation for wild-type *Sc*Mtr4 occurs at 800–1200 nM.

5. Fit the k_{unw}^{obs} vs protein concentration data to a binding isotherm equation:

$$k_{unw}^{obs} = k_{unw}^{max} * \frac{[E]}{K_{1/2} + [E]} \tag{3}$$

where k_{unw}^{obs} is the observed rate constant for a given substrate at a specific enzyme concentration, k_{unw}^{max} is the actual rate constant (or ratio of rate constants) for a given substrate at enzyme saturation, [E] is the concentration of enzyme (helicase), and $K_{1/2}$ is the functional affinity.

6. The actual rate constant k_{unw}^{max} is expressed in units of reciprocal time (min^{-1}).

5.7 Optimization and troubleshooting

5.7.1 Problem: Smeared samples, and/or inactive or slow helicase

Sometimes smearing is observed throughout the gel, giving the appearance of widespread degradation or substrate cleavage. Alternatively, no unwinding activity is observed or the helicase appears to be unexpectedly slow.

5.7.2 Potential solution to optimize the procedure

RNase contamination is the most likely cause of smeared bands. RNases can also cleave the 3′ end of the loading strand, making it inaccessible to the helicase.

Proper care should be taken to decontaminate and maintain RNase-free equipment, benchtops, and reagents. Excessive freeze/thaw of nucleic acid substrates can also compromise the substrate. In addition, samples may run poorly in gel if acrylamide-buffer solution is not degassed prior to casting.

5.7.3 Problem: Inconsistent or low intensity sample signal

Total band intensity on the gel is inconsistent from lane to lane or the overall signal is unexpectedly low.

5.7.4 Potential solution to optimize the procedure

Inconsistency between lanes is generally due to pipetting errors. Samples are clear and often difficult to load in a gel. While loading samples into the gel, it is easiest to use gel loading pipette tips that extend into the wells of the gel. This can prevent incomplete loading of the sample in the well (due to spillage or missing the well). A Sample Loading Guide may also be useful to load samples and count/check lanes.

Low intensity of the signal may be due to miscalculation of starting substrate concentration. Alternatively, it may be the result of photobleaching. Take precautions to protect the substrate from light throughout the experiment.

6. Conclusion

The protocols presented here reproducibly generate highly pure *Sc*Mtr4, *Sc*Trf4-Air2 and *Sc*TRAMP. A critical concern is the removal of nucleases during the prep. This approach enables productive assembly and analysis of TRAMP and a variety of TRAMP mutant complexes.

References

Bernstein, J., Patterson, D. N., Wilson, G. M., & Toth, E. A. (2008). Characterization of the essential activities of Saccharomyces cerevisiae Mtr4p, a 3′->5′ helicase partner of the nuclear exosome. *The Journal of Biological Chemistry*, *283*(8), 4930–4942.

Conrad, K. S., Hurley, J. M., Widom, J., Ringelberg, C. S., Loros, J. J., Dunlap, J. C., et al. (2016). Structure of the frequency-interacting RNA helicase: A protein interaction hub for the circadian clock. *The EMBO Journal*, *35*(15), 1707–1719.

de la Cruz, J., Kressler, D., Tollervey, D., & Linder, P. (1998). Dob1p (Mtr4p) is a putative ATP-dependent RNA helicase required for the 3′ end formation of 5.8S rRNA in Saccharomyces cerevisiae. *The EMBO Journal*, *17*(4), 1128–1140.

Delan-Forino, C., Spanos, C., Rappsilber, J., & Tollervey, D. (2020). Substrate specificity of the TRAMP nuclear surveillance complexes. *Nature Communications*, *11*(1), 3122.

Fiorini, F., Bonneau, F., & Le Hir, H. (2012). Chapter twelve - biochemical characterization of the RNA helicase UPF1 involved in nonsense-mediated mRNA decay. In E. Jankowsky (Ed.), *Vol. 511. Methods in enzymology* (pp. 255–274). Academic Press.

Gasteiger, E., Hoogland, C., Gattiker, A., Duvaud, S. E., Wilkins, M. R., Appel, R. D., et al. (2005). Protein identification and analysis tools on the ExPASy server. In J. M. Walker (Ed.), *The proteomics protocols handbook* (pp. 571–607). Totowa, NJ: Humana Press.

Houseley, J., Lacava, J., & Tollervey, D. (2006). RNA-quality control by the exosome. *Nature Reviews. Molecular Cell Biology*, 7(7), 529–539.

Jankowsky, E., & Fairman-Williams, M. E. (2010). Chapter 1 an introduction to RNA helicases: Superfamilies, families, and major themes. In *RNA helicases* (pp. 1–31). The Royal Society of Chemistry.

Januszyk, K., Weick, E. M., & Lima, C. D. (2020). Reconstitution of the human nuclear RNA exosome. *Methods in Molecular Biology*, *2062*, 467–489.

Jia, H., Wang, X., Anderson, J. T., & Jankowsky, E. (2012). RNA unwinding by the Trf4/Air2/Mtr4 polyadenylation (TRAMP) complex. *Proceedings of the National Academy of Sciences of the United States of America*, *109*(19), 7292–7297.

Jia, H., Wang, X., Liu, F., Guenther, U. P., Srinivasan, S., Anderson, J. T., et al. (2011). The RNA helicase Mtr4p modulates polyadenylation in the TRAMP complex. *Cell*, *145*(6), 890–901.

Johnson, S. J., & Jackson, R. N. (2013). Ski2-like RNA helicase structures: Common themes and complex assemblies. *RNA Biology*, *10*(1), 33–43.

LaCava, J., Houseley, J., Saveanu, C., Petfalski, E., Thompson, E., Jacquier, A., et al. (2005). RNA degradation by the exosome is promoted by a nuclear polyadenylation complex. *Cell*, *121*(5), 713–724.

Lim, J., Giri, P. K., Kazadi, D., Laffleur, B., Zhang, W., Grinstein, V., et al. (2017). Nuclear proximity of Mtr4 to RNA exosome restricts DNA mutational asymmetry. *Cell*, *169*(3), 523–537 e515.

Losh, J. S., King, A. K., Bakelar, J., Taylor, L., Loomis, J., Rosenzweig, J. A., et al. (2015). Interaction between the RNA-dependent ATPase and poly(A) polymerase subunits of the TRAMP complex is mediated by short peptides and important for snoRNA processing. *Nucleic Acids Research*.

Morales, Y., Olsen, K. J., Bulcher, J. M., & Johnson, S. J. (2018). Structure of frequency-interacting RNA helicase from neurospora crassa reveals high flexibility in a domain critical for circadian rhythm and RNA surveillance. *PLoS One*, *13*(5), e0196642.

Ogami, K., Chen, Y., & Manley, J. L. (2018). RNA surveillance by the nuclear RNA exosome: Mechanisms and significance. *Noncoding RNA*, *4*(1).

Olsen, K. J., & Johnson, S. J. (2021). Mtr4 RNA helicase structures and interactions. *Biological Chemistry*, *402*(5), 605–616.

Patrick, E. M., Srinivasan, S., Jankowsky, E., & Comstock, M. J. (2017). The RNA helicase Mtr4p is a duplex-sensing translocase. *Nature Chemical Biology*, *13*(1), 99–104.

Puno, M. R., & Lima, C. D. (2018). Structural basis for MTR4-ZCCHC8 interactions that stimulate the MTR4 helicase in the nuclear exosome-targeting complex. *Proceedings of the National Academy of Sciences of the United States of America*, *115*(24), E5506–e5515.

Schmid, M., & Jensen, T. H. (2019). The nuclear RNA exosome and its cofactors. In M. Oeffinger, & D. Zenklusen (Eds.), *The biology of mRNA: Structure and function* (pp. 113–132). Cham: Springer International Publishing.

Schuch, B., Feigenbutz, M., Makino, D. L., Falk, S., Basquin, C., Mitchell, P., et al. (2014). The exosome-binding factors Rrp6 and Rrp47 form a composite surface for recruiting the Mtr4 helicase. *The EMBO Journal*, *33*(23), 2829–2846.

Schuller, J. M., Sebastian, F., Lisa, F., Ed, H., & Conti, E. (2018). Structure of the nuclear exosome captured on a maturing preribosome. *Science*, *360*(6385), 219–222.

Studier, F. W. (2005). Protein production by auto-induction in high density shaking cultures. *Protein Expression and Purification*, *41*(1), 207–234.

Taylor, L. L., Jackson, R. N., Rexhepaj, M., King, A. K., Lott, L. K., van Hoof, A., et al. (2014). The Mtr4 ratchet helix and arch domain both function to promote RNA unwinding. *Nucleic Acids Research*, *42*(22), 13861–13872.

Vanacova, S., Wolf, J., Martin, G., Blank, D., Dettwiler, S., Friedlein, A., et al. (2005). A new yeast poly(a) polymerase complex involved in RNA quality control. *PLoS Biology*, *3*(6), e189.

Wang, X., Jia, H., Jankowsky, E., & Anderson, J. T. (2008). Degradation of hypomodified tRNA(iMet) in vivo involves RNA-dependent ATPase activity of the DExH helicase Mtr4p. *RNA*, *14*(1), 107–116.

Weick, E.-M., & Lima, C. D. (2021). RNA helicases are hubs that orchestrate exosome-dependent 3′–5′ decay. *Current Opinion in Structural Biology*, *67*, 86–94.

Weick, E. M., Puno, M. R., Januszyk, K., Zinder, J. C., DiMattia, M. A., & Lima, C. D. (2018). Helicase-dependent RNA decay illuminated by a Cryo-EM structure of a human nuclear RNA exosome-MTR4 complex. *Cell*, *173*(7), 1663–1677 e1621.

Wyers, F., Rougemaille, M., Badis, G., Rousselle, J. C., Dufour, M. E., Boulay, J., et al. (2005). Cryptic pol II transcripts are degraded by a nuclear quality control pathway involving a new poly(A) polymerase. *Cell*, *121*(5), 725–737.

Yang, Q., & Jankowsky, E. (2005). ATP- and ADP-dependent modulation of RNA unwinding and strand annealing activities by the DEAD-box protein DED1. *Biochemistry*, *44*(41), 13591–13601.

Yang, Q., & Jankowsky, E. (2006). The DEAD-box protein Ded1 unwinds RNA duplexes by a mode distinct from translocating helicases. *Nature Structural & Molecular Biology*, *13*(11), 981–986.

Zhang, N., Olsen, K. J., Ball, D., Johnson, S. J., & D'Arcy, S. (2022). Hydrogen-deuterium exchange mass spectrometry of Mtr4 with diverse RNAs reveals substrate-dependent dynamics and interfaces in the arch. *Nucleic Acids Research*. https://doi.org/10.1093/nar/gkac170.

CHAPTER SIXTEEN

Methods to assess helicase and translocation activities of human nuclear RNA exosome and RNA adaptor complexes

M. Rhyan Puno[a] and Christopher D. Lima[a,b,*]

[a]Structural Biology Program, Sloan Kettering Institute, Memorial Sloan Kettering Cancer Center, New York, NY, United States

[b]Howard Hughes Medical Institute, New York, NY, United States

[*]Corresponding author: e-mail address: limac@mskcc.org

Contents

Abstract

The nuclear RNA exosome collaborates with the MTR4 helicase and RNA adaptor complexes to process, surveil, and degrade RNA. Here we outline methods to characterize RNA translocation and strand displacement by exosome-associated helicases and adaptor complexes using fluorescence-based strand displacement assays. The design and preparation of substrates suitable for analysis of helicase and decay activities of

Methods in Enzymology, Volume 673
ISSN 0076-6879
https://doi.org/10.1016/bs.mie.2022.03.060

reconstituted MTR4–exosome complexes are described. To aid structural and biophysical studies, we present strategies for engineering substrates that can stall helicases during translocation, providing a means to capture snapshots of interactions and molecular steps involved in substrate translocation and delivery to the exosome.

1. Introduction

The RNA exosome is a conserved, eukaryotic 3′-to-5′ ribonuclease protein complex that is responsible for several aspects of nuclear and cytoplasmic RNA metabolism, ranging from general RNA turnover to processing and quality control of various coding and noncoding transcripts (Houseley, LaCava, & Tollervey, 2006; Puno, Weick, Das, & Lima, 2019). The human nuclear RNA exosome is composed of 9 core subunits with three S1/KH-domain proteins (EXOSC1–3) atop a hexameric ring of PH-domain proteins (EXOSC4–9) (Fig. 1). The exosome core contains a central channel that guides single-stranded RNA to two RNA-degrading enzymes: DIS3, a processive 3′-to-5′ RNAse II-like exoribonuclease and EXOSC10, a distributive 3′-to-5′ RNase D-like exoribonuclease that associate with the exosome core (Gerlach et al., 2018; Liu, Greimann, & Lima, 2006; Makino, Halbach, & Conti, 2013; Wasmuth, Januszyk, & Lima, 2014; Weick et al., 2018; Zinder & Lima, 2017).

Several functions of the nuclear RNA exosome require the helicase MTR4 (also known as SKIV2L2) (Weick & Lima, 2021). MTR4 is a multi-domain protein with a core structure composed of two RecA-like ATPase modules (RecA1 and RecA2), a winged helix (WH) fold and a C-terminal helical bundle (HB) domain (Gerlach et al., 2018; Jackson et al., 2010; Weick et al., 2018; Weir, Bonneau, Hentschel, & Conti, 2010). Inserted within WH is an arch-shaped protrusion consisting of an elongated stalk and a globular Kyprides-Ouzounis-Woese (KOW) domain that interacts with structured RNA and/or protein adaptors for exosome

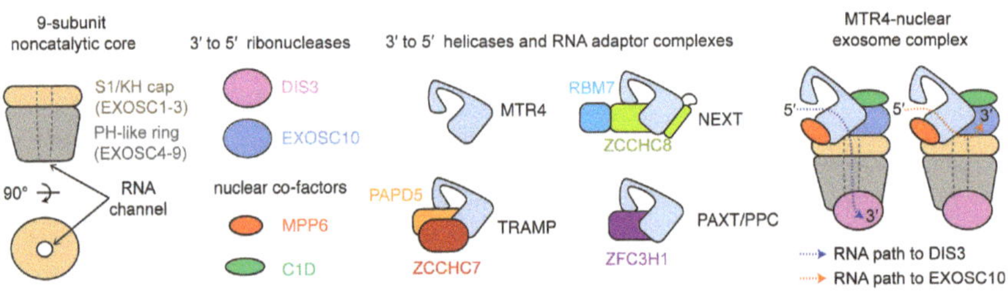

Fig. 1 Human RNA exosome subunits, co-factors, and RNA adaptor complexes.

recruitment (Jackson et al., 2010; Taylor et al., 2014; Thoms et al., 2015). Nuclear cofactors MPP6 and/or C1D/EXOSC10 tether MTR4 to the exosome (Fig. 1) (Gerlach et al., 2018; Wasmuth, Zinder, Zattas, Das, & Lima, 2017; Weick et al., 2018).

MTR4 is a central component of several nuclear RNA adaptor complexes that identify, capture and prime exosome substrates for decay. In human, these include the Nuclear Exosome Targeting (NEXT) complex, the TRF4-2/PAPD5-ZCCHC7-MTR4 polyadenylation (TRAMP) complex, and the Poly(A) Exosome Targeting (PAXT) connection or Polysome Protector complex (PPC) (Fig. 1) (Lubas et al., 2011; Meola et al., 2016; Ogami et al., 2017). NEXT, composed of an RNA binding protein RBM7, a zinc-knuckle protein ZCCHC8, and the MTR4 helicase, functions in turnover of unwanted transcripts produced by pervasive transcription and surveillance of aberrant RNA species that emerge from faulty transcription. PAXT/PPC core is a binary complex between MTR4 and ZFC3H1 and is loosely associated other cofactors that include poly(A) binding protein PABPN1. PAXT/PPC primarily targets processed polyadenylated transcripts. In contrast to its *S. cerevisiae* counterpart, less is known about endogenous targets of human TRAMP.

The ability of MTR4 and RNA adaptor complexes to disrupt RNA structures and displace proteins deposited on RNA is key to their function in promoting exosome-mediated decay (Puno et al., 2019; Puno & Lima, 2018; Weick & Lima, 2021). In this chapter, we present methods used in our laboratory to characterize helicase activities of exosome-associated helicases and RNA adaptor complexes to determine mechanisms and kinetic models underpinning substrate selection, translocation, unwinding, and targeting to the exosome. These assays may also prove useful in studies aimed at assessing helicases and RNA adaptor complexes for their ability to displace proteins bound to RNA. We also provide a protocol for analyzing helicase-dependent RNA decay by MTR4–exosome complexes. Finally, we describe strategies in designing substrates for trapping helicases during translocation to generate samples for biophysical and structural analysis.

2. RNA helicase assays with MTR4 and RNA adaptor complexes

Assays that detect strand displacement are widely used to query translocation and/or helicase activity in vitro. These assays measure the ability of a helicase to displace and release an oligonucleotide strand (reporter strand) from a duplex region of a substrate. Unwinding polarity and requirements for a single-stranded translocation strand can be assessed by designing duplex substrates with single-stranded regions extending from either 5′ or 3′ ends.

Several approaches are employed to monitor strand displacement using radiolabeled or fluorescently modified oligonucleotides. In a gel shift assay, the displaced reporter strand is captured by a complementary DNA oligonucleotide to prevent reannealing to the translocation strand while capturing the product for detection using native polyacrylamide gel electrophoresis (Fig. 2A and B). This is a common and accessible approach, since it utilizes materials and equipment available in most molecular biology laboratories, but it is limited by the number of samples and individual time points that can be conveniently taken manually. Throughput and temporal resolution are enhanced by using Fluorescence Resonance Energy Transfer (FRET)-based assays that enable detection of strand displacement in real time (Özeş, Feoktistova, Avanzino, Baldwin, & Fraser, 2014). Kinetic data can be obtained in minutes instead of hours as is often required for shift assays where assays are followed by gel electrophoresis and scanning. FRET substrates typically consist of a fluorescent reporter strand annealed

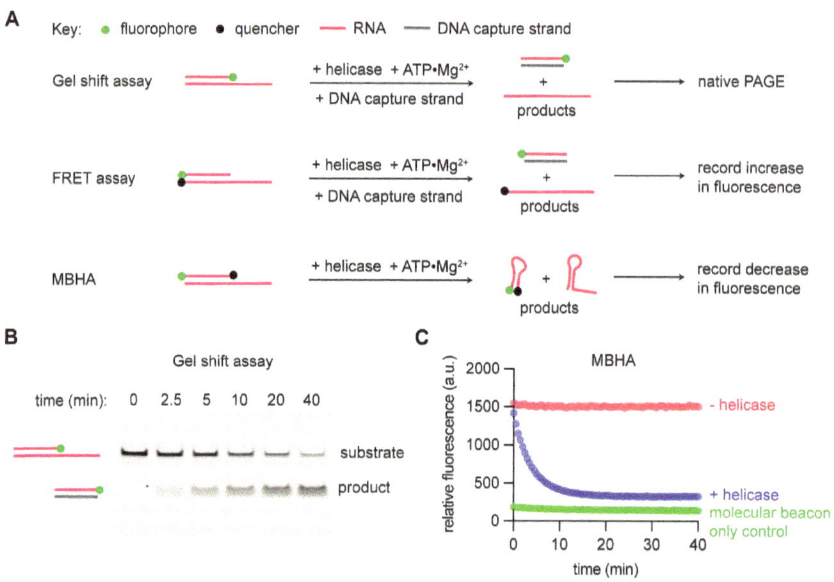

Fig. 2 Strand displacement assays. (A) Schematics comparing gel shift, FRET, and MBHA assays. (B) Representative native PAGE showing time courses of helicase unwinding reaction on a 3′ tailed RNA duplex substrate. (C) MBHA fluorescence change in the presence or absence of helicase.

to a translocation strand conjugated with a quencher dye (Fig. 2A). Upon strand displacement, the fluorophore is now distant from the quencher resulting in an increase in total reporter fluorescence.

A variant of the FRET-based method, the molecular Beacon Helicase Assay (MBHA), employs a displacement strand that forms a stem loop after being released from the translocation strand. The displacement strand (molecular beacon) is modified with the fluorophore and quencher moieties at opposite ends (Belon & Frick, 2008; Özeş et al., 2014). Strand displacement results in a decrease in total fluorescence that can be detected in real time (Fig. 2A and C). The molecular beacon cannot reanneal to the translocation strand, so this approach effectively eliminates the need for a capture strand in the reaction. It also bypasses the synthesis of two modified oligonucleotides, reducing cost and expanding the size range of synthetic translocation strands that can be used.

The assays described in this section are performed under multiple turnover conditions whereby the helicase sequentially binds multiple substrates. For single turnover conditions, a helicase trap oligonucleotide is included in the reactions to prevent rebinding of dissociated helicase to the substrate (Pang, Jankowsky, Planet, & Pyle, 2002). Enzyme titrations are performed with concentrations generally ranging from 10-fold above and below the estimated RNA binding constant and can be used to calculate maximal rate and apparent substrate affinity (half maximal rate concentration).

2.1 Equipment

- C1000 Touch Thermal Cycler (Bio-Rad)
- Nanodrop 2000 (Thermo Fisher Scientific)
- SpectraMax M5 microplate reader (Molecular Devices)
- ThermoMixer (Eppendorf)
- Typhoon FLA 9500 scanner (GE Healthcare)
- XCell SureLock Mini-Cell tank (Thermo Fisher Scientific)

2.2 Materials

- Annealing buffer (10 mM Tris–HCl pH 7.0, 100 mM potassium acetate)
- 10 × Assay buffer (100 mM Tris pH 7.0, 50% v/v glycerol, 15 mM $MgCl_2$)
- ATP·$MgCl_2$ solution (200 mM ATP, 200 mM $MgCl_2$, pH adjusted to 7.0 using KOH and supplemented with 1 mM Tris-HCl pH 7.0) (see Section 2.4 for details)

- Corning 3693 half-area 96-well white plate
- DNA capture strand (see Table 1, 8 μM Capture1 in annealing buffer)
- Lyophilized RNA (see Table 1; synthetic oligos obtained from IDT or Dharmacon)
- 2 × Helicase solution (purified helicase in 20 mM Tris pH 7.0, 100 mM NaCl, 0.1 mM TCEP)
- Helicase diluent (20 mM Tris pH 7.0, 100 mM NaCl, 0.1 mM TCEP)
- Nuclease-free water
- 2 × Quench solution (1% w/v sodium dodecyl sulfate, 10 mM ethylenediaminetetraacetic acid (EDTA), 10% v/v glycerol, 80 units/mL Proteinase K, 0.01% w/v xylene cyanol)
- Human placenta, RNAse inhibitor (New England Biolabs)
- β-mercaptoethanol (BME)
- 1 × Tris-Borate-EDTA (TBE) running buffer (89 mM Tris-borate, 2 mM EDTA, pH 8.3)
- Novex 12-well 20% TBE polyacrylamide gel

2.3 Procedures

2.3.1 Preparation of RNA substrates

1. Resuspend lyophilized oligonucleotides (GS1, GS2, MBHA1, and MBHA2 in Table 1) in annealing buffer to an estimated final concentration of 100 μM.
2. Incubate tube with resuspended RNA in ThermoMixer at 25 °C with a shake speed of 1000 rpm for 5 min. Transfer resuspended RNA into a fresh microcentrifuge tube.
3. Determine RNA concentration (in μM) by measuring absorbance at 260 nm (A_{260}) using Nanodrop 2000. Calculate concentration using Eq. (1):

$$A_{260} = \varepsilon b C \tag{1}$$

where ε is the molecular extinction coefficient ($M^{-1}\ cm^{-1}$), b is the path length, and C is the concentration.

4. Prepare 8 μM of each RNA strand in annealing buffer.
5. Mix 25 μL of 8 μM reporter strand and 25 μL of 8 μM translocation strand in 0.2 mL PCR tube.
   (a) For the gel shift assay, mix oligonucleotide pairs GS1 and GS2 to generate the 3′ A_{20} tailed RNA duplex substrate. Prepare a 10 × gel shift unwound control that will be used as a marker for captured

Table 1 List of oligonucleotides for substrate preparation.

Oligo ID	Type	Sequence (5′-to-3′)	Label	Manufacturer	Assay
GS1	RNA	AGCACCGUAAAGACGC[a]	5′ 6-FAM	IDT	Gel shift (reporter strand)
GS2	RNA	GCGUCUUUACGGUGCUAAAAAAAAAAAAAAAAAAAA[a]	None	IDT	Gel shift (translocation strand)
MBHA1	RNA	AGUGCGCUGUAUCUUCAAGGCCACU[b]	5′ Iowa Black RQ 3′ Cy5	IDT	MBHA (molecular beacon strand)
MBHA2	RNA	AGUGGCCUUGAAGAUACAGCGCACUAAAAAAAAAAAAAAAAAAA[b]	None	IDT	MBHA (translocation strand)
StrandA1	RNA	GCGTCTTTACGGTGCTCACCACACCACACCACACCACACCACACCACACCACACAAAAAAAA	None	Dharmacon	Gel shift (strand A)
StrandA2	RNA	GCGTCTTTACGGTGCTCACCACACCACACCACACCACACCACACCACACCACACAAAAAAAA	5′ Fluorescein	Dharmacon	Decay (strand A)
StrandA3	DNA/RNA	GCGTCTTTACGGTGCTCACCACACCACACCACACCACACCACACCACACCACACAAAAAAAA[c]	None	Dharmacon	Exosome loading (strand A)
StrandB1	DNA	GTGTGGTGTGGTGTGGTGTGGTGTGGTGTGGTGTGGT	None	Dharmacon	Decay (strand B)
StrandB2	DNA	GTGTGGTGTGGTGTGGTGTGGTGTGGTGTGGTGTGGT	5′ 6-FAM	IDT	Gel shift (strand B)
StrandB3	DNA	GTGTGGTGTGGTGTGGT	None	IDT	Exosome loading (strand B)
StrandC1	RNA	AGCACCGUAAAGACGC	None	Dharmacon	Gel shift/Decay/Exosome loading (strand C)
Capture1	DNA	GCGTCTTTACGGTGCT	None	IDT	Gel shift (capture strand)
Capture2	DNA	ACCACACCACACCACAC	None	IDT	Gel shift/Exosome loading (capture strand)

These oligonucleotide sequences are derived from [a]Jia, Wang, Anderson, and Jankowsky (2012) or [b]Belon and Frick (2008). [c]DNA sequences are underlined.

reporter strand product: mix 19.5 μL annealing buffer, 0.5 μL of 8 μM GS1 and 20 μL 8 μM Capture1.

(b) For the molecular beacon assay, mix oligonucleotide pairs MBHA1 and MBHA2 to generate the 3′ A_{20} tailed RNA duplex substrate. Prepare a 100× molecular beacon control: mix 35 μL annealing buffer and 5 μL of 8 μM MBHA1.

6. Using Bio-Rad C1000 Touch Thermal Cycler, heat the RNA mixture to 95 °C for 5 min followed by cooling to 16 °C for 10 min and 4 °C overnight. Transfer RNA mixture to a microcentrifuge tube and store at −80 °C until needed.

7. Analyze annealed RNA substrates using native polyacrylamide gel electrophoresis. Prepare 10 nM RNA substrate or reporter strand controls (GS1 or MBHA1) in 1× assay buffer. Mix 10 μL 10 nM RNA substrate or 10 nM reporter strand control with 10 μL of 2× quench solution. Load 10 μL sample into 12-well 20% Novex TBE polyacrylamide gel. Run gel in 1× TBE (pre-chilled to 4 °C) under constant voltage (200 V) for 1 h at 4 °C. Scan gel for fluorescence using Typhoon FLA 9500 scanner. For 6-carboxyfluorescein (6-FAM) labeled substrates, scan the gel using FAM settings (473 nm laser, LPB filter). For cyanine 5 (Cy5), scan the gel using Cy5 settings (635 nm, LPR filter). More than 95% of the reporter strand should be annealed.

2.3.2 Gel shift strand displacement assay

1. For each reaction, mix 13.5 μL nuclease-free water, 15 μL 10× assay buffer, 15 μL 50 mM BME and 1.5 μL 40 units/μL human placenta RNAse inhibitor.
2. Add 15 μL of 100 nM RNA solution to reaction tube. Mix well.
3. Prepare a 2× helicase solution in 20 mM Tris pH 7.0, 100 mM NaCl, 0.1 mM TCEP. Add 75 μL of 2× helicase solution in reaction mixture. Mix and incubate for 5 min at 22 °C. A protein titration is recommended with concentrations approximately 10-fold below and above the estimated equilibrium binding constant.
4. For the zero time point, pipet 18 μL reaction mixture into a microcentrifuge tube and add 2 μL 4 μM capture strand in annealing buffer and 20 μL quench solution.
5. Transfer 90 μL of the reaction mixture to a fresh microcentrifuge tube.
6. Prepare a fresh 10× ATP-capture solution. For 50 μL solution, mix 40 μL nuclease-free water, 5 μL 200 mM ATP·$MgCl_2$ solution and 2.5 μL 8 μM DNA capture strand.

7. Initiate the reaction by adding 10 μL of 10 × ATP-capture solution. Incubate the reaction tube in a ThermoMixer at 30 °C. Optional: Set shake speed to 30 rpm.
8. Take 18 μL aliquot of reaction mixture per time point and immediately transfer to a tube containing 18 μL 2 × quench solution. Mix well.
9. Incubate quenched samples at 30 °C for 10 min to complete proteinase K digestion.
10. Resolve samples in Novex 12-well 20% TBE polyacrylamide gel. Rinse gel cassette with water. Place the gel cassette in XCell SureLock Mini-Cell tank and fill the upper and lower chambers with 1 × TBE running buffer pre-chilled to 4 °C. Flush each well with 1 × TBE running buffer. Load 10 μL sample into each well. For captured reporter strand marker, mix 13.5 μL nuclease-free water, 15 μL 10 × assay buffer, 15 μL 50 mM BME and 1.5 μL 40 units/μL human placenta RNAse inhibitor, 15 μL 10 × gel shift unwound control, 75 μL helicase diluent, 15 μL 20 mM ATP·$MgCl_2$ solution. Load 10 μL captured reporter strand marker into a well. Run samples under constant voltage (200 V) for 1 h at 4 °C.
11. Remove the gel from the cassette and scan for 6-FAM fluorescence (473 nm laser, LPB filter) using Typhoon FLA 9500 scanner. Save the scanned image as .tif file.
12. Open .tif file in ImageJ and quantify band intensities of unreacted substrate ($I_{unreacted}$) and unwound product ($I_{unwound}$). Subtract background fluorescence from each intensities.
13. Calculate fraction unwound (f) with background corrected (^{bc}) intensities using Eq. (2):

$$f = \frac{I^{bc}_{unwound}}{I^{bc}_{unreacted} + I^{bc}_{unwound}} \tag{2}$$

14. Normalize data by subtracting fraction unwound at time = 0 due to nonenzymatic strand displacement from fraction unwound at each time point.
15. By plotting normalized fraction unwound (f_{norm}) versus time (t), the data can be fitted to Eq. (3) (one-phase association equation in GraphPad Prism) to determine the observed unwinding rate constant k_{obs} and reaction amplitude (A):

$$f_{norm} = A\,[1 - \exp(-k_{obs} \bullet t)] \tag{3}$$

16. Initial unwinding rate can be calculated by taking the product of observed rate constant and reaction amplitude or by the taking the slope of fitted linear curve of data at early time points.

2.3.3 Molecular beacon helicase assay

1. Prepare MBHA master mix: 162 μL nuclease-free water, 180 μL10 × assay buffer, 180 μL 50 mM BME, 18 μL human placenta RNAse inhibitor, 180 μL 100 nM RNA substrate (3′ A_{20} tailed MBHA substrate). This master mix is sufficient for 8 reactions with 3 technical replicates and can be used to assay unwinding at various helicase concentrations.

2. Pipet 20 μL MBHA master mix per reaction well into Corning 3693 half-area 96-well white plate. Use a multichannel pipette when conducting multiple reactions

3. Prepare a 2 × helicase solution in 20 mM Tris pH 7.0, 100 mM NaCl, 0.1 mM TCEP. Add 25 μL 2 × helicase solution or helicase diluent (no protein control) into each reaction well. Mix by pipetting.

4. Prepare molecular beacon control solution: 69 μL nuclease-free water, 10 μL 10 × assay buffer, 10 μL 50 mM BME, 1 μL human placenta RNAse inhibitor, 10 μL 100 × molecular beacon control. Add 45 μL molecular beacon control solution into 3 replicate wells.

5. Insert plate into SpectraMax M5 microplate reader pre-heated to 30 °C and record fluorescence (excitation/emission wavelength = 643 nm/ 667 nm, medium photomultiplier tube (PMT) setting, 6 flashes per read) every 30 s for 5–10 min or until the fluorescence stabilizes.

6. Initiate the reaction by adding 5 μL 20 mM ATP·$MgCl_2$ solution into each well using a multi-channel pipette.

7. Record fluorescence every 30 s for 40 min at 30 °C.

8. Calculate normalized fraction unwound for each time point using Eq. (4):

$$f_{norm} = \frac{S_t - S_0}{S_0 - M_t} - \frac{C_t - C_0}{C_0 - M_t} \qquad (4)$$

where S_t is the sample fluorescence at time t, S_0 is the sample fluorescence at time 0, M_t is the fluorescence of the molecular beacon control at time t, C_t is the fluorescence of no helicase control at time t and C_0 is the fluorescence of no protein/helicase control at time 0.

9. Data are fitted to Eq. (3) using GraphPad Prism to determine the observed unwinding rate constant k_{obs} and the extent of the reaction (A).

10. Initial rate can be calculated taking the product of the observed rate constant and reaction amplitude or by the taking the slope of fitted linear curve of data at early time points.

2.4 Notes

- Buffer conditions (buffer, pH, ionic strength) should be optimized for every helicase assayed.
- ATP·$MgCl_2$ stock solution (200 mM) is prepared using ATP disodium salt hydrate lyophilized powder (Sigma). Hydrate content should be checked in the accompanying analysis certificate. Adjust the pH using potassium hydroxide and supplement with 1 mM Tris-HCl pH 7.0. Add stoichiometric amount of $MgCl_2$ prior to storage at −80 °C.
- Fluorescence intensity of specific dyes can be reduced when placed adjacent to a guanine nucleotide.
- The influence of the position (5′ vs 3′ end of the reporter strand) of the fluorophore on helicase activity should be tested.
- Strand displacement assays are typically accompanied with measurements of nucleotide triphosphate (NTP) consumption to determine how energy expenditure is coupled to translocation. For some helicases, NTP hydrolysis is dispensable for strand displacement but is required for enzyme recycling (Liu, Putnam, & Jankowsky, 2008). The number of NTP used per nucleotide translocation also vary among helicases. Several methods (Bradley & De La Cruz, 2012; Brune, Hunter, Corrie, & Webb, 2002; Özeş et al., 2014; Rajagopal & Lorsch, 2013) and commercial kits are available to monitor NTPase activity.

3. RNA unwinding and decay assays of MTR4-exosome complexes

RNA unwinding by MTR4 and RNA adaptor complexes ultimately leads to 3′ end trimming or decay by the exosome. These processes are coupled through physical interaction of MTR4 with the exosome via nuclear co-factors MPP6 and/or the C1D/EXOSC10 heterodimer. To analyze RNA unwinding and decay activities of recombinant human MTR4-exosome complexes, we designed fluorophore-labeled tripartite substrates (Fig. 3, Table 1) that can be used for gel-based assays (Weick et al., 2018). For decay assays, the translocation strand (strand A) is labeled with a fluorescent moiety whereas for helicase assays, a DNA oligonucleotide (strand B) annealed to the translocation strand is conjugated with a fluorophore (Fig. 3). A short poly(A)$_8$ overhang is included for interaction

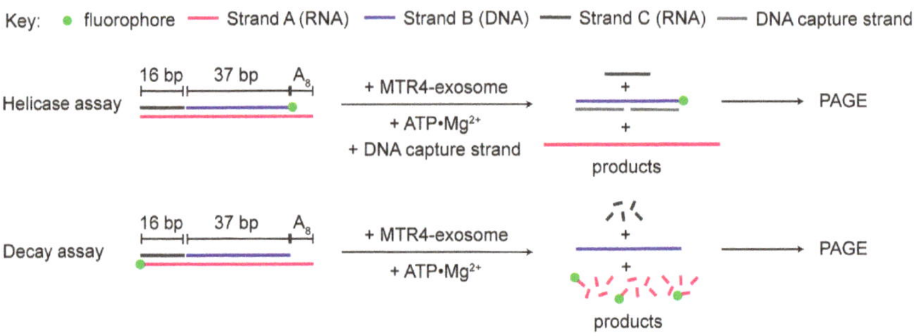

Fig. 3 Substrate design and reaction scheme for RNA helicase and decay assay of human MTR4-exosome complex.

with MTR4 (Jia et al., 2011). These substrates can serve as templates for exploring conditions or modifications that alter MTR4-exosome activities. The following procedures describe substrate preparation and protocols for helicase and decay assays of MTR4-exosome complexes.

3.1 Equipment

- C1000 Touch Thermal Cycler (Bio-Rad)
- Nanodrop 2000 (Thermo Fisher Scientific)
- ThermoMixer (Eppendorf)
- Typhoon FLA 9500 scanner (GE Healthcare)
- Xcell SureLock Mini-Cell tank (Thermo Fisher Scientific)

3.2 Materials

- Annealing buffer (10 mM Tris pH 7.0, 100 mM potassium acetate)
- ATP·$MgCl_2$ solution (200 mM ATP, 200 mM $MgCl_2$, pH adjusted to 7.0 using KOH and supplemented with 1 mM Tris-HCl pH 7.0) (see Section 2.4 for details)
- 50 mM ß-mercaptoethanol (BME)
- DNA capture strand (see Table 1, 3 μM Capture2 in annealing buffer)
- Lyophilized RNA (see Table 1; synthetic oligos from IDT or Dharmacon)
- Assay buffer (20 mM HEPES-KOH pH 7.5, 50 mM potassium acetate, 0.5 mM magnesium acetate, 2.5 mM DTT, 0.01% IGEPAL CA-630)
- Human placenta RNAse inhibitor (40 U/μL, New England Biolabs)
- Human DIS3/EXOSC10/MTR4–exosome protein complexes in assay buffer engineered with or without mutations in the DIS3 and/or

EXOSC10 active sites that disrupt hydrolysis and catalytic activity for helicase or decay assays, respectively (Weick et al., 2018)

- 2 × Quench solution (1% w/v sodium dodecyl sulfate, 10 mM ethylenediaminetetraacetic acid (EDTA), 10% *v*/v glycerol, 80 units/mL Proteinase K, 0.01% w/v xylene cyanol)
- 0.5 × TBE running buffer (44.5 mM Tris-borate, 1 mM EDTA, pH 8.3)
- Novex 12-well 4%–20% TBE polyacrylamide gel (Thermo Fisher Scientific)

3.3 Protocols

3.3.1 Preparation of RNA substrates

1. Resuspend and prepare oligonucleotides (StrandA1, StrandA2, StrandB1, StrandB2, and StrandC1) as in steps 1–3 of Section 2.3.1.
2. Prepare a stock solution of 32 μM for each oligonucleotide.
3. For helicase substrate, mix 10 μL of 32 μM StrandA1, 10 μL of 32 μM StrandB2, 10 μL of 32 μM StrandC1 and 10 μL annealing buffer in 0.2 mL PCR tube. For decay substrate, mix 10 μL of 32 μM StrandA2, 10 μL of 32 μM StrandB1, 10 μL of 32 μM StrandC1 and 10 μL annealing buffer in 0.2 mL PCR tube. Prepare a 10 × gel shift unwound control that will be used as a marker for captured reporter strand product: mix 35.5 μL annealing buffer, 0.5 μL of 8 μM StrandB2 and 4 μL 3 μM Capture2.
4. Using Bio-Rad C1000 Touch Thermal Cycler, heat the RNA mixture to 95 °C for 5 min followed by cooling to 16 °C for 10 min and 4 °C overnight. Transfer RNA mixture to a microcentrifuge tube and store at −80 °C until needed.
5. Analyze annealed RNA substrate using native polyacrylamide gel electrophoresis. Prepare 10 nM RNA substrate or fluorescent reporter strand control (StrandA2 or StrandB2) in 1 × assay buffer. Mix 10 μL 10 nM RNA substrate or 10 nM reporter strand control with 10 μL of 2 × quench solution. Load 10 μL sample into 12-well 20% Novex TBE polyacrylamide gel. Run gel in 1 × TBE (pre-chilled to 4 °C) under constant voltage (200 V) for 1 h at 4 °C. Scan gel for 6-FAM/fluorescein fluorescence (473 nm laser, LPB filter). More than 95% of the fluorescent strand should be annealed.

3.3.2 RNA unwinding assay of MTR4–exosome complexes

1. For each reaction, mix 69 μL assay buffer, 1 μL 40 units/μL human placenta RNAse inhibitor, 10 μL 100 nM RNA and 10 μL 200 nM

MTR4-exosome complexes with catalytically inert ribonucleases (Weick et al., 2018) in a microcentrifuge tube.
2. Incubate at 20 °C for 5 min.
3. For sample at 0 time point, take 18 μL reaction mixture and mix with 2 μL assay buffer and 10 μL quench solution.
4. Transfer 54 μL reaction mixture to a fresh microcentrifuge tube.
5. Prepare a fresh 10 × ATP-capture solution. For 50 μL solution, mix 42.5 μL nuclease-free water, 2.5 μL 200 mM ATP·$MgCl_2$ solution and 5 μL 3 μM DNA capture strand.
6. Initiate the reaction by adding 6 μL ATP-capture solution.
7. Take 10 μL aliquot of reaction mixture at various time points and terminate the reaction by adding 5 μL quench solution.
8. Incubate quenched samples at 30 °C overnight to complete proteinase K digestion.
9. Rinse each well of a pre-cast Novex 12-well 4%–20% TBE polyacrylamide gel with water.
10. Place the gel cassette in XCell SureLock Mini-Cell tank and fill the upper and lower chambers with 0.5 × TBE running buffer (pre-chilled to 4 °C). Flush each well with 0.5 × TBE running buffer.
11. Load 3.5 μL sample into each lane. For captured reporter strand marker, mix 39.5 μL assay buffer, 0.5 μL 40 units/μL human placenta RNAse inhibitor, 5 μL 10 × gel shift unwound control, 5 μL 20 mM ATP·$MgCl_2$, 50 μL quench solution. Load 3.5 μL captured reporter strand marker into a well.
12. Run samples under constant voltage (200 V) for 45 min at 4 °C.
13. Remove the gel from the cassette and scan for 6-FAM fluorescence (473 nm laser, LPB filter) using Typhoon FLA 9500 scanner.

3.3.3 RNA decay assays of MTR4–exosome complexes

1. For each reaction, mix 69 μL assay buffer, 1 μL 40 units/μL human placenta RNAse inhibitor, 10 μL 100 nM RNA and 10 μL 20 mM ATP·$MgCl_2$ of MTR4-exosome complexes in a microcentrifuge tube with or without mutations in EXOSC10 that disrupt its activity to facilitate analysis of DIS3 activities (Weick et al., 2018).
2. Incubate at 20 °C for 5 min.
3. For sample at 0 time point, take 18 μL aliquot of reaction mixture and mix with 2 μL assay buffer and 10 μL quench solution.
4. Pipet 54 μL reaction mixture to a fresh microcentrifuge tube.
5. Initiate the reaction by adding 6 μL 1 μM MTR4–exosome complex.

6. Take 10 μL aliquot of reaction mixture at various time points and stop the reaction by adding 5 μL quench solution.
7. Incubate quenched samples at 30 °C overnight to complete proteinase K digestion.
8. Rinse each well of a pre-cast Novex 12-well 4%–20% TBE polyacrylamide gel with water.
9. Place the gel cassette in XCell SureLock Mini-Cell tank and fill the upper and lower chambers with 0.5 × TBE running buffer. Flush each well with 0.5 × TBE running buffer.
10. Load 3.5 μL sample into each lane.
11. Run samples under constant voltage (200 V) for 45 min at 22 °C.
12. Remove the gel from the cassette and scan for fluorescein fluorescence (473 nm laser, LPB filter) using Typhoon FLA 9500 scanner.

3.4 Notes

- Assay buffer can be replaced with 20 mM Tris-HCl pH 7.0, 50 mM NaCl, 0.5 mM $MgCl_2$, 2.5 mM DTT, 0.01% IGEPAL CA-630 to avoid precipitation of SDS by potassium salt.

4. Preparation of RNA-loaded human MTR4-exosome complex for structural studies

To understand how MTR4 and other exosome-associated helicases unwind and feed RNA into the exosome, we engineered substrates to capture the helicase during the course of RNA translocation. We previously reported use of a substrate with a chimeric DNA-RNA translocation strand (Fig. 4A) to determine the cryogenic electron microscopy (cryo-EM) structure of RNA-loaded human MTR4–exosome complex in the process of unwinding and delivering RNA to the exosome (Weick et al., 2018). The DNA-RNA translocation strand allows MTR4 tracking on the RNA segment of the substrate but induces stalling as the helicase reaches the DNA-RNA junction (Fig. 4b). We designed the length of the translocation strand based on the number of nucleotides accommodated by MTR4, exosome core, and DIS3 in reported crystal structures of *S. cerevisiae* homologs (Weir et al., 2010; Zinder, Wasmuth, & Lima, 2016). The optimal length was refined by performing a helicase reaction followed by a crosslinking assay (Weick et al., 2018). In this section, we describe the preparation and loading of a tripartite DNA-RNA chimera substate to MTR4-exosome complex for cryo-EM analysis.

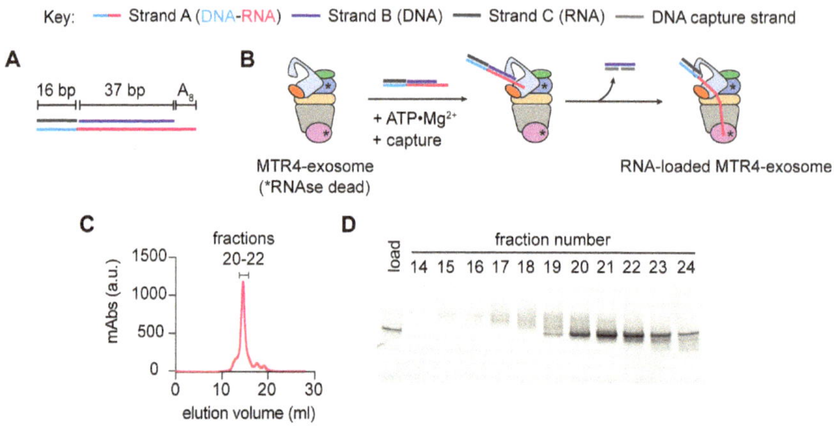

Fig. 4 RNA loading to human MTR4–exosome complex. (A) DNA-RNA chimera tripartite substrate. (B) Schematics showing the RNA loading strategy for human MTR4-exosome complex using the DNA-RNA chimera substrate. (C) Superdex S200 Increase 10/300 GL chromatogram of annealed DNA-RNA chimera tripartite substrate. (D) SYBR Gold-stained native PAGE showing Superdex S200 Increase fractions of purified DNA-RNA chimera tripartite substrate.

4.1 Equipment

- ÄKTA pure 25 L chromatography system (GE Healthcare)
- Bio-Rad ChemiDoc XRS+
- C1000 Touch Thermal Cycler (Bio-Rad)
- Nanodrop 2000 (Thermo Fisher Scientific)
- ThermoMixer (Eppendorf)
- Typhoon FLA 9500 scanner (GE Healthcare)
- XCell SureLock Mini-Cell tank (Thermo Fisher Scientific)

4.2 Materials

- Lyophilized RNA (see Table 1; synthetic oligos from IDT or Dharmacon)
- Annealing buffer (10 mM Tris-HCl pH 7.0, 100 mM potassium acetate)
- ATP·$MgCl_2$ (200 mM ATP, 200 mM $MgCl_2$, pH adjusted to 7.0 using KOH and supplemented with 1 mM Tris-HCl pH 7.0) (see Section 2.4 for details)
- DNA capture strand (see Table 1, 90 μM Capture2 in annealing buffer)
- β-mercaptoethanol (BME)
- Human placenta RNAse inhibitor (New England Biolabs)

- Human DIS3/EXOSC10/MTR4-exosome complex with mutations that disrupt DIS3 and EXOSC10 exoribonuclease activity (Weick et al., 2018; protein complex in 20 mM Tris-HCl pH 8.0, 50 mM NaCl, 0.5 mM $MgCl_2$, 2.5 mM BME)
- 4 × Lithium dodecyl sulfate (LDS) sample buffer (Thermo Fisher Scientific)
- Loading Buffer (20 mM Tris-HCl pH 8.0, 100 mM NaCl, 2.5 mM BME, 0.5 mM $MgCl_2$, 20% v/v glycerol, 10 mM EDTA, 0.2% v/v SDS, 50 units/mL Proteinase K)
- Lyophilized oligonucleotides (see Table 1, synthetic oligos from IDT or Dharmacon)
- MES running buffer (Thermo Fisher Scientific)
- Novex 12-well 4%–20% TBE polyacrylamide gel (Thermo Fisher Scientific)
- NuPage 4%–12% Bis-Tris polyacrylamide gel (Thermo Fisher Scientific)
- 2 × Quench solution (20 mM Tris-HCl pH 8.0, 50 mM NaCl, 6 mM AMPPNP, 1 mM BME)
- Superdex S200 Increase 10/300 GL (Thermo Fisher Scientific)
- SYBR Gold Gel Stain (Thermo Fisher Scientific)
- 1 × TBE running buffer (89 mM Tris-borate, 2 mM EDTA, pH 8.3)

4.3 Protocol

4.3.1 Substrate preparation

1. Resuspend lyophilized, HPLC-purified oligonucleotides in annealing buffer (10 mM Tris pH 7.0, 100 mM potassium acetate) to an estimated final concentration of 400 μM.
2. Follow steps 2–3 in Section 3.3.1.
3. Prepare 200 μL mixture of 100 μM StrandA3, 200 μM StrandB3, and 100 μM StrandC1. Aliquot 50 μL into a PCR 8-tube strip.
4. Using a thermocycler, incubate the RNA mixture at 95 °C for 5 min, 16 °C for 10 min then 4 °C overnight. Pool annealed RNA solution into a 1.7 mL microcentrifuge tube.
5. Purify annealed RNA using Superdex S200 Increase 10/300 GL pre-equilibrated with annealing buffer. Representative chromatogram is shown in Fig. 4C.
6. Analyze fractions for nucleic acid content using native PAGE. Take 1 μL aliquot and add 19 μL of 1 × TBE sample buffer. Resolve samples (10 μL/lane) in Novex 4%–20% TBE polyacrylamide gel using pre-chilled 1 × TBE running buffer (200 V for 1 h at 4 °C). Stain gel with

1 × SYBR Gold dissolved in 1 × TBE for 10 min. Wash gel with 1 × TBE for 10 min. Scan gel for fluorescence using Bio-Rad ChemiDoc XRS+. Representative native PAGE analysis of nucleic acids in Superdex S200 Increase fractions is shown in Fig. 4D.

7. Take peak fractions, determine RNA concentration, and store at −80 °C until needed.

4.3.2 Preparation of RNA-loaded human MTR4–exosome complex

1. Mix 50 μL 9 μM human MTR4-exosome complex with 50 μL 9 μM substrate.
2. Prepare fresh ATP-capture solution. For 50 μL solution, mix 20 μL nuclease-free water, 5 μL 200 mM ATP·$MgCl_2$ solution and 25 μL 90 μM DNA capture strand. Add 10 μL of 20 mM ATP-capture mix to initiate reaction.
3. Incubate reaction mixture at 22 °C for 2 h.
4. Add 50 μL of quench solution.
5. Spin at 14,000 × *g* using a microcentrifuge for 5 min at 4 °C.
6. Purify using Superdex 200 Increase 10/300 GL pre-equilibrated with SEC buffer.
7. Analyze fractions for protein content using SDS-PAGE. Take 15 μL aliquot and add 5 μL 4 × LDS sample buffer supplemented with 5% v/v BME (710 mM BME). Heat at 95 °C for 3 min. Resolve samples in 4%–12% NuPage Bis-Tris gel using 1 × MES running buffer. Stain gel with Coomassie staining solution. Destain with destaining solution.
8. Analyze fractions for co-purified substrate. Take 10 μL aliquot, add 10 μL of loading buffer and incubate at 37 °C for 3 h. Resolve samples (10 μL/lane) in Novex 4%–20% TBE polyacrylamide gel using 1 × TBE running buffer pre-chilled to 4 °C (200 V for 1 h at 4 °C). Stain gel with 1 × SYBR Gold dissolved in 1 × TBE for 10 min. Wash gel with 1 × TBE for 10 min. Scan gel for SYBR Gold fluorescence using Bio-Rad ChemiDoc XRS+.
9. Take peak fractions and supplement with 2 mM AMPPNP using quenching buffer. RNA loaded human MTR4-exosome complex can be used for subsequent cryo-EM grid preparation and analysis.

4.4 Notes

- Bulky chemical moieties can be conjugated to a nucleotide in the translocation strand to block MTR4 tracking on the RNA substrate.

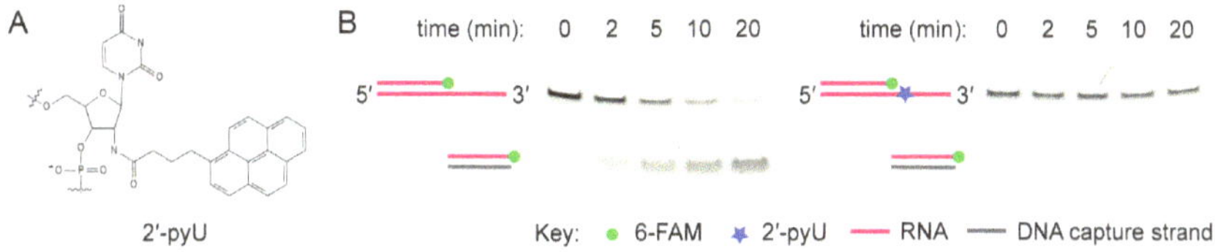

Fig. 5 2′-pyU-modified substrate inhibits helicase activity. (A) Chemical structure of 2′-amino-butyryl-pyrene uridine modification. (B) Native PAGE showing time courses of NEXT unwinding reaction on a 3′ tailed RNA duplex substrate with (right panel) or without (left panel) 2′-pyU-modification.

For example, we found that incorporation of 2′-amino-butyryl-pyrene-modified uridine (2'pyU) in the translocation strand inhibits unwinding by NEXT (Fig. 5).

Acknowledgments

We thank Ha An Nguyen and Lucas Repeta for sharing their critical review of the manuscript. This research was supported in part by NIH National Institute of General Medical Sciences (NIGMS) Grants R35 GM118080 (to C.D.L.). The content is solely the responsibility of the authors and does not represent the official views of the National Institutes of Health. C.D.L. is an investigator of the Howard Hughes Medical Institute.

References

Belon, C. A., & Frick, D. N. (2008). Monitoring helicase activity with molecular beacons. *BioTechniques*, *45*(4), 433–442. https://doi.org/10.2144/000112834.

Bradley, M. J., & De La Cruz, E. M. (2012). Analyzing ATP utilization by DEAD-box RNA helicases using kinetic and equilibrium methods. *Methods in Enzymology*, *511*, 29–63. https://doi.org/10.1016/B978-0-12-396546-2.00002-4.

Brune, M., Hunter, J. L., Corrie, J. E. T., & Webb, M. R. (2002). Direct, real-time measurement of rapid inorganic phosphate release using a novel fluorescent probe and its application to actomyosin subfragment 1 ATPase. *Biochemistry*, *33*(27), 8262–8271. https://doi.org/10.1021/BI00193A013.

Gerlach, P., Schuller, J. M., Bonneau, F., Basquin, J., Reichelt, P., Falk, S., et al. (2018). Distinct and evolutionary conserved structural features of the human nuclear exosome complex. *eLife*, 7, e38686. https://doi.org/10.7554/eLife.38686.

Houseley, J., LaCava, J., & Tollervey, D. (2006). RNA-quality control by the exosome. *Nature Reviews Molecular Cell Biology*, 7(7), 529–539. https://doi.org/10.1038/nrm1964.

Jackson, R. N., Klauer, A. A., Hintze, B. J., Robinson, H., Van Hoof, A., & Johnson, S. J. (2010). The crystal structure of Mtr4 reveals a novel arch domain required for rRNA processing. *The EMBO Journal*, *29*(13), 2205–2216. https://doi.org/10.1038/emboj.2010.107.

Jia, H., Wang, X., Anderson, J. T., & Jankowsky, E. (2012). RNA unwinding by the Trf4/Air2/Mtr4 polyadenylation (TRAMP) complex. *Proceedings of the National Academy of Sciences of the United States of America*, *109*(19), 7292–7297. https://doi.org/10.1073/pnas.1201085109.

Jia, H., Wang, X., Liu, F., Guenther, U. P., Srinivasan, S., Anderson, J. T., et al. (2011). The RNA helicase Mtr4p modulates polyadenylation in the TRAMP complex. *Cell*, *145*(6), 890–901. https://doi.org/10.1016/J.CELL.2011.05.010.

Liu, Q., Greimann, J. C., & Lima, C. D. (2006). Reconstitution, activities, and structure of the eukaryotic RNA exosome. *Cell*, *127*(6), 1223–1237. https://doi.org/10.1016/J.CELL.2006.10.037.

Liu, F., Putnam, A., & Jankowsky, E. (2008). ATP hydrolysis is required for DEAD-box protein recycling but not for duplex unwinding. *Proceedings of the National Academy of Sciences of the United States of America*, *105*(51), 20209–20214. https://doi.org/10.1073/PNAS.0811115106/SUPPL_FILE/0811115106SI.PDF.

Lubas, M., Christensen, M. S., Kristiansen, M. S., Domanski, M., Falkenby, L. G., Lykke-Andersen, S., et al. (2011). Interaction profiling identifies the human nuclear exosome targeting complex. *Molecular Cell*, *43*(4), 624–637. https://doi.org/10.1016/J.MOLCEL.2011.06.028.

Makino, D. L., Halbach, F., & Conti, E. (2013). The RNA exosome and proteasome: Common principles of degradation control. *Nature Reviews Molecular Cell Biology*, *14*(10), 654–660. https://doi.org/10.1038/nrm3657.

Meola, N., Domanski, M., Karadoulama, E., Chen, Y., Gentil, C., Pultz, D., et al. (2016). Identification of a nuclear exosome decay pathway for processed transcripts. *Molecular Cell*, *64*(3), 520–533. https://doi.org/10.1016/j.molcel.2016.09.025.

Ogami, K., Richard, P., Chen, Y., Hoque, M., Li, W., Moresco, J. J., et al. (2017). An Mtr4/ZFC3H1 complex facilitates turnover of unstable nuclear RNAs to prevent their cytoplasmic transport and global translational repression. *Genes & Development*, *31*(12), 1257–1271. https://doi.org/10.1101/gad.302604.117.

Özeş, A. R., Feoktistova, K., Avanzino, B. C., Baldwin, E. P., & Fraser, C. S. (2014). Real-time fluorescence assays to monitor duplex unwinding and ATPase activities of helicases. *Nature Protocols*, *9*(7), 1645–1661. https://doi.org/10.1038/nprot.2014.112.

Pang, P. S., Jankowsky, E., Planet, P. J., & Pyle, A. M. (2002). The hepatitis C viral NS3 protein is a processive DNA helicase with cofactor enhanced RNA unwinding. *The EMBO Journal*, *21*(5), 1168–1176. https://doi.org/10.1093/emboj/21.5.1168.

Puno, M. R., & Lima, C. D. (2018). Structural basis for MTR4-ZCCHC8 interactions that stimulate the MTR4 helicase in the nuclear exosome-targeting complex. *Proceedings of the National Academy of Sciences of the United States of America*, *115*(24), E5506–E5515. https://doi.org/10.1073/pnas.1803530115.

Puno, M. R., Weick, E.-M., Das, M., & Lima, C. D. (2019). SnapShot: The RNA exosome. *Cell*, *179*, 282–282.e1. https://doi.org/10.1016/j.cell.2019.09.005.

Rajagopal, V., & Lorsch, J. R. (2013). ATP and GTP hydrolysis assays (TLC). *Methods in Enzymology*, *533*, 325–334. https://doi.org/10.1016/B978-0-12-420067-8.00025-8.

Taylor, L. L., Jackson, R. N., Rexhepaj, M., King, A. K., Lott, L. K., van Hoof, A., et al. (2014). The Mtr4 ratchet helix and arch domain both function to promote RNA unwinding. *Nucleic Acids Research*, *42*(22), 13861–13872. https://doi.org/10.1093/nar/gku1208.

Thoms, M., Thomson, E., Baßler, J., Gnädig, M., Griesel, S., & Hurt, E. (2015). The exosome is recruited to RNA substrates through specific adaptor proteins. *Cell*, *162*(5), 1029–1038. https://doi.org/10.1016/j.cell.2015.07.060.

Wasmuth, E. V., Januszyk, K., & Lima, C. D. (2014). Structure of an Rrp6-RNA exosome complex bound to poly(A) RNA. *Nature*, *511*(7510), 435–439. https://doi.org/10.1038/nature13406.

Wasmuth, E. V., Zinder, J. C., Zattas, D., Das, M., & Lima, C. D. (2017). Structure and reconstitution of yeast Mpp6-nuclear exosome complexes reveals that Mpp6 stimulates RNA decay and recruits the Mtr4 helicase. *eLife*, *6*, e29062. https://doi.org/10.7554/eLife.29062.

Weick, E. M., & Lima, C. D. (2021). RNA helicases are hubs that orchestrate exosome-dependent 3′–5′ decay. *Current Opinion in Structural Biology, 67*, 86–94. https://doi.org/10.1016/J.SBI.2020.09.010.

Weick, E.-M., Puno, M. R., Januszyk, K., Zinder, J. C., Dimattia, M. A., & Lima, C. D. (2018). Helicase-dependent RNA decay illuminated by a cryo-EM structure of a human nuclear RNA exosome-MTR4 complex. *Cell, 173*, 1–15. https://doi.org/10.1016/j.cell.2018.05.041.

Weir, J. R., Bonneau, F., Hentschel, J., & Conti, E. (2010). Structural analysis reveals the characteristic features of Mtr4, a DExH helicase involved in nuclear RNA processing and surveillance. *Proceedings of the National Academy of Sciences of the United States of America, 107*(27), 12139–12144. https://doi.org/10.1073/pnas.1004953107.

Zinder, J. C., & Lima, C. D. (2017). Targeting RNA for processing or destruction by the eukaryotic RNA exosome and its cofactors. In *Vol. 31. Genes and development* (pp. 88–100). https://doi.org/10.1101/gad.294769.116.

Zinder, J. C., Wasmuth, E. V., & Lima, C. D. (2016). Nuclear RNA exosome at 3.1 Å reveals substrate specificities, RNA paths, and allosteric inhibition of Rrp44/Dis3. *Molecular Cell, 64*(4), 734–745. https://doi.org/10.1016/J.MOLCEL.2016.09.038.

CHAPTER SEVENTEEN

Using hydrogen-deuterium exchange mass spectrometry to characterize Mtr4 interactions with RNA

Darby Ball, Trang Nguyen, Naifu Zhang†, and Sheena D'Arcy*
Department of Chemistry and Biochemistry, The University of Texas at Dallas, Richardson, TX, United States
*Corresponding author: e-mail address: sheena.darcy@utdallas.edu

Contents

Abstract

Hydrogen deuterium exchange coupled to mass spectrometry (HDX-MS) is a valuable technique to investigate the dynamics of protein systems. The approach compares the deuterium uptake of protein backbone amides under multiple conditions to

† Current address: Institute for Cancer Genetics, Columbia University Irving Medical Center, New York, NY, United States.

Methods in Enzymology, Volume 673
ISSN 0076-6879
https://doi.org/10.1016/bs.mie.2022.04.002

characterize protein conformation and interaction. HDX-MS is versatile and can be applied to diverse ligands, however, challenges remain when it comes to exploring complexes containing nucleic acids. In this chapter, we present procedures for the optimization and application of HDX-MS to studying RNA-binding proteins and use the RNA helicase Mtr4 as a demonstrative example. We highlight considerations in designing on-exchange, bottom-up, comparative studies on proteins with RNA. Our protocol details preliminary testing and optimization of experimental parameters. Difficulties arising from the inclusion of RNA, such as signal repression and sample carryover, are addressed. We discuss how chromatography parameters can be adjusted depending on the issues presented by the RNA, emphasizing reproducible peptide recovery in the absence and presence of RNA. Methods for visualization of HDX data integrated with statistical analysis are also reviewed with examples. These protocols can be applied to future studies of various RNA-protein complexes.

Abbreviations

HDX hydrogen deuterium exchange
LC liquid chromatography
MS mass spectrometer or spectrometry
Mtr4 mRNA transport protein 4
TIC total ion chromatogram

1. Introduction

Hydrogen-deuterium exchange coupled to mass spectrometry (HDX-MS) is an analytical approach for studying proteins and their interactions with ligands. It monitors deuterium exchange into (on-exchange) or out of (off-exchange) amide groups in the backbone of a protein (Englander & Kallenbach, 1984; Katta, Chait, & Carr, 1991). As deuterium (2.01 amu) is heavier than hydrogen (1.01 amu), isotopic exchange can be measured in a mass spectrometer (MS) at either the protein or peptide scale. Major influences on the rate of exchange are pH, temperature, primary structure, and backbone amide hydrogen bonding and solvent accessibility (Bai, Milne, Mayne, & Englander, 1993). Exchange is minimized around pH 2.5, depending on ionic strength and peptide sequence, and low temperature (Hamuro, 2021; Kim & Baldwin, 1982; Rosenberg & Chakravarti, 1968), with the rate increasing 10-fold for each pH unit or for every increase of 22 °C (Englander, 2006). In most cases, pH, temperature, and primary structure are tightly controlled between samples, making differences in exchange reflective of changes in hydrogen bonding or solvent accessibility. These are readouts of protein structure at the secondary, tertiary, and quaternary levels. Fast exchanging amides are associated with a dynamic protein backbone or solvent exposure, while slow exchanging amides suggest a rigid protein backbone or burial in the protein core (Engen, Wales, & Shi, 2011; Marcsisin & Engen, 2010).

HDX-MS can be used to study protein interactions as ligands often cause a change in exchange rate (Woods & Hamuro, 2001). The rate can be enhanced or diminished and be a result of either a direct binding or a binding-induced conformational change (Chalmers, Busby, Pascal, West, & Griffin, 2011). Complementary experiments are required to distinguish between the latter two causes. Disadvantages of using HDX-MS to study protein interactions include limited spatial resolution, as sufficient peptides to provide single amino acid resolution for an entire protein are not typically recovered, and the analytical challenge associated with an inability to completely stop the exchange reaction. In an on-exchange set-up, deuterium in the protein is lost during sample processing in hydrogen-rich solvents. This is called back-exchange and is minimized primarily by handling exchanged samples rapidly and at low pH and temperature (Bai et al., 1993; Hamuro, 2021; Rand, Lund, Amon, & Jørgensen, 2011; Walters, Ricciuti, Mayne, & Englander, 2012). Advantages of using HDX-MS on the other hand include analysis in an aqueous environment, a modest sample requirement, and compatibility with diverse types of ligands. HDX-MS can study binding of proteins to other proteins, nucleic acids, small molecules, and even lipids (Engen et al., 2011; Narang, Lento, & Wilson, 2020; Pirrone, Iacob, & Engen, 2014).

While HDX-MS can be used to study protein binding to various types of ligands, some ligands are underrepresented in the literature. This is the case for nucleic acids and may be attributed to an incompatibility with the typical HDX-MS workflow (Fig. 1A) (Sperry, Wilcox, & Gross, 2008). Nucleic acids and their protein partners tend to aggregate at the low pH required to minimize back-exchange (Chalmers et al., 2011;

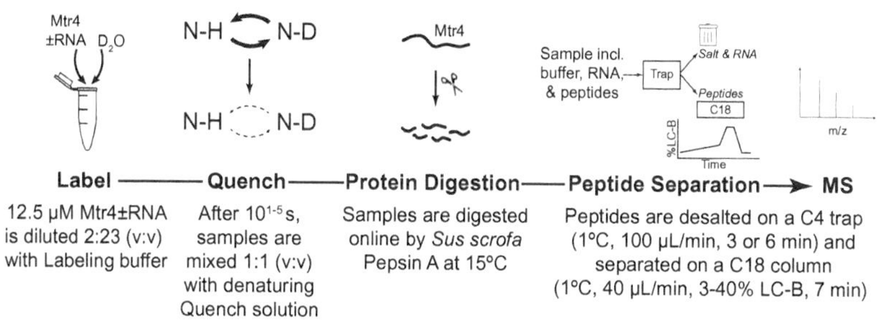

Fig. 1 Typical steps in bottom-up, on-exchange HDX-MS workflow. Experimental details for Mtr4 are included.

Poliakov, Jardine, & Prevelige, 2008; Sorokin, Gladchenko, & Valeev, 1986). Nucleic acids can also cause issues in liquid chromatography (LC) steps that digest protein and separate resulting peptides after the exchange reaction (Filandrova, Kavan, Kadek, Novak, & Man, 2021). Nucleic acids can reduce protein digestion and cause protein or peptide accumulation on columns. When accumulated peptides contaminate subsequent samples, it is described as carryover (Fang, Rand, Beuning, & Engen, 2011; Haeckel, 1988). It is not yet possible to predict if a particular nucleic acid will be problematic. The inherent instability of RNA compared to DNA may also make it more likely to be troublesome. When working with nucleic acids as ligands one should thus anticipate that alterations may need to be made to a standard HDX-MS workflow.

We have recently used HDX-MS to investigate the interaction between the Ski-2-like helicase Mtr4 and an assortment of RNA ligands (Zhang, Olsen, Ball, Johnson, & D'Arcy, 2022). Mtr4 is an activator of the nuclear exosome that utilizes ATP to unwind various types of RNA, including tRNA, rRNA, snoRNA, and mRNA (Allmang et al., 1999; van Hoof, Lennertz, & Parker, 2000; Wang, Jia, Jankowsky, & Anderson, 2008). It prefers substrates with single-stranded 3' overhang of 4 to 6 adenines (Bernstein, Patterson, Wilson, & Toth, 2008; Jia, Wang, Anderson, & Jankowsky, 2012; Jia et al., 2011). There are several high-resolution structures of Mtr4 showing a helicase core with two RecA domains, a winged-helix domain, and a helical bundle (Conrad et al., 2016; Jackson et al., 2010; Taylor et al., 2014; Wang et al., 2019). Mtr4 also contains a less-conserved arch domain that our HDX-MS studies implicate in dictating RNA preference (Zhang et al., 2022). During our HDX-MS experiments with Mtr4 and RNA, we encountered several technical challenges. In this chapter we describe our approach to predicting (Section 2) and troubleshooting (Section 3) such challenges and detail a final protocol for investigating RNA-binding by Mtr4 (Section 4). The procedures described herein can be applied to investigate nucleic acid binding of Mtr4, other helicases, or any other protein.

2. Designing an HDX-MS experiment for a protein-RNA complex

In this section, we broadly discuss the initial considerations in the design of an HDX-MS experiment that aims to characterize protein binding to RNA. Most of these aspects also apply to the analysis of DNA-binding proteins, which have been discussed elsewhere (Filandrova et al., 2021). We outline workflow choices and discuss the qualities of the protein and protein-RNA complex that impact experimental design. We give general guidance for any soluble protein with an RNA ligand and then include specific details for Mtr4.

2.1 Definition of the workflow

One of the initial considerations for any HDX-MS experiment is to choose between a top-down or bottom-up approach (Section 2.3 Note 1). Top-down measures the mass of the intact deuterated protein, while bottom-up measures the mass of peptides produced by digestion of the deuterated protein. A choice must also be made between monitoring the on-exchange or the off-exchange reaction. For Mtr4, we employ a bottom-up, on-exchange experiment. We select bottom-up as peptide-level data allows us to distinguish exchange of the different regions of Mtr4. At >120kDa, Mtr4 is a large protein that would be difficult to fragment and obtain peptide- or residue-level data in a top-down system (Pan, Han, Borchers, & Konermann, 2009). We select on-exchange to avoid difficulties in preparing Mtr4 that is both natively folded and fully-deuterated (Woods & Hamuro, 2001).

The general bottom-up, on-exchange workflow (Fig. 1) is well-established and is commonly reported in literature. Briefly, the main stages of the workflow are as follows:

1. *Labeling*: exchange reaction is initiated by diluting the protein with deuterium at a controlled ratio (volume to volume, v:v) of hydrogen to deuterium, pH, and temperature.
2. *Quenching*: at precise timepoints the exchange reaction is minimized, and the protein unfolded, by lowering the pH and temperature through mixing with a cooled solution that often contains a chaotrope.
3. *Digestion*: unfolded protein is cut into peptides by proteases as quickly as possible at low pH and temperature.
4. *Peptide Separation*: peptide mixtures are desalted by trapping and separated on an analytical LC column in line with an MS.
5. *Mass Measurement*: mass, and in some cases the ion mobility, of each peptide is measured in the MS.
6. *Data Analysis*: deuterated peptides are identified based on matching properties (retention time, ion mobility, and mass range) to non-deuterated controls that have sequence information from tandem MS data.

Spectra are manually inspected, and deuterium uptake calculated as the mass difference between deuterated and non-deuterated peptides.

For all samples submitted to the workflow, it is imperative that time, pH, and temperature are tightly controlled as they influence the intrinsic chemical exchange rate (Section 2.3 Note 2). This rate must be minimized at stages 2–4 to avoid back-exchange. Further details and recommendations for all stages can be found elsewhere (Chalmers et al., 2011; Engen & Wales, 2015; Kochert, Iacob, Wales, Makriyannis, & Engen, 2018; Konermann, Pan, & Liu, 2011; Masson et al., 2019; Walters et al., 2012; Weis, 2016). In our experience, we find that each protein system requires tweaks to some or all stages especially when nucleic acids are involved. We describe how we identify these tweaks (Section 3) before embarking on a complete analysis (Section 4).

It is also worth noting that the bottom-up, on-exchange workflow is frequently used for HDX-MS experiments that are comparative. Rather than interpreting absolute deuterium uptake (Section 2.3 Note 3), these experiments interpret the deuterium uptake difference when a protein is in various conditions. For comparative experiments, it is essential to identify the exact variable and keep all other parameters consistent. The nature of the variable will depend on the question being asked, but could be the presence of a ligand, a buffer additive, a post-translational modification, to name but a few. For Mtr4 our single variable is the presence of RNA, and we compare the deuterium uptake of Mtr4 alone to that of the Mtr4-RNA complex. Our Mtr4 workflow is thus a bottom-up, on-exchange, comparative, HDX-MS experiment.

2.2 Sample properties

Prior to performing any HDX-MS experiments, the properties of the protein and protein-RNA complex must be considered. Important properties are as follows:

1. Quantity of protein and RNA
   - The total protein required for an experiment can be estimated as follows:
   - *total protein = protein per injection x no. timepoints x no. complexes x no. replicates*
   - Typical values for each variable are:
   - *protein per injection* – 25 pmol with a range of 10 to 100 pmol (Guttman & Lee, 2016).

     no. timepoints – 5 including controls, 10^1 s, 10^2 s, 10^3 s, and 10^4 s (Section 2.3 Note 4)

- *no. complexes* – 2 for the protein alone and the protein in complex; add 1 for every extra RNA
- *no. replicates* – 3 if considering only technical replicates (Section 2.3 Note 5)
- Typical values give a total protein requirement of 750 pmol. This is approximately 40 μg for a 50 kDa protein or 90 μg for a 120 kDa protein like Mtr4. This is a lower limit estimate, as the calculation does not include protein for troubleshooting (Section 3), the best-practice of preparing volumes 10% greater than those being pipetted (Section 4), the inclusion of a biological replicate or several RNA ligands. The total protein required is likely closer to 1500 pmol. This is a modest amount of protein compared to other analytical tools such as X-ray crystallography (Iacob & Engen, 2012).
- If <1500 pmol are available, adjustments must be made to maintain feasibility. Easy adjustments are reducing the number of timepoints and injecting less protein per injection. The former means a change in deuterium uptake may be missed, while the latter may reduce the number of peptides recovered.
- The total RNA required can be calculated using the same equation as for protein, although typical values differ. The *RNA per injection* may be >25 pmol as excess RNA is needed to near-saturate the protein (Kochert et al., 2018) and the *no. of complexes* is 1 as RNA alone is not analyzed.
- Sufficient Mtr4 is available as approximately 30 nmol (3.5 mg) of recombinant protein can be purified per liter of *Escherichia coli* culture. Sufficient RNA quantities can be made in *in vitro* transcription systems (namely $\text{tRNA}_\text{i}^\text{Met}$) (Easton, Shibata, & Lukavsky, 2010) or purchased commercially (e.g., Integrated DNA Technologies) (Taylor et al., 2014).

2. *Purity of protein and RNA*

- Pure protein is free from contamination with other proteins. High purity (>90%) is not an absolute requirement but does make LC-MS more straightforward as there are less peptides from contaminating proteins. Peptides from contaminating proteins can have spectral overlap with peptides from the protein-of-interest and increase the likelihood of false positive peptide identifications. Impurity is less of an issue with a smaller protein, a higher-resolution MS, or a sub-zero LC system (Section 2.3 Note 6).
- Purity also relates to the presence of a non-native tag. If the tag is small (such as a poly-histidine, flag, or strep tag), it does not need

to be removed as it will contribute very few peptides to the sample. If the tag is large (such as glutathione *S*-transferase or maltose-binding protein), it may need to be removed depending on the extent of peptide spectral overlap. If the tag interferes with protein folding or activity in any biochemical assay, it should be removed regardless.

- RNA does not have to be pure assuming that the contaminating species do not interact with the protein-of-interest and do not interfere with measuring concentration. The biological relevance and structure predictions of the RNA should be considered.
- Recombinant Mtr4 purified with standard chromatographic techniques results in a sample with >95% purity (Jackson et al., 2010).

3. *Homogeneity of protein and protein-RNA complex*
   - Homogeneity means all copies of the protein are chemically and conformationally identical. Heterogeneity, on the other hand, means copies of the protein are distributed between populations with distinct chemical composition and/or conformational preferences. Heterogeneity impacts HDX-MS experiments, with as little as 5% heterogeneity resulting in a significant change in deuteration (Hageman & Weis, 2019a).
   - The most likely source of chemical heterogeneity is variable post-translational modification, especially on proteins expressed in eukaryotes (Baumann & Meri, 2004). How such modifications are best handled depends on their biological relevance and impact on peptide recovery. A modification can be added to the peptide search parameters or removed pre- or post-deuteration (Jensen et al., 2016). Removal post-deuteration would need to be done rapidly with pH and temperature conditions that minimize back-exchange.
   - Conformational heterogeneity can be difficult to foresee but can be due to slow domain motions (Miranker, Robinson, Radford, Aplin, & Dobson, 1993); constitutive, asymmetric homo-oligomerization (Kheterpal & Wetzel, 2006); unsaturated binding of cofactors; or heterogeneous disulfide bond formation. Conformational isomers will be detected by HDX if they are either non-interchanging or interchanging at a rate near to the exchange rate and have distinct deuterium uptake rates. Spectra will contain multi-modal distributions of isotopic peaks that can be difficult to deconvolute for protein alone or the protein-RNA complex (Section 2.3 Note 7).

- Recombinant Mtr4 is likely homogeneous as it is expressed in a prokaryotic system, is monomeric, and does not contain any disulfide bonds or cofactors.

4. *Stability of protein and protein-RNA complex*
   - A stable protein maintains its native primary, secondary, tertiary, and quaternary structure, while an unstable protein will degrade, unfold, and/or aggregate (Deller, Kong, & Rupp, 2016). Conditions for protein stability must be established prior to HDX-MS. Key environmental variables that influence protein stability and impact HDX-MS are light, temperature, pH, time, buffer additives, and protein concentration.
   - Temperature and pH alter the intrinsic chemical exchange rate (Section 2.3 Note 2) and should both be considered with time to identify shared conditions for stability and measurable exchange (Bai et al., 1993; Iacob & Engen, 2012). For example, if the protein is only stable at low temperature where the intrinsic chemical exchange rate is slow, this may be compensated by having the protein at a higher pH or exchanging for longer timepoints.
   - Buffer additives that improve stability are often compatible with HDX-MS. Common examples are salts such as NaCl, reducing agents such as TCEP (preferred due to its long half-life), chelating agents such as EDTA, and some lipids and detergents. Buffer components are removed from the sample at the desalting step, downstream of digestion and upstream of the MS. If a non-standard additive is required, the effect on the intrinsic chemical exchange rate (Section 2.3 Note 2), protein digestion, and peptide desalting, must be considered. If it is problematic, then additional steps may be required to remove the additive after quenching.
   - Protein must be stable both during exchange (0.5–1 μM or 25 pmol in 12.5–25 μL) and as a concentrated stock (at least 10–20 μM). Exchange is started by dilution with a deuterium buffer and the more concentrated the stock the higher the percentage deuterium in the final reaction, improving signal-to-noise by increasing deuterium incorporation.
   - Buffer additives and concentrations must be the same for the protein and the protein-RNA complex to ensure the only variable is the presence of the RNA.
   - Mtr4 can be considered stable as it does not degrade or aggregate at concentrations from 1 to >20 μM in an HDX-MS-compatible buffer

(Equilibration buffer: 50 mM HEPES, 100 mM NaCl, 0.5 mM $MgCl_2$, 0.2 mM TCEP, pH 6.5). A potential buffer additive to consider is an ATP analog if it is known to stabilize helicase conformations or states.

5. *Protein affinity for RNA*
   - The affinity between the protein and RNA must be low enough to reach a high percent-bound during exchange. The calculation for percent-bound is very clearly described elsewhere (Kochert et al., 2018). Percent-bound is usually maximized as it directly relates to the degree of change in deuterium uptake. There is no hard and fast rule for the minimum percent-bound, and as low as 5% may be detectable (Hageman & Weis, 2019b), but >50% is recommended.
   - Percent-bound is best increased by adding the RNA in stochiometric excess of the protein. This is possible as the RNA is not monitored in the MS as it is removed during upstream LC. Large excesses of RNA, however, can cause issues with complex solubility, online digestion, and LC (Section 3).
   - When comparing different RNA types, experiments are best set up at a common protein to RNA ratio, so the results reflect affinity differences.
   - The dissociation constants of Mtr4 for various RNA ligands are in the low micromolar range (Lingaraju et al., 2019; Puno & Lima, 2018). As exchange is normally performed at low micromolar concentrations, the RNA ligands should be added in stoichiometric excess (such as 2:1 molar ratio of RNA:protein) to ensure greater than 50% of the protein is RNA-bound (Section 2.3 Note 8).

2.3 Notes

1. In between bottom-up and top-down, there is the option of middle-down. This involves partial digestion of the protein to large fragments rather than peptides, followed by fragmentation by electron transfer dissociation for amino acid resolution. (Karch et al., 2018; Pan, Zhang, Chou, & Borchers, 2016).
2. Intrinsic chemical exchange rate refers to the rate of exchange of hydrogen to deuterium irrespective of the influence of solvent accessibility and hydrogen bonding. It can be measured by small molecule standards (Murphree, Vorauer, Brzoska, & Guttman, 2020) or peptides that lack structure (Toth et al., 2017; Zhang, Zhang, & Xiao, 2012). The intrinsic

chemical exchange rate is known to be influenced by several factors (for example, temperature, pH, and amino acid sequence) and these factors must be absolutely controlled for comparative samples.

3. The rate of deuterium uptake can be used to identify regions of relative disorder in a protein. The deuterium uptake of a peptide from one region of the protein can be qualitatively compared to the deuterium uptake of a peptide from another region. Peptides with greater or faster uptake likely represent loops or disordered domains. It is not possible, however, to go from deuterium uptake to fine structural details such as α-helix versus β-sheet.
4. One must decide the timepoints most useful for their protein. This is usually done empirically, although we typically use 10^1 s, 10^2 s, 10^3 s, and 10^4 s for the first experiment to cover as much of the exchange curve as possible without changing temperature and pH (Hamuro, 2017; Hoofnagle, Resing, & Ahn, 2004). When doing long timepoints ($>10^3$ s) it is important to include a 10 s pulsed-labeled control to ensure integrity of the protein (Masson et al., 2019).
5. Technical replicates report on the reproducibility of the exchange reactions and mass measurement (Masson et al., 2019). We use three replicates for data calculations but prepare four in case a sample is lost during processing. Biological replicates report on the biological variability of the system. For example, a biological replicate may involve repeating the entire experiment with a different protein preparation (Engen & Wales, 2015; Masson et al., 2019).
6. Sub-zero LC systems can minimize back-exchange over extended gradients that improve peptide separation (Hudgens, 2020; Wales, Fadgen, Eggertson, & Engen, 2017). In the mobile phase of sub-zero LC systems, solvent modifiers such as methanol or ethylene glycol are added to prevent freezing at temperatures below 0 °C. It should be noted that utilizing sub-zero systems does not improve the ion intensity of fragmented peptides (Wales et al., 2017).
7. A multi-modal distribution of isotopic peaks can have various causes and it is important to unambiguously diagnose the cause. Beyond conformational heterogeneity, it can be due to EX1 exchange kinetics (Englander & Kallenbach, 1984; Weis, Wales, Engen, Hotchko, & ten Eyck, 2006), StepWave settings (Guttman et al., 2016), or carryover on columns (Fang et al., 2011).
8. It is important to note that the experiments described herein monitor RNA binding, rather than any enzymatic turnover of the RNA.

To monitor enzymatic turnover, exchange would need to be performed under steady-state conditions with sufficient nucleotide and RNA through the entire exchange time course. One, however, could consider adding a non-hydrolysable nucleotide analog in the binding experiment, as well as $MgCl_2$ to facilitate its binding (Robinson, 1980). If this is done, the percentage of protein bound by nucleotide should be considered as already described for RNA binding.

3. Definition of experimental parameters for HDX-MS of an RNA-protein complex

In this section, we describe how to identify optimal parameters for a bottom-up, on-exchange, comparative HDX-MS experiment looking at protein binding to RNA. Parameters are sequentially optimized through a series of checkpoints, all of which must be satisfied before a complete experiment is attempted (Fig. 2). Checkpoints 1–3 focus on non-deuterated protein and generally apply to any HDX-MS experiment. RNA is included at checkpoints 2, 4 and 5, with deuterated samples being introduced at checkpoint 6. Particularly challenging checkpoints for protein with RNA relate to sample solubility, peptide recovery, and carryover. Our metrics to assess peptide recovery are intensity and shape of the Total Ion Chromatogram (TIC), percent sequence coverage, and peptide redundancy. These are defined as follows:

1. *TIC*: lot of the summed intensity of all ions across the entire mass range as a function of retention time on the LC system.

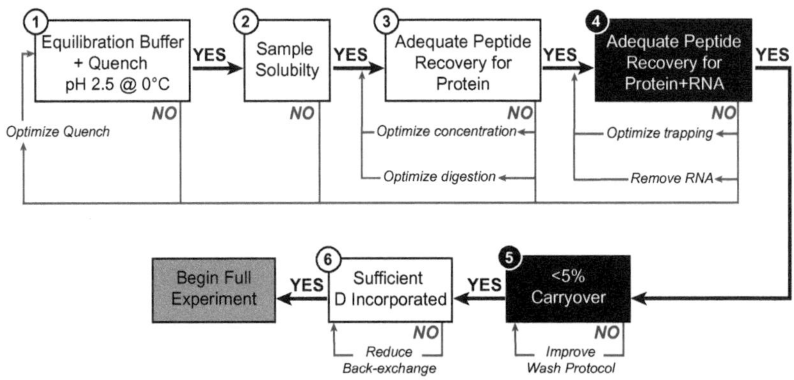

Fig. 2 Six major checkpoints in establishing parameters for HDX-MS of a protein-RNA complex. Black boxes highlight crucial checkpoints for samples with RNA. Black arrows indicate the most direct path, while grey arrows indicate troubleshooting.

2. *Percent sequence coverage*: number of residues covered by peptides divided by total number of residues in the protein multiplied by 100.
3. *Peptide redundancy*: sum of peptide lengths divided by the total number of residues covered by peptides.

The described procedure is designed for an LC setup with online digestion as published previously (Fig. 3) (Ahn, Jung, Wyndham, Yu, & Engen, 2012). The LC component (Section 3.5, Note 1) starts with injected protein samples being pushed by an acidic, aqueous buffer across a digestion column containing an immobilized protease that is active at low pH (e.g., pepsin). Resulting peptides flow directly to a short, reverse phase, trap column for desalting. Anything not retained on the trap flows to waste. Flow over the digestion column and trap is performed by an isocratic pump (Pump A in Fig. 3) with an aqueous mobile phase (LC-A). A valve switch then reverses the flow over the trap such that it flows directly to a longer, reverse phase, analytical column. Flow over the trap and analytical column is performed by a binary pump (Pump B in Fig. 3) with aqueous (LC-A) and organic (LC-B) mobile phases that are combined to create an elution gradient. In the final step, peptides are eluted into the MS with an increasing acetonitrile gradient via increasing the percent of LC-B. Starting values for key parameters of the LC method are listed in Table 1.

The described procedure also requires an MS run in positive-ion mode with an electrospray source and fragmentation capabilities, typically collision-induced dissociation. We use a Synapt G2-Si (Waters™) equipped

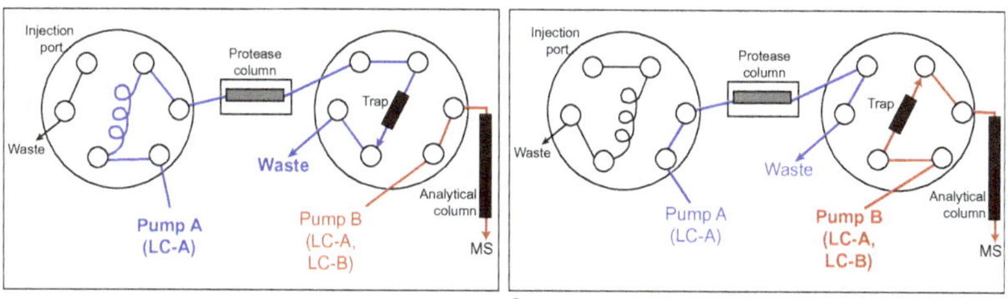

First valve position: isocratic pump A flows LC-A over the protease column and trap as sample is injected.

Second valve position: binary pump B flows a gradient of LC-B (mixed with LC-A) over the trap and analytical column.

Fig. 3 Two LC valve positions used during analysis of HDX samples. Blue represents flow path from the isocratic pump A that flows LC-A. Red indicates flow path from binary pump B that flows a gradient of LC-A and LC-B. The protease column is depicted in light gray, while the trap and analytical columns are depicted in black. *Based on Ahn, J., Jung, M. C., Wyndham, K., Yu, Y. Q., & Engen, J. R. (2012). Pepsin immobilized on high-strength hybrid particles for continuous flow online digestion at 10,000 psi.* Analytical Chemistry, *84(16), 7256. doi:10.1021/AC301749H.*

Table 1 Starting parameters for liquid chromatography.

Sample loop	50 μL
Digestion column	Waters™ Enzymate Protein Pepsin Column, 300 Å, 5 μm, 2.1 mm × 30 mm
Trap column	Waters™ ACQUITY UPLC Protein BEH C18 VanGuard Pre-column, 130 Å, 1.7 μm, 2.1 mm × 5 mm
Analytical column	Waters™ ACQUITY UPLC BEH C18, 130 Å, 1.7 μm, 2.1 mm × 100 mm
LC temperature	1 °C
Digestion temperature	15 °C
LC-A	0.1% (v/v) formic acid in H_2O
LC-B	0.1% (v/v) formic acid in ACN
Trapping time	3 min
Trapping flow rate	100 μL/min
Analytical flow rate	40 μL/min
Elution gradient (%LC-B)	3–40% in 7 min, 40–75% in 1 min, 75% for 2 min, 75–3% in 1 min, 3% for 1 min
Wash gradient (%LC-B)	3–85% in 1 min, 85–3% in 1 min, repeat six times
Protease column wash	50 μL
	1.6 M GuHCl, 4% (v/v) ACN, 0.8% (v/v) formic acid, 1% (w/v) n-Octyl-β-D-glucopyranoside

with an 8K quadrupole, collision-induced dissociation fragmentation, and ion mobility capabilities. Ion mobility data can reduce peptide overlap, especially for different charge states (Iacob, Murphy, & Engen, 2008). Pertinent MS-related parameters are listed in Table 2. Prior to starting optimization, the combined LC-MS system should be validated for HDX-MS using a standard or control protein, such as the commercially available Phosphorylase B or cytochrome *C*. Protocols for these are described elsewhere (Hoofnagle et al., 2004; Wales & Engen, 2006). Validation with both deuterated and non-deuterated samples ensures system integrity and permits an estimation of back-exchange (ideally <30%) after optimization of MS parameters (Guttman et al., 2016). When working on a validated system, the procedure described herein may take anywhere from two days if no issues are

Table 2 Starting parameters for mass spectrometry.

Source	Electrospray
Polarity	Positive mode
Analyzer	Resolution mode
Function type	$HDMS^E$ Continuum (DIA MS/MS with ion mobility)
m/z range	100–2000
Scan time	0.4 s
Low collison energy	6 V at Trap (off at Transfer)
High collision energy	22–44 V at Transfer (off at Trap)
Trap gas	2.0 mL/min
Helium gas	180 mL/min
IMS gas	90 mL/min
Lockspray	0.1 pmol/μL Glu-1-Fib B, 1 ng/μL Leu-enkephalin in 50:50 ACN/H_2O, 0.1% (v/v) formic acid, 5 μL/min
Lockspray flow rate	5 μL/min
Lockspray capillary	2.0 kV
MSMS mass	785.8426 m/z (Glu-1-Fib 2^+)
Capillary	2.8 kV
Sampling source	30 V
Source offset	30 V
Source temperature	80 °C
Desolvation	175 °C
Cone gas	90 L/h
Desolvation gas	650 L/h
Nebulizer	6.5 Bar

encountered, up to several weeks if extensive troubleshooting is required. Starting variables for sample-related parameters are listed in Table 3. These may be altered before the procedure based on specific knowledge about the protein or protein with RNA (Section 2).

Table 3 Starting parameters for sample preparation.

Stock concentration	25 μM
D_2O dilution v:v	1:24
Protein per injection	25 pmol
Quench solution	1.6 M GuHCl, 0.8% (v/v) formic acid, pH 2.3
Sample:Quench	1:1
Quenched sample	100 μL

3.1 Equipment

- General

  Dry bath, 1.5 mL tube block

  Ice bucket

  Liquid nitrogen

  Microcentrifuge tubes, 1.5 mL [for concentrated stocks and exchange reactions]

  Mini-centrifuge

  Nano-drop UV-Vis Spectrophotometer

  SevenCompact pH meter with InLab Micro Pro-ISM (Mettler Toledo)

  TempAssure colorful (five colors) PCR tubes, 0.2 mL (USA Scientific Inc) [for quenched samples]

  Timer, 4-channel

  Tube Rack, 0.2 mL and 1.5 mL

- LC-MS

  ACQUITY UPLC BEH C18, 130 Å, 1.7 μm, 2.1 mm X 100 mm (Waters™) [analytical column]

  ACQUITY UPLC Protein BEH C18 VanGuard Pre-column, 130 Å, 1.7 μm, 2.1 mm X 5 mm (Waters™) [trap column]

  Enzymate Protein Pepsin Column, 300 Å, 5 μm, 2.1 mm X 30 mm (Waters™) [protease column]

  ACQUITY UPLC M-Class System with HDX Automation (Waters™) [LC system]

  Synapt G2-Si HDMS, 8K quadrupole, Lockspray source with ESI probe (Waters™) [MS]

Injection syringe, needle point style 3, 22-gauge, 2-in needle length, 250 μL (Hamilton)

- Software

  MassLynx 4.2 (Waters™) [control of LC-MS]

  ProteinLynx Global Server (PLGS) 3.0.3 (Waters™) [peptides list(s) from non-deuterated controls]

  DynamX 3.0 (Waters™) [search for and refine spectra of deuterated peptides]

3.2 Reagents

Acetonitrile, LC-MS grade
DCl, 99.96% D (Cambridge Isotope Laboratories)
D_2O, 99.96% D (Cambridge Isotope Laboratories)
Formic acid, LC-MS grade
Guanidinium-HCl
H_2O, LC-MS grade
HCl
HEPES [4-(2-hydroxyethyl)-1-piperazineethanesulfonic acid]
n-Octyl-β-D-glucopyranoside (β-OG)
NaCl
NaOD, 99.5% D (Cambridge Isotope Laboratories)
NaOH
TCEP [tris(2-carboxyethyl)phosphine]

3.3 Buffers and samples

- LC-A: 0.1% (v/v) formic acid in LC-MS grade H_2O [500 mL]

  Pump A (isocratic pump that directs flow over digestion column and trap) and pump B (binary pump that directs flow over trap and analytical column) both use LC-A (Fig. 3).
- LC-B: 0.1% (v/v) formic acid in LC-MS grade acetonitrile [500 mL]

  Pump B mixes LC-B with LC-A to create an acetonitrile gradient over the trap and analytical column (Fig. 3).
- Protease Column wash: 1.6 M guanidinium-HCl, 4% (v/v) acetonitrile, 0.8% (v/v) formic acid, 1% (w/v) n-Octyl-β-D-glucopyranoside [10 mL] (Section 3.5 Note 2)

After each sample, 50 µL of Protease Column Wash is injected onto the digestion column at 25 µL/min followed by washing with LC-A. The digestion column should be flowing directly to waste rather than over the trap.

- Equilibration buffer: 20 mM HEPES, 150 mM NaCl, 0.2 mM TCEP, pH 7.5 [10 mL]

  The Equilibration buffer should permit the formation of a stable protein-RNA complex. It will vary depending on the protein-of-interest.
- Labeling buffer: 20 mM HEPES, 150 mM NaCl, 0.2 mM TCEP in D_2O, pD = 7.5 [2 mL] (Section 3.5 Note 3)

  The Labeling buffer should have the same composition as the Equilibration buffer but be made with D_2O instead of H_2O. Dilution of samples with Labeling buffer will start the exchange reaction. Labeling buffer percent deuterium, pD, and temperature must be tightly controlled and recorded.
- Quench solution: 1.6 M guanidinium-HCl, 1% (v/v) formic acid, pH 2.3 [10 mL]

  The Quench solution is used to minimize the exchange reaction and unfold the protein to improve digestion. It typically contains a chaotrope such as urea or guanidinium-HCl, and acid to reduce pH. If the protein has a lot of cysteines or disulfides, reducing agents such as TCEP can also be added (Hamuro & Coales, 2018; Zhang et al., 2010). If the Quench solution contains urea or reducing agent, prepare it fresh on the day of use. The pH of this Quench solution should be adjusted to be 2.3 and it is used to reduce pH by being mixed 1:1 (v:v) with the sample. It is not a buffer (pK_a of formic acid is 3.75), although a buffered Quench (such as with phosphate) can be used instead.
- The Protease Column wash, Equilibration buffer, Labeling buffer, and Quench solution are all optimized in the method. Unless other information about the protein-RNA complex is known (Section 2), these recipes are good starting points.
- Purified protein-of-interest

  The protein should be concentrated enough to be diluted to 25 µM with Equilibration buffer such that the final concentrations of buffer components are as close as possible to the desired Equilibration buffer.
- RNA ligand

  The RNA should be concentrated enough to be added to the protein at an equal or excess molar ratio dependent on the binding affinity (Section 2). RNA is normally stored in RNAase-free H_2O.

3.4 Procedure

1. *Checkpoint 1: Check that the mix of Equilibration buffer and Quench solution has a pH near 2.5 at 0 °C.*

   Simulate experimental conditions by mixing 100 μL of Equilibration buffer with 100 μL of Quench solution. Use the micro pH electrode to measure pH at 0 °C.

   If the pH is significantly above/below 2.5 revise the Quench solution or Equilibration buffer recipes. The relative volume of Quench solution can also be increased, although this would reduce the protein concentration of an actual sample.
2. Prepare stocks of 25 μM protein and 25 μM protein with RNA in Equilibration buffer. Use a protein-to-RNA ratio that ensures a high percent-bound (Section 2). If the protein and RNA are initially in different buffers, add the RNA buffer to the protein alone sample in place of the RNA to ensure the final buffers match exactly. Give the samples time to equilibrate (at least 10 min) before use.
3. Dilute 2 μL of the 25 μM stock with 48 μL of Equilibration buffer. This 1:24 dilution would give 96% D_2O and 1 μM protein during an exchange reaction.
4. *Checkpoint 2: Check that quenched samples are free from precipitate or suspended aggregates.*

   Add 50 μL of pre-cooled Quench solution to the 50 μL of diluted protein. Look for cloudiness indicative of protein precipitation. If the diluted sample is clear, repeat for protein with RNA. Perform an additional check on protein with RNA to detect suspended aggregates in clear samples. Use a nano-drop UV-vis spectrophotometer to measure sample absorbance at 280 nm (protein concentration) and 340 nm (turbidity) before and after a 1-min spin in a minicentrifuge. Blank the spectrophotometer with a matched buffer and record replicate absorbance readings. Decreased absorbance after spinning suggests the sample contains suspended aggregates.

   If cloudiness or aggregation are detected, repeat step 4 with a new Quench solution (see step 9 for guidance). If only the RNA sample is aggregating, the pH of the Quench solution should be increased; albeit at the cost of increased back-exchange when the pH is >2.5.
5. Prepare the LC-MS system for sample injection and analysis (Tables 1 and 2). Two methods files (prepared in MassLynx) are needed; one for non-deuterated samples that collect MS and MS/MS data, and another for deuterated samples that collect only MS data. Always perform a

wash run directly prior to analysis of each sample to ensure the system is clean (Section 3.5 Note 2).

6. Inject 80 μL of the quenched sample into a 50 μL sample loop and run the method. Wait until the valve switch to remove the syringe. We inject at least 120% of the loop volume to ensure the loop is completely full. If the quenched sample is deuterated and frozen in liquid nitrogen, thaw on ice for 45 s (timed) and draw 80 μL into a pre-cooled syringe before injection. Trouble injecting the sample may indicate a clogged injection port filter (Section 3.5 Note 4).
7. Assess the signal intensity and shape of the TIC (Fig. 4). For our system, intensity should have a maximum of 3E7-1E8 (depending on protein size and purity) and be distributed within the 7-min, 3–40% LC-B gradient.

   If the TIC looks as expected, skip to step 11.

   If the TIC intensity is high, go to step 8 and reduce protein per injection. High intensity can cause spectral artifacts such as peak ringing (Mahoney, Ray, & Hieftje, 1997)

   If the TIC intensity is low, go to step 8 and increase protein per injection. Low intensity restricts the recovery of peptide fragmentation products essential for unambiguous peptide sequencing, yielding reduced peptide identification and sequence coverage.

   If the TIC intensity is not evenly distributed within the LC-B gradient, return to step 5 and alter the gradient in the method file. Hints for gradient optimization can be found elsewhere (Wales, Fadgen, Gerhardt, & Engen, 2008; Walters et al., 2012), however, smaller retention time ranges can increase spectral overlap.

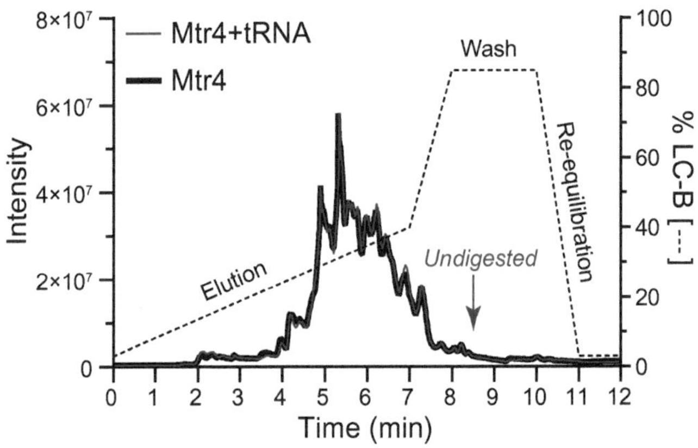

Fig. 4 Suitable total ion chromatograms. The traces for Mtr4 (black) and Mtr4 with tRNA (grey) superpose. The elution time for undigested protein and the LC-B elution gradient (dashes) are indicated.

If the TIC contains a peak of intensity after the LC-B gradient, and this peak is higher than the peaks during the gradient (Section 3.5 Note 5), skip to step 9 and optimize quenching. The post-gradient peak is undigested or partially-digested protein.

8. Repeat steps 3-7 using a different dilution at step 3. Use a 1:49 to reduce protein per injection to 10 pmol, and 2:23 to increase protein per injection to 50 pmol.

If 50 pmol still does not give sufficient TIC intensity, go to step 9.

9. Repeat steps 4–7 using different quench conditions. Variables include chaotrope type and concentration; duration and temperature of sample mixing; pH; and additives such as reducing agents. Hints for quench optimization can be found elsewhere (Guttman & Lee, 2016; Hamuro & Coales, 2018; Zhang et al., 2010).

If undigested or partially-digested protein persists in the TIC, go to step 10.

10. Repeat steps 5–7 using the most-promising quench conditions from step 9, but with different digestion parameters. Easy parameters to change in the method are temperature (increase up to 15 °C); time and pressure (changing flow rate or adding a flow restrictor to increase pressure); and the column (Ahn et al., 2012). Various protease columns are commercially available or can be made and may be more suitable depending on the protein sequence (Kadek et al., 2014; Papanastasiou et al., 2019; Zhang et al., 2008). If the protein alone fails to produce an appropriate TIC, a more drastic change (such as moving digestion off-line (Ahn, Cao, Yu, & Engen, 2013)) or innovative strategy will need to be developed.

11. Identify recovered peptides using MS and MS/MS data. We use PLGS with a curated database containing the protein and the protease (Masson et al., 2019). The amino acid sequences must exactly match the sample and include any tags. If using pepsin as the protease select "non-specific" as the digestion reagent (Hamuro & Coales, 2018; Hamuro, Coales, Molnar, Tuske, & Morrow, 2008). Also, be sure to allow for any relevant post-translational modifications. The output from PLGS is a .csv file containing a list of peptides with information about their fragmentation, ion mobility, and retention time.

12. *Checkpoint 3: Check that recovered peptides span the protein sequence with sufficient redundancy.*

Filter the list to estimate peptide coverage and redundancy across the protein sequence. We apply filters, namely 0.3 fragmentation products per amino acid and 1 consecutive product, using

DynamX. Ideally, peptides should span >80% of the sequence with >3 redundancy, but there are no strict requirements. Visualize peptides using a 'protein coverage map' that shows each peptide as a bar below the entire protein sequence. To assess the data more completely, we often apply a higher intensity threshold (5000 counts) and examine individual spectra.

If coverage and redundancy are as desired, go to step 13.

If coverage or redundancy need further optimization, return to step 10.

13. *Prepare and analyze (steps 5-7 and 11) at least 2 replicate samples of protein using optimized parameters. The run sequence should be "sample, wash, blank, sample, wash, blank" (Section 3.5 Notes 2 and 6). The blank runs will be used at step 17. The sample TICs should be superimposable. Re-assess peptide recovery for sample runs (like step 12) considering only peptides recovered in both replicates (2/2 file threshold).*
14. *Checkpoint 4: Check that parameters optimized for protein are compatible for protein with RNA. Repeat step 13 for protein with RNA.*

    If RNA has little effect on the TIC and peptide recovery, skip to step 16.

    If RNA reduces TIC intensity and/or peptide recovery dramatically, go to step 15. We have observed this for several RNAs due to insufficient RNA removal during trapping.
15. Repeat step 14 with altered trapping parameters. We suggest increasing trapping volume to >300 μL by increasing flow rate up to 200 μL/min and lengthening trapping time. Increasing flow rate part-way through trapping minimizes adverse effects on digestion. A starting point is 100 μL/min for 1 min, followed by 200 μL/min for 2 min, to give a total trapping volume of 500 μL. Additional minutes can be added incrementally to determine the minimum trapping volume required to remove the RNA. A less-retentive C4 trap with the same particle size can also be tried in place of the C18 trap.

    If TIC intensity cannot be recovered, add steps between quenching and digestion to remove the RNA. To minimize back-exchange, these steps must be rapid and at as low a pH and temperature as possible.
16. *Prepare and analyze (steps 5–7 and 11) at least 2 replicate samples of protein with RNA using optimized parameters (like step 13). Check that coverage is like that obtained at step 13.*
17. *Checkpoint 5: Check that blank runs and wash conditions prevent problematic peptide carryover.*

Use peptide lists from steps 13 and 16 to search the blank and sample runs collected at steps 13 and 16 in DynamX. Manually check and refine spectra for a minimal subset of peptides that cover all possible segments of the protein sequence. For example, if multiple peptides redundantly cover the same sequence, just refine one or two of them. Compare the intensity of peptide spectra in the blank runs to that in sample runs.

If the intensity of blank run spectra is >5% of the sample spectra for many peptides, go to step 18 to reduce carryover. Spectra from peptides with various levels of carryover are shown in Fig. 5.

If the blank runs contain few peptides or have only low-signal spectra, skip to step 19.

18. Return to step 16 and use different wash and blank conditions between runs. Solutions to minimize carryover are clearly described elsewhere (Fang et al., 2011; Hamuro & Coales, 2018; Majumdar et al., 2012).

19. *For this and following steps, use optimized parameters. Pre-incubate Labeling buffer and concentrated stocks of protein, and protein with RNA, at 25°C (the exchange temperature; Section 3.5 Note 7) for 10min.*

20. *Generate highly-deuterated samples by diluting concentrated stocks with Labeling buffer at the precise ratio. Use a final volume of at least 100μL for each exchange reaction so that samples can be analyzed in at least duplicate.*

21. *After 10^5 s, mix exchange reactions with pre-cooled Quench solution at the precise ratio, make 100μL aliquots, and flash freeze with liquid nitrogen. If required, samples can be stored at −80°C indefinitely.*

22. *Checkpoint 6: Check that deuteration can be measured with optimized parameters.*

Repeat steps 5–7 with highly-deuterated samples. Use peptide lists from steps 13 and 16 to search the data in DynamX (like in step 17). Look for any carryover or multi-modal spectra. Compare the deuterium uptake to the theoretical maximum deuterium uptake (Section 3.5 Note 8) to qualitatively judge back-exchange levels and suitability of the final timepoint. For a region of the protein known to be disordered or solvent-exposed, we hope to detect at least 60% of theoretical maximum deuterium uptake.

If multi-modal spectra are detected for most peptides, be sure to check step 4 and step 17.

If deuterium uptake is low, tweak the protocol to reduce back-exchange. This will likely be done by reducing time and temperature where possible.

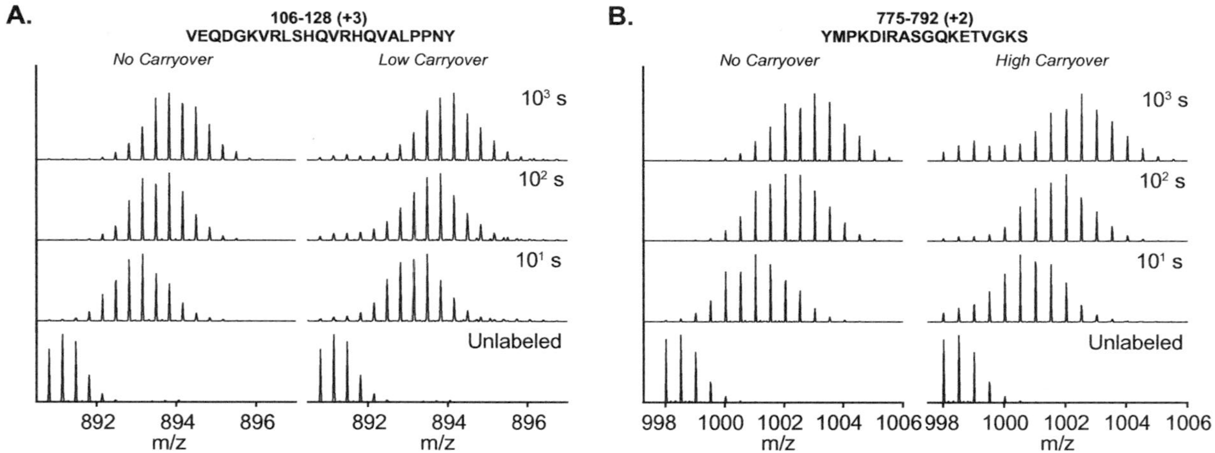

Fig. 5 Example peptide spectra showing different levels of carryover. Mtr4 peptide 106–128 (A) has low carryover. Mtr4 peptide 775–792 (B) has high carryover.

23. *Proceed with a complete analysis for the specific protein and RNA tested (Section 4).*
24. *If experiments with new RNA ligands are desired for the same protein, return to step 4 and use the new RNA with the optimized parameters.*

3.5 Notes

1. Our LC uses two six-port valves in a cooled chamber to direct flow across three columns (Fig. 3) (Ahn et al., 2012). The injection port goes to the loading valve, which contains the 50 μL sample loop and protease column; flows to waste; and is driven by pump A. The protease column sits in a separate chamber with independent temperature control. The loading valve also connects to the analytical valve, which contains the trap and analytical columns; flows directly to the MS; and is driven by pump B. During injection, pump A pushes flow through the loop, over the protease column and trap (forward direction) to waste. Pump B pushes flow over just the analytical column to the MS. After trapping, both valves switch so that pump A pushes flow directly to waste, while pump B pushes flow over both the trap (reverse direction) and analytical column to the MS. Pump B can then execute the acetonitrile elution gradient. The loading valve may be actuated independently during the gradient to manually wash the sample loop and protease column. These wash steps can be automated and in the reverse direction in other systems. Pressures of pump A and B should be closely monitored for the system to help identify leaks or clogs.
2. Cleaning the LC-MS between each sample run is required for almost all HDX-MS experiments. It is especially required if samples are not recovered during troubleshooting or when the sample contains RNA. Sample is often retained on the protease column, and protease-safe wash protocols to address this issue have been developed (Majumdar et al., 2012). We clean the protease column with injection(s) of Protease Column Wash and flow of LC-A. We clean the syringe and loop with a manual LC-A rinse. We clean the trap and analytical columns with a 15-min wash method. This method pumps LC-A over the protease and trap at 100 μL/min for 3 min before the valve is switched. It then executes a repeating LC-B gradient of 3% to 85% over 1 min and 85% to 3% LC-B over 1 min at 40 μL/min, for a total of 12 min.
3. The pD of a deuterated buffer is measured using a standard pH meter and adding a constant of 0.4 to the measured pH (Glasoe & Long, 1960).

To prepare Labeling buffer with a pD of 7.5, for example, adjust to a final measured pH of 7.1. To minimize addition of protons, do this using LC-MS grade DCl or NaOD. We also make the buffer from concentrated stocks of HEPES and NaCl made with deuterium. The only H_2O in the buffer comes from the 0.5 M TCEP stock that is diluted 1 in 2500 to give 0.2 mM TCEP.

4. Our system has an injection port filter that captures particulate material, preventing its injection onto a column. The filters are 2 μm stainless-steel frit filters. Pressure can be felt during sample injection if the filter is starting to be clogged. We change the filter regularly, approximately after every 30 injections and before starting a complete experiment.
5. The elution of undigested or partially digested protein after the gradient is not problematic if sufficient peptides can be recovered during the gradient. It does however suggest that digestion is inefficient and that perhaps less protein could be injected per sample. The intensity of the peak can also be used as an internal indicator of protease life. If the peak intensity increases through successive replicate runs, it may mean the protease column is losing activity.
6. Blank runs are mock injections of Equilibration buffer and Quench solution mixed exactly as would be done for an actual protein-containing sample. The LC-MS method used should match the one used for a protein-containing sample. Data collected in blank runs is used to assess the extent of carryover between samples.
7. The exchange temperature is an important variable as it influences the intrinsic chemical exchange rate (Bai et al., 1993; Hvidt & Nielsen, 1966). We typically start with 25 °C.
8. In most cases, theoretical maximum deuterium uptake can be calculated as follows:

   maximum deuterium = *percent deuterium x* (*peptide length* – *1* – *no. prolines not at N-terminus*)

   The *percent deuterium* will depend on the dilution of concentrated stock with deuterium at step 3. The *-1* accounts for fast back-exchange of the N-terminal residue. This depends on peptide sequence, with some choosing to use *-2* instead (Bai et al., 1993; Hamuro, 2021; Wales, Eggertson, & Engen, 2013). The *no. of proline not at N-terminus* accounts for proline not having a backbone amide that can participate in exchange. We specify *not at N-terminus* as an N-terminal proline would already be subtracted by the *-1*. DynamX 3.0 does not use the *not at N-terminus* qualifier and by default will incorrectly subtract the

N-terminal proline twice. Peptides with N-terminal prolines thus require manual correction to their maximum deuterium uptake value in the peptide editor. DynamX also does not multiply by the *percent deuterium*. DynamX will create uptake plots with the maximum deuterium uptake value as the highest value on the y-axis.

4. Using HDX-MS to identify molecular determinants of RNA-binding by Mtr4

In this section, we detail an HDX-MS protocol for Mtr4 with various RNA ligands. The protocol has been applied to several RNA ligands including the yeast initiator methionine tRNA ($tRNA_i^{Met}$); a 42-bp single-stranded RNA ($ssRNA^{42}$); a 32-bp double-stranded RNA containing a hairpin and a nick ($dsRNA^{32}$); and a 16-bp double-stranded RNA ($dsRNA^{16}$). The $tRNA_i^{Met}$ is purified from an in vitro transcription system (Easton et al., 2010), while the others are synthesized commercially (Integrated DNA Technologies, Inc) and annealed where relevant. The RNA ligands all contain a 3' poly-adenine overhang and bind Mtr4 with affinities in the μM range (0.3–1.1 μM) (Lingaraju et al., 2019; Puno & Lima, 2018; Zhang et al., 2022). The protocol has also been applied to an Mtr4 mutant missing residues 667–813 (Jackson et al., 2010)

The workflow was developed using the theoretical considerations and optimization steps already discussed (Sections 2 and 3) (Fig. 2). The Equilibration buffer was selected to match the buffer in an Mtr4 helicase assay (Weir, Bonneau, Hentschel, & Conti, 2010). We did not have any challenges with solubility of Mtr4 either with or without RNA at any stage, but we did need to optimize Mtr4 digestion to reduce peptide length and the amount of undigested Mtr4 (*checkpoint 3*). Elevating digestion temperature to 15 °C and increasing chaotrope concentration in the Quench proved sufficient. Addition of RNA to the Mtr4 samples was also problematic as it caused an almost complete loss of signal in the TIC in all cases (*checkpoint 4*). For $tRNA_i^{Met}$, signal was recovered by replacing the C18 trap with a C4 trap. Subsequent checkpoints were satisfied, and a complete experiment could be performed with the $tRNA_i^{Met}$ at a 5.4-fold molar excess to Mtr4 [>95% bound; 81% sequence coverage; 5.14 peptide redundancy; back-exchange 30%].

For the other RNAs, the C4 trap alone was insufficient to recover TIC signal for Mtr4 with RNA, and carryover was also detected (*checkpoint 5*). Signal was only recovered after reducing RNA concentration to a two-fold molar excess of Mtr4 and doubling trapping time/volume. Despite trying a few washing protocols, carryover of many peptides could not be mitigated in a reasonable time frame, so these were excluded from analysis. The trade-offs resulted in a lower percent of Mtr4 bound to RNA [81% for $tRNA_i^{Met}$, 75% for $ssRNA^{42}$, 66% for $dsRNA^{32}$, 56% $dsRNA^{16}$], increased back-exchange [38%], and less peptide recovery [58% coverage, 2.82 redundancy], compared to a complete experiment with only $tRNA_i^{Met}$.

The protocol we detail for HDX-MS of Mtr4 does not include any direct measures of the intrinsic chemical exchange or back-exchange rates. This can be done using small molecule standards or through the inclusion of a fully or maximally deuterated control (Hoofnagle et al., 2004; Murphree et al., 2020; Zhang & Smith, 1993). These allow scaling of the data such that deuterium incorporation can be compared between samples prepared or analyzed on different occasions (Toth et al., 2017; Zhang & Smith, 1993). Rather than rely on these controls, we perform large experiments where all samples that need to be compared are prepared and analyzed simultaneously. This is not always the most practical or efficient approach and is a point of improvement. The protocol also performs exchange reactions by hand, but validated sample-handling systems for HDX are available (Chalmers et al., 2006; Cummins et al., 2016). The duration of the protocol will depend on the number of RNA ligands and/or Mtr4 constructs with most time being spent manually inspecting spectra after data collection. The protocol is written for a single RNA ligand at a two-fold molar ratio to Mtr4 with 92% deuterium, a 1:1 mix with Quench solution, four technical replicates and five timepoints.

4.1 Equipment and reagents

Equipment and reagents required beyond those listed in Sections 3.1 and 3.2.

- $MgCl_2$
- ACQUITY UPLC Protein BEH C4 VanGuard Pre-column, 300 Å, 1.7 μm, 2.1 mm X 5 mm (Waters™)
- Multichannel pipettes (Thermo Scientific E1-ClipTip 200 and 1250)
- PyMOL 2.5 (Schrödinger)

4.2 Buffers and samples

Prepare buffers and samples listed in Section 3.3. The following have a different recipe.

- Equilibration buffer: 50 mM HEPES, 100 mM NaCl, 0.5 mM $MgCl_2$, 0.2 mM TCEP, pH 6.5 [1 mL].
- Labeling buffer: 50 mM HEPES, 100 mM NaCl, 0.5 mM $MgCl_2$, 0.2 mM TCEP in D_2O, pD = 6.5 (pH_{read} = 6.1) (Section 2.3 Note 2) [2.5 mL].
- Quench solution: 3.84 M guanidinium-HCl, 1% (v/v) formic acid, pH 1.75 [3 mL].

4.3 Procedure

- Exchange reactions
  1. Prepare concentrated stocks of 12.5 μM Mtr4 and 12.5 μM Mtr4 with 25 μM RNA (hereafter Mtr4-RNA) in Equilibration buffer. Mtr4 should be dialyzed into Equilibration buffer after purification and RNA should be in RNase-free H_2O. Be sure to add RNase-free H_2O to the Mtr4 alone sample in place of the RNA to ensure buffers match exactly. 132 μL of each stock is needed based on the following calculation.

     Volume = [(4 *μL per injection x* 4 *replicates x* 5 *timepoints*) + (4 *μL x* 10 *controls*)] *x* 1.1.

     The 4 μL of stock per sample is determined from a 50 μL sample volume with a 2:23 dilution with Labeling or Equilibration buffer; the replicates are technical; the timepoints are 10^1, 10^2, 10^3, 10^4, and 10^5 s; the controls are six non-deuterated samples and four pulse-labeled samples at 10^5 s; and the 1.1 adds 10% extra volume for ease of pipetting.
  2. Equilibrate aliquots of concentrated stocks and buffers in a dry bath at 25 °C (exchange temperature) for at least 10 min. Aliquots should be in 1.5 mL tubes and include:

     Exchange reactions
     - Four tubes with 22 μL of Mtr4
     - Four tubes with 22 μL of Mtr4-RNA
     - Four tubes with 560 μL of Labeling buffer

     Non-deuterated controls
     - One tube with 26.4 μL of Mtr4
     - One tube with 26.4 μL of Mtr4-RNA
     - Two tubes with 330 μL of Equilibration buffer

     Pulse-labeled controls
     - One tube with 17.6 μL of Mtr4

- One tube with 17.6 μL of Mtr4-RNA
- Two tubes with 210 μL of Labeling buffer

3. Prepare a series of 0.2 mL tubes containing 50 μL of Quench solution. Label the tubes with the sample composition, timepoint, and replicate number. We use five different tube colors to distinguish the four replicates and controls. A total of 60 tubes are required. Store quench tubes on ice with lids closed.
4. Start the exchange reactions of a single replicate of Mtr4 and Mtr4-RNA simultaneously. Use a multichannel pipette to dispense 253 μL of Labeling buffer into the two 22 μL samples (prepared in step 2). Be sure to start a timer as soon as Labeling buffer is dispensed into the samples.
5. Mix both reactions with an extra aspiration and dispense.
6. Return the reactions to the 25 °C dry bath as quickly as possible (certainly within first 3 s).
7. Leave the exchange reactions untouched in the 25 °C dry bath.
8. Immediately before each timepoint (10^1, 10^2, 10^3, 10^4, 10^5 s), draw 52 μL of both exchange reactions into a multichannel pipette.
9. At each timepoint, dispense 50 μL of exchange reactions into the correctly-labeled 0.2 mL tubes containing Quench solution (prepared in step 3). Make sure the samples mix with the Quench solution during pipetting.
10. Quickly cap the 0.2 mL tubes and flash-freeze the samples in liquid nitrogen.
11. Store samples at −80 °C until analysis by LC-MS. Long-term storage at temperatures higher than −60 °C may result in notable back-exchange (Watson et al., 2021).
12. Return to step 4 to start replicate exchange reactions. For efficiency, replicate reactions can be staggered. We typically start replicate reactions after the 10^2 s timepoint has been quenched for the prior replicate. We pipette non-deuterated controls (step 13) after the 10^3 s timepoint for all replicates, and pulse-labeled controls (step 14) after the 10^5 s timepoint for all replicates.
13. Prepare 6 non-deuterated controls each for Mtr4 and Mtr4-RNA. It is not essential to prepare non-deuterated controls of Mtr4-RNA, but they can be useful to confirm the effect of RNA on peptide recovery at any stage of the experiment. Dispense 303.6 μL of Equilibration buffer into the two 26.4 μL samples (prepared in step 2). Process samples immediately (no incubation required) as described in steps 8–11.

14. Prepare 4 pulse-labeled controls each for Mtr4 and Mtr4-RNA. The 17.6 μL stocks (prepared in step 2) should be incubated at 25 °C for 10^5 s alongside exchange reactions. After the 10^5 s at 25 °C, dispense 202.4 μL of Labeling buffer into the two 17.6 μL samples (prepared in step 2). Process samples as described in steps 5–11, quenching after 10^1 s exchange. Deuterium uptake of our pulse-labeled control (10 s deuteration after incubation at 25 °C for 10^5 s) of Mtr4 matched the 10 s deuterated sample showing Mtr4 stability through the experiment.

- Data collection and spectra analysis
  1. Prepare the LC-MS system for sample injection and analysis (Section 3.4, step 5). Method parameters are summarized in Fig. 1 and Tables 1 and 2 and may vary depending on RNA ligand.
  2. Transport samples from sub-zero storage to near the LC-MS in liquid nitrogen.
  3. Identify a sample for analysis and enter the file name and any notes in the data collection software (MassLynx). The file name should at a minimum include the date, sample composition, timepoint, and replicate number. It could also include the initials of the person injecting the sample and the injection number.
  4. Pre-chill a 250 μL syringe in an ice-box. A paper towel can be used to insulate the syringe from body temperature during handling.
  5. Use the syringe to partially aspirate and dispense the sample multiple times while the sample is thawed on ice for 50 s (timed).
  6. Draw 80 μL of sample into the syringe, dispense it into the injection port and sample loop, and run the method.
  7. Wash the system (Section 3.4, step 5). During the wash, verify the quality of the previous run. A simple yet powerful metric is TIC shape and intensity (Fig. 4). Variability in TIC indicates inconsistent sample concentration, potentially due to aggregation or insufficient RNA removal. Increased intensity of the undigested or partially digested protein peak may also suggest problems with the protease column. See also step 9 onwards once several samples have been injected.
  8. Repeat steps 3–7 until all samples have been injected. Samples from all replicates should be injected in a random order and in as short a timeframe as possible without interruption with other samples. Use a single set of columns for all samples and be sure to space unlabeled controls throughout labeled samples. Be sure that data are backed-up in real time.

9. Generate peptide lists from the non-deuterated controls (Section 3.4, step 11).
10. Import and filter peptide lists (Section 3.4, step 12). A file threshold greater than 2 is recommended to eliminate many false positive identifications.
11. Search raw data files for non-deuterated controls with the filtered peptide list (typically <1 h in DynamX). Manually check and refine spectra. Most manual corrections involve selecting a different retention time window. Any obviously misassigned peptides should be deleted from the peptide list, while the few poorly assigned peptides should be hidden. We suggest removing redundant charge states of a peptide that have poor data quality (such as too low/high intensity or spectra overlap with other peptides) before step 12.
12. Search raw data files for deuterated samples with the filtered peptide list (typically several hours). Manually check and refine spectra. This can take several days or weeks depending on user experience and data quality. Be sure to note any peptides with multi-model isotopic distributions that are not expected for Mtr4 alone.
13. Identify peptides requiring further attention using deuterium uptake plots (automatically generated in DynamX as spectra are refined) (Fig. 6). Uptake plots show deuterium uptake (Da) over time for each peptide. The y-axis maximum should be the theoretical maximum deuterium uptake (Section 3.5 Note 7) based on the specific peptide length and sequence. If this y-axis maximum is

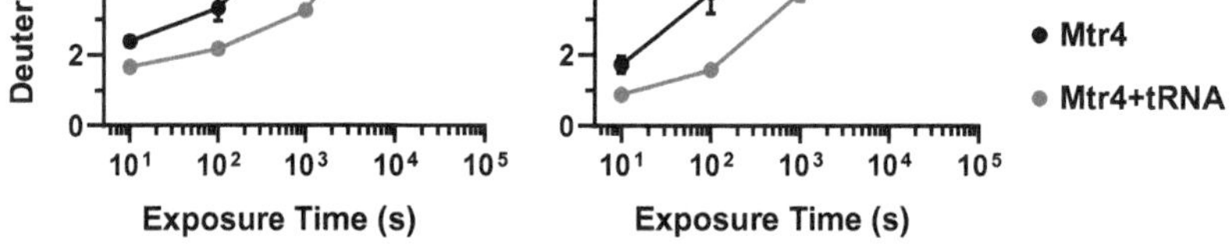

Fig. 6 Example deuterium-uptake plots. Deuterium uptake of Mtr4 peptides 184–212 (left) and 261–275 (right) comparing Mtr4 alone (black) and Mtr4with tRNA (grey). Error bars are plus/minus two standard deviations of triplicate measurements. Some error bars are too small to be seen. Y-axes are 80% of theoretical maximum deuteration.

altered, it should be done so consistently for all peptides (e.g., uptake plots for each peptide scaled to 80% of maximum uptake) and noted in a figure legend. The trajectory of an uptake plot relates to protein structure (Morgan & Engen, 2009). A peptide should be more closely evaluated if the uptake plot has a dramatically different trajectory to overlapping peptides; if the standard deviation of at least three replicates for any timepoint is >0.2 Da; and if deuterium uptake decreases over time. The latter is not possible and can indicate aggregation, carryover, or other technical issues. Return to step 12 until all peptides have acceptable uptake plots.

14. Prepare a summary table and export the data in preparation for publication (Masson et al., 2019). See Section 4.4 for a brief discussion on data visualization, beyond spectra and uptake plots. While there are many great options available, we tend to use heat maps or volcano plots.

4.4 Data visualization

When performing a comparative HDX-MS experiment, it is important to use significance testing to identify peptides with differential uptake between samples, in our case, Mtr4 and Mtr4-RNA. We follow guidelines clearly justified elsewhere that use a dual-significance filter (Hageman & Weis, 2019b). The first filter is a *P*-value less than 0.01, calculated using three replicates and a two-tailed Welch's t-test (Welch, 1947). The second filter is a minimum absolute average uptake difference set to be greater than the pooled standard deviation of the data. We typically use 0.2–0.5 Da. We apply both filters and represent the deuterium uptake difference in heat maps (aka chiclets) and volcano plots (Fig. 7A and B). We do this using a free web-based application called HD-eXplosion (http://hd-explosion.utdallas.edu/) (Zhang, Yu, Zhang and D'Arcy, 2021) or in-house Python scripts (available upon request). Note that peptides that have a reduced number of replicates will have an artificially small *P*-value as currently, HD-eXplosion applies the same replicate number to all peptides. We consistently do our comparisons as Mtr4 alone minus Mtr4 with RNA, and differences can be in absolute Daltons or as a percentage of theoretical maximum deuterium uptake (Section 3.5 Note 7). The latter is advisable if peptide lengths have a range greater than 10 residues. Finally, kinetic information can be compressed by summing differences across timepoints, allowing visualization of multiple timepoints on a single structure, although we do not do this

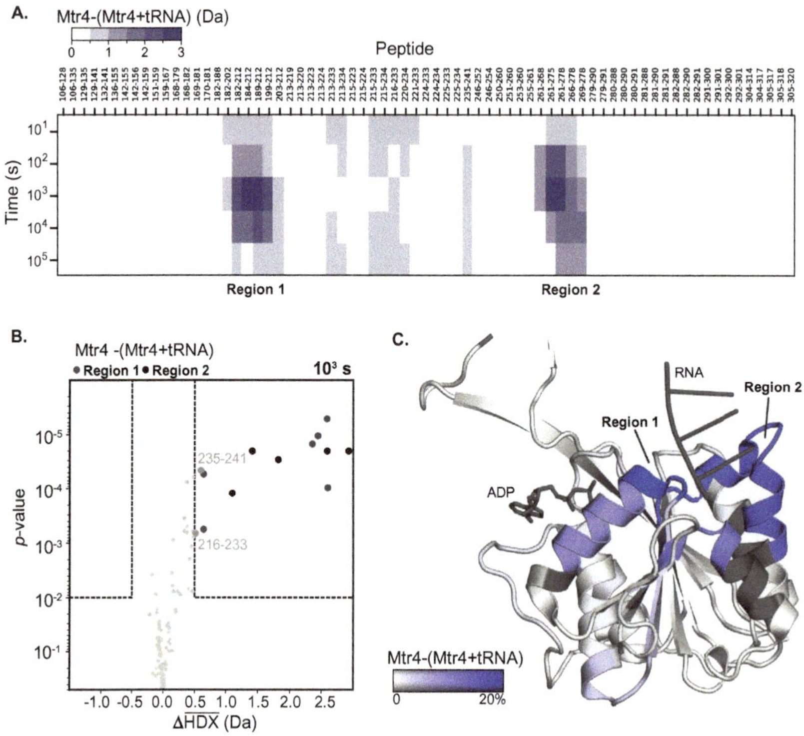

Fig. 7 Example data visualization formats for HDX-MS experiments. Difference in deuterium uptake between Mtr4 alone and Mtr4 with tRNA for the RecA1 domain only can be shown as a chiclet plot (A), volcano plot (B), or projected onto the structure (C). Statistical cut-offs in A and B are *P*-value <0.01 ($n=3$) and a difference greater than 0.5 Da. Data for 10^3 s only is shown in (B) and (C). The RecA1 domain is from PDB 2XGJ with RNA, ADP, and regions of no coverage shown in gray. Regions 1 and 2 correspond to RNA-binding elements in RecA1 that show protection upon addition of tRNA.

for Mtr4. This summing can be useful when there isn't a single timepoint that includes all observed differences.

Chiclets display the difference in average deuterium uptake between two conditions over time for each peptide. One axis is the peptide list (consistently ordered by either the first, middle, or last residue) and the other axis is the exposure time (left to right or top to bottom increasing). Average deuterium uptake difference is indicated by the color of each cell in the grid. Typically, white indicates no significant difference, blue indicates Mtr4 greater than Mtr4 with RNA (protection), and red indicates Mtr4 less than Mtr4 with RNA (deprotection). The graduations of color change should be equal in magnitude and number above and below zero, and a scale bar included. If there are multiple RNA ligands, it can be effective to align multiple chiclets for comparison. Chiclets allow for visual grouping of peptides and rapid identification of outliers but can be too large for publication when many peptides are recovered.

Volcano plots display the difference in average deuterium uptake between two conditions as a function of *p*-value. A single point is plotted for each peptide at each exposure time. Cut-offs can be clearly indicated by dotted lines such that only points in the top right (protection) or top left (deprotection) areas have significance. To avoid extensive point overlap, volcano plots are often created for a single timepoint so that peptides can be labeled. This loses information especially if significant differences are not all represented in a single timepoint. Volcano plots allow for quick identification of significant differences and rapid identification of outliers but do not allow for visual grouping of peptides. We create both chiclets and volcano plots when first assessing and making conclusions from the data.

A final element of data visualization is projecting the data onto structures from either the protein data bank or AlphaFold database (Jumper et al., 2021). This highlights the spatial proximity of peptides in the folded protein. This can be done for the deuterium uptake of Mtr4 alone to check it matches the known structure, insofar as surface-exposed loop regions have more deuterium uptake than buried, folded regions. Such projection is qualitative as it does not account for side-chain effects on the intrinsic chemical exchange rate (Bai et al., 1993; Hamuro, 2021). More importantly, the difference in average deuterium uptake between two conditions can also be projected (Fig. 7C). In both cases, projection requires that peptide-level data be represented at the residue level. DynamX does this by averaging deuterium uptake values (calculated as a percentage of theoretical maximum deuterium uptake; (Section 3.5 Note 7) of all peptides that include the residue not at the N-terminal position. This approach fails to account for prolines (and in fact assigns them an uptake value), and in the case of a comparison, fails to incorporate any error or significance test. The residue-level data can be exported as a script that can be visualized with precise color and numerical scales via "spectrum properties" in the molecular viewer PyMOL. It is imperative to make sure the sequence numbering in the script and structure file match exactly and to color the structure the "no coverage" color (typically grey) before showing the spectrum properties. We prefer the scale to be equal above and below zero (with zero shown as white), and to use a consistent scale that is reported for all projections within a project.

5. Summary

This chapter covered the main stages of an HDX-MS experiment to characterize RNA-binding of a protein-of-interest. We outlined important considerations in the design of an HDX-MS experiment, including workflow selection and sample quality. We described a checkpoint system for defining the optimal experimental parameters for all stages of an on-exchange, comparative workflow. We also detailed a specific protocol for the ATP-dependent, RNA-helicase Mtr4 that encompasses the exchange reactions to data collection, analysis, and visualization. The introduction of some types of RNA can be problematic for HDX-MS, with symptoms being sample aggregation upon quenching, reduced peptide recovery, and enhanced peptide carryover. Mitigation of these challenges, as done for Mtr4, can result in a successful, informative HDX-MS analysis. HDX-MS analysis can identify regions of the RNA-binding protein that either directly contact the RNA ligand, or alternatively undergo a related conformational change. It is essential to use complementary biochemical and structural approaches to absolutely differentiate between these two possibilities. The protocols in this chapter serve as a starting point for HDX-MS of any RNA-binding protein.

Acknowledgments

This work was supported by the National Institute of General Medical Sciences, National Institutes of Health (R35 GM133751-01 to S.D.) and The Welch Foundation (AT-2059-20210327 to S.D.). We thank current and past members of the D'Arcy laboratory for useful discussions. We thank the group of Sean Johnson (Utah State University) for materials and useful discussions. We thank John Engen and Thomas Wales (Northeastern University), and Stephen Coales (Trajan Scientific and Medical), for input and suggestions.

References

Ahn, J., Cao, M. J., Yu, Y. Q., & Engen, J. R. (2013). Accessing the reproducibility and specificity of pepsin and other aspartic proteases. *Biochimica et Biophysica Acta (BBA) - Proteins and Proteomics*, *1834*(6), 1222–1229. https://doi.org/10.1016/J.BBAPAP.2012.10.003.

Ahn, J., Jung, M. C., Wyndham, K., Yu, Y. Q., & Engen, J. R. (2012). Pepsin immobilized on high-strength hybrid particles for continuous flow online digestion at 10,000 psi. *Analytical Chemistry*, *84*(16), 7256. https://doi.org/10.1021/AC301749H.

Allmang, C., Kufel, J., Chanfreau, G., Mitchell, P., Petfalski, E., & Tollervey, D. (1999). Functions of the exosome in rRNA, snoRNA and snRNA synthesis. *The EMBO Journal*, *18*(19), 5399–5410. https://doi.org/10.1093/EMBOJ/18.19.5399.

Bai, Y., Milne, J. S., Mayne, L., & Englander, S. W. (1993). Primary structure effects on peptide group hydrogen exchange. *Proteins*, *17*(1), 75–86. https://doi.org/10.1002/PROT.340170110.

Baumann, M., & Meri, S. (2004). Techniques for studying protein heterogeneity and post-translational modifications. *Expert Review of Proteomics*, *1*(2), 207–217. https://doi.org/10.1586/14789450.1.2.207.

Bernstein, J., Patterson, D. N., Wilson, G. M., & Toth, E. A. (2008). Characterization of the essential activities of Saccharomyces cerevisiae Mtr4p, a 3' -> 5' helicase partner of the nuclear exosome. *Journal of Biological Chemistry*, *283*, 4930–4942. https://doi.org/10.1074/jbc.M706677200.

Chalmers, M. J., Busby, S. A., Pascal, B. D., He, Y., Hendrickson, C. L., Marshall, A. G., et al. (2006). Probing protein ligand interactions by automated hydrogen/deuterium exchange mass spectrometry. *Analytical Chemistry*, *78*(4), 1005–1014. https://doi.org/10.1021/AC051294F.

Chalmers, M. J., Busby, S. A., Pascal, B. D., West, G. M., & Griffin, P. R. (2011). Differential hydrogen/deuterium exchange mass spectrometry analysis of protein-ligand interactions. *Expert Review of Proteomics*, *8*, 43–59. https://doi.org/10.1586/epr.10.109.

Conrad, K. S., Hurley, J. M., Widom, J., Ringelberg, C. S., Loros, J. J., Dunlap, J. C., et al. (2016). Structure of the frequency-interacting RNA helicase: A protein interaction hub for the circadian clock. *The EMBO Journal*, *35*(15), 1707–1719. https://doi.org/10.15252/EMBJ.201694327.

Cummins, D. J., Espada, A., Novick, S. J., Molina-Martin, M., Stites, R. E., Espinosa, J. F., et al. (2016). Two-site evaluation of the repeatability and precision of an automated dual-column hydrogen/deuterium exchange mass spectrometry platform. *Analytical Chemistry*, *88*(12), 6607–6614. https://doi.org/10.1021/ACS.ANALCHEM.6B01650/SUPPL_FILE/AC6B01650_SI_001.PDF.

Deller, M. C., Kong, L., & Rupp, B. (2016). Protein stability: A crystallographer's perspective. *Acta Crystallographica. Section F, Structural Biology Communications*, *72*(Pt. 2), 72. https://doi.org/10.1107/S2053230X15024619.

Easton, L. E., Shibata, Y., & Lukavsky, P. J. (2010). Rapid, nondenaturing RNA purification using weak anion-exchange fast performance liquid chromatography. *RNA*, *16*(3), 647. https://doi.org/10.1261/RNA.1862210.

Engen, J. R., & Wales, T. E. (2015). Analytical aspects of hydrogen exchange mass spectrometry. *Annual Review of Analytical Chemistry*, *8*, 127–148. Annual Reviews Inc https://doi.org/10.1146/annurev-anchem-062011-143113.

Engen, J. R., Wales, T. E., & Shi, X. (2011). Hydrogen exchange mass spectrometry for conformational analysis of proteins. In *Encyclopedia of Analytical Chemistry* John Wiley & Sons, Ltd. https://doi.org/10.1002/9780470027318.a9201.

Englander, S. W. (2006). Hydrogen exchange and mass spectrometry: A historical perspective. *Journal of the American Society for Mass Spectrometry*, *17*(11), 1481. https://doi.org/10.1016/J.JASMS.2006.06.006.

Englander, S. W., & Kallenbach, N. R. (1984). Hydrogen exchange and structural dynamics of proteins and nucleic acids. *Quarterly Review of Biophysics*, *16*, 1. https://doi.org/10.1017/S0033583500005217.

Fang, J., Rand, K. D., Beuning, P. J., & Engen, J. R. (2011). False EX1 signatures caused by sample carryover during HX MS analyses. *International Journal of Mass Spectrometry*, *302*(1–3), 19–25. https://doi.org/10.1016/j.ijms.2010.06.039.

Filandrova, R., Kavan, D., Kadek, A., Novak, P., & Man, P. (2021). Studying protein–DNA interactions by hydrogen/deuterium exchange mass spectrometry. *Methods in Molecular Biology*, *2247*, 193–219. https://doi.org/10.1007/978-1-0716-1126-5_11.

Glasoe, P. K., & Long, F. A. (1960). Use of glass electrodes to measure acidities in deuterium oxide. *Journal of Physical Chemistry*, *64*(1), 188–190. https://doi.org/10.1021/J100830A521.

Guttman, M., & Lee, K. K. (2016). Isotope labeling of biomolecules: Structural analysis of viruses by HDX-MS. *Methods in Enzymology*, *566*, 405. https://doi.org/10.1016/BS.MIE.2015.05.021.

Guttman, M., Wales, T. E., Whittington, D., Engen, J. R., Brown, J. M., & Lee, K. K. (2016). Tuning a high transmission ion guide to prevent gas-phase proton exchange during H/D exchange MS analysis. *Journal of the American Society for Mass Spectrometry*, *27*(4), 662–668. https://doi.org/10.1007/S13361-015-1330-8.

Haeckel, R. (1988). Recommendations for definition and determination of carry-over effects. *The Journal of Automatic Chemistry*, *10*(4), 181–183. https://doi.org/10.1155/S1463924688000380.

Hageman, T. S., & Weis, D. D. (2019a). A structural variant approach for establishing a detection limit in differential hydrogen exchange-mass spectrometry measurements. *Analytical Chemistry*, *91*(13), 8017–8024. https://doi.org/10.1021/ACS.ANALCHEM.9B01326/SUPPL_FILE/AC9B01326_SI_001.PDF.

Hageman, T. S., & Weis, D. D. (2019b). Reliable identification of significant differences in differential hydrogen exchange-mass spectrometry measurements using a hybrid significance testing approach. *Analytical Chemistry*, *91*(13), 8008–8016. https://doi.org/10.1021/ACS.ANALCHEM.9B01325/SUPPL_FILE/AC9B01325_SI_001.PDF.

Hamuro, Y. (2017). Determination of equine cytochrome c backbone amide hydrogen/deuterium exchange rates by mass spectrometry using a wider time window and isotope envelope. *Journal of the American Society for Mass Spectrometry*, *28*(3), 486–497. https://doi.org/10.1007/S13361-016-1571-1/SUPPL_FILE/JS8B05485_SI_001.DOC.

Hamuro, Y. (2021). Tutorial: Chemistry of hydrogen/deuterium exchange mass spectrometry. *Journal of the American Society for Mass Spectrometry*, *32*(1), 133–151. https://doi.org/10.1021/JASMS.0C00260.

Hamuro, Y., & Coales, S. J. (2018). Optimization of feasibility stage for hydrogen/deuterium exchange mass spectrometry. *Journal of the American Society for Mass Spectrometry*, *29*(3), 623–629. https://doi.org/10.1007/s13361-017-1860-3.

Hamuro, Y., Coales, S. J., Molnar, K. S., Tuske, S. J., & Morrow, J. A. (2008). Specificity of immobilized porcine pepsin in H/D exchange compatible conditions. *Rapid Communications in Mass Spectrometry*, *22*(7), 1041–1046. https://doi.org/10.1002/RCM.3467.

Hoofnagle, A. N., Resing, K. A., & Ahn, N. G. (2004). Practical methods for deuterium exchange/mass spectrometry. *Methods in Molecular Biology (Clifton, NJ)*, *250*, 283–298. https://doi.org/10.1385/1-59259-671-1:283.

Hudgens, J. W. (2020). Construction of a dual protease column, subzero (-30 °C) chromatography system and multi-channel precision temperature controller for hydrogen-deuterium exchange mass spectrometry. *Journal of Research of the National Institute of Standards and Technology*, *125*. https://doi.org/10.6028/JRES.125.025.

Hvidt, A., & Nielsen, S. O. (1966). Hydrogen exchange in proteins. *Advances in Protein Chemistry*, *21*(C), 287–386. https://doi.org/10.1016/S0065-3233(08)60129-1.

Iacob, R. E., & Engen, J. R. (2012). Hydrogen exchange mass spectrometry: Are we out of the quicksand? *Journal of the American Society for Mass Spectrometry*, *23*, 1003–1010. https://doi.org/10.1007/s13361-012-0377-z.

Iacob, R. E., Murphy, J. P., & Engen, J. R. (2008). Ion mobility adds an additional dimension to mass spectrometric analysis of solution-phase hydrogen/deuterium exchange. *Rapid Communications in Mass Spectrometry: RCM*, *22*(18), 2898–2904. https://doi.org/10.1002/RCM.3688.

Jackson, R. N., Klauer, A. A., Hintze, B. J., Robinson, H., van Hoof, A., & Johnson, S. J. (2010). The crystal structure of Mtr4 reveals a novel arch domain required for rRNA processing. *The EMBO Journal*, *29*(13), 2205–2216. https://doi.org/10.1038/EMBOJ.2010.107.

Jensen, P. F., Comamala, G., Trelle, M. B., Madsen, J. B., Jørgensen, T. J. D., & Rand, K. D. (2016). Removal of N-linked glycosylations at acidic pH by PNGase A facilitates hydrogen/deuterium exchange mass spectrometry analysis of N-linked glycoproteins. *Analytical Chemistry*, *88*(24), 12479–12488. https://doi.org/10.1021/ACS.ANALCHEM.6B03951.

Jia, H., Wang, X., Anderson, J. T., & Jankowsky, E. (2012). RNA unwinding by the Trf4/Air2/Mtr4 polyadenylation (TRAMP) complex. *Proceedings of the National Academy of Sciences of the United States of America*, *109*(19), 7292–7297. https://doi.org/10.1073/PNAS.1201085109/SUPPL_FILE/PNAS.201201085SI.PDF.

Jia, H., Wang, X., Liu, F., Guenther, U. P., Srinivasan, S., Anderson, J. T., et al. (2011). The RNA helicase Mtr4p modulates polyadenylation in the TRAMP complex. *Cell*, *145*(6), 890–901. https://doi.org/10.1016/J.CELL.2011.05.010.

Jumper, J., Evans, R., Pritzel, A., Green, T., Figurnov, M., Ronneberger, O., et al. (2021). Highly accurate protein structure prediction with AlphaFold. *Nature*, *596*(7873), 583–589. https://doi.org/10.1038/s41586-021-03819-2.

Kadek, A., Mrazek, H., Halada, P., Rey, M., Schriemer, D. C., & Man, P. (2014). Aspartic protease nepenthesin-1 as a tool for digestion in hydrogen/deuterium exchange mass spectrometry. *Analytical Chemistry*, *86*(9), 4287–4294. https://doi.org/10.1021/AC404076J/SUPPL_FILE/AC404076J_SI_003.ZIP.

Karch, K. R., Coradin, M., Zandarashvili, L., Kan, Z. Y., Gerace, M., Englander, S. W., et al. (2018). Hydrogen-deuterium exchange coupled to top- and middle-down mass spectrometry reveals histone tail dynamics before and after nucleosome assembly. *Structure*, *26*(12), 1651–1663.e3. https://doi.org/10.1016/J.STR.2018.08.006/ATTACHMENT/9D1DC48C-5A93-40D4-BDE8-49FBD831E932/MMC3.XLSX.

Katta, V., Chait, B. T., & Carr, S. (1991). Conformational changes in proteins probed by hydrogen-exchange electrospray-ionization mass spectrometry. *Rapid Communications in Mass Spectrometry: RCM*, *5*(4), 214–217. https://doi.org/10.1002/RCM.1290050415.

Kheterpal, I., & Wetzel, R. (2006). Hydrogen/deuterium exchange mass spectrometry A window into amyloid structure. *Accounts of Chemical Research*, *39*(9), 584–593. https://doi.org/10.1021/AR050057W.

Kim, P. S., & Baldwin, R. L. (1982). Influence of charge on the rate of amide proton exchange. *Biochemistry*, *21*(1), 1–5. Retrieved from https://pubs.acs.org/sharingguidelines.

Kochert, B. A., Iacob, R. E., Wales, T. E., Makriyannis, A., & Engen, J. R. (2018). Hydrogen-deuterium exchange mass spectrometry to study protein complexes. *Methods in Molecular Biology*, *1764*, 153–171. https://doi.org/10.1007/978-1-4939-7759-8_10.

Konermann, L., Pan, J., & Liu, Y.-H. (2011). Hydrogen exchange mass spectrometry for studying protein structure and dynamics. *Chemical Society Reviews*, *40*, 1224–1234. https://doi.org/10.1039/c0cs00113a.

Lingaraju, M., Johnsen, D., Schlundt, A., Langer, L. M., Basquin, J., Sattler, M., et al. (2019). The MTR4 helicase recruits nuclear adaptors of the human RNA exosome using distinct arch-interacting motifs. *Nature Communications*, *10*(1), 1–11. https://doi.org/10.1038/s41467-019-11339-x.

Mahoney, P. P., Ray, S. J., & Hieftje, G. M. (1997). Time-of-flight mass spectrometry for elemental analysis. *Applied Spectroscopy*, *51*(1), 16A–28A.

Majumdar, R., Manikwar, P., Hickey, J. M., Arora, J., Middaugh, C. R., Volkin, D. B., et al. (2012). Minimizing carry-over in an online pepsin digestion system used for the H/D exchange mass spectrometric analysis of an IgG1 monoclonal antibody. *Journal of the American Society for Mass Spectrometry*, *23*(12), 2140–2148. https://doi.org/10.1007/S13361-012-0485-9/FIGURES/6.

Marcsisin, S. R., & Engen, J. R. (2010). Hydrogen exchange mass spectrometry: What is it and what can it tell us? *Analytical and Bioanalytical Chemistry*, *397*(3), 967–972. https://doi.org/10.1007/s00216-010-3556-4.

Masson, G. R., Burke, J. E., Ahn, N. G., Anand, G. S., Borchers, C., Brier, S., et al. (2019). Recommendations for performing, interpreting and reporting hydrogen deuterium exchange mass spectrometry (HDX-MS) experiments. *Nature Methods*, *16*(7), 595–602. https://doi.org/10.1038/s41592-019-0459-y.

Miranker, A., Robinson, C. V., Radford, S. E., Aplin, R. T., & Dobson, C. M. (1993). Detection of transient protein folding populations by mass spectrometry. *Science (New York, N.Y.)*, *262*(5135), 896–900. https://doi.org/10.1126/SCIENCE.8235611.

Morgan, C. R., & Engen, J. R. (2009). Investigating solution-phase protein structure and dynamics by hydrogen exchange mass spectrometry. *Current Protocols in Protein Science*, *58*(1), 17.6.1–17.6.17. https://doi.org/10.1002/0471140864.PS1706S58.

Murphree, T. A., Vorauer, C., Brzoska, M., & Guttman, M. (2020). Imidazolium compounds as internal exchange reporters for hydrogen/deuterium exchange by mass spectrometry. *Analytical Chemistry*, *92*(14), 9830. https://doi.org/10.1021/ACS.ANALCHEM.0C01328.

Narang, D., Lento, C., & Wilson, D. J. (2020). HDX-MS: An analytical tool to capture protein motion in action. *Biomedicines*, *8*. MDPI AG https://doi.org/10.3390/BIOMEDICINES8070224.

Pan, J., Han, J., Borchers, C. H., & Konermann, L. (2009). Hydrogen/deuterium exchange mass spectrometry with top-down electron capture dissociation for characterizing structural transitions of a 17 kDa protein. *Journal of the American Chemical Society*, *131*(35), 12801–12808. https://doi.org/10.1021/JA904379W.

Pan, J., Zhang, S., Chou, A., & Borchers, C. H. (2016). Higher-order structural interrogation of antibodies using middle-down hydrogen/deuterium exchange mass spectrometry. *Chemical Science*, 7(2), 1480–1486. https://doi.org/10.1039/C5SC03420E.

Papanastasiou, M., Mullahoo, J., DeRuff, K. C., Bajrami, B., Karageorgos, I., Johnston, S. E., & Jaffe, J. D. (2019). Chasing tails: Cathepsin-L improves structural analysis of histones by HX-MS *[S]. *Molecular & Cellular Proteomics*, *18*(10), 2089–2098. https://doi.org/10.1074/MCP.RA119.001325.

Pirrone, G. F., Iacob, R. E., & Engen, J. R. (2014). Applications of hydrogen/deuterium exchange MS from 2012 to 2014. *Analytical Chemistry*, *87*(1), 99–118. https://doi.org/10.1021/AC5040242.

Poliakov, A., Jardine, P., & Prevelige, P. E. (2008). Hydrogen/deuterium exchange on protein solutions containing nucleic acids: Utility of protamine sulfate. *Rapid Communications in Mass Spectrometry*, *22*(16), 2423–2428. https://doi.org/10.1002/rcm.3627.

Puno, M. R., & Lima, C. D. (2018). Structural basis for MTR4-ZCCHC8 interactions that stimulate the MTR4 helicase in the nuclear exosome-targeting complex. *Proceedings of the National Academy of Sciences of the United States of America*, *115*(24), E5506–E5515. https://doi.org/10.1073/PNAS.1803530115/SUPPL_FILE/PNAS.1803530115.SAPP.PDF.

Rand, K. D., Lund, F. W., Amon, S., & Jørgensen, T. J. D. (2011). Investigation of amide hydrogen back-exchange in Asp and His repeats measured by hydrogen (1H/2H) exchange mass spectrometry. *International Journal of Mass Spectrometry*, *1–3*(302), 110–115. https://doi.org/10.1016/J.IJMS.2010.08.027.

Robinson, J. D. (1980). Binding to the high-affinity substrate site of the (Na + + K +)-dependent ATPase. *Journal of Bioenergetics and Biomembranes*, *12*(3), 165–174. https://doi.org/10.1007/BF00744681.

Rosenberg, A., & Chakravarti, K. (1968). Studies of hydrogen exchange in proteins. *The Journal of Biological Chemistry*, *243*(19), 5193–5201. https://doi.org/10.1016/S0021-9258(18)92009-0.

Sorokin, V. A., Gladchenko, G. O., & Valeev, V. A. (1986). DNA protonation at low ionic strength of solution. *Die Makromolekulare Chemie*, *187*(5), 1053–1063. https://doi.org/10.1002/MACP.1986.021870502.

Sperry, J. B., Wilcox, J. M., & Gross, M. L. (2008). Strong anion exchange for studying protein-DNA interactions by H/D exchange mass spectrometry. *Journal of the American Society for Mass Spectrometry*, *19*(6), 887–890. https://doi.org/10.1016/j.jasms.2008.03.003.

Taylor, L. L., Jackson, R. N., Rexhepaj, M., Klauer King, A., Lott, L. K., van Hoof, A., et al. (2014). The Mtr4 ratchet helix and arch domain both function to promote RNA unwinding. *Nucleic Acids Research*, *42*, 13861–13872. https://doi.org/10.1093/nar/gku1208.

Toth, R. T., Mills, B. J., Joshi, S. B., Esfandiary, R., Bishop, S. M., Middaugh, C. R., et al. (2017). Correction to empirical correction for differences in chemical exchange rates in hydrogen exchange-mass spectrometry measurements. *Analytical Chemistry*, *89*(24), 13673. https://doi.org/10.1021/ACS.ANALCHEM.7B04690.

van Hoof, A., Lennertz, P., & Parker, R. (2000). Yeast exosome mutants accumulate 3′-extended polyadenylated forms of U4 small nuclear RNA and small nucleolar RNAs. *Molecular and Cellular Biology*, *20*(2), 441–452. https://doi.org/10.1128/MCB.20.2.441-452.2000/ASSET/45F1E2C9-5A17-4A87-BDD5-85364A98DD0F/ASSETS/GRAPHIC/MB0201360007.JPEG.

Wales, T. E., Eggertson, M. J., & Engen, J. R. (2013). Considerations in the analysis of hydrogen exchange mass spectrometry data. *Methods in Molecular Biology (Clifton, N.J.)*, *1007*, 263. https://doi.org/10.1007/978-1-62703-392-3_11.

Wales, T. E., & Engen, J. R. (2006). Hydrogen exchange mass spectrometry for the analysis of protein dynamics. *Mass Spectrometry Reviews*, *25*, 158–170. https://doi.org/10.1002/mas.20064.

Wales, T. E., Fadgen, K. E., Eggertson, M. J., & Engen, J. R. (2017). Subzero celsius separations in three-zone temperature controlled hydrogen deuterium exchange mass spectrometry. *Journal of Chromatography A*, *1523*, 275–282. https://doi.org/10.1016/J.CHROMA.2017.05.067.

Wales, T. E., Fadgen, K. E., Gerhardt, G. C., & Engen, J. R. (2008). High-speed and high-resolution UPLC separation at zero degrees Celsius. *Analytical Chemistry*, *80*(17), 6815. https://doi.org/10.1021/AC8008862.

Walters, B. T., Ricciuti, A., Mayne, L., & Englander, S. W. (2012). Minimizing back exchange in the hydrogen exchange-mass spectrometry experiment. *Journal of the American Society for Mass Spectrometry*, *23*(12), 2132–2139. https://doi.org/10.1007/S13361-012-0476-X.

Wang, J., Chen, J., Wu, G., Zhang, H., Du, X., Chen, S., et al. (2019). NRDE2 negatively regulates exosome functions by inhibiting MTR4 recruitment and exosome interaction. *Genes and Development*, *33*(9–10), 536–549. https://doi.org/10.1101/GAD.322602.118/-/DC1.

Wang, X., Jia, H., Jankowsky, E., & Anderson, J. T. (2008). Degradation of hypomodified tRNAiMet in vivo involves RNA-dependent ATPase activity of the DExH helicase Mtr4p. *RNA*, *14*(1), 107–116. https://doi.org/10.1261/RNA.808608.

Watson, M. J., Harkewicz, R., Hodge, E. A., Vorauer, C., Palmer, J., Lee, K. K., et al. (2021). Simple platform for automating decoupled LC-MS analysis of hydrogen/deuterium exchange samples. *Journal of the American Society for Mass Spectrometry*, *32*(2), 597–600. https://doi.org/10.1021/JASMS.0C00341/SUPPL_FILE/JS0C00341_SI_007.ZIP.

Weir, J. R., Bonneau, F., Hentschel, J., & Conti, E. (2010). Structural analysis reveals the characteristic features of Mtr4, a DExH helicase involved in nuclear RNA processing and surveillance. *Proceedings of the National Academy of Sciences of the United States of America*, *107*(27), 12139–12144. https://doi.org/10.1073/PNAS.1004953107/SUPPL_FILE/PNAS.1004953107_SI.PDF.

Weis, D. D. (2016). Hydrogen exchange mass spectrometry of proteins: fundamentals, methods, and applications. In *Hydrogen exchange mass spectrometry of proteins* John Wiley & Sons, Ltd. https://doi.org/10.1002/9781118703748.

Weis, D. D., Wales, T. E., Engen, J. R., Hotchko, M., & ten Eyck, L. F. (2006). Identification and characterization of EX1 kinetics in H/D exchange mass spectrometry by peak width analysis. *Journal of the American Society for Mass Spectrometry*, *17*(11), 1498–1509. https://doi.org/10.1016/J.JASMS.2006.05.014.

Welch, B. L. (1947). The generalisation of student's problems when several different population variances are involved. *Biometrika*, *34*(1–2), 28–35. https://doi.org/10.1093/BIOMET/34.1-2.28.

Woods, V. L., & Hamuro, Y. (2001). High resolution, high-throughput amide deuterium exchange-mass spectrometry (DXMS) determination of protein binding site structure and dynamics: Utility in pharmaceutical design. *Journal of Cellular Biochemistry*, *84*(S37), 89–98. https://doi.org/10.1002/JCB.10069.

Zhang, H.-M., Kazazic, S., Schaub, T. M., Tipton, J. D., Emmett, M. R., & Marshall, A. G. (2008). Enhanced digestion efficiency, peptide ionization efficiency, and sequence resolution for protein hydrogen/deuterium exchange monitored by Fourier transform ion cyclotron resonance mass spectrometry. *Analytical Chemistry*, *80*(23), 9034–9041. Retrieved from http://www.ncbi.nlm.nih.gov/pubmed/19551977.

Zhang, H. M., McLoughlin, S. M., Frausto, S. D., Tang, H., Emmett, M. R., & Marshall, A. G. (2010). Simultaneous reduction and digestion of proteins with disulfide bonds for hydrogen/deuterium exchange monitored by mass spectrometry. *Analytical Chemistry*, *82*(4), 1450. https://doi.org/10.1021/AC902550N.

Zhang, Z., & Smith, D. L. (1993). Determination of amide hydrogen exchange by mass spectrometry: A new tool for protein structure elucidation. *Protein Science: A Publication of the Protein Society*, *2*(4), 522–531. https://doi.org/10.1002/PRO.5560020404.

Zhang, N., Olsen, K. J., Ball, D., Johnson, S. J., & D'Arcy, S. (2022). Hydrogen-deuterium exchange mass spectrometry of Mtr4 with diverse RNAs reveals substrate-dependent dynamics and interfaces in the arch. *Nucleic Acids Research*. https://doi.org/10.1093/NAR/GKAC170.

Zhang, N., Yu, X., Zhang, X., & D'Arcy, S. (2021). HD-eXplosion: Visualization of hydrogen-deuterium exchange data as chiclet and volcano plots with statistical filtering. *Bioinformatics (Oxford, England)*, *37*(13), 1926–1927. https://doi.org/10.1093/BIOINFORMATICS/BTAA892.

Zhang, Z., Zhang, A., & Xiao, G. (2012). Improved protein hydrogen/deuterium exchange mass spectrometry platform with fully automated data processing. *Analytical Chemistry*, *84*(11), 4942–4949. https://doi.org/10.1021/AC300535R.

Kolibri 360 GmbH
Ferdinand-Jühlke-Straße 7
99095 Erfurt, Deutschland
produktsicherheit@kolibri360.de